Books are to be returned on or before
the last ~~~

Forensic Geoscience: Principles, Techniques and Applications

It is recommended that reference to all or part of this book should be made in one of the following ways:

PYE, K. & CROFT, D. J. (eds) 2004. *Forensic Geoscience: Principles, Techniques and Applications.* Geological Society, London, Special Publications, **232**.

SCOTT, J. & HUNTER, J. R. 2004. Environmental influences on resistivity mapping for the location of clandestine graves. *In*: PYE, K. & CROFT, D. J. (eds) 2004. *Forensic Geoscience: Principles, Techniques and Applications.* Geological Society, London, Special Publications, **232**, 33–38.

GEOLOGICAL SOCIETY SPECIAL PUBLICATION NO. 232

Forensic Geoscience:
Principles, Techniques and Applications

EDITED BY

K. PYE

Kenneth Pye Associates Ltd & Royal Holloway, University of London, UK

and

D. J. CROFT

Croft Scientific and Technical & Kenneth Pye Associates Ltd, UK

2004
Published by
The Geological Society
London

THE GEOLOGICAL SOCIETY

The Geological Society of London (GSL) was founded in 1807. It is the oldest national geological society in the world and the largest in Europe. It was incorporated under Royal Charter in 1825 and is Registered Charity 210161.

The Society is the UK national learned and professional society for geology with a worldwide Fellowship (FGS) of 9000. The Society has the power to confer Chartered status on suitably qualified Fellows, and about 2000 of the Fellowship carry the title (CGeol). Chartered Geologists may also obtain the equivalent European title, European Geologist (EurGeol). One fifth of the Society's fellowship resides outside the UK. To find out more about the Society, log on to www.geolsoc.org.uk.

The Geological Society Publishing House (Bath, UK) produces the Society's international journals and books, and acts as European distributor for selected publications of the American Association of Petroleum Geologists (AAPG), the American Geological Institute (AGI), the Indonesian Petroleum Association (IPA), the Geological Society of America (GSA), the Society for Sedimentary Geology (SEPM) and the Geologists' Association (GA). Joint marketing agreements ensure that GSL Fellows may purchase these societies' publications at a discount. The Society's online bookshop (accessible from www.geolsoc.org.uk) offers secure book purchasing with your credit or debit card.

To find out about joining the Society and benefiting from substantial discounts on publications of GSL and other societies worldwide, consult www.geolsoc.org.uk, or contact the Fellowship Department at: The Geological Society, Burlington House, Piccadilly, London W1J 0BG: Tel. +44 (0)20 7434 9944; Fax +44 (0)20 7439 8975; E-mail: enquiries@geolsoc.org.uk.

For information about the Society's meetings, consult *Events* on www.geolsoc.org.uk. To find out more about the Society's Corporate Affiliates Scheme, write to enquiries@geolsoc.org.uk.

Published by The Geological Society from:
The Geological Society Publishing House
Unit 7, Brassmill Enterprise Centre
Brassmill Lane
Bath BA1 3JN, UK
(*Orders*:   Tel. +44 (0)1225 445046
        Fax +44 (0)1225 442836)

Online bookshop: http://bookshop.geolsoc.org.uk

British Library Cataloguing in Publication Data
A catalogue record for this book is available from the British Library.

ISBN 1-86239-161-0

Typeset by Servis Filmsetting Limited, Manchester, UK

Printed by Cromwell Press, Trowbridge, UK

Distributors

USA
    AAPG Bookstore
    PO Box 979
    Tulsa
    OK 74101-0979
    USA
Orders: Tel. + 1 918 584-2555
        Fax + 1 918 560-2652
        E-mail bookstore@aapg.org

India
    Affiliated East-West Press PVT Ltd
    G-1/16 Ansari Road, Darya Ganj,
    New Delhi 110 002
    India
Orders: Tel. + 91 11 2327-9113/2326-4180
        Fax + 91 11 2326-0538
        E-mail affiliat@vsnl.com

Japan
    Kanda Book Trading Company
    Cityhouse Tama 204
    Tsurumaki 1-3-10
    Tama-shi, Tokyo 206-0034
    Japan
Orders: Tel. + 81 (0)423 57-7650
        Fax + 81 (0)423 57-7651
        E-mail geokanda@ma.kcom.ne.jp

Contents

Preface

This book arises from a 2-day international conference held at the Geological Society of London in March 2003. The meeting sought to bring together forensic scientists, geologists, sedimentologists, environmental scientists, archaeologists and related practitioners from academia, industry and the criminal justice system to discuss the history, principles, development of techniques, and recent research in and the application of geosciences in a forensic context.

The meeting was attended by over 120 delegates, with 25 oral papers given and 16 posters presented. It was sponsored by Hitachi High Technologies, Leica Microsystems and Kenneth Pye Associates Ltd, and was also supported by the Forensic Science Society. The editors would like to thank these organizations, together with staff at the Geological Society, and those who agreed to chair the sessions over the 2 days.

This publication contains 29 papers based on a selection of these oral papers and poster presentations. The wide range of topics addressed and the diversity of the contributions is reflected in the papers' content and style. The chapters have been arranged to reflect a continuum from fieldwork, through techniques, laboratory analyses and data handling, to case-based papers and teaching.

The editors would also like to extend their thanks to a host of reviewers, who include: J. R. L. Allen, A. Aspinall, J. R. Bacon, N. Branch, M. Burchell, N. J. Cassidy, P. Chamberlain, R. G. Cowell, G. C. Davenport, H. Demmelmeyer, R.-M. di Maggio, C. Doherty, M. J. Faulkner, P. Finch, R. Goodacre, A. Jackson, R. C. Janaway, C. J. Jeans, M. Lark, N. Linford, S. Mays, A. Mazurek, A. Neal, R. Ogilvy, R. D. Pancost, S. Parry, L. Pierce, P. Potts, D. Ryves, S. Roskams, M. Thompson, D. Thornley, E. Valsami-Jones, N. Walsh, N. Wells, R. Withnall, A. Wolf, D. Wray.

K. Pye
D. J. Croft

Forensic geoscience: introduction and overview

KENNETH PYE[1,2] & DEBRA J. CROFT[1,3]

[1] *Kenneth Pye Associates Ltd, Crowthorne Enterprise Centre, Crowthorne Business Estate, Crowthorne RG45 6AW, UK (e-mail: kpye@kpal.co.uk)*
[2] *Department of Geology, Royal Holloway, University of London, Egham Hill, Egham TW20 0EX, UK*
[3] *Croft Scientific & Technical, Blaen-y-Waun, Llanafan, Ceredigion SY23 4BD, UK*

The nature of forensic geoscience

Forensic geoscience may be defined as a subdiscipline of geoscience that is concerned with the application of geological and wider environmental science information and methods to investigations which may come before a court of law. The scientific boundaries of forensic geoscience are not clearly defined, and there are significant overlaps with other, related subdisciplines such as forensic archaeology (Hunter *et al.* 1987), forensic anthropology, forensic botany (Hall 2002; Horrocks & Walsh 1998), forensic engineering (Shuirman & Slosson 1992) and even forensic medicine and forensic pathology (Knight 1997; DiMaio & DiMaio 2001). Forensic geoscience is concerned with all aspects of earth materials, including rocks, sediments, soil, air and water, and with a wide range of natural phenomena and processes. Since modern sediments and soil also often contain objects and particles of human origin, man-made materials such as brick, concrete, ceramics, glass and various other industrial products and raw materials are also sometimes of interest. These may be of relatively modern origin or of archaeological importance (e.g. Henderson 2002).

Forensic geology (Murray & Tedrow 1975, 1992) may be regarded as a subset of forensic geoscience and is principally concerned with studies of rocks, sediments, minerals, soils and dusts. *Environmental forensics* (Morrison 2000; Murphy & Morrison 2002), on the other hand, has somewhat wider scope than forensic geoscience, with much stronger links to disciplines such as chemical engineering, and with a greater concern with such issues as groundwater contamination and air pollution modelling.

Forensic geoscience is by nature an integrative subdiscipline that draws directly on other subdisciplines in earth science such as geophysics, geochemistry, sedimentology, engineering geology and geoarchaeology. These, in turn, draw on the fundamental sciences (physics, chemistry and biology) and other derivative disciplines such as engineering, medicine, archaeology and anthropology (Fig 1).

Contribution of geoscience to forensic investigations

The potential contribution of studies of rocks, minerals and sediments to criminal investigations was recognized more than a century ago by the American criminologist Professor Hans Gross, and other early advocates in the early part of this century included the German forensic scientist Georg Popp and the French forensic science pioneer Edmond Locard (see Murray 2004). However, it was not until 1975 that the first textbook on forensic geology was published (Murray & Tedrow 1975), and there is still only a very limited dedicated literature in the field. Many general texts on crime scene investigation and criminalistics make little or no mention of forensic geology or soil evidence (e.g. Fisher 2000; Saferstein 2001), and there is widespread ignorance among the legal profession and police forces about its potential (e.g. White 1998; Townley & Ede 2004). Even among forensic practitioners, there are still many who believe that 'soil is soil, sand is sand, and mud is mud'.

In actual fact, there is an enormous diversity among earth surface materials, and the fact that modern techniques are capable of very detailed characterization and discrimination makes such material potentially highly useful in a forensic context. Forensic geoscience techniques and information can be applied to a wide range of civil law and criminal law issues, relating to such problems as environmental accidents, construction failures, pollution and, of course, serious crimes such as murder, terrorism, genocide, arson, drug smuggling and rape.

Forensic geoscience information may be used simply for 'intelligence' purposes within the framework of an ongoing police investigation or as evidence for presentation in court, depending on the quality of data and strength of the conclusions which can be drawn. By combining geological information with data obtained from related subdisciplines, such as forensic botany, useful information about provenance and geographical location can be provided. This type of approach is sometimes referred to as 'environmental profiling' and is analogous to the

From: PYE, K. & CROFT, D. J. (eds) 2004. *Forensic Geoscience: Principles, Techniques and Applications*. Geological Society, London, Special Publications, **232**, 1–5. © The Geological Society of London, 2004.

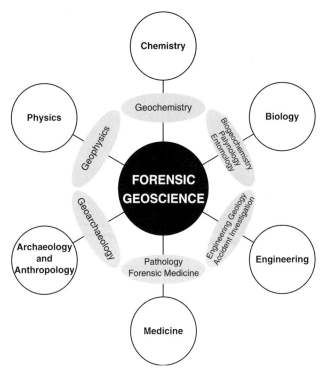

Fig. 1. The relationship of forensic geoscience to some other disciplines and subdisciplines.

geographical profiling and psychological profiling undertaken by behavioural scientists. Geoscientists can also assist in tasks such as body location through the application of geophysical survey and other remote-sensing techniques (Fenning & Donnelly 2004; Watters & Hunter 2004; Scott & Hunter 2004).

In recent years, soil and geological evidence has been increasingly used in investigations and both criminal and civil law trials in the UK, but worldwide its acceptance in the courts varies from country to country. This partly reflects the different legal traditions, the greatly varying degree of sophistication displayed in the investigation and analysis methods between different countries. However, professional and public awareness of the potential power of forensic geoscience evidence is increasing, and there can be little doubt that it will become increasingly used on a global scale over the coming decades.

Key questions and approaches

In serious crime cases, one of the most common tasks is to determine whether a suspect (or his means of transport, such as a car or motorcycle) was present at the scene of a crime. Items seized from the suspect, including footwear, clothing, digging implements or car, are routinely examined for traces of blood, DNA, fibres or hair that might link the suspect to the victim or the scene, but increasingly they are also being examined for traces of soil, mud, dust or pollen that might also indicate a link. Increasing forensic awareness among criminals, many of whom now take precautions not to leave DNA or similar traces, means that new types of potential evidence need to be found.

In such a case, the investigator may well be presented initially with an exhibit such as that shown in Figure 2, a pair of shoes with relatively small amounts of mud on the soles. In view of the relatively small amount of material available for analysis, the investigator needs to draw up an examination and analysis strategy in order to maximize the amount of information that can be obtained from the removed material and to obtain the maximum evidential value possible. This requires selection of the most appropriate analytical techniques, as well as the collection of appropriate background data against which the analytical data can be compared.

Two key questions are typically asked by investigating police officers: (1) is there a 'match' between the mud on the shoes and the crime scene, and, if so, (2) what is the evidential value of such a 'match'? In

Fig. 2. Footwear with mud deposits on the soles belonging to a murder suspect, later convicted. Attempts had been made to scrape mud off the sole of the shoe on the right.

order to address the first question the material must be examined and analysed in as many ways as possible, in order to determine whether any significant differences exist between the questioned sample and the known samples taken from the crime scene, the initial objective being to establish whether the possibility can be excluded that the questioned sample originated from the crime scene. The examination and analysis techniques used should be as non-destructive as possible, both to maximize the potential for as many tests as possible to be carried out and to preserve evidence for possible later defence examination. A wide range of techniques is available (as discussed by Jarvis *et al.* 2004; Blott *et al.* 2004; Pye 2004*a*, *b*; Edwards 2004; and others in later chapters in this volume) and considerable experience is required in order to assess which techniques are likely to provide the most significant information in any particular situation.

If test results show that the questioned mud cannot be excluded as having come from the crime scene, the next question is one of the *degree* of similarity, or 'match', and its evidential significance. No two natural soil or sediment samples are ever absolutely identical, and the key issue is therefore to decide how much similarity, and how much dis-similarity exists, and with what degree of probability it can be stated that the questioned sample did originate, in whole or in part, from the crime scene.

In some, relatively rare, circumstances it may be possible to say with virtual certainty that the crime scene was the source. This normally arises where there is some form of physical fit, for example between two halves of a broken rock, or where soil is taken from a characteristic shoe impression in the ground. It may also be possible where one or, more commonly, several, highly unusual particle types (sometimes referred to as 'exotics') are found in common between the questioned soil sample and the crime scene control samples. In many cases, however, no such evidence is available, and assessment of degree of match and its significance must be based on comparison of several different bulk sample properties, which are evaluated using a combination of statistical, graphical and other methods. Assessment of evidential value involves both quantification of the degree of similarity between two samples and comparison with data in some form of database which can provide information about similar samples in other locations. The quality of database information varies from place to place, and in many investigations it is necessary to collect and analyse additional samples for comparison purposes specifically in the context of the investigation in hand. It goes without saying that the better the database and quality of contextual information available, the greater the confidence that can be placed in any conclusions reached. Contextual information includes such factors as the timing of the alleged offence and seizure of the exhibit (e.g. footwear), method of storage and nature of any previous examinations, and whether the suspect lived close to the crime scene or had any legitimate reason to visit it. Investigating officers have a duty to provide the investigator with as much relevant information as possible without deliberately trying to influence acquisition or interpretation of the results.

It is always advantageous if the forensic geoscientist has an opportunity to examine critical exhibits first hand and to undertake any sampling of soil and mud that may be necessary. In this way, the sequence of mud acquisition may be identified and appropriate types of samples collected for analysis. Scenes of crime officers and police often are not fully aware of the requirements for geological sample collection, storage and handling, although the situation is improving. Similarly, it is of great benefit if the forensic geoscientist is able to visit the crime scene, preferably as soon as possible after the crime has been discovered. By doing so, hypotheses for investigation are generated and suitable 'control' samples can be taken.

The fact that geoscience, like all natural sciences, is not an exact science does not undermine its usefulness in a court of law. Geological and soil trace evidence is in many respects similar to glass and fibre evidence, where probabilities of occurrence and significance of findings are best considered within a

Bayesian type framework rather than solely through attempts to apply conventional statistics (Small *et al.* 2004). Inevitably, however, there will be situations when it is not possible for the geoscientific expert witness to offer anything more than an expert opinion. Guidance on the duties and responsibilities of the expert witness, both in report writing and in presenting evidence in court, is now widely available through publications and training programmes (e.g. Rothwell 1998; Bond *et al.* 1999; Townley & Ede 2004).

Wider benefits and the way ahead

Involvement in forensic casework often leads to wider benefits, both for the investigating forensic scientist and for wider geoscience as a whole. The specific requirements of forensic work, relating to sample continuity, documentation and critical review before a court of law, provide good general scientific discipline. Moreover, new questions and hypotheses often arise within the context of a specific case, which can lead to significant developments in analytical techniques and procedures or generate a major research programme lasting several years! The fact that much forensic science work involves small samples, which have to be preserved as far as possible, stimulates the need to develop ever more sensitive but reproducible techniques of analysis.

In the future, the promising developments are likely to occur as a result of closer interdisciplinary collaboration between different forensically related subdisciplines. For example, the issue of human identification and provenance determination offers a major research frontier which invites cooperation between anatomists, osteologists, forensic odontologists, geoscientists and archaeologists (Pye 2004; Evans & Tatham 2004). Similarly, issues such as geotraceability related to foodstuffs, drugs and other products is already fostering new collaborations between geologists, food scientists, biochemists and law enforcement agencies. The next decade promises to generate many other exciting collaborations and developments.

References

BOND, C. SOLON, M. & HARPER, P. 1997. *The Expert Witness in Court. A Practical Guide.* Shaw & Sons, Crayford, Kent, 162pp.

BLOTT, S. J., CROFT, D. J., PYE, K., SAYE, S. E. & WILSON, H. E. 2004. Particle size analysis by laser diffraction., 2004. *In:* PYE, K. & CROFT, D. J. (eds) *Forensic Geoscience: Principles, Techniques and Applications.* Geological Society, London, Special Publications, **232**, 63–73.

DIMAIO, V. J. & DIMAIO, D. 2001. *Forensic Pathology.* 2nd edition. CRC Press, Boca Raton.

EDWARDS, H. G. M. 2004. Forensic applications of Raman spectroscopy to the non-destructive analysis of bio-materials and their degradation. *In:* PYE, K. & CROFT, D. J. (eds) *Forensic Geoscience: Principles, Techniques and Applications.* Geological Society, London, Special Publications, **232**, 159–170.

EVANS, J. A. & TATHAM, S. 2004. Defining 'local signature' in terms of Sr isotope composition using a tenth- to twelfth-century Anglo-Saxon population living on a Jurassic clay-carbonate terrain, Rutland, UK. *In:* PYE, K. & CROFT, D. J. (eds) *Forensic Geoscience: Principles, Techniques and Applications.* Geological Society, London, Special Publications, **232**, 237–248.

FENNING, P. J. & DONNELLY, L. J. 2004. Geophysical techniques for forensic investigation. *In:* PYE, K. & CROFT, D. J. (eds) *Forensic Geoscience: Principles, Techniques and Applications.* Geological Society, London, Special Publications, **232**, 11–20.

FISHER, B. A. J. 2000. *Techniques of Crime Scene Investigation.* 6th edition. CRC Press, Boca Raton.

HALL, D. W. 2002. Forensic botany. *In:* HAGLUND. W. D. & SORG. M. (eds), *Forensic Taphonomy.* CRC Press, Boca Raton, 353–363.

HENDERSON, J. 2002. *The Science and Archaeology of Materials.* Routledge, London & New York.

HORROCKS, M. & WALSH, K. A. J. 1998. Forensic palynology: assessing the value of the evidence. *Review of Paleobotany and Palynology,* **103**, 69–74.

HUNTER, J. R., ROBERTS, C. & MARTIN, A. 1997. *Studies in Crime: An Introduction to Forensic Archaeology.* Routledge, London.

JARVIS, K. E., WILSON, H. E. & JAMES, S. L. 2004. Assessing element variability in small soil samples taken during forensic investigations. *In:* PYE, K. & CROFT, D. J. (eds) *Forensic Geoscience: Principles, Techniques and Applications.* Geological Society, London, Special Publications, **232**, 171–182.

KNIGHT, B. 1997. *Simpson's Forensic Medicine.* 11th edition. Arnold, London.

MORRISON, R. D. 2000. *Environmental Forensics: Principles and Applications.* CRC Press, Boca Raton.

MURPHY, B. L. & MORRISON, R. D. 2002. *Introduction to Environmental Forensics.* Academic Press, San Diego, 560pp.

MURRAY, R. C. & TEDROW, J. F. C. 1975. *Forensic Geology.* Rutgers University Press, New York.

MURRAY, R. C. & TEDROW, J. F. C. 1992. *Forensic Geology.* 2nd edition. Prentice Hall, New Jersey.

MURRAY, R. C. 2004. Forensic geology: yesterday, today and tomorrow. *In:* CROFT, D. J & PYE, K. 2004. *In:* PYE, K. & CROFT, D. J. (eds) *Forensic Geoscience: Principles, Techniques and Applications.* Geological Society, London, Special Publications, **232**, 7–9.

PYE, K. 2004a. Forensic examination of rocks, sediments, soils and dust using scanning electron microscopy and X-ray chemical analysis. *In:* PYE, K. & CROFT, D. J. (eds) *Forensic Geoscience: Principles, Techniques and Applications.* Geological Society, London, Special Publications, **232**, 103–122.

PYE, K. 2004b. Isotope and trace element analysis of human teeth and bones for forensic purposes. *In:* PYE,

K. & CROFT, D. J. (eds) *Forensic Geoscience: Principles, Techniques and Applications*. Geological Society, London, Special Publications, **232**, 215–236.

ROTHWELL, D.K. 1998. Presentation of expert forensic evidence in court. *In*: WHITE, P. (ed.) *Crime Scene To Court: The Essentials of Forensic Science*. Royal Society of Chemistry, London, 327–351.

SAFERSTEIN, R. 2001. *Criminalistics*. 7th edition. Prentice Hall, New Jersey.

SCOTT, J. & HUNTER, J. R. 2004. Environmental influences on resistivity mapping for the location of clandestine graves. *In*: PYE, K. & CROFT, D. J. (eds) *Forensic Geoscience: Principles, Techniques and Applications*. Geological Society, London, Special Publications, **232**, 33–38.

SHUIRMAN, G. & SLOSSON, J. E. 1992. *Forensic Engineering*. Academic Press, San Diego.

SMALL, I. F., ROWAN, J. S., FRANKS, S. W. WYATT, A. & DUCK, R. W. 2004. Bayesian sediment fingerprinting provides a robust tool for forensic geoscience applications. *In*: PYE, K. & CROFT, D. J. (eds) *Forensic Geoscience: Principles, Techniques and Applications*. Geological Society, London, Special Publications, **232**, 207–213.

TOWNLEY, L. & EDE, R. 2004. *Forensic Practice in Criminal Cases*. The Law Society, London.

WHITE, P. (ed.) 1998, *Crime Scene to Court: The Essentials of Forensic Science*. Royal Society of Chemistry, Cambridge.

WATTERS, M. & HUNTER, J. R. 2004. Geophysics and burials: field experience and software development. *In*: PYE, K. & CROFT, D. J. (eds) *Forensic Geoscience: Principles, Techniques and Applications*. Geological Society, London, Special Publications, **232**, 21–31.

Forensic geology: yesterday, today and tomorrow

RAYMOND C. MURRAY

Department of Geology, University of Montana, Missoula, MT 59812, USA
(e-mail: rcm@selway.umt.edu; www.forensicgeology.net)

Abstract: Sir Arthur Conan Doyle and Hans Gross suggested the possibility of using soil and related material as physical evidence. Edmond Locard provided the intellectual basis for the use of the evidence. High-visibility cases, such as the work of the Federal Bureau of Investigation (FBI) in the Camarena case, the laboratory of the Garda Siochana in the Lord Mountbatten case and G. Lombardi in the Aldo Moro case, contributed to the general recognition that geological evidence could make an important contribution to justice. The value of geological evidence results from the almost unlimited number of rock, mineral, soil and related kinds of material combined with our ability to use instruments that characterize these materials. Forensic examinations involve identification of earth materials, comparison of samples to determine common source, studies that aid an investigation and intelligence studies. The future will see increased use of the evidence, new automated methods of examination, improved training of those who collect samples, and research on the diversity of soils and how, when and what parts of soils are transferred during various types of contact. The microscope will remain important because it allows the examiner to find the rare and unusual particle.

The use of geological materials as trace evidence in criminal cases has existed for approximately 100 years. Murray 2004 provides an overview and reminds us that it began, as with so many of the other types of evidence, with the writings of Sir Arthur Conan Doyle. Doyle wrote the Sherlock Holmes series between 1887 and 1893. He was a physician who apparently had two motives: writing saleable literature and using his scientific expertise to encourage the use of science as evidence (Murray & Tedrow 1992). In 1893 Hans Gross wrote *Handbook for Examining Magistrates*, in which he suggested that perhaps more could be told about where someone had last been from the dirt on their shoes than from toilsome inquiries. In 1908, a German chemist, Georg Popp, examined the evidence in the Margarethe Filbert case. In this homicide a suspect had been identified by many of his neighbours and friends because he was known to be a poacher. The suspect's wife testified that she had dutifully cleaned his best shoes the day before the crime. Those shoes had three layers of soil adhering to the leather in front of the heel. Popp, using the methods available at that time, said that the uppermost layer, and thus the oldest, contained goose droppings and other earth materials comparable to samples from the walkway outside the suspect's home. The second layer contained red sandstone fragments and other particles comparable to samples from the scene where the body had been found. The lowest layer, and thus the youngest, contained brick, coal dust, cement and a whole series of other materials comparable to samples from a location outside a castle where the suspect's gun and clothing had been found. The suspect said that he had walked only in his fields on the day of the crime. Those fields were underlain by porphyry with milky quartz. Popp found no such material on the shoe, although the soil had been wet on that day. In this case, Popp had developed most of the elements involved in present-day forensic soil examination. He had compared two sets of samples and identified them with two of the scenes associated with the crime. He had confirmed a sequence of events consistent with the theory of the crime, and he had found no evidence supporting the alibi.

Rocks, minerals, soils and related materials have evidential value. The value lies in the almost unlimited number of kinds of materials and the large number of measurements and observations that can be made on these materials. For example, the number of sizes and size distributions of grains combined with colours, shapes and mineralogy is almost unlimited. There is an almost unlimited number of kinds of minerals, rocks and fossils. These are identifiable, recognizable and can be characterized. It is this diversity of earth materials, combined with the ability to measure and observe the different kinds, which provides the forensic discriminating power.

There have been many contributions to the discipline over the last 100 years. Many have been made by the Laboratory of the Federal Bureau of Investigation, in Washington DC, McCrone Associates in Chicago, the Centre for Forensic Sciences in Toronto, Microtrace in Elgin, Illinois, the former Central Research Establishment at Aldermaston, Kenneth Pye Associates Ltd in the UK, the Japanese National Research Institute of Police Science and the Netherlands Forensic Institute, as well as other government, private and academic researchers.

From: PYE, K. & CROFT, D. J. (eds) 2004. *Forensic Geoscience: Principles, Techniques and Applications.* Geological Society, London, Special Publications, **232**, 7–9. © The Geological Society of London, 2004.

Because much of the evidential value of earth materials lies in the diversity of, and the differences between the minerals and particles, microscopic examination at all levels of instrumentation is the most powerful tool. In addition, such examination provides an opportunity to search for man-made artefact grains and other kinds of physical evidence.

Individualization, that is the unique association of samples from the crime scene with those of the suspect to the exclusion of all other samples, is not possible in most cases. In this sense earth material evidence is not similar to DNA, fingerprints and some forms of firearms and tool-mark evidence. However, in a South Dakota homicide case, soil from the scene where the body was found and from the suspect's vehicle both contained similar material, including grains of the zinc spinel, gahnite. This mineral had never before been reported from South Dakota. Such evidence provides a very high level of confidence and reliability.

One of the most interesting types of studies is the *aid to an investigation.* There are many examples of cases where a valuable cargo in transit is removed and rocks or bags of sand of the same weight are substituted. If the original source of the rocks or sand can be determined, then the investigation can be focused at that place. In a high-visibility case, Enrique Camarena of the Drug Enforcement Agency was murdered in Mexico (McPhee 1997). His body was exhumed as part of a cover-up staged by members of the Mexican Federal Judicial Police. When the body was later found, it contained rock fragments that differed from the country rock of the locality and represented the rocks from the original burial site. From the combination of petrographic examination of these rocks and a detailed literature search of Mexican volcanic rock descriptions, the original burial location was found and the cover-up exposed.

Most examinations involve *comparison*. Comparison aims to establish a high probability that two samples have a common source or, conversely, that they do not have similar properties and thus are unlikely to have come from the same source. In comparative studies of soils, it is difficult to overestimate the value of finding artefacts in the soil or some other unusual type of evidence. In an Upper Michigan rape case, three flowerpots had been tipped over and spilled onto the floor during the struggle. It was shown that potting soil on the suspect's shoe had a high degree of similarity to a sample collected from the floor and represented soil from one of the pots. In addition, small clippings of blue thread existed both in the flowerpot sample and on the shoe of the suspect. The thread provided additional trace evidence that supplemented the soil evidence.

In a New Jersey rape case, the suspect had soil samples in the turn-ups of his trousers. In addition to glacial sand grains that showed a similarity to those in soil samples collected from the crime scene, the soil contained fragments of clean Pennsylvania anthracite. Such coal fragments are not uncommon in the soils of most of the older cities in eastern North America. However, this sample there was found to contain too much coal when compared with samples collected in the surrounding area. Further investigation showed that, some 60 years earlier, the crime scene had been the location of a coal pile for a coal-burning laundry. Again, the combining of soil evidence with an investigation of an artefact and local industrial history increased the evidential value.

A new and evolving type of study is one done for the purpose of *intelligence gathering.* An example might involve identifying mineral material on an individual who had claimed to have recently visited a particular location. In such a case the question would be asked whether the mineral material supports the claim and could have come from that location. *Identification* of the mineral material alone can be useful in the case of mine fraud, gem fraud and art fraud by providing information that demonstrates the fraud.

The alertness of those who collect samples, and the quality of collection, is critical to the success of any examination. If appropriate samples are not collected during the initial evidence gathering, they will never be studied and never provide assistance to the court. There was a case in which an alert police officer happened to look at an individual arrested for a minor crime. He observed: 'That is the worst case of dandruff I have ever seen.' It was not dandruff but diatomaceous earth, which was essentially identical with the insulating material of a safe that had been broken into the previous day.

The future of forensic geology holds much promise. However, that future will see many changes and new opportunities. New methods are being developed that take advantage of the discriminating power inherent in earth materials. Quantitative X-ray diffraction could possibly revolutionize forensic soil examination. When it has been developed to the point that this or similar methods become routine laboratory techniques, it will be possible to do a quantitative mineralogical analysis that is easily reproducible. However, the microscope will remain an important tool in the search for the unusual grain or artefact. Sampling methods, plus the thorough and complete training of those people who collect samples for forensic purposes, will be improved. Soils are extremely sensitive to change over short distances, both horizontally and vertically. Soil sampling in many cases is the search for a sample that matches. The collection of all the other samples serves only the purpose of demonstrating the range of local differences. In collecting soil samples for comparison, we are searching for one that affords to the possibility of matching. Screening techniques applied during sampling eliminated samples that are totally different are often appropriate. For example, a surface sample offers little possibility of matching with material collected at a depth of 1.2m in a grave.

Studies that demonstrate the diversity of soils are important. One approach is to take an area normally assumed to be fairly homogenous in its soil character and collect 100 samples on a grid. The pairs of samples are then compared with each other until all the pairs are shown to be different. Starting with colour and moving on to size distribution and mineralogy, different methods are used to eliminate all of the pairs that appear similar. Junger (1996) performed several such studies and suggested methods for soil examination.

The qualifications and competence of examiners are a very major problem. How do you learn to do forensic soil examinations? This requires a thorough knowledge of mineralogy and the ability to use effectively a microscope and the other techniques used in earth material examination. It is also important that examiners are familiar with the other kinds of trace evidence, as well as the law and practice of forensic examination.

References

JUNGER, E. P. 1996. Assessing the unique characteristics of close-proximity soil samples: just how useful is soil evidence? *Journal of Forensic Sciences*, **41**, 27–34.

MCPHEE, J. 1997. *Irons in the Fire*. Farrar, Straus and Giroux, New York.

MURRAY, R. C. 2004. *Evidence from the Earth*. Mountain Press, Montana.

MURRAY, R. C. & TEDROW, J. 1992. *Forensic Geology*. Prentice Hall, Englewood Cliffs, New Jersey

Geophysical techniques for forensic investigation

PETER J. FENNING[1] & LAURANCE J. DONNELLY[2]

[1]*VJ GeoConsultants, 9 Avon Wharf, Bridge Street, Christchurch, BH23 1DJ, UK*
(e-mail: VJ_Consultants@hotmail.com)
[2]*British Geological Survey, Kingsley Dunham Centre, Nicker Hill, Keyworth, Nottingham,*
NG12 5GG, UK

Abstract: A wide range of geophysical techniques is applied in forensic investigations where the target objects are frequently buried under ground and are often small in size. These include targets which are only a few centimetres in diameter but located within a large search area are often of several hectares. The application of a specific geophysical technique may be governed by the physical properties of the target object and its local surroundings. However, operational and financial costs are important. Topographical and geological variations or presence of man-made structures may hinder the application of the most cost-effective technique. Additionally, site area, logistics and weather conditions are also important factors.

Generally, to overcome these difficulties it is recommended that forensic geophysics be carried out along conventional geophysical guidelines as used in civil-engineering site investigation. On occasion, departure from these conventional guidelines may be beneficial in that field survey data acquisition can be complemented by simultaneous direct invasive assessment of geophysical anomalies instead of waiting until office reporting has been completed. Three case studies are presented: one relates to a search for a buried metal target located using a scanning magnetometer with simultaneous excavation, and two relate to searches for graves and buried wooden coffins

Forensic geology (also known as *geoforensics*) is a specialist branch of geology that is concerned with the applications of geology to solving crimes (Donnelly 2002*a, b*). Forensic geophysics is not formally defined or listed in the encyclopaedic dictionary of exploration geophysics (Sheriff, 1994) but we define it simply as '*the application of geophysical methods related to legal investigations*'. This could equally apply to both criminal and civil investigations, to which might be added military applications, for example one early military application was the detection of the location of enemy artillery guns during the First World War war using acoustic and seismic techniques. A description of these techniques is given in Lawyer *et al.* (2001). More recently, in the last 20 years, the application of geophysical methods in archaeological surveying, plus advances in geophysical instrumentation and computing technology, has allowed geophysicists to conduct high-resolution surveys of the top 1–2 m below ground surface. A geophysical methodology is proposed and some survey results reviewed.

Geophysical background

McCann *et al.* (1997) proposed a methodology for geophysical surveys in site investigations. A similar approach can be adopted in forensic geophysics.

The first stage consists of an initial desk or background study of the survey area, in which all available relevant information about a site is collated. This includes:

- Present and historical topographical maps, usually from the Ordnance Survey;
- Geomorphological research studies and reports;
- Present and historical geological survey maps and associated descriptive memoirs from the British Geological Survey;
- Aerial and satellite photography, both current and historical;
- Present and historical soil survey maps with surface vegetation detail;
- Web search of the English Heritage database of geophysical survey results related to archaeological investigations (Linford 2002);
- Library/web search for relevant scientific and press publications and photographs, including university research papers (this involves access to specialized libraries and local and county records offices.
- Information on the nature and physical properties of the survey target (e.g. buried metallic weapon, victim's discarded clothing, on buried human remains).

In addition, a visit to the site with a detailed 'walkover' survey is necessary in order to obtain what can only be described as a 'feel' of the site (i.e. putting the desk study into context). The site visit should include a trial of geophysical instruments and an

From: PYE, K. & CROFT, D. J. (eds) 2004. *Forensic Geoscience: Principles, Techniques and Applications*. Geological Society, London, Special Publications, **232**, 11–20. © The Geological Society of London, 2004.

Table 1. *Summary of geophysical survey methods. After Kearey and Brooks (1994).*

Method	Measured parameter	'Operative' physical parameter
Seismic	Travel times of waves Reflected/refracted seismic waves	Density and elastic moduli which determine the propagation velocity of seismic waves
Gravity	Spatial variations in the strength of gravitational field of the Earth	Density
Magnetic	Spatial variations in the strength of geomagnetic field	Magnetic susceptibility and remanence
Electrical resistivity induced polarization	Earth resistance, polarization voltages or frequency-dependent ground resistance	Electrical conductivity Electrical capacitance
Self-potential	Electrical potentials	Electrical conductivity
Electromagnetic	Response to electromagnetic radiation	Electrical conductivity and inductance
Ground-penetrating radar	Response to high-frequency electromagnetic radiation	Electrical conductivity and dielectric constant

assessment of the physical properties of soils and rocks, including a review of recent publications.

These are all time-consuming tasks but are essential if a satisfactory geophysical survey is to be carried out.

The second stage is to consider the possible success of applying a particular geophysical method (or methods) to a specified target which is to be located in an area of which the surface topography, geological setting, soils distribution and physical properties have been collated during the desk study. The results of the trial of geophysical instruments at site are important in this consideration.

In addition to an understanding of the capabilities of geophysical methods, it is essential that the limitations of these methods be clearly understood. MacDougall *et al.* (2002) point out various limitations which can make areas unattractive for geophysical surveys. Some of these are summarized below.

- Presence of man-made metallic features at ground surface e.g. gates, buildings, fences, overhead power cables, parked or moving vehicles and machinery;
- Presence of man-made metallic and non-metallic features below ground surface e.g. cables, pipes, sewers, reinforced oncrete;
- Severe ground topography;
- Access problems e.g. bushes and vegetation;
- Presence of farm animals;
- Current construction or farming activities;
- Electrical interference e.g. mobile phones, electrical machinery, power cables;
- Seasonal factors e.g. tourists, weather.

Most of these limitations can be assessed during the 'site walk-over' that is which should be part of the desk study.

The physical property contrast between a target and the surrounding host material of soil and rocks is essential if geophysical methods are to be effective. The main geophysical methods and their operative physical parameters have been summarized by Kearey and Brooks (1984) and are reproduced in Table 1 with the addition of the ground-penetrating radar (GPR) method.

A further consideration relates to the important factor of detection probability and spatial survey measurement density. Firstly an estimate has to be made of the likely target area and the volume and depth of burial. Next, boundaries must be placed upon the surface area to be surveyed. Benson and Yuhr (1996) have investigated this problem and discuss the quantity of geophysical measurements required to locate a known area of surface target within an overall search area. With reference to their simple area plan (see Fig. 1) the authors define the target area as At, which is considered to be located in the search area As, allowing a search to target area of As/At to be calculated. In the example given, a site to target ratio As/At of 10 is shown. If we assume that a buried body exhibits a surface area of $c.$ 1.0 m^2, and that it is located within a survey area of $20 \times 50m$ ($1000\,m^2$), then As/At = 1000. For a 90% probability of target detection 1300 geophysical station measurements are required. To approach a probability of detection of 100% at least 1600 station measurements should be taken.

It is important to assess the As/At ratio and the likely number of required geophysical measurements to obtain a good probability of target detection. Obviously, for a good probability a regular survey grid pattern is required. This may be obtained either by individual station grid measurements or by closely spaced profile lines using a 'walking' (mobile continuously reading) magnetometer, GPR or inductive conductivity (IC) unit.

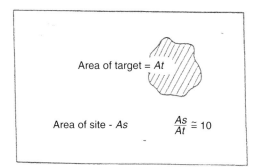

Area of target = At

Area of site - As $\frac{As}{At} \cong 10$

Fig. 1. Detection probability example. After Benson and Yuhr (1996).

By no means least is the capital purchase or rental costs (and availability) of geophysical instruments and the availability of experienced geophysicists to carry out the survey work and interpretation. Typical geophysical daily survey rates for these various geophysical methods have been summarized by MacDougall *et al.* (2002).

Geophysical methods and their applications

The applicability of geophysical methods in forensic geophysical surveys can therefore be summarized as follows.

Seismic method

Generally, the seismic method is utilized in refraction and reflection mode and is either applied to layered geological structures or used to determine depth of bedrock beneath superficial deposits. As such, it is rarely applicable in forensic geophysics where a distinct target, such as a buried human body, is the survey objective. However, Hildebrand *et al.* (2002) compared seismic reflection and GPR imaging over a dead pig buried in a wooden coffin at a test site in Illinois, USA. They were able to detect this target at a depth of 1.6–2.4 m below surface by both methods. It would appear that the geophysical anomalies were related to the wooden coffin rather than the dead pig which it contained. In terms of survey and interpretation duration, the GPR survey was many times faster than the seismic reflection survey.

Gravity method

The gravity method is a costly and time-consuming technique, usually employed to detect subsurface cavities, such as caves, graves and disused mine shafts. The physical property contrast is the density difference between local rocks and an air-filled cavity. Emsley and Bishop (1997) describe the application in cavity location but this method has been rarely applied in forensic geophysics.

Magnetic method

The naked human body has virtually no associated magnetic anomaly and, when buried, is very unlikely to be detected by a magnetic survey. However, a fully clothed body is a different matter. Clothing may include metal buttons, zip-fasteners, shoe eyelets and belt buckles, while pockets may contain spectacles, keys, pens and other ferrous metallic objects. The magnetic response of any such items will be small, usually a positive anomaly of 1–5 nT (nanotesla) at a buried depth of 50 cm. In comparison, 1 kg of iron buried at a depth of 100 cm gives a positive response of *c.* 50 nT. Both responses are measured at ground level. Magnetic anomalies such as those described are easily within the measuring capability of modern proton, optically pumped and fluxgate magnetometers/gradiometers, provided that the soil and rock surrounding the target consists of uniformly low-magnetic material. However, the possibility of a negative magnetic anomaly associated with a human body devoid of any ferrous objects and located in a highly magnetic topsoil should be considered.

Magnetometer surveys may be carried out with continuous recording at a walking pace, but survey lines must be very closely spaced in order to detect personal magnetic items and scanning fluxgate magnetometers with sensors 10 cm above ground surface may be most effective. Breiner (1973) gives a comprehensive study of the survey applications of portable magnetometers.

Resistivity method

Contrasts in the electrical resistivity between a target and its surroundings can be delineated using a range of resistivity depth-sounding and profiling techniques. All involve inserting four steel electrodes into the ground and measuring vertical and horizontal variation in resistivity. Recently, multi-electrode arrays have been utilized which involve up to 80 electrodes producing a resistivity cross-section or image of the subsurface. The application of this array configuration to shallow-depth investigations are summarized by Clark (1996) and Noël (1992). Barker (1997) describes a number of case histories in which this resistivity-imaging method was applied to ground-engineering investigations. It

should be stressed that resistivity is influenced by soil water content and chemistry, soil grain structure and pore space. A target resistivity contrast in a host material may vary depending upon whether the host material is dry or water-saturated. In the case of a buried body, no published case histories appear to be available that describe direct location by resistivity methods.

Induced polarization method

This method is similar to resistivity profiling and is used in disseminated metallic ore prospecting. The induced polarization (IP) effect is a transient voltage which is observed after current flow ceases in a resistivity array. The method has been investigated in archaeological surveying by Aspinall and Lynam (1970). These authors found the method to be slower than resistivity surveys and less effective. No published case histories utilizing the IP method in buried body location are available.

Self-potential method

This is a naturally occurring ground potential due to electrochemical reactions between different rocks and groundwater levels and flow. It is a very simple and inexpensive technique that utilizes two non-polarizing ground electrodes and a millivoltmeter. Typically it has been used in the location of metallic sulphides but it is being used increasingly in the mapping of geological boundaries and cavity features. No relevant publications are available with reference to buried bodies, but the method may have an application in mapping disturbed soils over a buried body. Corwin (1990) describes the self-potential (SP) method used in environmental and civil engineering applications.

Electromagnetic methods

An effective and rapid surveying alternative to resistivity profiling is the electromagnetic inductive conductivity (IC) profiling method, which allows continuous recording of the subsurface conductivity at a walking pace. Electrical conductivity is the reciprocal of electrical resistivity. Two instrument types are available; one measuring conductivity to c. 1.5 m subsurface, the other to c. 7 m. Clark (1996) discusses their application. These instruments are one-man, portable and operate in the frequency range of 10–15 kHz. No case histories relating to direct detection of buried human remains are known but the application of the method in defining archaeological features such as graves and tombs are given

in Frohlic and Lancaster (1986) and Dalan (1991). As with resistivity surveying, if a buried body and the associated soil disturbance caused by burial give rise to an electrical resistivity contrast against the local host material this would be an effective method.

The much higher frequency range of 25 MHz to 2 GHz is the realm of GPR which has received substantial publicity for its ability to produce high-resolution cross-sections of the subsurface. GPR applications have received major publicity in the case of buried murder victims in Cromwell Street, Gloucestershire (the so-called 'Fred West Murders'), and in locating a buried cache of ransom money in the case of the kidnapped estate agent Stephanie Slater in 1992. No technical publications of these applications appear to have been presented.

Certainly GPR is very effective at locating buried graves that are lined 'cavity' structures. Examples are given in Bevan (1991). The direct location of a buried body by GPR in a search area is more problematical. Certainly GPR experiments by Hildebrand over buried pigs at test sites indicate that GPR can locate such objects. Additionally, more realistic tests have been carried out at the University of Tennessee Forensic Anthropology Unit, where donated human cadavers are buried. A recent GPR study at this facility by Miller (2002) investigated the effects of buried decomposing human body targets using GPR over a period of time. Changes in GPR anomaly response are compared with stages of decomposition.

However, in most test area GPR studies, the investigator is aware of the target location, and additionally, test sites are frequently constructed in uniform ground. Care must be taken in extrapolating test area results to actual field survey conditions. Mellett (1992) is credited with the first discovery of a buried human body in the USA using GPR. It should be noted that other evidence had considerably reduced the search area and that the victim wore synthetic clothing and was buried at a depth of 0.5 m. Mellett (1996) discusses eight different GPR searches for human body remains in which only one, that described above, was successfully located. GPR may be more successful in the indirect location of a buried body by delineating the change in physical properties of disturbed soil overlying the cadaver. Similarly, GPR surveys over a concrete floor may detect anomalies resulting from a void beneath the concrete being caused by compaction of underlying disturbed soil or by the space occupied by the body beneath.

Metal detector methods

These instruments are based on the pulse induction or time-domain principle and are one-man, portable hand-held scanning devices with an audible signal or meter output. More sophisticated units embody interchangeable search heads and can locate metal coins at depths up to 0.5 m while larger targets, such as metal spades, may be located at depths of 1 m. This is usually the limiting depth of this type of metal detector unless a very large metal object, such as military ordnance, is involved. Gaffney and Gater (2003) describe the application of the metal detector in archaeological investigations while Das *et al.* (2001) carried out a series of comparison tests of the application of metal detectors in the detection of buried land-mines. A wide selection of metal detectors is now available because of the large number of amateur users, and details can be found in popular magazines such as *Treasure Hunting*.

Standards/procedures

No formal standards or codes of practice exist for forensic geophysical investigations but useful guidelines for high-resolution, shallow-depth geophysical surveys have been published by McCann *et al.* (1997) and Milsom (2003). Additionally, David (1995) details basic guidelines and good practice in archaeological geophysical surveys. Gaffney *et al.* (2002) also provide detailed information on the use of geophysical techniques in archaeological evaluations and give good practice guidance.

Physical properties

The physical properties of soils and rocks have been investigated and compiled over many years, and a series of laboratory and field tests for measuring resistivity, density, magnetic susceptibility, dielectric constant and seismic velocity have evolved. These are detailed by Hallenburg (1997). The physical properties of decaying human bodies are not well known but research on actual cadavers is being carried out at the Tennessee, USA, test site. Killam (1990) reports on this and other research. Davenport *et al.* (1992) report on geophysical investigations to locate dead pigs buried in order to simulate human cadavers. In the UK, Home Office sponsored research has been conducted over buried pigs but the results are not in the public domain. Bournemouth University has carried out resistivity experiments over buried pigs (Cheetham 2003 unpublished observations).

Physical properties of a decaying body, such as resistivity (conductivity), density and dielectric constant will vary with state of body decay. Additionally, the physical properties of soil and weathered rock will vary according to soil type, degree of weathering and moisture content.

The above considerations are essential in deciding whether a geophysical method will be effective in a body search at a particular location. Theoretical modelling of possible responses from various geophysical methods is also impeded by this lack of physical property knowledge, but useful theoretical simulation studies have been carried out. For example, theoretical GPR responses from buried human remains have been researched by Hammon *et al.* (2000).

Multisensor applications

In attempting to survey large site areas more cost effectively and to ensure a uniform survey coverage, geophysicists have developed arrays of geophysical sensors which can be man-carried or vehicle-mounted and walked/towed across site. Frequently, differential global-positioning systems (DGPS) are integral to such an array in order to provide a location for the geophysical readings and to control survey traverse progress. Such systems involving GPR, magnetic and electromagnetics have been developed for unexploded ordnance (UXO) location and, to a minor extent, archaeological surveys. Fera *et al.* (2003) describe a magnetic survey carried out in Austria using five magnetic sensors mounted on a non-magnetic vehicle. Erkul *et al.* (2003) describe a motorized multisensor system involving five fluxgate magnetometer sensors. The advantages of using mobile multi-sensor systems in forensic studies are clearly apparent where there are large tracts of survey and copious detailed data points are required.

Case histories

We now present three case histories of the application of geophysical surveys in forensic investigations in which we were personally involved. These examples have been selected to illustrate the complex nature and variety of circumstances encountered in forensic geophysical investigations.

1. Vault detection using GPR, Hampshire

This investigation was carried out by Structural Testing Services (UK) Ltd, (Fenning, 1988) to detect concealed graves in the form of stone vaults within a cemetery measuring 120×70 m. The locality was several kilometres north of Portsmouth, UK. A cavity, resuting from a collapse of the vault ceiling, had occurred beneath a pathway across the cemetery and constituted a hazard. The possibility of similar collapses from other concealed vaults had to be investigated. Examination of the collapsed vaults

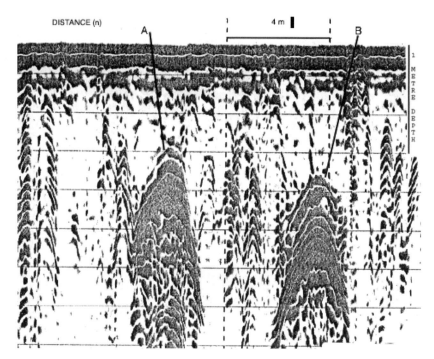

Fig. 2. GPR profile over buried graves.

indicated a depth of 1 m. Given a typical vault surface area (At) of 3.5m² and a survey area (As) of 13600 m², then the As/At ratio is c. 4000.

The situation was complicated by surface stonework structures, such as other graves and vaults, and the presence of numerous decorative metal enclosure rails was evident. The surface topography was reasonably flat, and local investigations indicated that the vaults were located within topsoil and drift with no evidence of bedrock. Selecting a geophysical method to detect near-surface voids could involve most geophysical methods.

The relevant physical property contrast in this case is the air-filled void, possibly with stone vertical margins within topsoil, and drift, with the collapsed vault being above the water table and possibly containing relict ferrous objects. The magnetic method was rejected because of the surface metallic objects while galvanic resistivity profiling would have been difficult to perform due to stone surface obstructions. Although the gravity method was considered initially, it was ruled out on cost grounds. It was decided to conduct trials using both IC measurements with a Geonics EM-31 and GPR with a Geological Survey Systems, Inc. (GSSI) SIR3 unit with a range of frequency antennas. Trial surveys indicated that both methods could detect voids. The IC anomaly gave a negative read-out across the void while the GPR was most effective at a frequency of 500 MHz and yielded a pronounced hyperbolic anomaly.

During the course of the survey it became apparent that the IC measurements were being adversely affected by the presence of decorative ironworks but the GPR was very effective. A total of 33 GPR anomalies was interpreted as being indicative of vault voids, and these were marked at site for future invasive investigations. The results of these invasive investigations are not known. Figure 2 depicts a typical GPR section 12 m in profile length with a number of distinctive hyperbolic reflectors. The two major reflectors are beneath point A at c. 1 m below surface and point B at 1.2 m subsurface. Other minor hyperbolic reflectors are interpreted as relict metal objects or buried items of stonework

2. The Moors Murders, Saddleworth Moors, northern England

A second case history involved the search for the buried body of Keith Bennett, a murder victim in the so-called 'Moors Murders', who disappeared in July 1964. His body is believed to be buried at Saddleworth Moor, West Yorkshire. The background history is detailed by Topping and Ritchie (1989), who indicated that the body target was likely to be found buried at a depth of 0.5–1.0 m below surface,

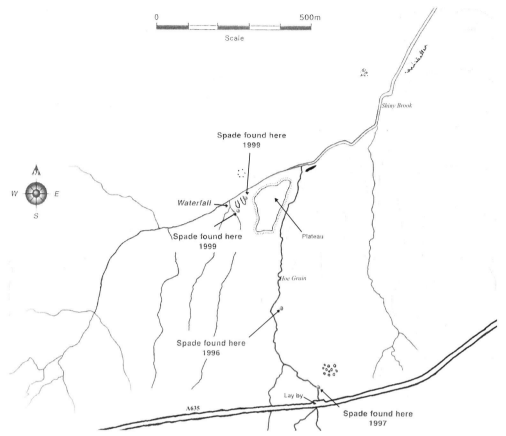

Fig. 3. Sketch map of survey site, Saddleworth Moor, West Yorkshire.

within the peat and possibly adjacent to a stream. There was a possibility that a metal spade used to bury the victim was also buried at this site.

The body has been buried for almost 40 years and the search area is at least 5km². The *As/At* target area ratio is estimated to be in the realms of several million. The site is inhospitable, with surface peat bogs, bog bursts and outcrops of rock and generally covered in rough heather and grass. Surface topography is very uneven, with concealed gullies and circular depressions. The solid geology is Carboniferous, with shale, thin mudstones and siltstone, plus occasional exposures of more durable millstone grit and sandstones known locally as Kinderscout grit. In the summer months the surface peat cover varied from dry on the stream slopes to water-saturated adjoining the stream, with a consequent wide variation in other physical properties. Hobbs (1986) comments on the wide density variation of peat with varying moisture content. Likewise, resistivity at site varies similarly, according to the degree of water content and compaction. A site plan is shown in Figure 3.

An assessment of the likely physical contrast between the target and host peat or alluvial is difficult. After almost 40 years the target could be well preserved if it was buried in the peat but little might remain if buried in alluvial deposits. The physical properties of peat, particularly its resistivity and dielectric constant, will vary considerably, depending upon the degree of water saturation. Given this problem and the nature of the surface topography, a trial of various geophysical methods was carried out.

Firstly, any mechanically pulled or towed instrument proved inoperable due to the nature of the surface terrain. This eliminated GPR and wheeled time-domain electromagnetics. Where GPR could be deployed, radar profiles exhibited many anomalous subsurface reflective features, probably due to buried sandstone boulders, variable peat thickness and infilled gullies. GPR, even if it could be deployed, would yield many anomalies that would have to be investigated by digging.

Secondly, inductive electromagnetics such as the Geonics EM-38 (penetration depth 1 m) again revealed many anomalous features similar to those revealed by GPR. The gravity and seismic methods were rejected due to cost and time implications and also the nature of the target. Resistivity profiling was also investigated but yielded results similar to the IC. It was concluded that, given the vast search area, lack of a diagnostic physical property contrast, and with no definitive background information to limit the search area, there was little chance of success in directly locating an anomalous human body or decomposed remains. However, evidence was available which indicated that a spade might be buried with the body or nearby. Additionally the victim might have been using metal-rimmed spectacles or wearing clothing with metal zips. A test loop of metal in the shape of spectacles was buried and checked using a scanning fluxgate magnetometer and a metal detector. These instruments successfully located the test source from 0.2m to 0.3 m distance while a buried spade could be detected up to 0.7 m. For operational reasons a Schoenstedt scanning fluxgate magnetometer and GeoInstruments Explorer metal detector were selected.

The magnetometer and metal detector both give an audible output which varies with an increase in either target magnetic susceptibility, (the magnetometer), or in conductivity (the metal detector).

In any search area it is important to ensure 100% ground coverage, and this was achieved by outlining 2 m wide swathes, up to 50 m long, with rope markers. Each swathe was systematically covered in a scanning mode using both instruments. As soon as an anomalous change in signal was obtained the location was marked with a wooden peg and detailed excavation with spade and trowel immediately followed. The site map (Fig. 3) shows the location of four buried spades found during the detailed survey. Additionally a volume of discarded metal objects was found in streams and below ground surface. As far as is known, no definite correlation was made between any excavated metal object and the victim.

In conclusion, the methodology of rapid scanning of a large area using scanning magnetometer and time-domain metal detector followed by almost simultaneous intrusive excavation was employed as an indirect method of buried victim search. However, to date, this search has not been successful in locating the target victim.

3. Search for wooden coffins, southeastern England

The third case history involves a search for buried wooden coffins at a construction site near Chatham, in southeastern England (Hasan & Fenning 1990). During the course of ground excavations, which

involved removing a 2 m thickness of superficial and industrial deposits across a 20 ha site, several small wooden coffins containing human bones were located embedded in the underlying blue-clay strata. A desk study had indicated that a significant number of these coffins, relating to Napoleonic prisoners of war, was buried across this area of approximately 20 ha. A typical coffin measured $2 \times 0.75 \times 0.75$ m, which represents an As/At in excess of 2 million. The survey objective was to investigate the most cost-effective method of detecting buried wooden coffins across this large area.

The physical properties of the clay were initially assessed *in situ* and were found to show a resistivity of 3–10 Ω, while a magnetometer and metal detector scan over an excavated coffin showed virtually no response. This excavated wooden coffin was water-saturated. A large trench had been mechanically excavated to a depth of 2 m through the overlying fill to the top surface of the blue clay, and several coffins had been located by mechanical probing methods. A geophysical trial was conducted using the following methods:

- GPR: GSSI-SIR3 with 900 500 and 100 MHz antennas;
- Inductive conductivity: Geonics EM-31 and EM-38. Figure 4 depicts a field operator carrying out IC measurements using an EM-38 over the marked location of a buried coffin;
- Resistivity profiling: ABEM Terrameter SAS 300C with Wenner array traversing;
- Magnetics: EDA Omni-IV proton magnetometer in conjunction with a second unit deployed as a diurnal monitor;
- Metal detector: White's pulse induction type.

A series of composite geophysical profiles using the above methods was carried out along two test lines, each 12 m in length. One test profile was located over four known *in situ* coffins separated c. 1–2 m apart along the profile. The second test profile was conducted over two known *in situ* coffins located 1 m apart. GPR profiles at the three frequencies showed no discernable anomalies coincident with known coffin locations. Similarly, resistivity Wenner profiling with interelectrode spacings of 0.5 and 1 m revealed no resistivity anomalies. Total field magnetic profiles and metal detector scans also revealed no anomalies. Additional composite surveys over an area measuring 10×50 m, using an 0.5 m grid, again yielded no significant anomalies.

The results were inconclusive and no definitive geophysical anomalies were noted over the coffin locations using the above methods.

The explanation for this lack of success of the geophysical surveys was simply that the buried waterlogged coffins possessed similar physical

Fig. 4. Photograph showing field operator carrying out inductive conductivity measurements over the marked location of a buried coffin.

properties to the blue clay, and also any metal used in Napoleonic coffin construction was degraded by rust. The solution to this detection problem proved to be mechanical probing using metal rods pushed into the wet clay by hand and noting the variation in impact signal when the wooden coffin lid was encountered.

Conclusions

Conventional police searches for murder victims' graves or buried artefacts rely on recognizing obvious visual signs of ground disturbance and discarded items such as clothing. These are undertaken by large-scale ground searches and trial and error excavations. These are manpower intensive, cost prohibitive, often non-productive and can destroy criminal evidence, and cause subtle ground disturbances. Carefully designed and implemented geophysical surveys, when integrated with other scene of crime and forensic investigations, can provide alternative cost-effective methods for searching the ground. The following conclusions are drawn:

(1) The physical properties of buried human remains when contrasted with the variety of host geological environments of burial present a difficult problem that requires further research.
(2) The ratio of area of target burial to the likely search area should be considered prior to any geophysical area. Large ratios may preclude cost-effective geophysical surveys.
(3) Indirect detection is feasible if associated human artefacts such as a victim's personal effects are considered in addition to direct detection of human remains.

(4) The conventional methodology for deploying geophysical surveys in site investigations should be applied to forensic geophysical surveys.

The authors would like to thank A. Aspinall and an anonymous reviewer for constructive comments for manuscript improvement. We are grateful to V. Dennis for her technical assistance in the preparation of this manuscript and accompanying figures.

References

ASPINALL, A. & LYNAM, J. T. 1970. An induced polarization instrument for the detection of near surface features. *Prospezioni Archeologiche*, **5**, 67–75.

BARKER, R. D. 1997. Electrical imaging and its applications in engineering investigations. *In*: McCANN, D. M., EDDLESTON, M., FENNING, P. J. & REEVES, G. M. (eds) *Modern Geophysics in Engineering Geology*. Geological Society, London, Engineering Geology Special Publications, **12**, 37–43.

BENSON, R. C. & YUHR, L. 1996. *An Introduction to Geophysical Techniques and their Application for Engineers, Geologists and Project Managers*. Technos Inc., Miami.

BEVAN, B. W. 1991. The search for graves. *Geophysics*, **56**, 1310–1319.

BREINER, S. 1973. *Applications Manual for Portable Magnetometers*. Geometrics Inc., San Jose.

CLARK, A. 1996. *Seeing Beneath the Soil: Prospecting Methods in Archaeology*. Routledge, London.

CORWIN, R. F. 1990. The self potential method for environmental and engineering applications. *Geotechnical and Environmental Geophysics*, **1**, 127–145.

DALAN, R. A. 1991. Defining archaeological features with electromagnetic surveys at the Cahokia Mounds state historic site. *Geophysics*, **56**, 1280–1287.

DAS, Y., TOEWS, J. D., RUSSELL, K. & McFEE, J. E. 2001. Laboratory in-air testing of metal detectors. *In*: DUBEY,

A. C., HARVEY, J. F., BROACH, T. & GEORGE, V. (eds) Detection and remediation technologies for mines and mine like targets.? *Proceedings of the Society of Photo-optical Instrument Engineers* , **4394**, 8–19

DAVENPORT, G. C., FRANCE, D. L. *et al.* 1992. A multidisciplinary approach to the detection of clandestine graves. *Journal of Forensic Science,* **37**, 1445–1458

DAVID, A. 1995. *Geophysical Survey in Archaeological Field Evaluation. Research and Professional Services Guidelines, No. 1.*, English Heritage, London.

DONNELLY, L. J. 2002*a*. *Forensic Geology (Geoforensics): How Do Geologists Help Solve Crimes.* IMC Group Consulting Ltd & British Geological Survey, Special Report.

DONNELLY, L. J. 2002*b*. Finding the silent witness – how forensic geology helps solve crimes. All-Party Parliamentary Group for Earth Science, Geological Society of London. *Geoscientist,* **12**, 24.

EMSLEY, S. J. & BISHOP, I. 1997. Application of the microgravity technique to cavity location in the investigation for major civil engineering works. *Modern Geophysics in Engineering Geology*, Geological Society Engineering Group Special Publication, **12**, 183–192

ERKUL, E., RABBEL, W. & STUMPEL, H. 2003. Development of a mobile multi-sensor system: first results. *Archaeologia Polonia*, **41**, 159–160.

FENNING, P. J. 1988. *Geophysical Investigation to Detect Concealed Vaults at a Cemetery in Hampshire.* Structural Testing Service (UK) Ltd, Southampton, Internal Report 2004.

FERA, M., NEUBAUER, W., DONEUS, M. & EDER-HINTERLEITNER, A. 2003. Magnetic prospecting and targeted excavation of the prehistoric settlement platt – Reitlüsse, Austria. *Archaeologia Polonia*, **41**, 165–167.

FROHLIC, B. & LANCASTER, W. J. 1986. Electromagnetic surveying in current middle eastern archaeology: application and evaluation. *Geophysics*, **51**, 1414–1425.

GAFFNEY, C. & GATER, J. 2003. *Revealing the Buried Past: Geophysics for Archaeologists.* Tempus Publishing, Gloucester.

GAFFNEY, C., GATER, J. & OVENDEN, S. 2002. The use of geophysical techniques in archaeological evaluation. Institute of Field Archaeologists, University of Reading, Papers, **6**.

HAMMON, William, S., MCMECHAN, G. A. & ZENG Xiaoxian. 2000. Forensic GPR-finite-difference simulation of responses from buried remains. *Journal of Applied Geophysics*, **45**, 171–186.

HALLENBURG, J. K. 1997. *Non-hydrocarbon Methods of Geophysical Formation Evaluation.* Lewis Publishers, London.

HASAN, S. & FENNING, P. J. 1990. *Geophysical Investigations at Chatham Dockyard.* Internal Report 3239/1, Structural Testing Services (UK) Ltd, Southampton.

HILDEBRAND, J. A., WIGGINS, S. M., HENKART, P. C. & Conyers, L. B. 2002. Comparison of seismic reflection and ground penetrating radar imaging at the controlled archaeological test site, Champaign, Illinois. *Archaeological Prospection,* **9**, 9–21.

HOBBS, N. B. 1986. Mire morphology and the properties and behaviour of some British and foreign peats. *Quarterly Journal of Engineering Geology*, **19**, 7–80.

KEAREY, P. & BROOKS 1984. An Introduction to Geophysical Exploration. Blackwell Science Ltd, Oxford.

KILLAM, E. W. 1990. *The Detection of Human Remains.* Charles C. Thomas, Springfield.

LAWYER, L. C., BATES, C. C. & RICE, R. B. 2001. *Geophysics in the Affairs of Mankind – A Personalised History of Exploration Geophysics.* Society of Exploration Geophysicists, Tulsa.

LINFORD, N. 2002. The English Heritage Geophysical Survey Database. World Wide Web Address: http://www.eng-h.gov.uk/SDB/

MACDOUGALL, K. A., FENNING, P. J., COOKE, D. A., PRESTON, H., BROWN, A., HAZZARD, J. & SMITH, T. 2002. *Non-intrusive Investigation Techniques for Groundwater Pollution Studies.* Environment Agency, Bristol, Research & Development Technical Report P2-178/TR/10.

MCCANN, D., CULSHAW, M. G. & FENNING, P. J. 1997. Setting the standard for geophysical surveys in site investigation. *In:* MCCANN, D. M., EDDLESTON, M., FENNING, P. J. & REEVES, G. M. (eds) *Modern Geophysics in Engineering Geology.* Geological Society, London, Engineering Geology Special Publications, **12**, 3–34

MELLETT, J. S. 1992. Location of human remains with ground penetrating radar. *Fourth International Conference on GPR.* Geological Survey of Finland Special Papers, **16**, 359–365.

MELLETT, J. S. 1996. GPR in forensic and archaeological work: hits and misses. *In: Symposium on the Application of Geophysics to Environmental Engineering Problems,* (SAGEEP) 1991. Environmental & Engineering Geophysical Society Co., USA. 487–491.

MILLER, M. 2002. *Coupling ground penetrating radar applications with continually changing decomposing human targets: an effort to enhance search strategies of buried human remains.* MA thesis, University of Tennessee.

MILSOM, J. 2003. *Field Geophysics.* Open University Press, Milton Keynes.

NÖEL, M. 1992. Multielectrode resistivity tomography for imaging archaeology. *In:* SPOERRY, P. (ed.) *Geoprospection in the Archaeological Landscape.* Oxbow Monograph, **18**, 89–99, Oxbow Books, Oxford.

SHERIFF, R. E. 1994. *Encyclopedic Dictionary of Exploration Geophysics.* Society of Exploration Geophysics, Tulsa.

TOPPING, J. & RITCHIE, J. 1989. *Topping: The Autobiography of the Police Chief in the Moors Murder Case.* Angus & Robertson, London.

Geophysics and burials: field experience and software development

M. WATTERS & J. R. HUNTER

Institute of Archaeology and Antiquity, University of Birmingham, Birmingham B15 2TT, UK
(e-mail: Meg_watters@hotmail.com; j.r.hunter@bham.ac.uk)

Abstract: In the UK, geophysical survey methods have been increasingly applied in the search for clandestine burials and in the elimination of land from murder enquiries as part of a broader spectrum of systematic search techniques. This paper considers recent forensic search work and stresses the general resources and intelligence needed for a successful search, and the role played by geophysics. The main emphasis, however, is to discuss the applicability of ground-penetrating radar (GPR) in grave detection and to review recent developments in software, particularly the importance of high-resolution mapping. The paper also discusses the use of image analysis of both planes through blocks of buried landscape, together with predictive modelling and the value of 3-D analysis for data manipulation.

This paper combines the experience of a forensic archaeologist (JH) and an archaeological geophysicist (MW) in considering the applicability and potential of geophysical survey techniques, not only in locating buried human remains but also in looking at potential imaging of a burial itself for a more informed approach to the site. The archaeological background is based on some 60 individual murders and, more recently, work in the context of mass graves in the Balkans. The geophysical background comes from a wide exposure to different types of sites and features, non-invasive imaging and software development. Forensic geophysics is a relatively recent field of application and has little published literature (but see Hunter *et al.* 1996, p. 96*ff*). Although geophysics has persistently played a part in the wider repertoire of search techniques (Killam 1990; France *et al.* 1997; Davenport 2001) it is just now being examined and is becoming a forensics tool with considerable research potential (eg Nobes 2000; Hunter *et al.* 2001; Buck 2003).

The wider context of forensic search

Much of the experience which underpins this paper has been based on the search for buried remains in a variety of different scenarios and contexts in the UK. These searches also served to expose a distinct lack of awareness of systematic search techniques within UK police forces. They identified, for example, a general unfamiliarity with the underlying principles of various techniques, a misunderstanding of the value and limitations of particular methods, and a reluctance to understand the desirability of using complementary techniques. The use of ground-penetrating radar (GPR), for example, achieved a period of vogue after the well-publicized Cromwell Street burials in Gloucester in 1994 and thereafter was used

in a number of instances where it was clearly inappropriate, usually by operators drawn from the military or engineering sectors. A more in-depth and refined use of GRP is discussed below. There was a perception too, that search methods were looking for the body *per se*, whereas many methods responded purely to the disturbance brought about by burial rather than by the body itself, and this is still to be fully recognized.

The establishment of the Forensic Search Advisory Group (FSAG) in 1996 was a move to improve the information available on search techniques and is a facility supported by the National Crime and Operations Faculty (CENTREX) and offering a central advice service from a variety of experts, ranging in scope from aerial imagery, through geophysics and fieldcraft to cadaver dogs. The FSAG allows the user (usually a senior investigating officer, scientific support manager or crime scene investigator) access to impartial advice and recommends specific techniques, sequences of techniques and experienced specialists according to scenario. The FSAG also undertakes research and experimental work to further develop the effectiveness of forensics investigations in the UK.

A significant factor in searching for buried human remains is the fact that the buried human body is a relatively small, shallow, subsurface target that can present a challenge for detection through geophysical techniques. Geophysical mapping applications and methods (mainly addressing the method of GPR) have developed over the past 30 years to fulfil a variety of functions, including straightforward geological mapping, locating unexploded ordnance, analysing concrete reinforcements (http 1) and even providing treeroot imaging (United States Department of Agriculture Research, Triangle Park, North Carolina, USA). Since the 1940s, archaeological applications have provided the basic methodology

From: PYE, K. & CROFT, D. J. (eds) 2004. *Forensic Geoscience: Principles, Techniques and Applications.* Geological Society, London, Special Publications, **232**, 21–31. © The Geological Society of London, 2004.

for high-resolution data collection, processing and imaging (see Scollar *et al.* 1990; Clark 1996; Conyers & Goodman 1997; Gaffney *et al.* 2002). Forensic targets such as recent human remains, unlike typical archaeological targets, possess a unique decay dynamic (eg Haglund & Sorg 1997, 2002). This decay process can have a major effect not only on how the burial might best be detected, but also on the time frame in which detection might be either minimal or optimal. Some pioneering work in this field has been undertaken by the US group NecroSearch International (see Jackson 2002 for context) and the results widely disseminated (eg. France *et al.* 1992, 1997). Work has also been undertaken by the FSAG in the UK, as well as by other agencies in Europe and the USA where actual human remains have been used for geophysical purpose (http 2).

UK research has hitherto concentrated on controlled burials using pigs that have died from natural causes. These have been undertaken in a number of locations and have focused on specific objectives (Hunter 1999). In one experiment in a moorland environment resistivity measurements were taken over a number of pig graves for several months in order to assess the optimum time for detection in relation to a control grave. Resistivity survey measures the variation across a given area (see Clark 1996) and has the potential to identify disturbances in the ground on the assumption that the difference in moisture content between disturbed and undisturbed ground will provide an observable difference in electrical resistance. The graves were of consistent dimensions, depth and orientation and included a control grave, but the animals were wrapped using different materials. Results demonstrated that resistance change occurred most prominently at around 16 weeks and could be attributed to the onset of decay which caused the gravefill to become wetter. There was, however, some variation in this according to wrapping. In another experiment, using GPR and with pig burials under purpose-built concrete and hardcore, the burials were detected much earlier, and detection was enhanced as time progressed, possibly as a result of the collapse of the body creating a cavity. It was also observed that a lower frequency antenna was more effective in detecting the burials (Hunter 1999). GPR reads the basic contrast in electrical properties (dielectric permittivities) of the earth and features within it (see Daniels 1996 and Conyers & Goodman 1997.) Perhaps more significantly, this test stressed the difficulty of the interpretation of field data at certain points in the decay process, depending on the alignment of GPR traverse to the axis of the burial. This serves to underline the necessity for operators to have not only familiarity with the appropriate response and decay processes of the target, but also a knowledge of the geophysical technique that they are using and the most effective data collection methods. It may, therefore be difficult for operators familiar only with military or engineering applications to transfer into a forensic environment without a full appreciation of the specialist nature of the work.

A third experimental scenario, again in a different location, was designed to detect heat emitted during the decay process. This demonstrated not only an even shorter interval before decay set in, but also the narrowness of the decay window itself (approximately 3 weeks in this instance). Rates of decay vary significantly and reflect a number of extrinsic and intrinsic variables, all of which can act singly or in combination. An understanding of these variables, and hence the ability to postulate the interval between burial and decay, and the likely duration of the decay process itself, can therefore be a necessary prerequisite for geophysical survey. This information can help ensure that certain methods are deployed at the optimum time, for example the use of thermal imaging during the decay process when heat emission is at a maximum. There is no easy solution, or quick way, to finding buried bodies.

Forensic goals

Search work involves the elimination of suspected burial locations as much as finding the remains in question. Not finding buried remains is not a measure of failure, but a reminder that a forensic search is as much about negative evidence and elimination as it is of discovery. Increasingly, case studies have shown that search enquiries tend to fall broadly into two distinct camps (Hunter 2002): in one, there is typically a named victim and a suspect but the location of disposal tends to be unknown. In the other, there is usually *no* named victim nor even a missing person, and the allegation is typically made some considerable time after the event. Poor memory and alcohol or drug abuse is often a pertinent factor in the allegation. In the latter scenario, the location tends to be very specific, for example in the corner of a garden, under a tree, or below a shed or outbuilding. From a police point of view, the two types of scenario may require subtly different types of investigation. The former requires a positive, considered search in order to find the victim, perhaps using a combination of methods, including aerial photography, geophysical surveying and cadaver dogs. In the latter the exercise is normally geared towards elimination and might, for example, involve geophysics in the first instance, followed by earthmoving machinery to expose large areas of substrates within which disturbances might be evident. The methodologies are therefore not necessarily the same. Both need careful handling. The former is required not only to target specific locations, but also to ensure that areas *not* targeted can be eliminated for good, confident reasons. In either case, the worst possible outcome is one in which the victim is found within an area that has already been eliminated by searching. Geophysics has a serious role to play in both these scenarios.

Working in a search environment for buried human remains has also emphasised the importance of continuity – namely that while search for buried remains itself is a specialist area, it is not operationally separable from the recovery of those same remains. There is no hard defining line between the two processes, and the factors that affect one (e.g. stratigraphy, geology, hydrology or taphonomy etc) also affect the other. Even careful trial trenching – probably the most invasive part of any search process – can provide key data about the burial (e.g. its depth, the area of evidence potential and burial taphonomics.

In short, geophysical survey information can provide intelligence to support the subsequent recovery operation as part of a continuous process. For example in one typical garden search, geophysics (in this case resistivity) identified three possible targets. Each was carefully tested by trial section, the third locating the desired target. The process of cutting the trial section also had the advantage of identifying the grave edges, the depth of the grave and the depositional process involved in the burial, as well as the condition of remains. Geophysical survey results provided initial information on possible location of the burial, enabling very specific and careful excavation, hence supporting the overall effectiveness of the forensics investigation.

Identification of the victim and *conviction* of the offender also have to be viewed as part of the overall investigative process, and both rely heavily on the integrity of the recovery methodology. In evidentiary terms it becomes extremely difficult to separate the four elements, and the whole process, search, recovery, identification and conviction, has to be seen as continuous. It is also a process that is vulnerable to dysfunction and open to contamination of evidence, and this is why the individual specialists in any one area need to be familiar with the evidentiary requirements of those in the others. Successful completion of the total process requires the buried evidentiary resource not to be damaged or contaminated. To achieve this, the optimum search design is one that moves systematically from non-invasive to invasive methodologies.

Of course, the use of the most efficient non-invasive aids, notably aerial photography and geophysical survey, also brings about its own problems. Identifying anomalies through these methods involves a commitment: the anomalies will need to be investigated by intervention, even if they are of the wrong size, wrong character or located in an inappropriate place. Once these techniques have been applied the enquiry has no option but to follow them through and test the results of each one, irrespective of their perceived value.

In a further typical example, a garden was searched initially by GPR and a number of anomalies identified. All of these, however, were considered to be too small or shallow to have been relevant. Nevertheless, each one was vented for work with a cadaver dog (using a drill in hard surfaces), but the dog responded negatively throughout. After this the senior investigating officer (SIO) needed to make a decision according to the balance of other intelligence available as to whether further investigation was necessary. This depended on a wider range of evidence and on the SIO's confidence in the value of that information. One option was to close down the investigation of the garden at that stage, and another was to become more invasive and thereby increase the level of confidence in elimination. The latter option was chosen, the surface stripped by machine and each individual disturbance tested. The disturbances were found to be rubbish pits and pet burials. A cynical observer might have pointed out that this machine stripping could have been done in the first place without the need for either the GPR or the dog. But that would have meant working invasively from the outset, and it would have resulted in inevitable loss of evidence had the grave been there. The primary use of non-invasive techniques is an evidential safeguard within the sequence of search, recovery, identification and conviction – therein lies a key role for geophysics.

Clearly therefore, the greater the intelligence that can be obtained about a grave, the more likely it is that the evidence can be maximised. This is a matter not just of finding the grave in the first place (although sometimes that can be hard enough), but also of trying to understand the character of the grave – its depth, its profile, the nature of its infilling, or any depositional sequences – in fact any detail that might inform the recovery process and lead to maximization of buried evidence. The remainder of this short paper, in attempting to find some resolution to the very specific problems of forensic search, focuses on GPR applications for not only burial location but also for imaging individual features such as bones, objects and soil interfaces.

GPR and graves

GPR maps the form of contrasting electrical properties (dielectric permittivity and conductivity) of a soil or material below the ground surface. The stronger the difference between the electrical properties of two materials, the stronger the reflected signal in the GPR profile. The conductivity of soils and buried

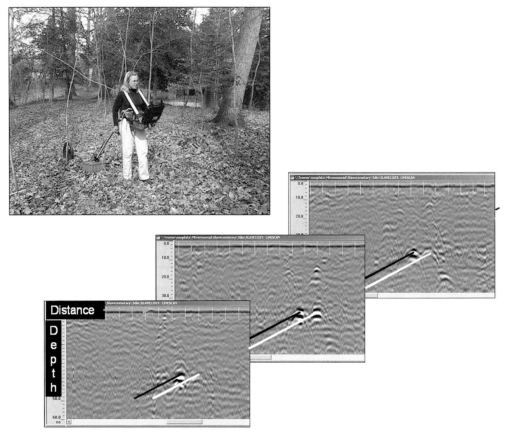

Fig. 1. GPR at Mt Vernon: the SIR2000 with a 400 MHz antenna was used to map slave burials from the time of George Washington. Three 2-D profiles are shown with *hyperbola*, or anomalies, that are identified as possible burials. On individual profiles, positive identification of burials is difficult, but with three adjacent profiles spaced at 1 m, lines can be drawn joining the *hyperbola* and establishing the location of probable burials.

features has the primary control on the attenuation, or loss, of the GPR signal that impacts the effectiveness of GPR survey. Though a highly conductive material will attenuate the GPR signal, it can also be an effective mapping tool, contributing information to the nature of the subsurface and features within it. (Daniels 1996; Conyers & Goodman 1997).

GPR records information on the amplitude, phase and time related to the capture and induction properties of the antenna in addition to the energy propagation, scattering and reflection of subsurface features. Unlike resistivity or other archaeologically-based geophysical methods, GPR data are collected as 2-D vertical profiles into the earth. The 2-D profiles are made up of a number of traces (or scans) at a particular location (x,y) that record the response of subsurface properties to the radar's electromagnetic wave at discrete points at a particular time (or depth) in the earth. The horizontal axis represents surface distance along the transect with the vertical

axis recording time (often referred to as two-way travel time). Time can be easily converted to depth in two ways: the first is by having a known dielectric permittivity value for the material in the survey area; the second is by having a known depth to a feature that appears in the radar profile. The latter is more accurate of these two methods but requires digging or coring. It must be kept in mind that earth properties are not constant and can change drastically over an area. Depth conversion should be checked across a site if possible (Fig. 1).

When considering burial resolution, differences are best viewed based on a relative scale. The outline of graves may be identified at the lowest resolution and individual features, such as bones and artefacts, within a grave may be imaged at the highest resolution (to date, no conclusive research has been conducted that has positively imaged bones within a grave, although this paper proposes a method by which this may be possible). GPR is easily adaptable

 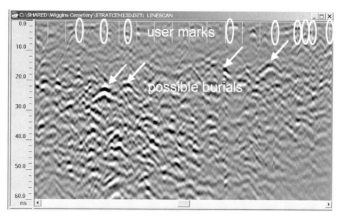

Fig. 2. Data collected at Wiggins Cemetery and house site at Stratham, New Hampshire, USA, with an SIR2 GPR unit and 400 MHz antenna. The 2-D profile displays the glacial till nature of the survey area, with many reflectors being recorded. User marks were inserted in the data when the antenna passed by visible head- and foot-stones for assistance in identification and calibration of GPR anomalies to burials. Possible burials are identified with arrows based on visual *hyperbola* recognition.

to these different scales because of its range of antenna frequencies. These frequencies range between 10 MHz and 1.5 GHz, with the lowest frequencies being used to map in geological and environmental targets with a typical penetration of approximately 30m. The highest frequency of 1.5 GHz, will effectively penetrate to about 1.5–2m in basic, dry loamy soils but often much less, particularly in wet, clay-rich material. The deeper penetration achieved with lower frequency antennas provides a coarser resolution, while the finest feature resolution is achieved with higher frequency antennas (but with a limited depth penetration). The most suitable antennas for grave location and detailed imaging are the 400, 900 and 1500 MHz antennas. This group provides a range of depth and resolution flexibility.

Some previous work with GPR in grave detection has met with mixed success (Bevan 1987, 1991, 1992; http 3). Radar data is complex and without experience and a full understanding of what is being recorded, consistent and effective work cannot be achieved. Even with experience and a familiarity with how a grave may be reflected in GPR data, any GPR expert will be challenged successfully to identify a particular anomaly as a grave. GPR is sensitive to changes in the subsurface electrical properties, and when mapping graves a number of phenomena are being interpreted, such as disrupted soil, solid reflections from the burial itself (metals, artefacts) or voids that may occur as a result of the degradation of the burial etc. Although the search is for the grave itself, GPR will map everything in the ground, for example tree roots and boles, stratigraphic changes, glacial deposits and rodent burrows.

One key factor for success in locating graves is software. Archaeological applications of GPR are increasingly employing time slicing, or plan view imaging, as an aid to data interpretation and, consequently, advanced processing and imaging software appears to be the optimum route to a more intuitive approach to the successful location and mapping of graves (Goodman *et al.* 1995; Goodman 1996, Conyers & Goodman 1997). A few software packages for 3-D imaging of GPR data are now commercially available, for example GPR-Slice (http 4), GPR_Process (http 5), 3D QuickDraw (http 6).

Traditional GPR data formats produce vertical profiles across a site. The normal way to interpret this data is to look at each profile individually, to attempt to classify anomalies based on their appearance and amplitude values, and to plot these anomalies on a site map (Fig. 2). This may work in some situations, but in others it would be impossible. One example of GPR used for grave mapping is at the Wiggins cemetery, a seventeenth-century colonial cemetery located in Stratham, New Hampshire, USA. The site soils are glacial till, very disturbed and filled with lenses of sands, clays, gravels and other glacial deposits, and the site is overgrown by trees and dense ground cover. Prior to GPR survey (approximately 13×18 m) some head- and foot-stones were visible on the site and were mapped. User marks were inserted into the data when the antenna passed either of these targets (Fig. 2).

GPR transects were spaced at 1m intervals: 33 scans per metre were collected along each transect. Because of the visible surface-markers (head- and foot-stones) and the historical background of the site, GPR transects were set up to cross perpendicu-

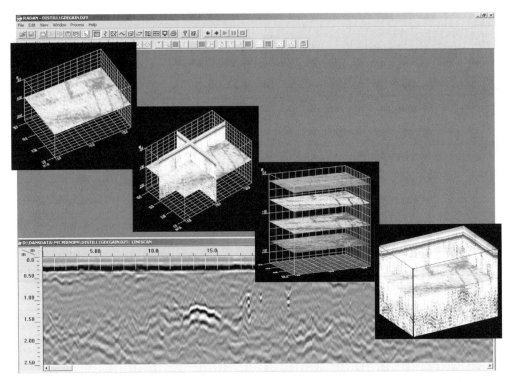

Fig. 3. An example of some of the display capabilities for 3-D imaging of GPR data. Data were collected at George Washington's distillery site at Mt Vernon using an SIR2000 and 400 MHz antenna. Data are processed and imaged in RADAN NT Main and 3D QuickDraw.

lar to the expected orientation of the burials. Typical transect spacing for graves should be 0.25– 0.5 m in order to obtain repeating anomalies in adjacent profiles that will provide a stronger record for positive feature identification. Optimally, GPR surveys should be done in an orthogonal grid (two overlapping grids, oriented perpendicularly to each other) and stitched together in the post-processing software to eliminate the risk of passing over a burial along an orientation that may not reflect the strongest potential signal (see pig burials above). Initial interpretations of graves have been attempted on the vertical profiles (Fig. 2). The anomalies identified in this way are based solely on the fact that they are the strongest anomalies in the profile (and may correspond to the user marks inserted at head- and foot-stone locations). Because of the nature of the site this method of feature identification was ineffective and the confidence level of burial identification was low.

GPR data interpretation for near-surface application has advanced since its beginnings in the early 1990s, when Goodman created the first GPR-Slice program (oil/gas industry excepted) and showed the archaeological community the potential of GPR imaging (Goodman & Nishimura 1992; Goodman et al. 1995; Goodman 1996). Despite this monumental

advance, the field of archaeology has been slow in adapting GPR data collection and processing to this relatively easily achievable level. The late 1990s and early 2000s have witnessed a more widespread use of 3-D imaging and the archaeological community is beginning to realize the great potential value of GPR as a tool for site imaging (Fig. 3). This application is far advanced from GPR's traditional role as 'check' on anomalies already identified from more commonly used archaeological mapping techniques, such as magnetometry or resistivity (Bradley & Fletcher 1996; Conyers & Goodman 1997, 1999; Conyers 1998, 2000; Conyers & Cameron 1998; Neubauer et al. 2003).

Looking at the Wiggins cemetery in a 3-D format, the identification of graves is based on a set of qualifiers that include feature size and amplitude (represented by a colour display). When investigating data in this 3-D format, it is necessary to keep in mind the processing steps that have been applied to the data from initial filtering to profile interpolation for the 'time slice' display. By slicing through the data, with plan views representing 0.5m slices of soil, individual anomalies appear that match the qualifiers set by the three presumed 'known' burials (Fig. 4). The locations of individuals burials can be seen across

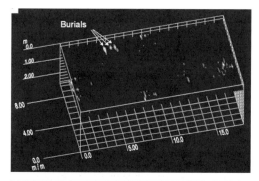

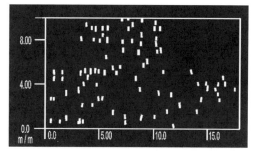

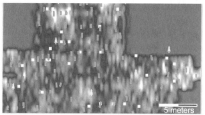

Fig. 5. Graves identified in 3D QuickDraw with on-screen annotation are exported as a .dxf file and displayed in Surfer (top image). The bottom image shows an exported GPR time slice (in .csv format), with overlain layers including identified burials (in top image) and observed surface features, including trees and head- and foot-stones.

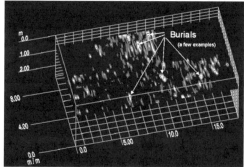

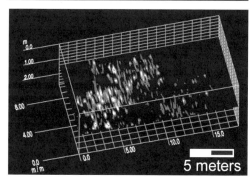

Fig. 4. Three time slices at different depths (from ~ 0.25 m to 1.75 m), displaying the distribution and spatial order of burials.

the site at various depths. This might give an insight into:

- how spatial distribution may relate to different periods of occupation and cemetery use;
- the season during which the body was buried; or, possibly,
- the culture of the burial, as on this site both colonial and native American burials are recorded (Wiggins, pers. comm.).

The diversity of the display options within the software enables the interpreter to 'get inside' the data, and to view it from all angles. Feature mapping can

be done on-screen and the results exported as .dxf files to any geological information system (GIS), computer-aided design (CAD) or other mapping program in order to compare different layers of information collected on the site (Fig. 5). This provides the project director with a number of different options to interpret and generates a comprehensive mapping and analytical facility to support decision making and further action.

There are a number of potential forensic applications for GPR survey:

- site assessment: identifying anomalies that may be graves, or have the potential to be graves;
- identifying an undisturbed area that is not related to a grave;
- grave location and mapping: actual positions for possible grave locations that can be further investigated with other methods;
- grave imaging: establishing the size of the possible grave (depth, width, shape and orientation).

All of these applications provide important information to the site investigators so that they can proceed with a more informed background on their site. Until now, GPR applications for grave location and mapping have not gone beyond this point, but recent product development in the GPR industry has

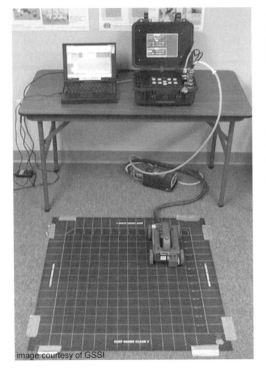

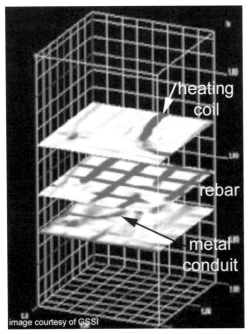

image courtesy of GSSI

Fig. 6. GPR configuration for concrete investigation. The set-up displayed is the StructureScan system with a 3-D cube displaying multiple time slices that identify features (heating coil, rebar mesh and metal conduit) within a concrete slab. Images courtesy of Geophysical Survey Systems, Inc.

opened new possibilities for even more detailed imaging and modelling of graves. One particular research area of forensic interest is the use of GPR in the non-invasive examination of mass graves (e.g. in the Balkans as a result of the civil war in the 1990s). These graves can contain anything from a handful to hundreds of victims, and many are machine-dug and typically 15–20 m × 3 m and over 2 m deep. The problem is not so much in locating these graves but in evaluating them in terms of depth, the number of individuals, their taphonomy and the logistics necessary for their recovery. Many graves can contain separate depositional events as a result of bodies being brought in by machine, often separated by discrete deposits of soil. The taphonomic variables are such that some depositions are skeletonized, some burnt or congealed, while other deposits may be saponified. Any intelligence that enables a better understanding of the character of configuration of these events is a major aid to their eventual investigation.

Advances are already moving in this direction. A previous example (Wiggins cemetery) illustrates the location and mapping of burials, while the subsequent example provides the potential for imaging features within a burial itself. This also illustrates the value of a background in geological and engineering products for archaeological application, and promotes the use of software and hardware development specifically to suit the challenges presented in archaeological or forensic mapping.

Advanced GPR imaging

One new and fast-growing GPR industry development over the past 2 years has been concrete investigation. GPR manufacturers have developed a specific package for this work that consists of a high-frequency antenna (1000–1500 MHz), a survey cart and a survey pad with 2 in (5 cm) grid (Fig. 6). Data are collected every 2 in (5 cm) across the pad at a fixed scan rate. During collection, each radar scan is coded and located geospatially within the 2 × 2 ft (60 × 60 cm) survey area.

Targets in concrete can include PVC piping, rebar mesh, cracks and deterioration, and pre- and post-tension cables (Fig. 6). From an archaeological perspective, this kit can be directly applied to archaeological questions. The first test was to see whether this radar system could locate artefacts and/or burials within pits under Incan house floors in Peru. Based on known information, a test site was constructed and a GPR survey conducted over the mapped buried artefacts. The results were very successful in locating each of the buried features, and the software engineer was able to make any necessary changes to the software program to help

Fig. 7. Data were collected at a Cathedral north of Valencia, Spain, with the GPR concrete configuration. Mapping over buried skeletons provided a good test case for the effectiveness of GPR in this application. (**a**) A Handyscan is used in the trench over a buried skeleton. (**b**) Time slices through the cube of data. (**c**) A femur was excavated corresponding to the reflector circled in the middle time slices. Images courtesy of Geophysical Survey Systems, Inc.

enhance the data. The application of this particular GPR configuration is also suitable for use within excavation trenches and similar contexts where high-resolution imaging is useful (e.g. standing structures, mosaic floors, exacavation balks and building facades).

Subsequently, in Valencia, Spain, another site presented itself for testing with the concrete-mapping GPR configuration. During the process of being structurally renovated, a cathedral courtyard was discovered to be a mass burial ground dating from a period of plague in the fourteenth or fifteenth centuries. Construction was imminent and the team of archaeologists was given a month to exhume and record all possible information on the burials. The concrete GPR kit was tested to see if it would be useful in helping the archaeologists in their work. Figure 7 shows the grid pad in the excavation trench collecting data. Once the data are transferred to a laptop the raw profile can be processed and a plan view of the area surveyed at varying depths. The pre-set processing techniques are aimed at concrete analysis, but changes can be made to make it more applicable for archaeological applications. As the data is scrolled down the 3-D survey cube (Fig. 7), anomalies appear and disappear according to their depth in the survey area. In the middle view of Figure 7, the anomaly on the top edge of the grid, in the centre, was positively confirmed as a femur through subsequent excavation. Further use of GPR in burial imaging will enable interpretations such as these to be developed and demonstrates the considerable potential of the application.

Conclusion and future paths

This has been a very brief overview of the problems experienced in forensic work and in some developments, particularly in GPR field applications and software. There is little doubt that geophysics, especially radar, has a major role to play in forensic search – either in locating graves, eliminating areas, and/or in the non-invasive analysis of buried anomalies. The time slice imaging presented in this paper is not new to the field of archaeology but deserves greater attention and development as a potentially powerful mapping tool in forensic contexts.

To date, little has been done with geophysical applications in forensics (with the exception of Davenport 2001 and occasional use by law enforcement agencies) in the context of fieldwork. Laboratory research by Freeland et al. (2003), NecroSearch (Davenport 2001) and FSAG (Hunter 1999) is beginning to establish a baseline for geophysical survey over burials in order to better understand the response of decaying bodies to various geophysical methods. In order to best utilize these valuable tools in forensic investigations, more work is seen to be necessary in both field application and training.

The authors have taken a step towards this process in Bosnia, in their work with the International Committee on Missing Persons. During a 2003 visit funded by the British Council, GPR equipment was used in an attempt to locate mass graves dating back to the civil conflict between Bosnian Serbs and Bosnian Muslims in the mid-1990s. Four different sites were visited and GPR was used in an attempt to pinpoint the location of graves. Each site provided a unique challenge to the GPR method (such as petrol station tanks and pipes, clay soils, massive surface disturbance, possible live ordnance etc.) for mapping the targeted features. The results of this project will be presented after further site investigation and excavations have been undertaken. This was the first stage of a possible extended forensic research project utilizing GPR not only to locate

graves but also to implement the proposed imaging presented in this paper. The purpose of this particular research was to provide investigators with detailed information to help in the planning and eventual recovery of the victims and, ultimately, in the conviction of war-crime perpetrators.

The second imperative step towards the successful use of geophysical survey in forensics work lies in training. The collection of valid data, the processing, the interpretation and the imaging of information can be done only after relevant training, similar perhaps to that offered by the US National Park Service (http 7) but tailored to forensic application using a hands-on test site and software workshops.

The potential for further forensic application is not only considerable but, at a time of growth in forensic science generally, is also opportune.

The authors are grateful to Geophysical Survey Systems Inc. and the British Council for supporting relevant fieldwork. They would also like to than R. Barker, N. Cassidy and C. Davenport for their helpful comments and advice given in the production of the text.

References

BEVAN, B. 1987. A Radar Search for Burials in Sandy Soil. Geosight, Pitman. Midwest Archeological Center, Lincoln, Nebraska.

BEVAN, B. 1991. The search for graves. Geophysics, 56, 1310–1319.

BEVAN, B. 1992. Testing for unmarked graves, In: BUTLER, D.K. (ed.) Proceedings of the Government Users Workshop on Ground Penetrating Radar, US Army Corps of Engineers, Miscellaneous Papers, GL-92-40, 34–38.

BRADLEY, J. & FLETCHER, M. 1996. Extraction and visualization of information for ground penetrating radar surveys. In: KAMERMAN, H. & FENNEMA, K. (eds) CAA95 – Interfacing the Past. Vol. 1. University of Leiden, Holland, 103–110.

BUCK, S. C. 2003. Searching for graves using geophysical technology: field tests with ground penetrating radar, magnetometry, and electrical resistivity. Journal of Forensic Sciences, 48, 5–11.

CLARK, A. J. 1996. Seeing Beneath the Soil: Prospecting Methods in Archaeology. Revised edition. Batsford, London.

CONYERS, L. B. 1998. Acquisition, processing and interpretation techniques for ground-penetrating radar mapping of buried pit-structures in the American Southwest. Proceedings of the 7th International Conference on Ground-penetrating Radar. Radar Systems and Remote Sensing Laboratory, University of Kansas, 53–59.

CONYERS, L. B. (2000) Subsurface mapping of a buried Paleoindian living surface, Lime Creek Site, Nebraska, USA. Geoarchaeology, 15, 799–817.

CONYERS, L. B. & CAMERON, C. M. 1998. Finding buried archaeological features in the American southwest: new ground-penetrating radar techniques and three-dimensional computer mapping. *Journal of Field Archaeology*, **25**:4, 417–430.

CONYERS, L.B. & GOODMAN, D. 1997. *Ground-penetrating Radar: An Introduction for Archaeologists*. Altamira Press, Walnut Creek.

CONYERS, L.B. & GOODMAN, D. 1999. Archaeology looks to new depths: ground-penetrating radar technology allows subsurface imaging of buried archaeological sites. *Discovering Archaeology*, **1**, 70–77.

DANIELS, D. J. 1996. *Surface-penetrating Radar*. Institution of Electrical Engineers Radar Series, ERA Technology, London, **6**.

DAVENPORT, G. C. 2001. Remote sensing applications in forensic investigations. *Journal of Historical Archaeology*, **35**, 87–100.

FRANCE, D. L., GRIFFIN, T. J. *ET AL.* 1992. A multidisciplinary approach to the detection of clandestine graves. *Journal of Forensic Sciences*, **37**, 1435–1750.

FRANCE, D. L., GRIFFIN, T. J. *ET AL.* 1997. Necrosearch revisited: further multidisciplinary approaches to the detection of clandestine graves. *In*: in HAGLUND, W. D. & SORG, M. H. (eds) *Forensic Taphonomy: The Postmortem Fate of Human Remains*. CRC Press, Boca Raton, 497–509.

FREELAND, R. S., MILLER M. L., YODER, R. E. & KOPPENJAN, S. K. 2003. Forensic application of FM-CW and pulse radar. *Journal of Environmental & Engineering Geophysics*, June 2003, **8**, 97-104.

GAFFNEY, C. F., GATER, J. A. & OVENDEN, S. M. 2002. *The Use of Geophysical Techniques in Archaeological Evaluations*. Insiitute of Field Archaeology, University of Reading, Papers, **6**. Reading, UK.

GEOPHYSICAL SURVEY SYSTEMS, INC. (GSSI). 2000. *3D Quickdraw User's Manual*.

GOODMAN, D. 1996. Comparison of GPR time slices and archaeological excavations. *In*: *Proceedings of the 6th International Conference on Ground Penetrating Radar*. Department of Geoscience and Technology, Tohoku University, Sendai, Japan, 77–82.

GOODMAN, D. & NISHIMURA, Y. 1992. Two-dimensional synthetic radargrams for use in archaeological investigation. *In*: HÄNNINEN, P. & AUTIO, S. (eds) *4th International Conference on Ground Penetrating Radar, 8–13 June 1992*. Finland Geological Survey, Rovaniemi, Special Papers, **16**, 339–343.

GOODMAN, D., NISHIMURA, Y. & ROGERS, J. D. 1995. GPR time slices in archaeological prospection. *Archaeological Prospection*, **2**, 85–89.

HAGLUND, W. D. & SORG, M. H. (eds) 1997. *Forensic Taphonomy: The Postmortem Fate of Human Remains*. CRC Press, Boca Raton.

HAGLUND, W. D. & SORG, M. H. (eds) 2002. *Advances in Forensic Taphonomy: Method, Theory and Archaeological Perspectives*. CRC Press, Boca Raton.

HUNTER, J. R. 1999. Research for the detection of clandestine graves using controlled animal burials. *Proceedings of the International Association of Forensic Sciences 15th Triennial Meeting, Los Angeles, 1999*, 45.

HUNTER, J. R. 2002. A pilgrim in forensic archaeology: a personal view. *In*: HAGLUND, W. D. & SORG, M. H (eds) Advances in *Forensic Taphonomy: Method, Theory and Archaeological Perspectives*. CRC Press, Boca Rajon, xxv – xxxii.

HUNTER, J. R., BRICKLEY, M. B *et al.* 2001. Forensic archaeology, forensic anthropology and human rights in Europe. *Science and Justice*, **41**, 173–178.

HUNTER, J. R, ROBERTS, C. A. & MARTIN, A. 1996. *Studies in Crime: An Introduction to Forensic Archaeology*. Routledge, London.

JACKSON, S. 2002. *No Stone Unturned*. Kensington, New York.

KILLAM, E. W.1990. *The Detection of Human Remains*. Charles C. Thomas, Springfield.

NEUBAUER W., EDER-HINTERLEITNER, A., SEREN, S. S. & MELICHAR, P. 2003. Integrated geophysical prospection of Roman villas in Austria. *In*: FASSBINDER, J. W. E. and IRLINGER, W. E. (eds) *Archaeological Prospection: Third International Conference on Archaeological Prospection*. Bayerisches Landesamt für Denkmalpflege, München, **41**, 239–241.

NOBES, D. C. 2000. The search for 'Yvonne': a case example of the delineation of a grave using near-surface geophysical methods. *Journal of Forensic Science* , **45**, 715–712.

SCOLLAR, I, TABBAGH, A., HESSE, A. & HERZOG, I. 1990. *Archaeological Geophysics and Remote Sensing*. Cambridge University Press, Cambridge.

World Wide Web Addresses:

http 1. http://www.geophysical.com/applications.htm

http 2. http://etd.ukt.edu/2002/MillerMichelle.pdf

http 3. NADAG: http://www.cast.uark.edu/nadag/ (search: archaeological site type/cemetery

http 4. GPR-Slice: http://www.gpr-survey.com/pages/362603/index.htm

http 5. GPR_Process: http://www.du.edu/~lconyers/gpr_data_processing_example.htm

http 6. 3D QuickDraw: http://www.geophysical.com/software.htm

http 7. StructureScan: http://www.geophysical.com/StructureScan.htm

Environmental influences on resistivity mapping for the location of clandestine graves

J. SCOTT[1] & J. R. HUNTER[2]

[1]School of Geography, Earth and Environmental Sciences, University of Birmingham,
Birmingham B15 2TT, UK (e-mail: j.b.t.scott@bham.ac.uk)
[2]Institute of Archaeology and Antiquity, University of Birmingham, Birmingham B15 2TT, UK
(e-mail: j.r.hunter@bham.ac.uk)

Abstract: Geophysical surveys are being increasingly applied for the detection of clandestine burials. Ground-penetrating radar has been used successfully but is not appropriate for many sites. For this reason, other techniques such as resistivity mapping are being tested. Resistivity is sensitive to many different changes in the ground that can be caused by a grave; however, the small physical extent of the target makes location extremely difficult. Other environmentally caused variations in resistivity may be of the same magnitude as those caused by a grave, and it is important to consider whether the resistivity anomaly caused by a grave can be detected over this background variation. This paper provides a summary of the method and discusses two case studies in which resistivity mapping was employed to locate a clandestine grave. The resistivity results of a survey over moorland were dominated by the thickness of the peat coverage while the resistivity of a survey in a river valley was dominated by various environmental variations, including a medieval furrow system. Both sites were eliminated from enquiries after extensive cadaver dog searches and some excavation, although it was considered that the environmentally caused resistivity variations at both sites would have made location of a grave by resistivity alone extremely difficult. However, the success of resistivity for landfill locations suggests that resistivity may be more appropriate for the location and delineation of larger features such as mass graves.

Geophysical methods are now being used regularly for the location of clandestine graves (Buck 2003). One geophysical technique that has been used successfully to locate graves is ground-penetrating radar (GPR). However, radar is not appropriate for many sites and therefore other geophysical techniques such as magnetometry and resistivity are being developed for this purpose. The electromagnetic technique has been used with success to map changes in ground resistivity due to old landfill sites (Green *et al.* 1999; De Iaco *et al.* 2000) and should respond in a similar way to the resistivity for mass graves. However, a major problem with clandestine graves is the small physical size of the target and its associated geophysical anomaly. In addition, there is little published information on the nature and magnitude of the expected anomalies. Any change in the physical properties of the ground caused by a grave might be brought about by a wide variety of different causes, and it is important to be aware of these and their potential magnitude in order to assess the applicability of a technique to a particular site. Here we discuss the resistivity mapping technique and its potential for the location of clandestine graves using two case studies from two sites in order to highlight some of the difficulties involved.

Resistivity mapping

Measurement technique

The most commonly used ground resistivity measurement technique involves passing an electrical current (I) into the ground between two electrodes, and measuring the electrical potential (V) between two further electrodes. The electrical resistance (R) is then calculated by,

$$R = \frac{V}{I} \tag{1}$$

where V is in volts, I is in amps and R is given in ohms (Ω). Two electrode configurations often used are the Wenner and Pole–Pole arrays, which are shown in Figure 1. The Wenner array has four mobile electrodes that are moved with every reading while the Pole–Pole array has two mobile electrodes and two remote electrodes which are fixed away from the survey area at a distance much greater than the electrode separation. In resistivity mapping it is common to take readings of ground resistance at set spacings on a grid and then to contour the values obtained. However, different electrode arrays and electrode separations will give different resistance readings for the same ground. For a comparison

From: PYE, K. & CROFT, D. J. (eds) 2004. *Forensic Geoscience: Principles, Techniques and Applications*. Geological Society, London, Special Publications, **232**, 33–38. © The Geological Society of London, 2004.

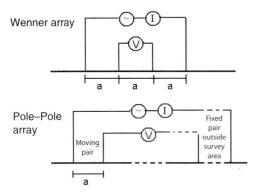

Fig. 1. The Wenner and pole–pole arrays.

Fig. 2. Resistivity mapping with the pole–pole array.

between different arrays it is necessary to convert the values to resistivity, ρ. Assuming a homogeneous ground, the resistivity for a Wenner array with an electrode separation of a is given by:

$$\rho = 2\pi a R \qquad (2)$$

and for a pole–pole array the resistivity is given by:

$$\rho = \pi a R \qquad (3)$$

where a is expressed in metres, and R is in ohms and ρ is in ohm metres (Ωm).

Strictly speaking, as the ground is never completely homogeneous, the resistivity computed in this way is referred to as 'apparent resistivity'.

The array used by Buck (2003) for the location of graves was the Wenner array with an electrode separation of 1 m. This is highly inappropriate for the detection of single grave features, which will have a spatial extent of less than the array length. Orlando et al. (1987) were able to locate relatively large tombs using a Wenner array with an electrode separation of 1 m, but found the results poor, and they claim to have better results with alternative 'focused' arrays. For our work the pole–pole array with an electrode separation of 50 cm was chosen. This array is commonly used in archaeological mapping as it is easier to use than the Wenner array. It also produces much simpler anomalies which can be more easily identified in a noisy environment. The equipment used incorporates two 'mobile' electrodes fixed to a frame with the resistance meter attached. Figure 2 shows a survey being carried out with this equipment on moorland. Readings were taken on a grid at 50 cm intervals, along profiles that were separated by 50 cm. This electrode separation and grid spacing was necessary to ensure ground resistance readings would characterize any anomaly caused by a feature as small as a single grave with a

spatial extent of less than 1 m². It may be necessary to have even smaller spacings, although smaller electrode separations are highly affected by near-surface ground variation and smaller grid spacing will increase the survey time. Less sensitive electrode arrays that can be used at smaller electrode separations are currently being developed and tested.

The resistivity of a grave

One of the most difficult problems in grave location using electrical resistivity is knowing the type of anomaly that a grave will produce. A buried body and the disturbance caused in burying the body both have the potential to change the resistivity of the ground. However, this change may be an increase or decrease in the actual resistivity. Some of the key factors that may change the resistivity of the ground in a grave environment are the following:

- Size of target
- Area of disturbance
- Depth of burial
- Body wrapping, including clothes and sacks
- State of decomposition
- Nature of substrates into which grave has been cut
- Climate
- Ground saturation.

All these factors could be extremely important on their own and could significantly change the magnitude, character and distribution of the ground resistivity. For this reason it is not possible to define a 'typical' anomaly. However, it is useful to make an estimate of the magnitude of a resistivity change caused by a burial in order to assess whether the grave could be detected by the resistivity-mapping technique. The magnitude of the anomaly will

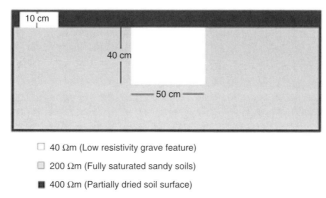

☐ 40 Ωm (Low resistivity grave feature)

☐ 200 Ωm (Fully saturated sandy soils)

■ 400 Ωm (Partially dried soil surface)

Fig. 3. Simple resistivity model of a grave.

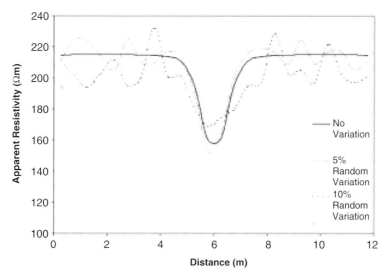

Fig. 4. Resistivity variation caused by a model grave.

depend on the contrast in properties between the grave and the surrounding soil. This contrast should be compared to the resistivity variation produced by the surrounding environment in order to ascertain whether the grave anomaly could be located within its background.

A very simple grave model is shown in Figure 3. This grave model is used to estimate the magnitude of the resistivity change that is produced by variations in the porosity of the ground due to digging. In cross-section the grave model (Fig. 3) is considered to be 40 cm deep by 50 cm wide. It has been assumed that digging the grave has increased the porosity of the surrounding sandy soil from 20% to 40% and that the soil is fully saturated with water of 10 Ωm resistivity. Thin (10 cm) partially saturated soil covers the surface with a higher resistivity. The resis-

tivity of the fully saturated ground can be estimated using the Winsauer equation

$$\rho_{ground} = \frac{0.62 \times \rho_{water}}{Porosity^{2.15}} \qquad (4)$$

which has been proposed as a model for unconsolidated sediment (Winsauer *et al.* 1952). The apparent resistivity measured by using the pole–pole array at 50 cm spacing can be calculated by using the program Res2Dmod and the results of this are given in Figure 4. Some random noise has been added to the data in order to make the anomaly more realistic. It can be seen that the spatial distribution of the anomaly is small, around 1 m, and the anomaly is just visible above the 10% added noise. However, 20%

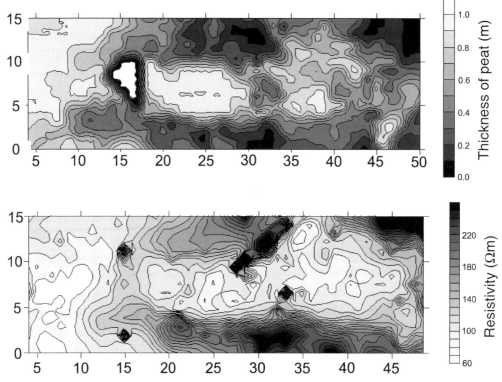

Fig. 5. Resistivity and thickness of peat maps from a moorland gully.

noise due to environmental variation would be of the same magnitude as the anomaly. The anomaly is of a similar magnitude to anomalies measured over a test site consisting of pig burials (Bray 1996; Hunter 1999). It is important to assess the resistivity variation over a site in order to determine whether a spatially small variation in resistivity of this magnitude would be visible. Further information on anomaly magnitudes of clandestine graves would be useful but not applicable to all general cases because of the many factors influencing the resistivity of a grave.

Case studies

A moorland environment

In the early 1960s four children were abducted, murdered and buried in the peat moors in the uplands between West Yorkshire and Lancashire (now Greater Manchester). The crimes came to be popularly known as 'The Moors Murders' (Harrison 1987). The landscape is hostile, disorienting and difficult to search. The surface vegetation consists of large, uneven tussocks, which are difficult to

walk through. Despite the expanse of land available, burial is only feasible in the areas of exposed peat found in gullies as well as on higher exposed ground. Police searching in the 1960s recovered two of the victims, and a third body was found during renewed investigation in 1988. This body was well preserved and partially mummified by the peat. The body of the final victim, a 12-year-old boy, is still missing. Renewed search in 2000/1 identified three gullies for investigation, in order either to discover the body or to eliminate them from any future work. Resistivity surveys were carried out along with auguring at 1 m intervals. It was clear that the high variation in moisture saturation and consolidation of the ground could make detecting a grave very difficult.

By carrying out the auguring it was found that the peat was generally less than 1 m in thickness and that it overlies sandstone. It was possible to make a contour map of the peat thickness, and the results for one of the gullies, along with the apparent resistivity map, are given in Figure 5. The most striking feature is the extremely close correlation between the apparent resistivity and the thickness of the peat. Peat generally has a low resistivity, and the sandstone has a much higher resistivity than the peat. The sandstone in the gully area was shallow enough to contribute to the measured resistivity value; therefore those places where the sandstone was shallow tended to give higher apparent resistivity readings. The resulting apparent resistivity changes (from 60 Ωm to > 200 Ωm) are far larger an anomaly than would be expected due to a grave feature. It would be extremely difficult to remove the effect of the varying depth of the sandstone on the resistivity to within the accuracy needed to locate a grave feature. Areas of resistivity variation that appeared not to be due to changes in the depth of sandstone were identified; where these occurred in places that were also considered to be deep enough for burial (around 40 cm) a cadaver dog search was used. The dog reacted to only one location, which was then excavated. However, no body was located and it is likely that the dog was reacting to methane released from the ground by auguring. The results suggest that resistivity mapping is inappropriate for this terrain, although the ability of the resistivity to give the depth of the peat could be important in eliminating areas too shallow for burial. Ground conductivity mapping using an electromagnetic system could be employed for faster mapping of the peat thickness.

A river valley environment

From witness statements police identified the corner of a field in Wales as the possible location of up to four child graves. The area, located next to a river, was flat, grass-covered and bordered by trees, an ideal location for access and ease in carrying out a resistivity survey. A resistivity-mapping survey was carried out during the summer of 2001 in order to locate the graves or to help eliminate the field from further police enquiries. The annotated apparent resistivity contour map of the field is shown in Figure 6. The resistivity of the main field area varied from between 30 Ωm to *c*. 120 Ωm, with some of the field edges increasing to over 200 Ωm resistivity. One of the main features is a banding of the resistivity in the X direction (Fig. 6), which occurs at approximately 2 m intervals and may be due to a previous, possibly medieval, furrow system. These bands are definitely not due to the grid sampling used because the survey lines were oriented in the Y direction and measurements were taken at 50 cm intervals. Areas of higher resistivity around the field edges could be due to drying of the ground from the tree and hedge root systems located at the edges of the field. Low-resistivity areas, particularly at 45 m in the Y direction (Fig. 6), could be due to increased ground saturation caused by shelter from the sun by tree canopies, or just to changes in the soil composition near the field edges. Single-point high-

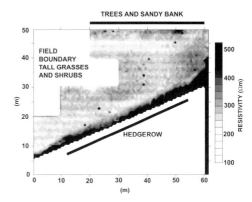

Fig. 6. Resistivity map from the corner of a field in Wales.

resistivity anomalies (black dots, Fig. 6) are erroneous readings, due to poor ground contact.

Based on the simple model of the resistivity of a grave feature given above, it would not be possible to distinguish an anomaly caused by a grave from the background environmental resistivity variation. This is particularly true around the field edges, where a clandestine grave is most likely to be located for maximum concealment. To completely eliminate the field from the investigation, cadaver dog searches were conducted across several of the larger anomalies and some trial machine excavations were carried out. However, these failed to ascertain the cause of the resistivity changes observed. Finally, a large band of topsoil was removed from a wider section of the field which, based on the evidence of witnesses, was most likely to contain graves, but this failed to locate any remains. The evidence, supported by no additional, strong resistivity anomalies suggested the elimination of the field from the enquiries.

Conclusions

The resistivity technique is sensitive to the type of variations in the ground that can be caused by a grave. However, the magnitude and spatial scale of a resistivity anomaly from a grave can only be estimated, on account of the many variables possible in grave type and location. Many other environmental factors are likely to produce anomalies of the same size and of potentially even greater magnitude than a grave. The case study of a flat, grass-covered field in Wales shows environmentally created resistivity variations that could occur anywhere, including gardens. The resistivity variation on the moorland site was dominated by the depth of the peat. These factors will make the location of single clandestine graves by the resistivity method extremely difficult.

Improved electrode arrays may make the resistivity technique more effective for mapping small anomalies such as graves although they will not remove the problems of the other environmental variations in resistivity that are regularly encountered. A very interesting potential use for the resistivity-mapping technique may be the location and delineation of mass graves, where the larger spatial extent of the anomaly will make the grave an easier target. Other techniques that map similar properties of the ground, such as the electromagnetic-mapping technique, may be quicker to use.

References

BRAY, E. 1996. *The use of geophysics for the detection of clandestine burials: some research and experimentation.* MSc dissertation, Bradford University.

BUCK, S. C. 2003. Searching for graves using geophysical technology: field tests with ground penetrating radar, magnetometry, and electrical resistivity. *Journal of Forensic Sciences*, **48**, 1–7.

DE IACO, R., GREEN, A. G. & HORSTMEYER, H. 2000. An integrated geophysical study of a landfill and its host sediments. *European Journal of Environmental and Engineering Geophysics*, **4**, 223–263.

GREEN, A., LANZ, E., MAURER, H. & BOERNER, D. 1999. A template for geophysical investigations of small landfills. *The Leading Edge*, **18**, 248–254.

HARRISON, F. 1987. *Brady and Hindley: Genesis of the Moors Murders.* Grafton Books.

HUNTER, J. R. 1999. Research for the detection of clandestine graves using controlled animal burials. *In: Proceedings of the International Association of Forensic Sciences 15th Triennial Meeting, Los Angeles, 1999*, 45.

ORLANDO, L., PIRO, S. & VERSINO, L. 1987. Location of subsurface geoelectric anomalies for archaeological work – a comparison between experimental arrays and interpretation using numerical methods. *Geoexploration*, **24**, 227–237.

WINSAUER, W. O., SHEARIN, H. M., MASON, P. H. & WILLIAMS, M. 1952. Resistivity of brine-saturated sands in relation to pore geometry. *Bulletin of the American Association of Petroleum Geologists*, **36**, 253–277.

The importance of stratigraphy in forensic investigation

IAN D. HANSON

*Centre of Forensic Science, Technology and Law, School of Conservation Sciences,
Bournemouth University, Talbot Campus, Fern Barrow, Poole, BH12 5BB, UK
(e-mail: idhanson@inforce.org.uk)*

Abstract: The laws of stratigraphy, developed in geology, have long been adopted for archaeological use. Archaeological excavation in the UK relies on the application of these principles to define, interpret and understand site history. The adaptation of archaeological methods to forensic settings has recently resulted in successful analysis of stratigraphy-defining complex series of events on murder burial scenes. The breadth of physical evidence that can be recovered through stratigraphic excavation is great, and that which can be lost without due attention to the buried surfaces forming an intrinsic part of the stratified deposits of a site is significant. Recent case examples demonstrate the importance of employing stratigraphical principles in the excavation and interpretion of buried cultural and natural deposits as part of multidisciplinary forensic investigation.

This paper focuses on stratigraphic principles in forensic investigations from an archaeological perspective. The development of forensic anthropology has led to the widespread application of archaeological methods to cases of homicide, human rights violations and war crimes in the last 20 years (e.g. Spennemann & Franke 1995a; Haglund 2002). Although many aspects of common archaeological practice, such as soil analysis, geophysical survey, recording methods and sampling, have positive applications for forensic investigation, this paper seeks to define and exemplify how stratigraphy has a central role in the methodology of forensic archaeological practice.

The laws of stratigraphy

The well-established laws of stratigraphy developed within geology have long been adapted for archaeological use, because, as Harris (1998, p. 2) describes it, 'much of the surface of the Earth is now blanketed with stratigraphical features and deposits made by people. Such structures represent a very complex history that destroyed much geological stratification in its creation'. Stratigraphy can be natural, cultural or a mixture of both, but this is not directly relevant to the definition of a stratigraphic unit (Lucas 2001). Harris (1979) argued for a set of laws for archaeological stratigraphy, and they have been widely adopted. These are based on the constant that, as with geological formations, all archaeological sites are stratified and are 'recurring phenomena', so the laws apply to all archaeological deposits. The first three laws are adapted from geology, the last being an archaeological development.

1. The law of superposition In a series of layers and interfacial features, as originally created, the upper units of stratification are younger and the lower are older, because each must have been deposited on, or created by the removal of, a pre-existing mass of archaeological stratification. In archaeological stratigraphy, the law of superposition must also take account of interfacial units of stratification which are not strata in a strict sense. These interfacial units of stratification may be seen as abstract layers and will have superpositional relationships with strata which lie above them or through which they were cut or 'lie above'.

2. The law of original horizontality Any archaeological layer deposited in an unconsolidated form will tend towards a horizontal position. Strata which are found with tilted surfaces were originally deposited that way, or lie in conformity with the contours of a pre-existing basin of deposition.

3. The law of original continuity Any archaeological deposit, as originally laid down, or any interfacial feature, as originally created, will be bounded by a basin of deposition, or may thin down to a feather edge. Therefore, if any edge of a deposit or interfacial feature is exposed in a vertical view, a part of its original extent must have been removed by excavation or erosion, and its continuity must be sought or its absence explained.

4. The law of stratigraphical succession A unit of archaeological stratification takes its place in the stratigraphic sequence of a site from its position between the undermost (or earliest) of the units which lie above it and the uppermost (or latest) of all the units which lie below it and with which the unit

From: PYE, K. & CROFT, D. J. (eds) 2004. *Forensic Geoscience: Principles, Techniques and Applications.* Geological Society, London, Special Publications, **232**, 39–47. © The Geological Society of London, 2004.

has a physical contact, all other superpositional relationships being redundant (Harris 1979, p. 30).

Stratigraphical analysis and the forensic sequence

Archaeological excavation in the UK relies on the application of these stratigraphical principles to define, interpret and understand site history. Stratigraphy provides a sequence (Gamble 2001, p. 63). The sequence of deposits of a site can, through excavation, be determined in four dimensions; those of 3-D space and that of time. Following the law of superposition, the archaeologist seeks to remove the stratified deposits of a defined area from the latest to the earliest, in a sequence that reverses the time line of deposition. Through the surviving stratification and the artefacts contained within, a history of events through space and time can be interpreted within the area of the excavated site. We can describe an event as a noticeable unit of experience, which may be tangible or intangible (Kind 1987, p. 36). In archaeological terms events are the physical remains of temporal phenomena, whether natural or cultural (Lucas 2001, p. 169), and are interpreted through the conditioned perceptions of the excavators.

Determining this sequence of events is exactly the outcome desired by criminal investigators at a crime scene. Examination of a crime scene can recover physical evidence that allows the interpretation and reconstruction of a sequence of events through time of (often) a single physical episode, such as a murder, and links that evidence to persons, other events and locations. Similarly, the aim of the forensic archaeologist is to recover physical evidence to allow a reconstruction of a time sequence of (often) a single event, usually in a buried context. As Kind (1987, p. 7) states, 'the thought processes of the archaeologist are very apposite to crime investigation. The great syntheses which archaeologists have made from fragmentary evidence enable us to draw an increasingly complete picture of man's past'. The principle of the reconstruction of a sequence of events through time from analysis of deposits, features and artefacts applies to surviving physical evidence, whether it is one day or 10000 years old, and for this reason gives archaeologists an intrinsic forensic capability.

Determination of the relevant event sequence in an environment that existed before and after a particular crime, and that is obscured through sheer complexity and taphonomic effects, is also a common aim of criminal and forensic archaeological investigators. The adaptation of archaeological methods to forensic settings has been successful in analysing stratification and its stratigraphy, determining quite complex sequences of events at burial sites in the UK and internationally.

Burial as a rapid deposition

Clandestine single graves, or mass graves, often take their place in the archaeological landscape as a single event of rapid deposition. In this respect, they differ from most archaeological sequences, which may be seen as an aggregate of small-scale, short-term acts (Foxhall 2000, p. 486). Many excavations reveal evidence of events, cultures and their sequence of change over extended periods of time, perhaps hundreds or thousands of years. The surviving physical deposits and features may have been slowly laid down, eroded, re-dug and subject to other long-term transformation processes, and it is less common to find buried strata that can be interpreted as reflecting single events in time, such as a murder. As Gowlett states (1997, p. 166), as archaeologists examining the 'internal timescale of sites . . . generally we do not have moments in time'.

Rapid deposition does reflect single, short-term events. These may be catastrophic events, such as earthquakes, which cause a roof to collapse, burying the contents of the floor below, as identified at the excavation of the temple at Archanes–Anemospilia, or the series of pyroclastic flows of volcanic ash erupting from Mount Vesuvius (from the well-known example of the burial of Pompeii in AD 79). While the eruption took some time, and many escaped the falling ash, excavations have revealed the positions at death of those in the streets who had not fled the eruption and were suffocated by heat and gases, and rapidly covered by flowing ash of the final 'pyroclastic surges' (Chamberlain & Parker Pearson 2001, p. 152). Rapid deposition may be accidental events, such as a ship sinking, as with the *Mary Rose*, or deliberate events, such as the burial of a coin hoard. They represent what Gamble (2001, p. 125) calls 'time capsules' and 'the archaeological equivalent of the Polaroid snap': events sealed by burial concerning a time frame of a single day or some hours.

Burials fit into this category of rapid deposition. At its most basic, a grave is excavated, a body or bodies placed within it and the excavated soil replaced. Further along the time line, the grave site may be transformed and disturbed, for example by robbing, truncation, burrowing animals or landscaping, creating a more complex stratigraphic sequence. However, compared with long-term settlement site stratigraphy, exemplified by excavations at Jericho by Kenyon (1957), it is still a simple sequence of deposits. Revealing and interpreting these rapid depositional events is the common experience of the archaeologist working in a forensic setting.

The most important forensic aspect of rapid deposition in burial is that of sealing the contents of a grave in their position at the time of backfilling. This 'freezing in a point of time' of the position of remains and evidence is very significant to an investigation. The original *in situ* position of items has great evidentiary value (Wiggins & Houck 2001, p. xii; Gerberth 1996, p. 3). Burial also protects and preserves aspects of a crime scene, especially in the short term, when decomposition and taphonomical change are not advanced. There is a tendency for evidence, of whatever kind, to disappear with the passage of time (Kind 1987). However, some evidence types have been shown to survive for long periods of time (in forensic terms) in an undisturbed buried environment, compared with above-ground placement, because of the protective and stabilizing environment and the limiting of decomposition (Mann *et al.* 1990; Spennemann & Franke 1995b; Rodriguez 1997). Not all evidence or artefacts in a burial context, especially if they are organic, such as body fluids, will survive more than a few days. Nor will those that are preserved, until exposure and recognition, be connected to a criminal event, but as with any crime scene, any item can and may constitute physical evidence (Gerbeth 1996, p. 3) and should be treated as such.

Recognition of evidence in the forensic landscape

Digging a clandestine grave creates a new surface – below ground – onto which a body is placed, and on which a perpetrator moves, works and leaves evidence. In the same way, as evidence of a murder may be left on the floor, walls, and furniture in a room in the form of a body, body materials, objects and impressions (Gerberth 1996, p. 505), so these evidence forms can be left on the subterranean surface (the walls and floor) of a grave during a body's burial.

The grave and its creation form the buried part of a wider sequence. The spoil from excavation left around a grave, dropped surface artefacts such as shell cases, footprints and vehicle tracks, crushed and broken vegetation and subsequent taphonomical alterations are all examples of the physical evidence (often stratified) in an environment associated with the creation of the grave. Together, they represent the sequence of events within the forensic landscape. The archaeologist, having a wide-ranging knowledge of archaeology, the environment and the natural sciences, is in a strong position to recognize such sequences and to call on the assistance of other specialists, such as botanists and palynologists, to recover and record evidence of this wider sequence. The archaeologist can also recognize the potential

short-term survival of many forms of physical evidence in such a sequence.

The concealment of burials by deliberate landscaping of the gravesite and surrounding area is common. A field containing several mass graves in Bosnia (excavated for the International Criminal Tribunal for the Former Yugoslavia, (or ICTY, in 1997–2000) typifies many such sites in the Balkans. Test trenches revealed the local stratigraphic sequence, including the original ground surface into which graves had been dug, covered by up to 3 m of imported material. The stratigraphic section of the trench showed a black line, representing the decayed turf line of the original ground surface, covered by soil layers from grave backfilling, then a layer of building debris, and two deposits of clay soil that had been used to level the site. Vegetation grew evenly across the site, which looked like a normal undisturbed meadow. Analysis of the strata allowed a sequence interpretation, the planning of the excavation, and removal of landscaping layers to expose the top surface of the graves in plan at the former ground surface. Analysis and excavation of the stratigraphic sequence in the wider landscape led to the uncovering of these graves. Excavation of the graves themselves showed that they had been robbed. Removal of a second layer of later backfilled soil revealed the tool marks of the machine used to rob the grave. The tool marks formed a surface separating the first layer of earlier backfilled soil and the remaining bodies in the grave.

For single graves it is normal to find one fill of soil as a backfilling, although it is not uncommon to find additional strata, formed by material such as straw or lime placed over bodies to hasten decomposition or mask odour (Mont 1950, p. 32; Hochrein 2002, p. 48). In mass graves, however, especially when used over time, there may be a number of soilfills separating dumps of bodies. Stratigraphic excavation reveals the depositional sequence that allows an interpretation of events. In one secondary grave in Bosnia, where bodies and soil from robbed graves were imported, a sequence of deposits was uncovered. Each deposit of soil and body parts was separated by a surface interface represented by vehicle tracks pressed into the clay soilfill. On excavation, four separate episodes of dumping and bulldozing could be determined in surface wheel tracks. This suggested that, as each deposit was dumped into the ramped grave, a frontloader type of vehicle drove into the grave, pushing the soil and remains into the far end, and then reversed out. The next deposit sealed the surface created by the vehicle track.

The surface in evidence recognition and recovery

An understanding of the laws of stratigraphy allows a logical, controlled exposure and recording of such buried evidence from the latest event to the earliest event in sequence. The acknowledgement of surfaces within this stratigraphic sequence is essential to their identification and exposure, and to maximise the potential for recovery of evidence from a grave. As Harris describes, the stratigraphic sequences are 'the outer expression of the remnants of pre-existing deposits that were partly destroyed in the creation of new surfaces' and 'are the great determinants of stratigraphic time, and share with time the fact that they do not exist, unless recorded in a diagrammatic form' (Harris pers. comm. 2002). Their recognition and careful recording by planning, scanning and photography should be considered a powerful tool in criminal evidence recovery.

What forensic evidence can be recovered from the carefully excavated stratified sequence of a burial and the wall and floor surfaces that are revealed?

Often overlooked in past investigations, wheel tracks and other machine tool marks are very common in machine-excavated mass graves. When these are dug, the excavating-bucket tooth marks (and wheel tracks in ramped graves) leave compressed impressions in the new surface of the excavation. Upon sequenced removal of the soilfill and bodies, these impressions are revealed. The quality of impression varies depending on the composition of the deposits through which the grave has been dug. For deep graves these will usually be 'natural' geological deposits. Sands and clays take bucket impressions easily, so that each tooth can be observed and measured. Gravel by its nature will not show smaller detail, but may retain the shape of a large machine bucket or general wheel shape. The excavation of a series of secondary graves along one road in Bosnia, excavated into soft degraded sandstone, showed that they had been dug by the same machine, with a specific bucket-tooth arrangement, linking the sites and demonstrating a degree of organization to the disposal.

As well as tool marks, trace evidence can be left on the grave wall and floor surfaces. This is especially significant in robbed graves. Stratigraphic excavation can reveal a sequence of activity when bodies have been deliberately removed (including a second cut truncating original backfilled soil). Removal of a body always leaves some evidence behind. The Locard exchange principle applies as much to buried evidence as it does to evidence elsewhere. Examples from Bosnia show that, in clay and sandy soils, impressions of body parts such as the head may remain in the walls and floor of graves. Hair, fibre, skull and ballistic fragments can be left in these, especially when loosened by decomposition at the time of movement.

Bone and tissue fragments from the backspatter of close-range gunshot wounds to the head heavily contaminate nearby surfaces (Burnett 1991). Small bone fragments can be recovered from surfaces if the soilfill adhering to the surface is levered, peeled and lifted away carefully with a trowel. Shots fired into a grave can pass through the walls and floor into the surrounding soil, especially in sand and clay, and careful examination of such surfaces has revealed the entry holes of ballistic paths in graves in Guatemala (Peccerelli pers. comm. 2000).

Seasonal and flowering plant material can survive between and under bodies, either because it fell into a grave at the time of burial, or because the plants were growing in depressions, such as quarries and pits utilized to conceal bodies; this can be used to date the time of burial. At an execution site in a shallow gravel quarry in Bosnia, several species of flowering plants growing at the time of the killings were found flattened under the bodies, indicating that the grave was made in high summer. The bodies acted as a stratigraphic layer, protecting the plants lying on the ground surface. The interface between grave surface and fill can also reveal the point at which plant roots were severed by the digging of the grave and also any subsequent growth.

In a robbed or disturbed burial, fluids and other by-products of decomposition left in the grave can signify the removal of a body. Rough movement of a body during robbing can detach adipocere or leave it adhering to the grave surface. Organic-rich soils forming the surfaces can be stained by decomposition (Mant 1950). Biomarkers such as volatile fatty acids from a decomposing body can penetrate the soil of the surface interface (Vass *et al.* 1992) and can also be helpful in determining the time elapsed since death. The ratio of trace elements in soil adjacent to the place of deposition of a skeleton may assist in the sourcing of bones that are found away from the burial site (Trueman pers. comm. 2003). The movement of fluids and elements from body to the soil of the interface can be investigated adequately only through the recognition and subtle examination of the grave surface.

The body as a stratigraphic deposit

A body or a mass of bodies (especially clothed and before extensive decomposition) can cover, protect and seal trace evidence, impressions, tool marks and other evidence found on the grave surface, separating it from the backfilled soil of the grave. Although Hunter *et al.* (1994) envisage the body within a grave as an artefact, it should be seen as a separate and separating deposit. While the body itself has artefactual qualities, such as aiding the dating of the grave through comparison of radio-isotope ratios with those of a relevant living population (Swift pers.comm. 2003), the body represents a depositional event, having horizontal and vertical dimensions and being differentiated from, and separating, surrounding deposits.

This is especially true in the short term of forensic time (before loss of organic remains through decomposition) and in mass burials, where bodies may completely cover the floor and walls of a grave. In this case, the body mass, a dense, contiguous aggregate of bodies (Haglund 2002, p. 247) may, by volume, be the largest deposit in the grave. There may also be separately deposited bodies or body masses in the same grave, so that bodies themselves may form a stratigraphic sequence as part of the filling of a grave. This has been a common event in mass graves in the Balkans.

Clothing and tissue can survive for many years in a body mass, where decomposition seems to slow and adipocere forms. This is especially true when graves are dug to beneath the water table and anaerobic conditions occur. Bodies excavated after 5 years from a grave in Bosnia, dating from the Srebrenica Massacre (July 1995), showed limited decomposition, and this simplifies the recognition and recovery of evidence from surfaces, sealed beneath and between a fatty, organic blanket of material. Recognizing the separation of a stratigraphic sequence of bodies during excavation is more difficult when complete skeletonization and loss of clothing has occurred and the skeletal elements press together from the weight of soil. However, it can be seen when the survey data of the adjacent body positions is analysed using 3-D imaging software (such as the freeware Rotate by Marijke van Gans, for which 3-D body point files were written by 'Bodrota' and developed by Richard Wright for Bosnian mass graves).

It may be possible to determine the sequence of deposit of each body in a mass grave when they have been separately placed, and even to recover evidence from the surface of each body. In practice, however, bodies in a grave intertwine and press together. Therefore it may not be possible to recognize any sequence of disposal, or to recover bodies in the correct depositional sequence. Bodies are often placed together, being dumped from trucks or bulldozed into a grave, and, although bodies are sampled and removed separately, they can give important information on the formation processes of the body mass as a single deposit.

The body itself therefore forms part of the stratigraphic sequence. The body and the grave surface form a forensic stratigraphy that often benefits (in terms of survival and recognition of evidence) from the short time line between site formation and investigation found in many forensic burial contexts.

Recovery of evidence from the surface interface

Careful excavation can reveal fine stratigraphic boundaries, allowing control of artefact recovery and stratigraphic recording. It should be remembered that, in the forensic short term, perhaps several years, the grave wall may extend visibly from the ground surface from which it was cut and be traceable, for example in a turf line. Over time, however, the clarity of a cut edge and the upper zone of a soil-fill may become obscured. This is caused by the physical and biochemical breakdown and movement of the near-surface organic material and soil layers, often designated the O and A horizons (Brady & Weil 2002). Plant root growth, animal burrowing, ploughing and the action of earthworms, for example, all cause extensive bioturbation of these organic soil horizons. A grave in Guatemala, originally dug in 1981 and containing five bodies, when investigated in 1999 had lost all traces of grave structure (both fill and grave cut) in the top 40 cm of the soil horizon because of the energetic nature of these organic soil layers. It was only visible where it cut into the distinct structure of the underlying B horizon, where there was less organic material and where clays and iron oxides had accumulated, providing a stable horizon that retained stratigraphic, intrusive distinctions such as gravefills and surfaces for centuries.

When the uppermost contexts of the grave have been exposed and identified, and the controlled removal of fill from a grave begins, care should be taken in the few centimetres where the fill has an interface with the grave wall. If 10 cm depth of fill is removed, the soil immediately against the surface should be left until last, so that it can be levered away and have room to 'fall' from the grave wall. Sometimes, in very wet or dry conditions, the soilfill will separate slightly from the grave wall, accentuating the stratigraphical boundary and forming a crack. A trowel can be used gently to pull or lever the fill away from the wall, revealing the intact surface. In 'sticky' clay soils, this levering action may remove clumps of soilfill that pull away trace evidence, such as hair or fibres, from the surface. The interface should be carefully observed during this action and any 'clump' so removed should be checked for adhering trace evidence. The soilfill will usually be looser than the truncated deposit it lies against and, in very loose and sandy fills, can simply be encouraged to fall away from the wall of a grave with the tip of a trowel. Once the surface is revealed, the traditional scraping action that archaeologists use to clean a surface should not be employed until a careful examination is made. This action will destroy trace evidence and fine tool marks, to ignore stratification is to ignore some of the most direct information of how the evidence came to be interred (Hochrein 2002, p. 48).

This technique should also be applied to removing soil from the surface of bodies, whether skeletonized or with soft-tissue preservation. Ballistic fragments, plant material and other trace evidence adhere to tissue, clothing and bone, and also adhere readily to the soil covering them; if this is removed without due care the evidence will go with it.

Loss of evidence: through destruction of the stratified sequences, and non-recognition of surface interface

For some archaeologists and anthropologists the surfaces between deposits are ignored or hold little stratigraphic importance. Many archaeologists outside the UK simply do not recognize the cut (and so the surface that represents it) as a significant stratigraphic entity (Lucas 2001, p. 154). With this lack of recognition, it is the deposits themselves that are considered revealing in terms of recovering evidence. In forensic burial terms, this places an emphasis on the soilfill of a grave that covers the body, and the body itself. It is normal to sieve this soil to recover any artefacts. This is important work in terms of evidentiary recovery, but should not be carried out in ignorance of the separate stratigraphic unit, the interfacial surface between the soilfill, and the deposits that have been cut through by the digging of the grave. Evidence on a surface is in its original *in situ* depositional location; the evidence within the soilfill is often not.

Some burial excavation techniques remove the backfilled soil within an identified grave, as well as the surrounding stratified deposits through which the grave has been dug, as a single horizontal arbitrary unit (also called levels, spits or planum). The gradual removal of these arbitrary units eventually leaves the body on a 'pedestal'. These methods have been widely used internationally and described (Mant 1950; Joukowsky 1980; United Nations 1991; Spennemann & Franke 1995a; Vanezis & Busuttil 1996; Ramey Burns 1999). This method gives a spatial and depth control to the removal of soil and to artefact recovery, and provides working space around bodies. However, it destroys and ignores the stratigraphic sequence of the burial even and, additionally, the artefact retrieved often has no known stratigraphic origin other than that of the arbitrary unit from which it was recovered (Figs 1 & 2).

There are only three occasions when horizontal arbitrary units should be removed during forensic excavation: (1) when test trenches are made (away from the immediate area of a grave) to observe the natural local stratigraphic sequence; (2) when removing uniform overburden or soil to locate a burial when the highest archaeological deposits cannot immediately be observed during excavation; (3) within the soilfill of a grave that appears to be a single, uniform context, as a control for spatial location of artefacts. However, Hochrein (2002, p. 48) regards each shovelful of soil returned to a grave at backfilling as theoretically representing a separate stratigraphic unit. This effect will be visible in clay soils, for example, where soil clods can retain tool marks. As with bodies in a body mass, such soil clods may perhaps best be described as part of one deposit, providing useful information on site formation processes (Roskams pers. comm. 2003) that should be sampled separately when appropriate.

If a grave is narrow, surrounding soils are unstable or waterlogged, or access within the grave is made difficult because of body positions, it may be necessary to removal the walls of a grave to assist in recovery. This should only be done with stratigraphic understanding after controlled removal of the grave-fill and recording of the exposed surface evidence of the wall. This wall removal provides working and drainage space and can help with lighting for photography, especially in deep graves. This can often be limited to one wall, or part of one wall.

However, in many cases the removal of walls is carried out as a standard procedure, whether or not it is necessary. There is a common feeling that removing walls allows adequate access so that a number of excavators can work on the burial, which also means that the work can be carried out in less time. It is quite normal for an access trench to be dug around all four sides of a grave during this process, removing and mixing together the soil from within the grave and from the soil horizons through which the grave was dug. There is also a common feeling that the grave itself is too constricting a space to work in, and that standing in the grave will damage the remains. The excavator, however, is working in the same space as the person who dug the grave. If a suitable amount of soil (some 20–30 cm) is left over a body in a grave while the wall surface is being examined, it will displace the excavator's concentrated weight and the remains will not be damaged. An excavator weighs less than the soil content of a grave; most deep graves can be successfully excavated within the structure of the grave itself or, in the case of shallow burials, from the ground surface beside the grave. Although Byers (2002) suggests 'when the excavation is down more than 8 inches, workers usually can no longer comfortably dig from the surface', in practice, working from the surface can be efficiently achieved to a depth of 0.7–0.8 m.

A simple calculation on a hypothetical hand-excavated single grave is illuminating concerning the space and time factors. If one person excavates a of grave dimensions $2 \times 1 \times 1$ m, removing fill from within the grave itself, 2 m^3 of soil can be removed,

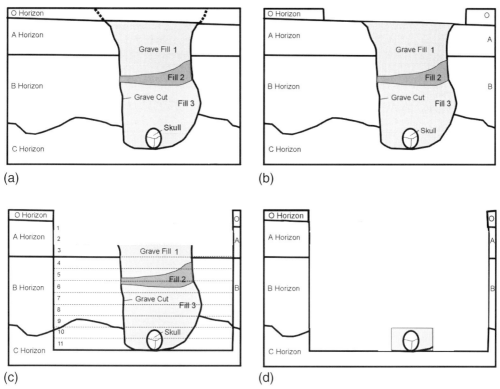

Fig. 1. Arbitrary excavation of graves. (**a**) Section across grave and natural stratigraphy. (**b**) Removal of O horizon leaf litter exposes grave in plan. (**c**) Removal of arbitrary levels (1–11) cuts through stratigraphic sequence and mixes soils. (**d**) Body and soil immediately above left intact surrounded by access trench. Grave surface removed.

retaining the intact grave surfaces. If a grave of the same dimensions with a 1 m wide trench on all four sides of it is excavated by five people, then 12m of soil can be excavated, or 2.4 m³ per person, destroying the grave walls. The soil will also have to be moved a greater distance from the grave to a place of storage. In many cases, excavating trenches (without machinery) around a grave is not economical in terms of saving effort, time or space, and is destructive to the structure of the grave. The feeling that there is always a need to remove the walls of a grave is therefore more of a mind-set than a practical necessity.

By arbitrarily stripping above and around a burial in this way, soils and artefacts from the gravefill and from separate stratified deposits, which both pre and post-date the burial event, become mixed together and lose their original contexts. This has implications not only for artefacts *in situ* and their location and association to a criminal event, but also to *terminus post quem* dating of associated deposits. More importantly, the surface forming the walls and (parts of) the floor of the grave is destroyed, losing the impressions and tool marks described previously.

Fragile trace evidence is lost and damaged by this

excavation process; at the very least there is loss of context. In forensic terms, 'the majority of forensic evidence is trace evidence'. Once a piece of evidence is moved, it can never be put back and its provenance is lost (Wiggins & Houck 2001, p. xii). Loss of ballistic tracks and the contextual location of embedded bullets also results from removal of the sides and floor of the grave.

Removal of the grave surface and the deposits immediately adjacent to it prevents effective sampling strategies of soils and soil water associated with the grave structure. This will affect the interpretation of trace element patterns in bone associated with the grave, and of diffusive passage of biomarkers from the body into the soil (Trueman pers. comm. 2003).

Given the sequence of events that can be determined and the evidence that can be recovered by exposing the surface of the walls and floor of a grave excavating stratigraphically, using the arbitrary/pedastalling method would seem to be flawed as a forensic archaeological technique. Archaeological excavation is formalized destruction (Praetzellis 1991), but this method is perhaps too formal and too

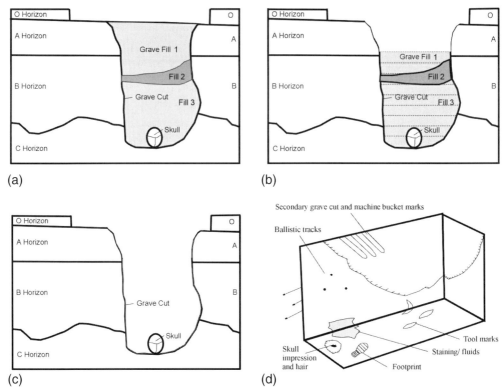

Fig. 2. Stratigraphic excavation. (**a**) Removal of O Horizon leaf litter exposes grave in plan. (**b**) Controlled removal of gravefill without destroying stratigraphic boundaries. Care taken not to damage grave surface. (**c**) Complete removal of gravefill, exposing body and grave surface for analysis. (**d**) Stratigraphic excavation exposes a variety of trace and other evidence on the walls and floor of the grave.

destructive, given the potential for loss of physical and stratigraphical evidence and the implications this has for forensic investigations.

Stratigraphic analysis: a link to separate locations

As well as the stratigraphic sequencing of graves, forensic investigations have been helped by analysis of natural stratigraphy. This has, for example, helped link soils moved from one grave to another through geological and palynological comparison. The Srebrenica Massacre saw the contents of primary mass graves robbed by heavy machinery and moved to remote secondary locations and reburied. A multi-disciplinary approach saw archaeologists identify exotic soils at these locations, which were then sampled by a palynologist. Soils from the backfilled soil of the secondary graves were compared with the natural stratigraphy of the secondary grave sites and primary grave sites and matched through soil structure and pollen content (United Nations 2000).

Microstratigraphic soil and debris layers can be deposited inside vehicle wheel arches from the turning of the wheel, building up over time as a vehicle moves between different locations. These have been analysed to determine their sequence, allowing comparison of their mineral and pollen ratios with those of sampled locations. On occasion, this has led to the positive determination that a certain vehicle was at a specific location, and was there in a sequence of movements between other locations (Brown pers. comm. 2003).

In the last 25 years, archaeological techniques in forensic investigations have developed using the influences of both US and UK archaeological theory. Recent forensic archaeological investigations, both nationally for the police and internationally for human rights organizations and war crimes tribunals, have successfully utilized stratigraphic methods of excavation and analysis to examine crime scenes, resulting in prosecutions (such as General Krstic at the ICTY) and investigations of cases of homicide, mass murder and burial In on site investigations and excavations, the archaeologist has always acted as a manager and utilizer of other scientific experts. On burial scenes, especially in mass grave and human rights cases, the archaeologist should act as a keystone in the building of a scientific multidisciplinary investigation and analysis. Stratigraphy is the tool that reveals the blueprint of the sequence of events through the time reflected in surviving buried deposits and features.

Many thanks for comments and assistance to A. Brown, E. Harris, F. Peccerelli, S. Roskams, B. Swift, C. Trueman. Thank you to M. Cox, M. Lewis, R. Wright and the reviewer for assistance and advice with the drafts of this paper.

References

BRADY, N. C. & WEIL, R. R. 2002. *The Nature and Properties of Soils*. 13th Edition. Prentice Hall, New Jersey.

BURNETT, B. R. 1991. Detection of bone and bone-plus-bullet particles in backspatter from close range shots to heads. *Journal of Forensic Sciences*, **36**, 1745–1752.

BURNS, K. R. 1999. *Forensic Anthropology Training Manual*. Prentice Hall, New Jersey.

BYERS, S. 2002. *Introduction to Forensic Anthropology*. Allyn and Bacon, Boston.

CHAMBERLAIN, A. T. & PARKER PEARSON, M. 2001. *Earthly Remains*. The British Museum Press, London.

FOXHALL, L. 2000. The running sands of time: archaeology and the short term. *World Archaeology*, **31**, 484–498.

GAMBLE, C. 2001. *Archaeology: The Basics*. Routledge, London.

GERBERTH, V. J. 1996. *Practical Homicide Investigation: Tactics, Procedures and Forensic Techniques*. CRC Press, Boca Raton.

GOWLETT, J. A. J. 1997. High definition archaeology: ideas and evaluation. *World Archaeology*, **29**, 152–171.

HAGLUND, W. D. 2002. Recent mass graves, an introduction. *In*: HAGLUND, W. D. & SORG, M. H. (eds) *Advances in Forensic Taphonomy: Method, Theory and Archaeological Perspectives*. CRC Press, Boca Raton, 243–261.

HARRIS, E. 1979. *Principles of Archaeological Stratigraphy*. Academic Press, London.

HARRIS, E. 1998. 25 years of the Harris matrix. Reprinted from the *Newsletter of the Society of Historical Archaeology*. World Wide Web Address: http://www.harrismatrix.com/history2.

HOCHREIN, M .J. 2002. An autopsy of the grave: recognizing, collecting, and preserving forensic geotaphonomic evidence. *In*: HAGLUND, W.D. & SORG, M.H. (eds) *Advances in Forensic Taphonomy: Method, Theory and Archaeological Perspectives*. CRC Press, Boca Raton, 45–70.

HUNTER, J. R., HERON, C. *ET AL*. 1994. Forensic archaeology in Britain. *Antiquity*, **68**, 758–769.

JOUKOWSKY, M. 1980. *A Complete Manual of Field Archaeology*. Prentice-Hall, New Jersey.

KENYON, K. M. 1957. *Digging up Jericho*. Ernest Benn, London.

KIND, S. 1987. *The Scientific Investigation of Crime*. Forensic Science Services Ltd, Harrogate.

LUCAS, G. 2001. *Critical Approaches to Fieldwork: Contemporary and Historical Archaeological Practice*. Routledge, London.

MANN, R. W., BASS, W. M. *ET AL*. 1990. Time since death and decomposition of the human body: variables and observations in case and experimental field studies. *Journal of Forensic Sciences*, **35**, 103–111.

MANT, A.K. 1950. *A study in exhumation data*. PhD thesis, University College London.

PRAETZELLIS, A. 1991. The limits of arbitrary excavation. *In*: E. C. HARRIS, BROWN III, M. R. & BROWN, G. J. (eds) *Practices of Archaeological Stratigraphy*. Academic Press, London, 68–87.

RAMEY BURNS, K. 1999. *Forensic Anthropology Training Manual*. Prentice Hall, New Jersey.

RODRIGUEZ, W. C. 1997. Decomposition of buried and submerged bodies. *In*: HAGLUND, W. D. & SORG, M. H. (eds) *Advances in Forensic Taphonomy: Method, Theory, and Archaeological Perspectives*. CRC Press, Boca Raton, 459–467.

RODRIGUEZ, W. C. & BASS, W. M. 1985. Decomposition of buried bodies and methods that may aid in their location. *Journal of Forensic Sciences*, **30**, 836–852.

SPENNEMANN, D. H. R. & FRANKE, B. 1995*a*. Archaeological techniques for exhumations: a unique data source for crime scene investigations. *Forensic Science International*, **74**, 5–15.

SPENNEMANN, D. H. R. & FRANKE, B. 1995*b*. Decomposition of buried human bodies and associated death scene materials on coral atolls in the tropical Pacific. *Journal of Forensic Sciences*, **40**, 356–367.

UNITED NATIONS. 1991. Model protocol for disinterment and analysis of skeletal remains. *In*: *Manual on the Effective Prevention and Investigation of Extra-Legal, Arbitrary and Summary Executions*. United Nations Office at Vienna, Centre for Social Development and Humanitarian Affairs, New York. New York, 30–41.

UNITED NATIONS. 2000. General Krstic trial transcripts 26/05/2000. World Wide Web Address: http://www.un.org/icty/ krstic/TrialC1/judgement/.

VANEZIS, P. & BUSUTTIL, A. 1996. *Suspicious Death Scene Investigation*. Arnold, London.

VASS, A. A., BASS, W. M. *ET AL*. 1992. Time since death determinations of human cadavers using soil solution. *Journal of Forensic Sciences*, **32**, 1264–1270.

WIGGINS, K. & HOUCK, M. M. 2001. Introduction. *In*: HOUCK, M. M. (ed.) *Mute Witnesses: Trace Evidence Analysis*. Academic Press, London, xi-xxxi.

Colour theory and the evaluation of an instrumental method of measurement using geological samples for forensic applications

DEBRA J. CROFT & KENNETH PYE

Kenneth Pye Associates Ltd, Crowthorne Enterprise Centre, Crowthorne Business Estate, Old Wokingham Road, Crowthorne RG45 6AW, UK (e-mail: d.croft@kpal.co.uk)

Abstract: Colour is a fundamental characteristic of many materials, including soils and sediments, and has been much used in geological, pedological and Quaternary science research. Traditionally, colour has been described qualitatively by visual comparison with standard charts, such as the *Munsell Soil Color Charts* or the Geological Society of America *Rock Color Chart*. Instrumental colour determination has been developed and used in industry for a variety of applications, including quality testing of paints, dyes and foodstuffs. In this paper, colour theory is outlined, and the Minolta® CM-2002® hand-held spectrophotometer is tested on geological samples to investigate reproducibility, discriminatory power, and accuracy in analysis. Standard methods for calibration, presentation and testing of a variety of (often small) samples have been developed. Examples are provided for the use of the method in forensic geoscience casework.

The importance of colour

Colour is a key attribute of many geological samples, including minerals, soil, dusts, sand and building stone. A number of general observations can be made, for example: the blacker the soil, often the greater the organic content; the darker the soil, the greater the moisture content; spotting or mottling indicates 'gleying' or reducing conditions; red and red–orange colours indicate oxidizing conditions and the presence of iron oxides. Red soils and sediments have attracted much research and debate (Van Houten 1961, 1968, 1973; Schwertmann & Taylor 1989; Turner 1980; Pye 1983; Schwertmann 1993). Schwertmann (1993) distinguishes between various iron oxides present in soils, haematite, geothite and ferrihydrite, based on the Munsell indices Hue (H), Value (V) and Chroma (C). Colour analysis has also been widely used to characterize soils and sediments in mapping and provenance studies (e.g. Walden *et al.* 1996; Pell *et al.* 2000; Wells 2002). Bigham & Ciolkosz (1993) provides an excellent summary of colour work to that date, including instrumental measurement.

In everyday life colour is ubiquitous and simple verbal descriptions such as 'red' or 'brown', with accompanying adjectives of 'bright', 'dull' or 'light', are given as adequate definitions. However, in determining and reporting colour as a definitive characteristic in scientific studies, it is necessary to consider a variety of factors that can affect the perception of colour. These include light sources and direction or angle of view, effects of background and contrast, grain size, compaction and the crystalline nature of the material being considered, as well as moisture content and temperature (Johnston 1967; Thornton 1997; Minolta 1998).

There are a number of documented systems for characterizing colour for a variety of applications, the most widespread being the Munsell® Color System. The *Munsell Soil Color Charts* (1994), used to determine colour in soils and geological materials, are a subset of the full Munsell Color System and are widely used in earth science research. However, the use of the comparison charts to make a visual match inevitably involves a degree of subjectivity, and different operators will frequently assign different values to the same sample. Indeed the same operator may interpret the same sample differently on separate occasions.

This paper presents the evaluation of a more objective instrumental method for colour determination using the Minolta CM-2002® spectrophotometer. The early work of Gangakhedkar (1981) provided a summary of instrumental methods including colorimeters, spectrophotometers and spectro-radiometers. His analysis of (then) 'new generation' spectrophotometers (those dedicated to colour determination, as opposed to adapted from chemical analysis systems) lists many factors that make them an ideal system, including small sample requirement, non-destructive nature, speed of analysis, portability, and repeatable and accurate results. McDonald's (1982) review of the relationship between instrumental and visual comparisons of colour from an industrial perspective is a valuable text and concluded that the ultimate versatility of visual comparisons has to be balanced against the subjectivity involved. This early work was further investigated and reported by a number of authors comparing instrumental methods with visual colour matching (e.g. Shields *et al.* 1996; Nagao & Nakashima 1991), together with work on laboratory method and accuracy (e.g. Torrent & Barron 1993),

From: PYE, K. & CROFT, D. J. (eds) 2004. *Forensic Geoscience: Principles, Techniques and Applications*. Geological Society, London, Special Publications, **232**, 49–62. © The Geological Society of London, 2004.

and the presentation and interpretation of results (Wells *et al.* 2002).

However, the instrumental methods have not been widely used by geologists and Quaternary scientists. This is also true in the forensic context where a number of authors have examined the use of colour (Dudley 1975; Murray & Tedrow 1991; Saferstein 1995; Sugita & Marumo 1996; Trujillo *et al.* 1996; Thornton 1997; Schafer 2001), but all have used the Munsell Color System.

Colour theory, characterization and notation systems

Colour is the perception by the brain of light of a particular wavelength falling onto the eye. For all practical purposes, this is limited to the visible spectrum between 400 nm (short wavelength, violet) and 700 nm (long wavelength, red). There are thought to be approximately 1100 distinguishable colours in soils (Saferstein 1995), making colour a potentially powerful method for the discrimination of samples. Sugita and Marumo (1996, p. 202), working on the forensic analysis of soil, stated that: 'Color is the most distinguishable property of soil in which the pedogenic environment and history are reflected.'

In the laboratory, observations can be made under controlled conditions. Some authors prescribe various pre-treatments for soil prior to testing for colour: Dudley (1975) suggested ashing and the comparisons of dry soil colour, Janssen (1983) suggested separation and testing of the clay fraction only, while Sugita and Marumo (1996) removed organic matter and iron oxide. However, even when strict protocols of preparation and presentation are followed, such as those set out in McCrone *et al.* (1973), Murray and Tedrow (1992) and Antoci and Petraco (1993), a degree of subjectivity in naming soil colour or in matching soil to a Munsell colour chip is inevitable. This can be overcome by the use of a spectrophotometer, which imposes further standardization on the variables involved in colour perception.

The development of *colour space*, by which it is possible to express colour in a notation system such as a combination of numbers, has been worked on for more than 100 years. These developments include the first Munsell Classification in 1905, which was subsequently updated to produce the current system based on H, V and C, as well as the XYZ system, L*a*b* colour space and the Hunter Lab colour space. More specialized systems also exist, such as the Japanese, *Göfo* system for measuring the colour of pork in order to indicate a degree of freshness and consumer acceptability. The three systems most widely used are outlined below.

Munsell Color System

The *Munsell Soil Color Charts* are widely available and contain the nine grouped colour cards particularly relevant to soil colour determination, a modified version of the full *Book of Color*. Each colour chip can be expressed by a standard name and by a combination of three parameters, Hue (H), Value (V) and Chroma (C), written as H V/C. A soil might be described as 'reddish brown' (5 YR 5/3). The classification system can be viewed as a wheel on an axle, which has the spectrum of colours around it (H), where the intensity or purity of the colour is an expression of the distance from the centre of the wheel to the outer edge (C), and the achromatic vertical axle, which the wheel travels up and down, gives relative lightness and darkness (V).

There are a number of cards within the chart which show similar colour chips and it can be difficult to assign a sample to one chip rather than another. For example, the charts for 2.5Y (yellow) and 5YR (yellow–red) both have chips called 'dark grey'. The full notation for each is 2.5Y 4/1 and 5YR 4/1, and distinguishing between the two chips is visually difficult.

Tristimulus XYZ system

The XYZ value system is based on tristimulus values, as defined by the CIE (Commission Internationale de l'Eclairage) in 1931. Its foundation is the three-colour component vision of the eye, where colours are perceived as combinations of red, green and blue. It is less continuous, more stepped, than some other systems of colour measurement. Although based on the human eye, the system has far less discriminatory power than the eye.

L*a*b* colour system

The L*a*b* system was devised in 1976 to provide a more continuous and uniform colour space for fine visual differentiation, particularly in manufacturing where there is a demand for low tolerances (such as the automotive paint industry). It is also known as CIELAB space, developed from earlier CIE systems. The system can be visualized as a colour solid, spherical in shape. L* represents Lightness, an achromatic measure running from zero (very dark) to 100 (very pale), through the sphere. A cross-section through the solid, at a value L*, gives a disc with two axes at right angles crossing in the centre at L*. The a* axis represents a green–red continuum, where green is a negative value and red is positive. The b* axis represents a blue–yellow continuum, with blue as a negative and yellow as positive. The centre of the disc is dull and achromatic, with vividness increasing towards the outside edge. A test colour can be fixed to a position, as a point in 3-D space, using the three coordinates.

Instrumental method for the spectrophotometry trials

The Minolta CM-2002® spectrophotometer is shown in Figure 1, together with standard colour tiles used for calibration. The instrument is initially calibrated negatively to a closed black box (zero reflectance) and positively to an International Standard white tile (100% reflectance) before each use. In our laboratory, the white tile has been analysed to assess drift over a period of 3 years and the instrument has returned values within $\pm 0.6\%$ of the certified values over the 400–700 nm wavelength range over the time period.

There are various illuminant types available within the instrument; standard illuminant D^{65} has been used throughout the work reported here: that is average daylight, including the ultraviolet wavelength region, at a temperature of 6504 °C (Minolta 1998). The spectrophotometer has the facility to exclude specular light during analysis, but in these trials this component has been included to reflect the crystalline nature of many geological materials.

Sample material was presented, in all cases, against a background of a cleaned white ceramic tile and measured against an appropriate certified CERAM II colour standard tile, six of which are shown in Figure 1. The 3 mm diameter aperture of

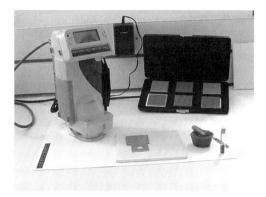

Fig. 1. Minolta CM-2002® spectrophotometer with CERAM II colour standard tiles.

the spectrophotometer was centred over the sample and the average of five measurements was taken for each sample tested. Three types of parameters were recorded: L*a*b* indices, Munsell indices (H, V and C), and the percentage light reflected over a range of 400 nm to 700 nm in 10 nm intervals, from which a reflectance curve can be constructed. The on-board computer can calculate mean and standard deviation figures for repeat measurements, and all results can be downloaded to a spreadsheet format on a standard PC.

The instrument was tested on a number of sand, dust and soil samples, as well as against a selection of colour chips from a *Munsell Soil Color Chart*. The range of geological samples used is listed in Table 1.

Table 1. *Geological samples tested for colour*

Group	Geographical source	Code	Material
Soils	Royal Holloway, Egham, Surrey, UK	CLC	Grassed soil, derived from London Clay
	Tanah Rata, Malaysia	MWG	Weathered granite regalith
	Tower Hamlets Cemetery, London, UK	THC	Grassed soil, Quaternary Drift derived
Dust deposits	Lanzhou, China	CLD	Loess
	Mt St Helens, USA	TVD	Volcanic ash
Coastal sediments	Blue Lagoon Beach, Australia	ABS	Siliceous beach sand
	Two Islands, Gt Barrier Reef, Australia	ACS	Coralline beach sand
	Beach, Nr Darwin, Australia	DBS	Coastal dune sand
	Denmark	DCS	Coastal dune sand
	North Kohala, Big Island, Hawaii	HBS	Volcanic beach sand
	Essaouira, Morocco	MDS	Coastal dune sand
	Berrow (Severn Estuary), Somerset, UK	SES	Coastal dune sand
Desert sediments	Nr Ayers Rock, Australia	ADS	Linear dune sand
	White Sands, New Mexico	GDS	Gypsum barchan dune sand
	Mauritania, S. Sahara	MRS	Desert flat sand
	El Facuar, Tunisia	SDS	Transverse dune sand

Table 2. *Variation in determined L*a*b* and Munsell indices (H, V/C) measurements repeated over 5 sequential days using five subsamples from sample THC (a Quaternary Drift derived soil). (Each subsample tested 5 times)*

Test parameters		L*	a*	b*	H	V/C
Day 1						
n = 25	Mean	28.70	2.73	8.11	0.94Y	2.8/1.4
	SD	0.95	0.08	0.25	0.09Y	0.11/0.04
	CV%	3.31	2.93	3.08		
Day 2						
n = 25	Mean	28.81	2.73	8.04	0.86Y	2.8/1.3
	SD	1.35	0.06	0.15	0.09Y	0.13/0.05
	CV%	4.69	2.20	1.87		
Day 3						
n = 25	Mean	28.23	2.63	7.90	1.00Y	2.8/1.3
	SD	1.34	0.14	0.34	0.14Y	0.15/0.04
	CV%	4.75	5.32	4.30		
Day 4						
n = 25	Mean	28.32	2.64	7.77	0.90Y	2.8/1.3
	SD	1.42	0.14	0.33	0.14Y	0.13/0.04
	CV%	5.01	5.30	4.25		
Day 5						
n = 25	Mean	27.78	2.55	7.68	1.04Y	2.7/1.3
	SD	2.19	0.09	0.30	0.11Y	0.24/0.00
	CV%	7.88	3.53	3.91		
Overall						
n = 125	Mean	28.37	2.66	7.90	0.95Y	2.8/1.3
	SD	1.42	0.12	0.30	0.13Y	0.01/0.05
	CV%	5.02	4.44	3.86		

Experimental work

A number of individual experiments were carried out to test the spectrophotometer under laboratory conditions, with the aim of establishing reproducibility within samples, comparability with respect to Munsell colour chips, discriminatory power between samples, variation in size fractions within a sample, variation with presentation method, and variation after heating to a range of temperatures.

Instrumental measurement for repeatability on natural geological samples

A single sample was tested on each of 5 days. The sample used was a <150 μm fraction from THC (a Quaternary Drift derived soil). Each test on each day consisted of five sub-samples, each tested five times. The mean and standard deviations for L*a*b* and for Munsell indices are summarized in Table 2, as measured day by day for the 5 days, and overall, as well as the coefficient of variation (CV%) for L*a*b* values.

Variation is extremely low for all parameters, showing excellent reproducibility over this short-time period. The exercise was repeated after 6 months on the same sample (which had been stored in an air-tight container, in the dark, at constant low temperature). The results summarized in Table 3 were again highly reproducible within themselves and in relation to the earlier set.

Repeat tests were also carried out on 4 of the 16 bulk geological samples outlined in Table 1. One sample from each type was chosen: SDS, ABS, CLD and CLC (see Table 4). Five repeat measurements were carried out on each of five subsamples. Table 4 shows the mean and standard deviation values for L*a*b* and Munsell indices, as well as CV% for the L*a*b* indices. Figure 2 shows the reflectance curves for these samples.

It can be seen from Table 4 and from the reflectance curves (Fig. 2) how reproducible the data are, even using the dried bulk samples with no pre-treatments. The most variation is seen in sample CLC (London Clay derived soil), supporting the theory that soils are inherently less homogenous than sand and loess samples. However, even for this sample, the CVs for L*a*b* are less than 7% and the Munsell H values all fall within 0.5 of a colour unit, making the method reliable even at this crude level of sample presentation.

Table 3. *Variation in determined L*a*b* and Munsell indices (H, V/C) measurements, repeated over 5 sequential days using five subsamples from sample THC (period 2)*

Test parameters	L*	a*	b*	H	V/C
Overall					
n = 125 Mean	28.67	2.63	7.92	0.98Y	2.8/1.4
SD	1.66	0.13	0.37	0.13Y	0.17/0.05
CV%	5.80	4.89	4.68		

Table 4. *Variation in L*a*b* and Munsell indices (H, V/C) measurements, as carried out on four contrasting bulk soil sample types*

Sample	Subsample	L*	a*	b*	H	V/C
SDS	Bulk 1	64.87	9.06	22.85	8.7YR	6.4/3.9
(Saharan desert sand)	Bulk 2	65.15	9.13	23.33	8.8YR	6.4/4.0
	Bulk 3	66.43	9.19	23.65	8.7YR	6.6/4.1
	Bulk 4	65.49	9.06	23.33	8.8YR	6.5/4.0
	Bulk 5	65.55	9.44	23.90	8.7YR	6.5/4.1
	Mean	**65.50**	**9.18**	**23.41**	**8.74YR**	**6.48/4.02**
	SD	**0.59**	**0.16**	**0.39**	**0.05YR**	**0.08/0.08**
	CV%	**0.90**	**1.71**	**1.67**		
ABS	Bulk 1	50.74	6.83	18.36	9.8YR	5.0/3.0
(Australian beach sand)	Bulk 2	51.73	6.03	17.28	10.0YR	5.1/2.8
	Bulk 3	50.73	6.47	17.49	9.8YR	5.0/2.9
	Bulk 4	51.55	6.94	18.23	9.6YR	5.1/3.0
	Bulk 5	52.18	6.71	18.04	9.7YR	5.1/3.0
	Mean	**51.39**	**6.60**	**17.88**	**9.78YR**	**5.06/2.94**
	SD	**0.64**	**0.36**	**0.48**	**0.15YR**	**0.05/0.09**
	CV%	**1.24**	**5.47**	**2.66**		
CLD	Bulk 1	55.40	6.77	18.73	9.6YR	5.4/3.1
(Chinese loess)	Bulk 2	53.78	6.57	18.18	9.6YR	5.3/3.0
	Bulk 3	54.34	6.67	18.38	9.6YR	5.3/3.0
	Bulk 4	54.64	6.95	18.87	9.5YR	5.4/3.1
	Bulk 5	55.06	7.02	19.04	9.5YR	5.4/3.2
	Mean	**54.64**	**6.80**	**18.64**	**9.56YR**	**5.36/3.08**
	SD	**0.63**	**0.19**	**0.35**	**0.05YR**	**0.05/0.08**
	CV%	**1.15**	**2.75**	**1.90**		
CLC	Bulk 1	43.65	6.43	19.18	0.6Y	4.3/3.0
(London Clay derived soil)	Bulk 2	42.63	6.50	18.64	0.5Y	4.2/2.9
	Bulk 3	46.59	6.83	20.46	0.5Y	4.6/3.3
	Bulk 4	45.94	7.56	20.96	0.1Y	4.5/3.4
	Bulk 5	43.48	6.64	19.33	0.5Y	4.3/3.0
	Mean	**44.46**	**6.79**	**19.71**	**0.44Y**	**4.38/3.19**
	SD	**1.71**	**0.46**	**0.96**	**0.19Y**	**0.16/0.22**
	CV%	**3.84**	**6.73**	**4.88**		

Comparison to Munsell Soil Color Chart chips

Five colour chips from each of the seven Hue cards used in the visual examinations of soil colour were tested by spectrophotometer to ascertain compatibility of method and accuracy. This might be necessary in forensic casework where the two sides in a court case might use the different methods. The cards tested were 10R, 2.5YR, 5YR, 7.5YR, 10YR, 2.5Y and 5Y. Each chip was tested five times and the results averaged by the on-board computer. Table 5 shows the indices and colour descriptions stated on the Munsell Hue cards, together with the mean (*n* = 5) indices returned by the spectrophotometer. The range of absolute differences and the mean and standard deviation for the total data set are also shown.

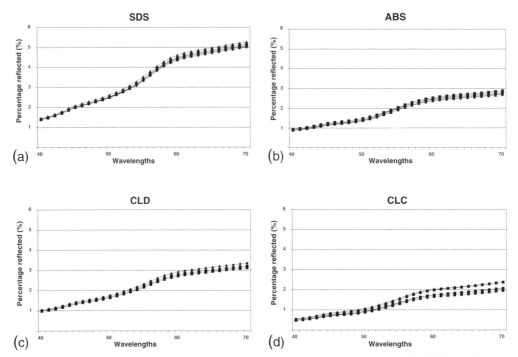

Fig. 2. Reflectance curves (400–700 nm range) for five subsamples taken from four samples: (**a**) SDS (Saharan desert sand), (**b**) ABS (Australian beach sand), (**c**) CLD (Chinese loess deposit), and (**d**) CLC (London Clay derived soil).

The returned V indices vary by 0.2 units or less from the published value, with the exception of 10YR 2.5/1 (black) which varies by 0.7 units (** in Table 5). The mean and standard deviation (SD) of the absolute differences for the dataset are both 0.1. The returned C indices vary by 1.1 units or less from published values, with the mean and SD of the absolute differences for the data both equal to 0.2. The returned H indices vary by 1.3 units or less, with the exception of the same black chip as above, which varies by 3.5 units. The mean and SD of the absolute differences for the dataset are both 0.6 units.

The largest variations occur in the black chip 10YR 2.5/1 and then, as a general observation, in the chips named as 'red'. Further investigation of all the chips named as 'red' returned V and C indices extremely close to the published ones and the H values varied by less then 1.0 units. All variations were well within the 2.5 units of Hue change found in the sequence of conventional Munsell cards, and varied by a maximum of one chip in direction for both the V and C components.

This exercise demonstrates that, when comparing the instrument-returned data with visually obtained data, care should be taken, particularly when dealing with very dark or black samples. The increased resolving power of the instrument compared with the standard card sets also has to be borne in mind, as well as the potential level of wear, fading and contamination on the soil chart being used.

Variation between natural geological samples

Testing of bulk dried samples was carried out on the eleven sands (seven coastal and four desert), two dusts and three soils listed in Table 1. Each sample was subsampled five times, and each subsample was tested five times and the mean values taken ($n = 25$). Table 6 summarizes the data obtained for L*a*b* and Munsell indices.

There is good discrimination in all indices between the different geological materials, even when they are visually similar, demonstrating the potentially powerful nature of the instrument.

Variation in colour of different size fractions is found in geological samples

Different size fractions were obtained from four of the samples (ADS, CLD, THC and SES) by wet sieving through stainless steel sieves. Sieves of mesh size 1 mm, 0.5 mm, 150 μm, 63 μm and 20 μm were

Table 5. *Summary of Munsell indices (H, V/C) and descriptions compared with the average of five spectrophotometer readings for the selected colour chips*

| | Munsell Chart Chips | | Returned indices | |
H	V/C	Description	H	V/C
10R	8/1	White	9.5R	7.9/1.3
10R	7/8	Light red	9.9R	6.9/8.1
10R	5/3	Weak red	9.6R	5.0/3.2
10R	4/8	Red	0.4YR	4.0/7.5
10R	2.5/1	Reddish black	9.8R	2.6/1.1
2.5YR	8/1	White	2.1YR	8.0/1.3
2.5YR	7/8	Light red	2.2YR	7.0/8.0
2.5YR	5/3	Reddish brown	1.3YR	5.0/3.1
2.5YR	4/8	Red	2.1YR	4.1/7.6
2.5YR	2.5/1	Reddish black	0.6YR	2.6/1.2
5YR	8/1	White	4.4YR	7.9/1.1
5YR	8/4	Pink	3.9YR	7.9/3.9
5YR	5/3	Reddish brown	4.4YR	5.0/2.9
5YR	5/8	Yellowish red	4.7YR	5.0/7.7
5YR	2.5/1	Black	4.3YR	2.6/0.9
7.5YR	8/1	White	6.9YR	7.9/0.9
7.5YR	8/6	Reddish yellow	7.0YR	8.0/6.0
7.5YR	5/3	Brown	6.6YR	5.0/3.0
7.5YR	5/8	Strong brown	7.3YR	5.0/7.4
7.5YR	2.5/1	Black	7.4YR	2.6/0.8
10YR	8/1	White	9.6YR	8.0/1.0
10YR	8/8	Yellow	9.7YR	8.0/7.8
10YR	5/3	Brown	8.9YR	5.0/2.7
10YR	5/8	Yellowish brown	9.3YR	5.1/6.9
**10YR	**2.5/1**	**Black**	**6.5YR**	**3.2/0.4**
2.5Y	8/1	White	1.2Y	7.9/1.1
2.5Y	8/8	Yellow	2.4Y	8.0/7.7
2.5Y	5/3	Light olive brown	2.3Y	5.0/3.0
2.5Y	5/6	Light olive brown	2.4Y	4.9/5.6
2.5Y	2.5/1	Black	3.0Y	2.7/0.9
5Y	8/1	White	5.3Y	8.0/1.0
5Y	8/8	Yellow	4.9Y	7.9/8.0
5Y	5/3	Olive	4.5Y	4.9/3.0
5Y	5/6	Olive	4.3Y	5.0/5.5
5Y	2.5/1	Black	4.0Y	2.6/0.9
Absolute		Min.	0.1	0.0/0.0
differences		Max.	3.5	0.7/1.1
		Mean	0.6	0.1/0.2
		SD	0.6	0.1/0.2

used. The fractions were then dried at low temperature (30 °C) and stored in air-tight containers in a cool and dark environment.

Table 7 shows the L*a*b* and Munsell indices for the four samples and size fractions tested. Figure 3 shows the variation found in the reflectance curves for the different size fractions found in each of the four samples.

From the graphs in Figure 3 and data in Table 7, it can be seen that variation does exist between size fractions in each sample. However, the general shapes of the reflectance curves obtained from the different size fractions in a single sample are broadly similar, the main difference being in reflectance intensity (position against the y-axis). These differences can be readily explained; for example, in sample CLD (Chinese loess deposit), the sieved sample largely fell into the >20 μm and the 20–63 μm fractions, and the curves for these and the bulk sample were closely related in both shape and position. The 63–150 μm fraction shows a curve with the same general shape and inflections but lower reflectance, particularly at the red end of the spectrum. On visual inspection, a large number of dark brown and black grains, mainly of biotite mica and other heavy minerals, could be seen in this size fraction. Other samples have similar explanations.

The variation found strongly suggests that the same size fraction should be used wherever possible in colour testing. Ideally colour should be determined on the bulk material and on a specific size fraction, or fractions, which are known to be characteristic of the particular material being analysed. Coarse sediments may need additional repeated analyses to compensate for the fact that the aperture of the instrument at 3 mm could easily be approaching the size of the individual grains.

Further investigation of the relationship of colour contribution by the size fractions to the colour of the bulk sediment was carried out using sediments from Monument Valley, Utah, USA. The sample was dry sieved to three size fractions: <250 μm, 250–500 μm and >500 μm. Results are presented in Table 8, with reflectance curves in Figure 4.

The bulk and the <250 μm subsample are very similar in colour numerically and in terms of their reflectance curves. These curves are close in position on the y-axis for all wavelengths and very similar in terms of detailed shape and inflection points. The 250–500 μm fraction curve is similar in shape and inflection, but offset on the vertical axis, particularly at the red end of the spectrum (i.e. this fraction is darker/less reflective and 'less red' than the bulk). The >500 μm fraction exhibits a different-shaped curve than the other three: darker than the bulk and flatter overall, indicating a colour closer to grey than the other curves. This latter fraction contains large amounts of dried organic matter, which largely represents dried sage-bush fragments and is also mineralogically different from the other samples. The <250 μm subsample is closest in all measures to the bulk, and this is supported by the fact that this size fraction contributes approximately 75% of the bulk sample, consisting of largely quartzo-feldspathic sand and sandstone fragments.

Table 6. *L*a*b* and Munsell indices (H, V/C) for 16 bulk geological samples*

Sample code	Sample group	L*	a*	b*	H	V/C
ABS	Coastal sediments	73.73	1.65	9.03	1.1Y	7.3/1.3
ACS		68.20	2.95	14.57	1.2Y	6.7/2.2
DBS		49.58	6.72	17.95	9.8YR	4.9/3.0
DCS		57.96	2.83	11.95	1.3Y	5.7/1.8
HBS		30.13	1.72	4.30	0.2Y	3.0/0.7
MDS		53.20	7.27	18.30	9.3YR	5.2/3.1
SES		47.08	5.31	16.58	0.6Y	4.6/2.6
ADS	Desert sediments	32.15	20.25	23.16	3.2YR	3.2/5.4
GDS		71.19	2.13	10.37	0.9Y	7.0/1.5
MRS		48.09	15.51	28.25	6.8YR	4.8/5.3
SDS		65.70	9.47	23.72	8.6YR	6.5/4.1
CLD	Dust deposits	51.25	6.75	18.63	9.7YR	5.0/3.1
TVD		64.27	1.54	8.59	1.5Y	6.3/1.2
CLC	Soils	44.54	6.78	20.34	0.6Y	4.4/3.2
MWG		57.87	10.63	28.49	9.4YR	5.7/4.8
THC		24.10	2.09	8.36	2.3Y	2.4/1.4

Effects of different sample presentation methods (using loess)

Six dried loess samples (A, B, C, E, F, G) from different global locations were tested in three ways, using:

(1)   A small quantity of sample which was dampened with de-ionized water and smeared on a white ceramic tile, then allowed to dry;
(2)   A sample reclaimed from (1) and presented as a dry powder, packed in a frame to a standard thickness (3 mm), on a white tile;
(3)   A split of the original dry bulk powder, packed in a frame to a standard thickness, on a white tile.

Figure 5 shows reflectance curves for samples tested in this way. The *y*-axis shows the percentage light reflected and has been kept constant at a maximum scale of 60%. The shapes of the curves remain consistent for the three testing methods, but the vertical positions across all wavelengths vary. The bulk dry sample (3) was the least reflective ('darkest'), the reclaimed powder (2) intermediate, and the moistened and dried sample (1), the most reflective ('lightest'). Figure 6 shows, on the same graph, the three reflectance curves for the three presentation methods of the loess sample A. The L* and Munsell V indices showed the same pattern. The differences demonstrate that variations can be expected if the sample is presented as a dry bulk powder or wetted and allowed to dry, probably due both to grain size separation effects and hydration processes. During wetting and drying, salts may migrate to the surface of the sample, increasing the reflectance of that sample.

These results indicate the importance of standardizing the way that samples are handled, pre-treated and presented to the instrument for analysis. Moisture content of the sample is critical, and it should be ensured that, where samples are submitted to a laboratory (as opposed to measurements made in the field), all samples in a batch are either air-dried, or oven-dried at a constant temperature and presented in a standard way.

Effects of heating on sample colour

The effects of heating on sample colour were investigated using four samples: RF-107A (a red tropical sandy soil) and C-KP1, C-KP3 and C-KP8 (forensic soil samples taken from adjacent fields in Wiltshire, UK). For RF-107A, a split of the bulk air-dried (unheated) sample was tested, together with subsamples which were heated to 400 °C for 6 hours and to 800 °C for 6 hours. For the other three samples, the wet-sieved <150 μm fraction was used: one air-dried (unheated) and subsamples which were heated to 400 °C for 6 hours and to 800 °C for 6 hours. Table 9 presents the L*a*b* and Munsell indices for these samples, and Figure 7 shows their reflectance curves.

The Munsell H index for each of the samples indicates an increase towards red with increased temperature, and this is reflected by the increasingly positive a* parameter (towards red, away from green) and the b* parameter (towards yellow, away from blue). The Munsell C parameter consistently increases with heating, indicating an increase in intensity of colour. The achromatic parameter on both systems (L* and V) shows no overall pattern.

Table 7. *L*a*b* and Munsell indices (H, V/C) for different size fractions obtained from four samples*

Sample	Size fractions	L*	a*	b*	H	V/C
ADS	Bulk sample	32.15	20.25	23.16	3.2YR	3.2/5.4
(Australian desert sand)	500 μm–1 mm	33.89	18.21	21.76	3.7YR	3.4/4.9
	150 μm–500 μm	35.00	21.58	24.84	3.1YR	3.5/5.8
	63 μm–150 μm	35.75	21.73	24.88	3.0YR	3.6/5.8
	20 μm–63 μm	40.13	21.09	26.85	3.7YR	4.0/5.9
	<20 μm	39.91	20.65	27.72	4.2YR	4.0/5.9
CLD	Bulk sample	51.25	6.75	18.63	9.7YR	5.0/3.1
(Chinese loess)	63 μm–150 μm	41.54	4.59	15.09	1.0Y	4.1/2.3
	20 μm–63 μm	50.19	6.81	18.92	9.8YR	4.9/3.1
	<20 μm	51.44	6.43	17.53	9.6YR	5.0/2.9
THC	Bulk sample	24.10	2.09	8.36	2.3Y	2.4/1.4
(Quaternary Drift	500 μm–1 mm	37.32	3.29	12.73	1.5Y	3.7/1.9
derived Soil)	150 μm–500 μm	37.90	2.64	10.25	1.7Y	3.7/1.6
	63 μm–150 μm	28.14	1.98	7.91	2.2Y	2.8/1.3
	20 μm–63 μm	27.95	2.56	8.94	1.6Y	2.8/1.5
	<20 μm	29.16	2.55	9.53	1.8Y	2.9/1.6
SES	Bulk sample	47.08	5.31	16.58	0.6Y	4.6/2.6
(Somerset coastal sand)	150 μm–500 μm	43.94	5.49	16.78	0.6Y	4.3/2.6
	63 μm–150 μm	47.89	5.13	17.05	0.9Y	4.7/2.7
	20 μm–63 μm	41.43	5.18	16.42	0.8Y	4.1/2.5
	<20 μm	44.10	5.59	17.60	0.6Y	4.3/2.7

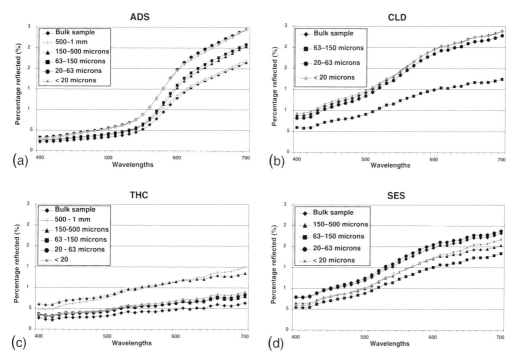

Fig. 3. Reflectance curves (400–700 nm range) for different size fractions obtained from four samples: (**a**) ADS (Australian desert sand), (**b**) CLD (Chinese loess deposit), (**c**) THC (Quaternary Drift derived soil), and (**d**) SES (Somerset coastal sand).

Table 8. *L*a*b* and Munsell indices (H, V/C) for bulk and size fractions obtained from a sample of Monument Valley sediment*

Size fractions	L*	a*	b*	H	V/C
Bulk	45.50	18.00	25.11	4.5YR	4.5/5.3
>500 μm	43.36	9.49	18.79	7.8YR	4.3/3.3
250–500 μm	34.92	10.04	16.47	6.9YR	3.5/3.1
<250 μm	45.71	19.11	26.25	4.3YR	4.5/5.6

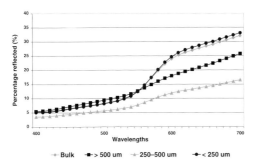

Fig. 4. Reflectance curves (400–700 nm range) for bulk and different size fractions obtained from a sample of Monument Valley sediment.

The reflectance curves show greatest difference between temperatures of 400 °C and 800 °C, and heating to 400 °C evidently has less impact on the soil colour for the soils analysed.

Effects of cling film over the instrument aperture

In order to prevent fine particles adhering to, or entering, the aperture of the instrument, cling film is often used as a protective screen. The effect of this was investigated by comparing values obtained with and without the film present, using the International Standard white tile. Reflectance values over the 400–700 nm range were compared with the certified values for the standard tile.

In all cases, the reflectance values for each wavelength are increased by less than 1%. The Munsell indices show a small increase in H (0.90–0.97 purple-blue), with negligible change in V and C. This slight increase towards the blue end is confirmed with the tests carried out on the CERAM certified standards, which each show small increases in the losses in reflectance from the red end (700 nm) and slight increases at the blue end (400 nm), as well as supporting variation in the Munsell H index. For the Munsell V index, the returned values are marginally lighter

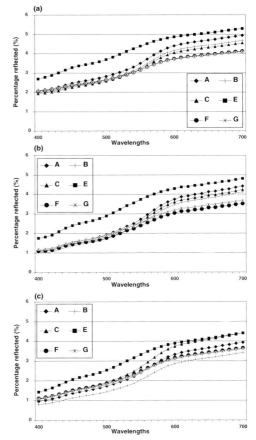

Fig. 5. Reflectance curves (400–700 nm range) for loess samples presented for analysis in three different ways on a white ceramic tile: (**a**) moistened and allowed to dry; (**b**) reclaimed dry powder scraped from (a) and presented in a frame; and (**c**) a split of the original dry bulk material presented in a frame.

(more light reflected), and for C values they are marginally lower, showing small losses in intensity. These changes were consistent across the colour spectrum tested (white, grey, blue, green, yellow, red and brown standard colour tiles) and are very small in magnitude. Use of cling film appears to make a minor difference to the results obtained for most samples, although its use should be stated where applicable, and all samples being compared should be measured either with or without this barrier.

Case examples

The spectrophotometer has been widely used in research (e.g. Croft 2003) and casework in our labo-

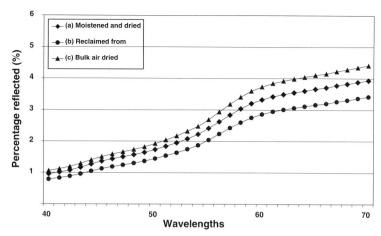

Fig. 6. Reflectance curves (400–700 nm range) for loess A presented in three different ways on a white ceramic tile: **(a)** moistened and allowed to dry, **(b)** reclaimed dry powder scraped from (a) and presented in a frame, and **(c)** a split of the original dry bulk material presented in a frame.

Table 9. *L*a*b* and Munsell indices (H, V/C) for four samples tested as air-dried (unheated), and after heating to 400 °C and 800 °C*

Samples	L*	a*	b*	H	V/C
RF-107A (unheated)	35.57	17.49	20.59	3.7YR	3.5/4.6
RF-107A (to 400 °C)	34.02	20.69	23.23	3.0YR	3.4/5.4
RF-107A (to 800 °C)	41.79	30.01	32.77	1.9YR	4.2/8.0
C-KP1 (unheated)	53.17	6.21	18.41	10.0YR	5.2/3.0
C-KP1 (to 400 °C)	49.87	11.40	22.41	7.5YR	4.9/4.1
C-KP1 (to 800 °C)	54.46	18.83	26.75	4.4YR	5.4/5.7
C-KP3 (unheated)	47.54	6.39	17.89	10.0YR	4.7/2.9
C-KP3 (to 400 °C)	43.11	13.00	23.15	6.9YR	4.3/4.3
C-KP3 (to 800 °C)	47.88	21.88	31.60	4.5YR	4.8/6.6
C-KP8 (unheated)	41.80	4.31	12.96	0.3Y	4.1/2.0
C-KP8 (to 400 °C)	42.64	8.19	17.59	8.6YR	4.2/3.0
C-KP8 (to 800 °C)	48.60	18.84	27.34	4.7YR	4.8/5.7

RF-107A: red tropical sandy soil
C-KP1, 3 and 8: soil from adjacent fields in Wiltshire, UK

ratory, and colour determinations obtained in this way have been found to provide a highly useful investigative tool. Two forensic casework examples have been chosen and are given here as a demonstration of the application of this instrument.

Example 1: fatal hit-and-run investigation, Wales

It was alleged that a vehicle had veered across a road, hitting both banks before hitting two pedestrians (one of whom subsequently died). The vehicle was seized and mud samples removed from the nearside rear tyre and the offside front area of the vehicle. Control samples were also taken from both banks. A number of techniques were used to examine these samples (see also Blott *et al.* 2004). The colour data are shown in Table 10 and the reflectance curves in Figure 8. The similarity between the samples from the two sides of the road and the samples from the respective vehicle sides can be clearly seen, both numerically and in the reflectance curves. This is a good example of the necessity of sampling carefully at the crime scene, where the samples from each side of the road exhibit very different characteristics.

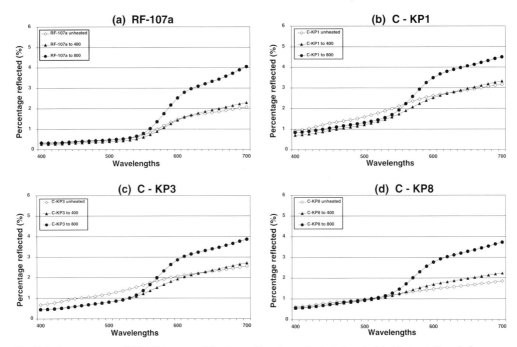

Fig. 7. Reflectance curves (400–700 nm range) for four sediment samples tested as air-dried (unheated), and after heating to 400°C and 800°C: (**a**) RF-107a (a red tropical sandy soil); and (**b**) C-KP1, (**c**) C-KP3 and (**d**) C-KP8 (forensic soil samples from adjacent fields in Wiltshire, UK).

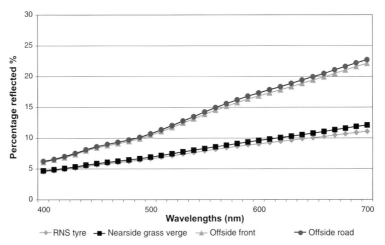

Fig. 8. Reflectance curves (400–700 nm range) for samples taken from a hit-and-run case: two vehicle samples and two control samples from the crime scene.

Example 2: illegal release of degus into a nature conservation area, Sefton coast, UK

Approximately 100 degus (exotic South American rodents) were released illegally into a conservation area, posing serious concern for indigenous wildlife.

A variety of techniques were used in the analysis of samples from the case (see also Saye & Pye 2004). Sand removed from a dustbin lid (which had been seized from a suspect's home because it was believed that the dustbin had been used to transport the rodents) was compared with sediments from the

Table 10. *L*a*b* and Munsell indices (H, V/C) for four samples from a 'hit and run' case: two vehicle samples and two control samples from the crime scene*

Samples	L*	a*	b*	H	V/C
Rear nearside tyre	33.78	2.70	9.02	1.1Y	3.3/1.4
Nearside grass verge	34.63	3.01	9.56	0.8Y	3.4/1.5
Offside front sample	43.99	5.23	15.84	0.4Y	4.3/2.5
Offside road bank	44.54	5.30	16.07	0.4Y	4.4/2.5

Table 11. *L*a*b* and Munsell indices (H, V/C) for samples from a case of eco-vandalism*

Samples	L*	a*	b*	H	V/C
Release site 1	43.24	5.78	15.72	0.0Y	4.2/2.5
Release site 2	44.09	5.43	15.27	0.1Y	4.3/2.4
Dustbin lid	58.14	3.31	13.05	0.7Y	5.7/2.0

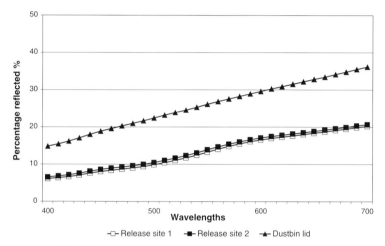

Fig. 9. Reflectance curves (400–700 nm range) for samples taken from a case of eco-vandalism.

release site and also to a database of UK sand dune samples. Table 11 presents the L*a*b* and Munsell indices data, and Figure 9 shows the reflectance curves for the samples concerned. In this case of eco-vandalism, the colour of the target sample (the sand from the dustbin lid) and the release site samples can be seen to be different, both numerically and in the reflectance curves, thereby excluding that piece of evidence from the investigation.

Conclusions

Quantitative measurement of the colour of soils and sediments can be a very useful tool both in forensic and wider geological investigations. Tests reported here show the Minolta CM-2002® spectrophotometer provides a precise and alternative measurement of colour that allows rapid comparison of different samples.

Care must be taken, however, to ensure that the different samples being compared are handled, pretreated and analysed in the same way. Our experi-

ments have shown that variations exist due to grain size, organic matter content, moisture content, and whether the sample is presented as a loose, dry powder or as moist slurry which is allowed to dry. Analysing before and after heat treatment may provide further discrimination. In many forensic investigations, it may be desirable to analyse both the bulk *in situ* or 'undisturbed' material as well as selected grain size fractions.

Thanks are given to two reviewers for their comments and helpful advice in the preparation of this paper.

References

ANTOCI, P. R. & PETRACO, N. 1993. A technique for comparing soil colors in the forensic laboratory. *Journal of the Forensic Science Society*, **38**, 437–441.

BIGHAM, J. M. & E. J. CIOLKOSZ (eds) 1993. *Soil Color*. Soil Science Society of America, Madison, Special Publications, **31**.

BLOTT, S. J., CROFT, D. J., PYE, K., SAYE, S. E. & WILSON, H. E. 2004. Particle size analysis by laser diffraction.

In: PYE, K. & CROFT, D. J. (eds) *Forensic Geoscience: Principles, Techniques and Applications*. Geological Society, London, Special Publications, **232**, 63–73.

CIE. 1931. *Proceedings of the 8th Session of the Commission Internationale de l'eclairage, Cambridge, UK*. Bureau Centrale de la CIE, Paris.

CROFT, D. J. 2003. *Forensic geoscience: development of techniques for soil analysis*. PhD thesis, Royal Holloway, University of London.

DUDLEY, R. J. 1975. The use of colour in the discrimination between soils. *Journal of the Forensic Science Society*, **15**, 209–218.

GANGAKHEDKAR, N. S. 1981. Instrumental measurement of colour. *PaintIndia Annual*, **1981**, 72–79.

GODDARD, E. N., TRASK, P. D., DEFORD, R. K., ROVE, O. N., SINGEWALD, J. T. & OVERBECK, R. M. 1948. [Also various reprints.] *Rock Color Chart*. Geological Society of America, Boulder.

JANSSEN, D. W., WILLIAMS, A. R. & PRITCHARD, W. W. 1983. The use of clay for soil color comparisons. *Journal of Forensic Sciences*, **28**, 773–776.

JOHNSTON, R. M. 1967. Spectrophotometry for the analysis and description of color. *Journal of Paint Technology*, **39**, 346–354.

MCCRONE, W. C., DELLY, J. G. & PALENIK, S. 1973. *The Particle Atlas: An Encyclopaedia of Techniques for Small Particle Identification*. Vols I–IV. Ann Arbor Science, Chicago.

MCDONALD, R. A. 1982. Review of the relationship between visual and instrumental assessment of colour difference. Part I. *Journal of the Oil Colour and Chemistry Association*, **65**, 43–53.

MINOLTA. 1998. *Precise Colour Communication*. Minolta, Tokyo. [Pamphlet]

MUNSELL COLOR. 1994. *Munsell Soil Color Charts*. Gretag Macbeth, New Windsor.

MURRAY, R. C. & TEDROW, J. C. F. 1992. *Forensic Geology*. Prentice Hall, New Jersey.

NAGAO, S. & NAKASHIMA, S. 1991. A convenient method of color measurement of marine sediments by colorimeter. *Geochemical Journal*, **25**, 188–197.

PELL, S. D., CHIVAS, A. R. & WILLIAMS, I. S. 2000. The Simpson, Strzelecki and Tirari Deserts: development and sand provenance. *Sedimentary Geology*, **29**, 517–525.

PYE, K. 1983. Red beds. *In*: PYE, K. & GOUDIE, A. S. (eds) *Chemical Sediments and Geomorphology*. Academic Press, London, 227–263.

SAFERSTEIN, R. 1995. Physical properties: glass and soil. *In*: SAFERSTEIN, R. *Criminalistics: An Introduction to Forensic Science*. Prentice Hall, New Jersey, 89–117.

SAYE, S. E. & PYE, K. 2004. Development of a coastal dune sediment database for England and Wales: forensic applications. *In*: PYE, K. & CROFT, D. J. (eds) *Forensic Geoscience: Principles, Techniques and Applications*. Geological Society, London, Special Publications, **232**, 75–95.

SCHAFER, A. Th. 2001. The colour of the human skull. *Forensic Science International*, **117**, 53–56.

SCHWERTMANN, U. 1993. Relations between iron oxides, soil colour and soil formation, *In*: BIGHAM, J. M. & CIOLKOSZ, E. J. (eds) *Soil Colour*. Soil Society of America, Madison, Special Publications, **31**, 51–69.

SCHWERTMANN, U. & TAYLOR, R. M. 1989. Iron oxides. *In*: DIXON, J. B. & WEED, S. B. (eds) *Minerals in Soil Environments*. Soil Science Society of America, Madison, Special Publications, **1**, 379–438.

SHIELDS, J. A., ST ARNAUD, R. J., PAUL, E. A. & CLAYTON, J. S. 1966. Spectrophotometric measurement of soil color. *Canadian Journal of Soil Science*, **46**, 83–90.

SUGITA, R. & MARUMO, Y. 1996. Validity of colour examinations for forensic soil identification. *Forensic Science International*, **83**, 201–210.

THORNTON, J. I. 1997. Visual colour comparisons in forensic science. *Forensic Science Review*, **9**, 37–57.

TORRENT, J. & BARRON, V. 1993. Laboratory measurement of soil color: theory and practice. *In*: BIGHAM, J. M. & CIOLKOSZ, E. J. (eds) *Soil Color*. Soil Science Society of America, Madison, Special Publication, **31**, 21–33.

TRUJILLO, O., VANEZIS, P. & CERMIGNANI, M. 1996. Photometric assessment of skin colour and lightness using a trisimulus colorimeter: reliability of inter- and intra- investigator observations in healthy adult volunteers. *Forensic Science International*, **81**, 1–10.

TURNER, P. 1980. *Continental Red Beds*. Elsevier, Amsterdam.

VAN HOUTEN, F. B. 1961. Climatic significance of red beds. *In*: NAIRN, A. E. M. (ed.) *Descriptive Paleo-climatology*. Interscience, New York, 83–139.

VAN HOUTEN, F. B. 1968. Iron oxides in red beds. *Geological Society of America Bulletin*, **79**, 399–416.

VAN HOUTEN, F. B. 1973. Origin of red beds: a review 1961–1972. *Annual Review of Earth Planetary Science*, **1**, 39–61.

WALDEN, J., WHITE, K. & DRAKE, N. A. 1996. Controls on dune colour in the Namib sand sea: preliminary results. *Journal of African Earth Sciences*, **22**, 349–353.

WELLS, N. A. 2002. Quantitative evaluation of color measurements: I. Triaxial stereoscopic scatter plots. *Sedimentary Geology*, **151**, 1–16.

WELLS, N. A., KONOWAL, M. & SUNDBACK, S. A. 2002. Quantitative evaluation of color measurements: II. Analysis of Munsell color values from the Colton and Green River Formations (Eocene, central Utah). *Sedimentary Geology*, **151**, 17–44.

Particle size analysis by laser diffraction

SIMON J. BLOTT[1], DEBRA J. CROFT[1], KENNETH PYE[1,2], SAMANTHA E. SAYE[1] & HELEN E. WILSON[2]

[1]Kenneth Pye Associates Ltd, Crowthorne Enterprise Centre, Crowthorne Business Estate, Old Wokingham Road, Crowthorne, RG45 6AW, UK (e-mail: s.blott@kpal.co.uk)
[2]Department of Geology, Royal Holloway, University of London, Egham Hill, Egham, Surrey, TW20 0EX, UK

Abstract: Particle size distribution is a fundamental property of any sediment or soil, and particle size determination can provide important clues to sediment provenance. For forensic work, the particle size distribution of sometimes very small samples requires precise determination using a rapid and reliable method with a high resolution. A protocol has been developed using a Coulter™ LS230 laser granulometer, which can analyse particles in the size range 0.04 μm–2000 μm. The technique is essentially non-destructive, permitting the recovery of critical samples, and has been demonstrated to have high precision for a range of soils, sediments and powders of interest in forensic investigations.

Size is a fundamental property of sediment particles, affecting their entrainment, transport and deposition. Particle size analysis therefore provides important clues to the sediment provenance, transport history and depositional conditions (e.g., Krumbein 1941; Friedman 1979; Sheridan *et al.* 1987; Bui *et al.* 1989) and, in a forensic context, offers a potentially useful means of 'fingerprinting' soils and sediments for purposes of comparison and environmental interpretation. The various techniques employed in particle size determination include direct measurement, dry and wet sieving, sedimentation, and measurement by laser granulometer, X-ray sedigraph and Coulter counter (e.g. Irani & Callis 1963; Komar & Cui 1984; Wanogho *et al.* 1987; McManus 1988; Molinaroli *et al.* 2000). All these methods involve the division of the sediment sample into a number of size fractions, enabling a particle size distribution to be constructed from the weight or volume percentage of sediment in each size fraction.

However, the various techniques describe widely different aspects of 'size', including *maximum calliper diameter*, *sieve diameter* and *equivalent spherical diameter*, and are to a greater or lesser extent influenced by variations in particle shape, density and optical properties. For this reason, the results obtained using different methods may not be directly comparable, and it can be difficult to assimilate size data obtained using more than one method (Pye 1994; Blott 2002). For many applications, including forensic casework, a single technique is required which can not only simultaneously measure the sand, silt and clay fractions within a sample, but also reliably determine the size distribution of very small volume samples rapidly and accurately (Wanogho *et al.* 1985, 1987; Chappell 1998).

This paper describes such a method using a laser diffraction instrument.

The Coulter™ LS230 laser granulometer

Particle size analysis using the principle of laser diffraction has been undertaken for several years, and a variety of instruments are available commercially (e.g. Coulter, Malvern, Fritsch, Retsch and Horiba instruments). The instruments vary in their sensitivity and sophistication, which influence the range of particle sizes that can be analysed, the ease with which data can be manipulated and summarized, and the overall quality of the results. In our laboratory a method of particle size determination has been developed and extensively tested using a Coulter™ LS230 laser granulometer, equipped with a variable-speed fluid module (Fig. 1). This instrument measures the size of waterborne particles by the diffraction of laser light of 750 nm wavelength,

Fig. 1. The Coulter™ LS230 laser granulometer.

From: Pye, K. & Croft, D. J. (eds) 2004. *Forensic Geoscience: Principles, Techniques and Applications.* Geological Society, London, Special Publications, **232**, 63–73. © The Geological Society of London, 2004.

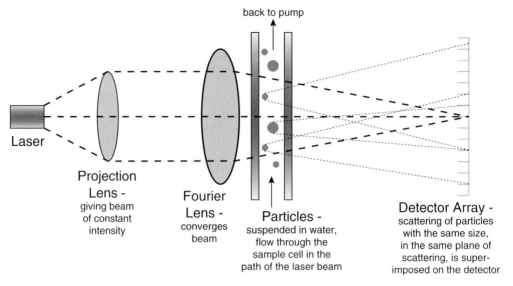

Fig. 2. Schematic diagram of the Coulter™ LS230 laser granulometer optical system.

based on Fraunhofer diffraction theory (de Boer *et al.* 1987). The instrument has a total of 126 photodiode detectors and routinely measures particles in the size range $0.4\,\mu m$–$2000\,\mu m$ (Fig. 2). This system uses a single Fourier lens, obviating the need to merge data using more than one lens, as required with some alternative instruments. An optional additional detector system is available, which measures the polarization intensity differential scattering (PIDS) of light. This system measures submicron particles down to $0.04\,\mu m$ using a single-frequency polarized light beam scattered onto banks of an additional six detectors. Data from the two systems are integrated providing an increased size range of $0.04\,\mu m$–$2000\,\mu m$, which is particularly useful for the analysis of soil with a potentially large clay fraction. The technique is non-destructive, permitting the recovery of critical samples after analysis, and rapid, with a sample turnaround time of 5 to 6 min. For very small samples a small volume module can also be used, and dry powders can be analysed using a further separate module.

The basic operating procedure requires a small subsample of the material under testing to be added to the water reservoir in the fluid module of the instrument. Although coarse sediments can be subsampled using a riffle box, in our experience combining numerous (>10) representative subsamples from a well-homogenized sample pot or bag provides consistent, repeatable results. These subsamples should then be disaggregated and dispersed in water to form a slurry prior to introduction into the fluid module. The granulometer is operated via a PC running specialized Particle Characterization soft-

ware under Microsoft Windows® operating system. A number of sample preparation and instrument run conditions can be varied by individual users, as follows:

(1) *Use of acid pre-treatment* Acids may be used to remove unwanted fractions in the sample. Hydrogen peroxide solution has been used in many studies to remove organic matter, and hydrochloric acid has been used to remove carbonates such as shell material.

(2) *Use of a dispersant solution* A dispersant, such as sodium hexametaphosphate (Calgon) can be added to the sample slurry prior to its introduction into the instrument in order to aid disaggregation and dispersion. The strength of the dispersant and stand times prior to introduction can be varied.

(3) *Use of ultrasonication* An ultrasonic treatment, either before or during the sample run, may aid disaggregation of the sample, although some particles may disintegrate if sonicated for too long. Also, certain types of sample experience agglomeration during ultrasonication.

(4) *Volume of sample* The amount of sample added to the instrument is measured by the reduction in laser intensity passing through the sample cell. As more sediment is added to the instrument the laser beam is increasingly obscured from the detectors. The recommended percentage of obscuration is between 8% and 12% for the Fourier system and between 45% and 55% for the PIDS system, which equates to between 0.2 g and 4 g for most soils and sediments. Finer

samples cause greater obscuration and therefore a smaller sample size is required. Where only a small amount of sample material is available, a special small volume module can be used.

(5) *Length of measurement time* Longer run times provide a more precise calculation of the particle size distribution, although some particles may begin to disintegrate with very long run times.

(6 *Instrument pump speed* The speed of fluid circulation through the sample cell can be varied to ensure that all particles are entrained into the flow while ensuring a random orientation of the particles. A very high pump speed may cause particles to orient themselves parallel (rod-shaped particles) or perpendicular (plate-shaped particles) to the direction of flow.

(7) *Optical model* A Fraunhofer mathematical model is used by default to calculate the size of particles based on the concentric ring pattern of lightscatter around spherical particles from the laser beam. If the optical properties (the real and imaginary refractive indices) of the sample are known, a customized optical model can be used, although results are only significantly different if the sample contains particles smaller than $10\,\mu m$.

(8) *Use of the PIDS system* The measurement of very fine particles with the PIDS system is optional. In general, less sample is required when the PIDS system is used. Disabling the PIDS system permits larger (and hence more representative) samples to be analysed, which is particularly suitable for coarser sediments.

A range of statistical parameters can be generated by the instrument software, including mean, median, mode, standard deviation, variance, skewness (degree of symmetry), kurtosis (degree of peakedness), and cumulative percentile values (the particle size at which a specified percentage of the particles are finer). Statistics are calculated using the method of moments either arithmetically (based on a normal distribution) or geometrically (based on a log-normal distribution, which is a good approximation for many well-sorted soils and sediments). Size distribution data can also be exported into other programs, such as the GRADISTAT statistical package (Blott & Pye 2001), which can additionally calculate the value of secondary and tertiary modes, express values in phi units, and determine particle size statistics using Folk and Ward (1957) equations. Selection of appropriate measures is partly a matter of user preference and partly determined by sample type (e.g. whether or not it is unimodal, multimodal or log-normally distributed).

As background to our forensic case work and research investigations, a series of experiments

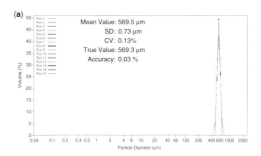

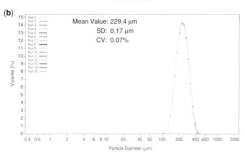

Fig. 3. Differential volume plots for 15 repeat runs of: (**a**) a glass bead standard, and (**b**) a coastal dune sand.

involving some 1100 sample analyses was conducted to assess the accuracy, precision and reproducibility of particle size analysis using this instrument, and to determine the optimum operating conditions and sample pre-treatments relevant to forensic and other applications. As with other techniques, reliable data comparison can only be made if the same parameters are used for the analysis of each sample in any given investigation. However, owing to differences in the properties and behaviour of different types of material, it is not always possible to use a single protocol for all sample types and every investigation.

Instrument accuracy and precision

The accuracy of the instrument in our laboratory is regularly checked by: (1) annual servicing and certification, (2) a proficiency-testing scheme which compares results from a group of laboratories four times a year, and (3) internal testing using glass bead standards and laboratory reference materials. Repeat testing with a certified glass bead standard has demonstrated an accuracy of 0.03% for the mean particle size when compared to the 'true' value (Fig. 3a).

Instrument precision was assessed by 15 repeat runs of the same subsample for both the glass bead standard and a well sorted dune sand from the Sefton coast, UK. The coefficient of variation (CV) for the mean particle size of these runs was 0.13% and 0.07% respectively (Fig. 3). The modal particle size

Table 1. *Comparison of variability data for 15 subsamples taken from a well-sorted dune sand from the Sefton coast, UK*

Sub-sample	Mean	Median (D_{50})	SD	Mode 1	Mode 2	D_{10}	D_{25}	D_{75}	D_{90}	$D_{90}-D_{10}$	$D_{50}/(D_{90}-D_{10})$	Sand	Silt	Clay
1	229.3	230.0	1.3	223.4	na	166.1	193.2	273.7	316.0	149.9	0.728	100.0	0.0	0.0
2	229.0	229.7	1.3	223.4	na	165.6	192.7	273.6	316.3	150.7	0.726	100.0	0.0	0.0
3	228.7	229.5	1.3	223.4	na	165.7	192.7	273.1	314.9	149.2	0.729	100.0	0.0	0.0
4	228.2	229.0	1.3	223.4	na	165.2	192.1	272.6	314.1	148.9	0.729	100.0	0.0	0.0
5	228.9	229.7	1.3	223.4	na	165.6	192.7	273.4	315.6	150.0	0.728	100.0	0.0	0.0
6	228.6	229.3	1.3	223.4	na	165.4	192.4	273.1	315.1	149.7	0.728	100.0	0.0	0.0
7	228.5	229.4	1.3	223.4	na	165.3	192.4	273.0	314.9	149.6	0.728	100.0	0.0	0.0
8	229.0	229.7	1.3	223.4	na	165.7	192.8	273.6	316.1	150.4	0.727	100.0	0.0	0.0
9	228.8	229.5	1.3	223.4	na	165.4	192.5	273.4	315.9	150.5	0.726	100.0	0.0	0.0
10	228.7	229.5	1.3	223.4	na	165.5	192.6	273.2	315.0	149.5	0.729	100.0	0.0	0.0
11	228.8	229.6	1.3	223.4	na	165.4	192.6	273.5	315.8	150.4	0.727	100.0	0.0	0.0
12	228.3	229.2	1.3	223.4	na	165.2	192.3	272.8	314.5	149.3	0.729	100.0	0.0	0.0
13	228.5	229.4	1.3	223.4	na	165.4	192.5	273.0	314.8	149.4	0.729	100.0	0.0	0.0
14	228.7	229.4	1.3	223.4	na	165.6	192.5	273.1	314.9	149.3	0.728	100.0	0.0	0.0
15	229.0	229.7	1.3	223.4	na	165.6	192.7	273.6	316.2	150.6	0.726	100.0	0.0	0.0
Mean	228.7	229.5	1.3	223.4	na	165.5	192.6	273.2	315.3	149.8	0.728	100.0	0.0	0.0
SD	0.3	0.2	0.0	0.0	na	0.2	0.3	0.3	0.7	0.6	0.001	0.0	0.0	0.0
CV%	0.1	0.1	0.1	0.0	na	0.1	0.1	0.1	0.2	0.4	0.1	0.0	0.0	0.0

was exactly the same for all runs, differences being less than the resolution between size classes measured by the instrument. It was not possible to assess precision by repeat runs of finer sediments due to agglomeration of silt and clay particles once the sample had been circulating for more than approximately 10 min.

Sampling precision

Variation due to combined subsampling and instrumental measurement error was determined by running 15 subsamples taken from each of three different bulk samples: (1) dune sand from the Sefton coast, UK; (2) loess from Lanzhou Province, China; and (3) parkland soil from Hampstead Heath, London, UK (Tables 1–3, Fig. 4). Sample heterogeneity was found to be the dominant factor influencing the variability of the results. The well-sorted Sefton coast sand was found to be the least variable, with a CV for the mean particle size of 0.13% (Table 1). The Chinese loess was less homogenous, with a CV for the mean of 3.2% (Table 2). The Hampstead Heath soil was found to be the most variable, with a CV for the mean of 6.5%. This can be attributed to greater heterogeneity within the sample, especially in the coarse fraction (Table 3), and less reliable dispersal of aggregates prior to analysis.

It can be seen from Tables 1–3 that, for the three sample types, the median and primary modes show lower CVs than the mean and, for inherently heterogeneous, poorly sorted soils, the median often provides a more consistent (robust) basis for comparison, when used together with the percentages of sand, silt and clay. In the case of well-sorted sediments (e.g. Fig. 4a & b) it may be sufficient to undertake a single analysis to obtain a representative result, whereas in the case of many soils (e.g. Fig. 4c) it is desirable to undertake several replicate analyses on three to five sub-samples. The results should be reported individually, together with an average and standard deviation (SD) for the runs, so that the full magnitude of inherent variability can be appreciated.

Protocol for sample pre-treatment and analysis

In order to develop a protocol for the use of laser granulometry for a variety of soils and sediments, a series of experiments was carried out to determine the optimum operating conditions and sample pre-treatments. Variables considered were:

- Soil texture (one clayey and one sandy-silt soil type)
- Operator variation (two different operators used)
- Size range analysed ($<2\,mm$, $<500\,\mu m$, $<150\,\mu m$)
- Use of hydrogen peroxide to remove organic matter (use and non-use)
- Sample size (sufficient to achieve recommended obscuration, 100 mg and 50 mg)
- Strength of dispersant used (0, 1, 3 and 5%)

Table 2. *Comparison of data for 15 subsamples taken from a sample of loess from Lanzhou Province, China (size values in micrometres)*

Sub-sample	Mean	Median (D_{50})	SD	Mode 1	Mode 2	D_{10}	D_{25}	D_{75}	D_{90}	$D_{90}-D_{10}$	$D_{50}/(D_{90}-D_{10})$	Sand %	Silt %	Clay %
1	25.6	33.3	3.3	50.2	na	4.8	15.6	56.7	84.3	79.6	0.418	20.5	74.4	5.1
2	25.0	32.6	3.3	50.2	na	4.6	15.1	55.9	82.3	77.7	0.420	19.8	75.0	5.2
3	25.4	33.1	3.2	50.2	na	4.8	15.5	56.2	82.3	77.5	0.427	20.0	74.9	5.1
4	25.2	32.7	3.3	50.2	na	4.7	15.2	56.1	83.3	78.6	0.415	20.0	74.8	5.2
5	24.3	31.8	3.3	50.2	na	4.3	14.3	55.4	82.3	78.0	0.407	19.6	75.1	5.3
6	26.2	34.0	3.3	50.2	na	4.9	16.0	57.6	86.0	81.1	0.419	21.2	73.8	5.0
7	25.5	33.2	3.2	50.2	na	4.7	15.6	56.3	82.5	77.8	0.427	20.1	74.7	5.1
8	24.4	32.2	3.3	50.2	na	4.4	14.7	55.1	80.2	75.8	0.425	19.1	75.4	5.4
9	24.1	32.2	3.3	50.2	na	4.2	14.5	54.9	79.5	75.2	0.427	18.9	75.6	5.5
10	24.8	32.5	3.3	50.2	na	4.5	14.9	56.1	82.7	78.2	0.416	20.0	74.7	5.3
11	25.3	32.8	3.3	50.2	na	4.6	15.1	56.6	84.2	79.5	0.413	20.5	74.4	5.1
12	25.1	32.9	3.3	50.2	na	4.4	15.0	56.5	83.4	79.0	0.416	20.3	74.4	5.3
13	24.5	32.0	3.3	50.2	na	4.3	14.2	56.2	84.2	79.9	0.400	20.2	74.4	5.4
14	22.8	30.7	3.2	50.2	na	4.0	13.7	52.4	74.3	70.2	0.436	16.7	77.6	5.6
15	25.1	32.9	3.3	50.2	na	4.5	15.2	56.5	83.5	79.0	0.416	20.3	74.5	5.3
Mean	24.9	32.6	3.3	50.2	na	4.5	15.0	55.9	82.3	77.8	0.419	19.8	74.9	5.3
SD	0.8	0.8	0.0	0.0	na	0.2	0.6	1.2	2.7	2.6	0.009	1.0	0.9	0.2
CV%	3.2	2.4	1.0	0.0	na	5.3	4.1	2.1	3.3	3.3	2.1	5.1	1.2	3.2

Table 3. *Comparison of data for 15 subsamples taken from a parkland soil sample from Hampstead Heath, London, UK (size values in micrometres)*

Sub-sample	Mean	Median (D_{50})	SD	Mode 1	Mode 2	D_{10}	D_{25}	D_{75}	D_{90}	$D_{90}-D_{10}$	$D_{50}/(D_{90}-D_{10})$	Sand %	Silt %	Clay %
1	40.2	56.8	4.1	87.9	684.2	5.3	18.5	97.3	149.4	144.1	0.394	46.6	49.6	3.8
2	38.4	56.4	3.8	87.9	223.4	5.3	18.6	95.0	136.1	130.8	0.431	46.2	49.9	3.8
3	38.9	57.4	3.5	87.9	269.2	6.2	21.4	91.9	122.7	116.5	0.492	46.5	50.1	3.4
4	32.1	46.0	3.5	87.9	269.2	5.0	16.0	83.6	110.4	105.4	0.437	39.5	56.8	3.7
5	35.0	50.4	3.6	87.9	269.2	5.2	17.2	89.5	123.2	118.0	0.427	42.5	53.7	3.8
6	35.3	51.5	3.7	87.9	295.5	5.1	17.1	90.5	125.1	120.0	0.429	43.4	52.8	3.8
7	32.9	44.9	3.7	87.9	295.5	4.8	15.2	87.6	125.6	120.8	0.372	39.8	56.4	3.9
8	38.0	53.8	3.9	87.9	295.5	5.4	18.3	93.3	133.2	127.8	0.421	44.8	51.5	3.7
9	38.9	55.7	3.8	87.9	356.1	5.5	18.8	95.7	137.5	132.0	0.422	45.9	50.5	3.7
10	37.9	53.1	3.9	87.9	269.2	5.4	18.0	92.7	132.4	127.0	0.418	44.2	52.1	3.7
11	36.6	52.4	4.1	87.9	269.2	4.7	16.1	93.5	136.5	131.8	0.398	44.2	51.6	4.2
12	36.6	48.4	4.0	87.9	269.2	5.1	16.4	91.7	140.3	135.2	0.358	41.7	54.5	3.8
13	37.9	53.0	3.9	87.9	269.2	5.3	17.9	92.1	131.6	126.3	0.420	44.2	52.0	3.8
14	40.4	53.2	4.2	87.9	269.2	5.4	18.3	94.3	146.2	140.8	0.378	44.4	52.0	3.6
15	37.7	53.6	3.8	87.9	269.2	5.3	18.1	91.8	129.6	124.3	0.431	44.4	52.0	3.7
Mean	37.1	52.4	3.8	87.9	304.9	5.3	17.7	92.0	132.0	126.7	0.415	43.9	52.4	3.7
SD	2.4	3.7	0.2	0.0	108.5	0.3	1.5	3.4	9.9	9.9	0.032	2.2	2.2	0.2

- Standing time prior to testing (5 min, 1 h and 24 h)
- Ultrasonication time (0, 30 and 60 s, 2 min and 5 min).
- Run time (30, 60 and 90 s)

The most reproducible results for both operators and for both soils were achieved by dispersing the soil in a small amount of 3% Calgon solution (*c.* 2 ml), leaving the dispersed sample to stand for 5 min before analysis, gentle physical disaggregation on a watchglass using a rubber pestle, and ultrasonication during sample introduction and during a measurement period of 90 s. In general, the shorter standing times gave the most reproducible results, with samples agglomerating when left for 24 h. Lower

(a)

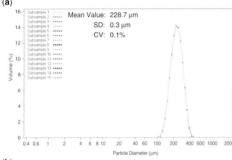

(b)

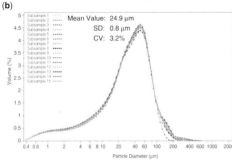

(c)

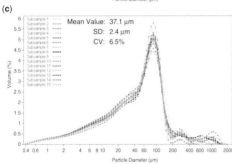

Fig. 4. Superimposed differential volume plots for 15 subsamples each of: (**a**) a well-sorted dune sand from the Sefton coast, UK; (**b**) a loess from Lanzhou Province, China; and (**c**) a parkland soil from Hampstead Heath, London, UK.

obtained using the manufacturer's recommended obscuration levels (8–12 %), although results were still adequate with levels as low as 3%, representing a sample size of approximately 50 mg depending on sample type. Useful indicative results could also be obtained using obscuration levels as low as 1% with the standard fluid module, although for very small samples there are often advantages in using the small-volume sample module.

The use of hydrogen peroxide to remove organic matter was only partially successful, since some samples required repeated treatment for a complete dissolution. There was also no improvement in measurement accuracy or reproducibility for the soils examined, a result confirmed by other studies (e.g. Beuselinck *et al.* 1998). The process is therefore considered too time-consuming for routine analysis and should be reserved for highly glutinous sediments, such as some lacustrine muds, where the organic fraction comprises the bulk of the sediment.

Although the development of a standard protocol is desirable for many applications, the tests reported here relate to sample reproducibility for just three sample types. Under certain circumstances and with certain types of materials (e.g. when dealing with sediments which are transported in a naturally aggregated state, or which show unusual chemical sensitivity), the sample preparation protocol may need to be modified in order to yield the most representative results. For example, additions of dispersants can sometimes cause flocculation in sediments with a high salt content (such as some intertidal muds). Highly cemented sediments may require lengthy standing times or stronger ultrasonic treatment to disaggregate the sample fully. Alternatively, dispersion should be avoided in analysis of samples such as windblown dusts, where the size of aggregates which have been transported by wind is of interest. Some aggregated samples, particularly those which are soluble or water-sensitive, are also more appropriately analysed in a dry powder form.

Summary parameters and intersample comparison

In terms of the statistical parameters generated, the most robust measures for discriminating between samples have been found to be the modal particle size(s), the median (D_{50}), and the measure of distribution spread defined by the difference between the tenth and ninetieth percentile values ($D_{90}-D_{10}$). When comparing samples, a number of statistical measures need to be used, including the percentages of sand (63–2000 μm), silt (2–63 μm) and clay (<2 μm), mean, median, primary mode, secondary

strength dispersants (0, 1 and 3%) also gave the most reproducible results, while soils tended to agglomerate when using the 5% strength over all standing time periods. Samples generally benefited from a short ultrasonic treatment, although ultrasonication for more than 2 min tended to alter the size distribution significantly, due to disintegration of some coarser particles. Reproducibility improved with longer run times, although not appreciably above 60 s.

Although the <500 μm size fraction gave the highest reproducibility, this was at the expense of limiting the potentially valuable and discriminatory information in the coarser part of the particle size distribution. The best reproducibility was also

Table 4. *Comparison of particle size distribution parameters for beach sands from Lincolnshire, UK, determined by dry sieving and laser granulometry*

Sample	Method	Statistical parameter (values in μm)				
		Mean	Median	Mode	$D_{90}-D_{10}$	SD
A	Sieve	389.9	379.1	302.5	448.9	1.52
	Laser	491.9	496.6	429.7	689.0	1.81
	Difference	+26%	+31%	+42%	+53%	+19%
B	Sieve	399.7	392.4	302.5	433.3	1.50
	Laser	486.9	487.5	429.7	653.4	1.75
	Difference	+22%	+24%	+42%	+51%	+17%
C	Sieve	241.2	233.2	215.0	149.5	1.28
	Laser	263.1	264.3	269.5	172.6	1.28
	Difference	+9%	+13%	+25%	+15%	0%
D	Sieve	220.4	218.7	215.0	159.6	1.36
	Laser	235.2	250.6	245.5	185.7	1.72
	Difference	+7%	+15%	+12%	+16%	+27%

and tertiary modes (if any), SD and $D_{90}-D_{10}$ (measure of sorting). The shape of the particle size distribution curve, plotted using a logarithmic size scale, is also extremely discriminatory when visually or mathematically described by the position and relative importance of the modal values and the position of inflection and turning points. Previous studies have successfully characterized sediments using just these parameters (e.g. Allen & Rippon 1997; Allen 2000). Great care is needed when interpreting open-ended distributions, or where the sediment is not unimodal, since many of the statistical parameters, including the mean, skewness and kurtosis, can become unreliable. In such instances, the sediment is usually best described by the position of the modal size values, defined percentile values, and inflection and turning points on the particle size curve.

Comparison with other techniques

Before the development of laser diffraction instruments, the grain size of sediments and soils was traditionally measured using a combination of sieving and sedimentation techniques. These techniques have a number of advantages and disadvantages compared with laser granulometry. Sieving has the disadvantage of being limited to analysis of the gravel, sand and coarse silt fractions, and measures the particle size distribution with a lower resolution (larger class size interval) than laser granulometry. The two techniques also measure size in different ways. Laser granulometry provides a measure of the volume of sediment particles based on their optical properties (which are related to size using an optical model). The technique measures the equivalent

spherical diameter by providing an average of all possible particle diameters and relies upon the random orientation of the particles presented to the laser beam, with particles passing before the laser beam many times during one analysis.

Sieving measures the intermediate 'calliper' diameter of individual particles (related to the size of the sieve mesh), and expresses particle abundances in terms of weight in each size class. The analysis is therefore affected by particle density, results effectively being a measure of weight distribution rather than size distribution. Laser granulometry is unaffected by sample density. Because sieving measures the intermediate diameter of particles, shape also influences the results, so that elongate or platy-shaped particles can pass through an aperture which spherical particles of an identical equivalent diameter cannot. Laser granulometry, on the other hand, tends to overestimate the size (and hence equivalent weight) of platy particles that appear (to the laser beam) to be larger than their equivalent spherical diameter. For this reason, the measured mean size for non-spherical particles by sieving is frequently less than the mean size determined by laser granulometry. As a consequence, the particle size distribution measured by sieving tends to be slightly compressed (i.e. better sorted) and fine-skewed. This is clearly demonstrated by comparison of four different beach sands analysed by the two techniques (Table 4). Differences between the techniques increase with larger proportions of non-spherical grains, such as samples containing shell or mica (Hayton *et al.* 2001).

Sedimentation has been the traditional method used to measure the fine fraction for many years, based on the settling velocity of particles through fluids. However, the platy form of fine silts and clays

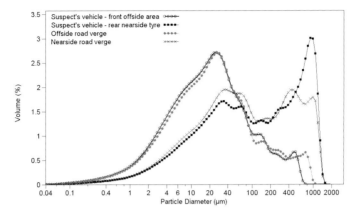

Fig. 5. Superimposed differential volume plots showing a close similarity between (**a**) mud from the front offside area of a hit-and-run suspect's vehicle and control samples from the offside verge; and (**b**) mud from the rear nearside tyre and control samples from the nearside verge.

tends to lead to an underestimation of the size of these particles due to their slower settling velocity compared with spheres of the same equivalent size, and therefore an overestimation of the volume of fine silt and clay (Konert & Vandenberghe 1997). In contrast, studies have shown that the laser diffraction technique frequently underestimates the volume of clay material due to a worsening approximation of the Fraunhofer diffraction theory for small particle sizes (Agrawal *et al.* 1991), particularly where the particle diameter is close to the wavelength of light ($0.75 \mu m$) when refraction becomes important (Loizeau *et al.* 1994). Experiments have also shown that the amount of clay is underestimated in inverse proportion to the actual clay content, with small amounts of clay being overlooked in poorly sorted sediments. Thus, results obtained by laser diffraction can show greater mean and median values, and lower sorting, when compared to sedimentation methods (McCave *et al.* 1986; Loizeau *et al.* 1994; Konert & Vandenberghe 1997).

Ultimately, the choice of technique is to a large extent governed by the type(s) of sediment under consideration. Coarse sand, mixtures of sand and gravel, or gravel samples are normally best analysed by dry sieving. Samples which are wholly composed of silts and clays can alternatively be measured by settling tube analysis. However, for samples which are predominantly composed of sizes ranging from clean medium sands to fine silt, and particularly poorly sorted muds and soils, laser granulometry provides the best method available based on a combination of instrument precision, size class resolution, small sample size and time/cost. Different manufacturers' instruments do, however, vary significantly in their capabilities (British Standards Institution 1999), and maintenance of high-quality

results requires a high standard of equipment maintenance and operator training.

Forensic case examples

Hit and run accident

In a fatal hit-and-run incident, it was alleged that a vehicle had veered violently across a main road, hitting banks on both sides before hitting two pedestrians walking in the opposite direction. The vehicle drove off without stopping. One of the pedestrians subsequently died of injuries sustained. As a result of police enquiries, a vehicle was seized and the driver arrested. Examination revealed the presence of mud on the nearside rear tyre and offside front area of the vehicle. Samples of these mud deposits were compared with control samples taken from the verges on each side of the road at the scene using a range of physical and chemical techniques, including laser granulometry. The particle size distribution curves of the four soil samples are displayed in Figure 5. It is clear from the differential volume plots that there is a close similarity between the front area of the vehicle and the offside verge, and between the nearside tyre and the nearside verge, in terms of the overall shapes of the curves and the positions and relative sizes of the modal values and inflection points. Although the actual numerical values of mean particle size and percentiles are not identical (Table 5), the values are very similar and typical of the range that can be expected due to sampling variation. The similarity between the suspect and control samples was confirmed by chemical tests and quantitative spectrophotometric colour determination (see Croft & Pye 2004).

Table 5. *Particle size summary statistics for mud from the front offside area and rear nearside tyre of a hit-and-run suspect's vehicle, and control samples from the offside and nearside road verges where impact marks were visible*

Statistical measure	Suspect's vehicle		Control samples from scene	
	Front offside area	Rear nearside tyre	Offside verge	Nearside verge
Mean (μm)	18.9	95.8	21.3	72.7
Median (μm)	19.6	100.2	21.6	69.9
Primary mode (μm)	26.2	906.1	26.2	38.0
Secondary mode (μm)	127.8	31.5	140.3	471.7
Tertiary mode (μm)	517.8	66.5	751.9	60.6
$D_{90}-D_{10}$ (μm)	142.4	903.0	198.1	739.0
SD (μm)	4.8	6.9	5.2	6.6
% Sand	21.0	57.4	23.5	52.2
% Silt	71.4	39.3	69.3	44.2
% Clay	7.6	3.2	7.2	3.5

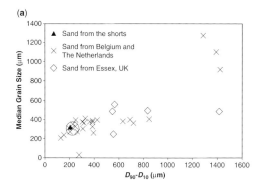
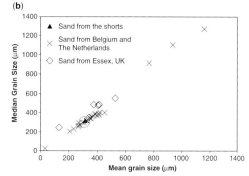

Fig. 6. Scatter plots showing: (**a**) $D_{90}-D_{10}$ against median grain size; and (**b**) mean against median grain size, for sand from the shorts and control beach samples from Belgium, The Netherlands and Essex, UK. Samples from the Dutch harbour where the boat departed are circled.

Drugs-smuggling operation

Following a Customs and Excise surveillance operation, a boat carrying a large consignment of drugs was intercepted at a small harbour on the east coast of England. The boat was met by two males in a car that was similar to one which had previously been seen near another small harbour on the Dutch coast, and from which a small boat was heard to depart in the early hours of the previous morning. Among items seized from the car were a pair of wet, sandy shorts. This sand was compared with control samples taken along the coast of The Netherlands and Belgium and from relevant locations on the east coast of England. Grain size statistics were compared using a variety of scatter plots (Fig. 6). Results showed a high degree of similarity between sand on the shorts and control samples from a localized area close to the harbour in The Netherlands from which the boat departed, both in terms of grain size and

chemical composition. Control samples taken from the harbour and adjoining area in England were significantly different to the sand on the shorts.

Conclusions

Particle size determination using the Coulter™ LS230 laser granulometer has been found to be very precise for a range of sample types. Measurement errors associated with the instrument are much less than those produced as a result of subsampling and pre-treatment procedural variations. Reproducible results can be obtained with obscuration values of 3% or less, although 8–12% is recommended. However, significant errors can be generated with samples that are too small to be representative, particularly of coarser materials.

Samples should be washed through a 2 mm stainless steel sieve into the instrument fluid module in order to remove particles which are too large to be analysed, including organic matter. In specific circumstances (such as comparing control samples to a mud smear on clothing where the coarse fraction has been lost), there may be advantages in limiting the distribution to a narrow size fraction (e.g. 250–500 μm), obtained by prior wet sieving.

It is recommended that samples be dispersed in a weak dispersant solution (2 ml of 3% Calgon is adequate) prior to measurement. Greater concentrations can cause sample agglomeration. For routine work with most sample types, removal of organic matter using hydrogen peroxide is not required since little improvement in measurement accuracy or reproducibility is obtained. It is recommended that samples be left to disperse fully before analysis (up to 5 min for samples containing more than 10% silt and clay). Longer standing times can again cause sample agglomeration.

A short ultrasonic treatment (up to 2 min) is recommended before analysis. Longer treatment times can cause disintegration of some particles. Discrimination between samples should be undertaken using a number of statistical measures, including the percentage of sand, silt and clay, mean, median, mode(s), SD and $D_{90}-D_{10}$, and a description of the shape of the particle size curve using the position and relative importance of modal values, inflection and turning points.

Particle size analysis alone is not usually sufficient to draw definitive conclusions about a potential 'match' between questioned samples and controls but should be regarded as one of a range of tools which can be applied in combination.

The authors would like to thank D. H. Thornley and J. R. L. Allen at the Postgraduate Research Institute for Sedimentology at the University of Reading, UK, for their helpful comments on an earlier version of this paper.

References

AGRAWAL, Y. C., MCCAVE, I. N. & RILEY, J. B. 1991. Laser diffraction size analysis. In: SYVITSKI, J. P. M. (ed.) *Principles, Methods and Applications of Particle Size Analysis.* Cambridge University Press, Cambridge, 119–128.

ALLEN, J. R. L. 2000. Late Flandrian (Holocene) tidal palaeochannels, Gwent Levels (Severn Estuary), SW Britain: character, evolution and relation to shore. *Marine Geology,* **162**, 353–380.

ALLEN, J. R. L. & RIPPON, S. 1997. Iron Age to early modern activity at Magor Pill and Palaeochannels, Gwent: an exercise in lowland coastal-zone geoarchaeology. *Antiquaries Journal,* **77**, 327–370.

BEUSELINCK, L., GOVERS, G., POESEN, J., DEGRAER, G. & FROYEN, L. 1998. Grain-size analysis by laser granulometry: comparison with the sieve-pipette method. *Catena,* **32**, 193–208.

BLOTT, S. J. 2002. *Morphological and sedimentological changes on artificially nourished beaches, Lincolnshire, UK.* PhD Thesis, University of London.

BLOTT, S. J. & PYE, K. 2001. GRADISTAT: a grain size distribution and statistics package for the analysis of unconsolidated sediments. *Earth Surface Processes and Landforms,* **26**, 1237–1248.

BRITISH STANDARDS INSTITUTION. 1999. *Particle Size Analysis – Laser Diffraction Methods.* British Standards Institution, London, ISO 13320.

BUI, E. N., MAZZULLO, J. M. & WILDING, L. P. 1989. Using quartz grain size and shape analysis to distinguish between aeolian and fluvial deposits in the Dallol Bosso of Niger (West Africa). *Earth Surface Processes and Landforms,* **14**, 157–166.

CHAPPELL, A. 1998. Dispersing sandy soil for the measurement of particle size distributions using laser diffraction. *Catena,* **31**, 271–281.

CROFT, D. J. & PYE, K. 2004. Colour theory and the evaluation of an instrumental method of measuring colour using geological samples for forensic applications. In: PYE, K. & CROFT, D. J. (eds) *Forensic Geoscience: Principles, Techniques and Applications.* Geological Society, London, Special Publications, 232, 49–62.

DE BOER, G. B. J., DE WEERD, C., THOENES, D. & GOOSSENS, H. W. J. 1987. Laser diffraction spectrometry: Fraunhofer versus Mie scattering. *Particle Characterisation,* **4**, 14–19.

FOLK, R. L. & WARD, W. C. 1957. Brazos River bar: a study in the significance of grain size parameters. *Journal of Sedimentary Petrology,* **27**, 3–26.

FRIEDMAN, G. M. 1979. Differences in size distributions of populations of particles among sands of various origins. *Sedimentology,* **26**, 3–32.

HAYTON, S., NELSON, C. S., RICKETTS, B. D., COOKE, S. & WEDD, M. W. 2001. Effect of mica on particle-size analyses using the laser diffraction technique. *Journal of Sedimentary Research,* **71**, 507–509.

IRANI, R. R. & CALLIS, C. F. 1963. *Particle Size: Measurement, Interpretation, and Application.* John Wiley & Sons, NewYork.

KOMAR, P. D. & CUI, B. 1984. The analysis of grain-size measurements by sieving and settling-tube techniques. *Journal of Sedimentary Petrology,* **54**, 603–614.

KONERT, M. & VANDENBERGHE, J. 1997. Comparison of laser grain size analysis with pipette and sieve analysis: a solution for the underestimation of the clay fraction. *Sedimentology,* **44**, 523–535.

KRUMBEIN, W. C. 1941. Measurement and geologic significance of shape and roundness of sedimentary particles. *Journal of Sedimentary Petrology,* **11**, 64–72.

LOIZEAU, J.-L., ARBOUILLE, D., SANTIAGO, S. & VERNET, J.-P. 1994. Evaluation of a wide range laser diffraction grain size analyser for use with sediments. *Sedimentology,* **41**, 353–361.

MCCAVE, I. N., BRYANT, R. S., COOK, H. F. & COUGHANOWR, C. A. 1986. Evaluation of a laser-diffraction-size analyser for use with natural sediments. *Journal of Sedimentary Petrology,* **56**, 561–564.

MCMANUS, J. 1988. Grain size determination and interpretation. *In:* TUCKER, M. E. (ed.) *Techniques in Sedimentology.* Blackwell, Oxford, 63–85.

MOLINAROLI, E., DE FALCO, G., RABITTI, S. & PORTARO, R.

A. 2000. Stream-scanning laser system, electric sensing counter and settling grain size analysis: a comparison using reference materials and marine sediments. *Sedimentary Geology*, **130**, 269–281.

PYE, K. 1994. Properties of sediment particles. *In*: PYE, K. (ed.) *Sediment Transport and Depositional Processes*. Blackwell, Oxford, 1–24.

SHERIDAN, M. F., WOHLETZ, K. H. & DEHN, J. 1987. Discrimination of grain-size sub-populations in pyroclastic deposits. *Geology*, **15**, 367–370.

WANOGHO, S., GETTINBY, G. & CADDY, B. 1987. Particle size distribution analysis of soils using laser diffraction. *Forensic Science International*, **33**, 117–128.

WANOGHO, S., GETTINBY, G., CADDY, B. & ROBERTSON, J. 1985. Determination of particle size distribution of soils in forensic science using classical and modern instrumental methods. *Journal of Forensic Sciences*, **34**, 823–835.

Development of a coastal dune sediment database for England and Wales: forensic applications

SAMANTHA E. SAYE & KENNETH PYE

Kenneth Pye Associates Limited, Crowthorne Enterprise Centre, Crowthorne Business Estate, Old Wokingham Road, Crowthorne, RG45 6AW, UK (e-mail: s.saye@kpal.co.uk)

Abstract: A database of coastal dune sediments in England and Wales has been developed with potential applications in forensic investigations. Coastal dunes are popular sites for criminal-related activities, including burial of drugs, weapons and murder victims. The coastal dunes and associated areas of windblown sand in England and Wales occupy an area of approximately 200 km^2 within 112 individual identified systems. Research has been undertaken to ascertain the spatial variation in sedimentological properties of the coastal dune sediments in England and Wales. Field sediment sampling has been undertaken at each dune system. More than 1500 sediment samples have been analysed by laser diffraction to determine particle size characteristics, and more than 500 have been analysed by inductively coupled plasma atomic emission spectrometry (ICP-AES) to determine chemical composition. Two examples illustrate how the database has been used in criminal investigation.

Coastal dunes are popular sites for criminal-related activities, including burial of drugs, weapons and murder victims, and traces of sand are frequently found on vehicles, footwear and items of clothing seized from suspects. Methods are therefore required: (1) to identify the likely environmental origin of the sand (e.g. coastal dune, beach, riverbank, inland coversand sheet, building sand), and (2) to determine its likely geographical origin. Task (1) requires specification of a number of textural and compositional criteria which are environmentally diagnostic, while task (2) requires the existence of a database of background information against which a questioned sample can be compared. While major progress has been made in the past 30 years in relation to environmental characterization and discrimination, database information is still patchy.

Previously existing data for coastal dune environments in England and Wales were spatially incomplete and often obtained in a non-standardized manner. Earlier sedimentological studies have largely involved analysis of particle size at specific sites to determine the relationships between dunes and adjoining beaches (e.g. Greenwood 1978; Pye 1991; Jay 1998) or calcium carbonate content determination and its significance for dune vegetation (e.g. Salisbury 1952; Wilson 1960), and no systematic comparisons between systems have been considered at either regional or national scale. For this reason, a preliminary national survey of the particle size and chemical composition of coastal sand dune systems has been undertaken to ascertain the degree and pattern of variation which exists, and to highlight areas which require more detailed investigation.

Particle size and distribution and chemical composition are fundamental properties of sediments which can provide important clues to provenance, transport pathways and depositional conditions (McLaren 1981; Pye 1982; 1994; McLaren & Bowles 1985; Gao & Collins 1992; 1994; Winspear & Pye 1995; 1996). Particle size distribution reflects the nature of the source materials and also prevailing energy conditions. Sediment chemical composition reflects the mineralogy of the geological source material, any anthropogenic contaminants which may be present, and the effects of post-depositional weathering and soil development. Although many other properties can also be of significance, particle size distribution and chemical composition have the advantage that they can be determined relatively quickly and inexpensively using modern analytical equipment such as laser granulometry (Blott *et al.* 2004) and inductively coupled plasma spectrometry (Pye & Blott 2004). As such they are suitable as screening criteria which allow selection of certain samples that require more detailed investigation.

Coastal sand dune systems in England and Wales

The coastal dunes and associated areas of windblown sand in England and Wales occupy an area of approximately 200 km^2 within 112 individual identified systems (Fig. 1). A dune system is defined here as an area of aeolian accumulation composed of individual dunes, with or without areas of sand sheets, which displays morphological or functional integrity. Each identified system has been allocated a dune 'site' number (Fig. 1). However, within some systems individual components were recognized and these 'sites' have been labelled a, b, c etc. Sites

From: PYE, K. & CROFT, D. J. (eds) 2004. *Forensic Geoscience: Principles, Techniques and Applications.* Geological Society, London, Special Publications, **232**, 75–95. © The Geological Society of London, 2004.

Fig. 1. Location of the major coastal dune systems in England and Wales. Coastal process cells (as defined by Motyka & Brampton 1993) are also shown.

within systems arise from either the dissection by urban and industrial development, and occasionally natural erosion into separate fragments of the former whole or natural spatial separation and/or character differences. A total of 150 sites has been identified in England and Wales. The morphology and factors influencing the distribution of the dune systems are discussed in Saye (2003).

Research has been undertaken to determine the nature and degree of spatial variation in the sedimentological properties of the dune systems. In addition to being of forensic interest, variations in particle size distribution potentially have an impact on aeolian processes and dune morphology, while chemical and mineralogical characteristics provide information about sediment provenance and supply rates. These aspects are of importance in terms of coastal engineering and sustainable habitat conservation management (Saye 2003).

Sediment sampling and analysis methods

Sediment samples were collected along transects perpendicular to the coastline at regular intervals along the dune frontage at each of the 150 dune sites. Generally, samples were taken at 1 km intervals but at smaller sites the distance between samples was reduced to obtain a sufficient quantity of samples for statistical analysis. Samples were taken from the frontal dune crest (the most seaward dune) and each hind dune ridge crest inland of the frontal dune, where present. An assumption was made that sediment samples from the dune crest are representative of the dune unit as a whole.

However, it is known that particle size can vary with depth depending on the type of dune and the sedimentary processes of ripple formation, grainfall and grainflow (Pye 1982; Neal 1993; Al-Enezi 2001; Clemmensen *et al.* 2001). In order to test the likely

Fig. 2. Depth sampling using an auger within the Sefton coastal dune system.

The samples were collected over a 2-year period, between 1999 and 2001, which was the time taken to complete the ground surveys at all 150 sites. While seasonal and interannual variation in particle size characteristics is possible as a result of variation in transport processes on these time scales, it is unlikely substantially to alter average trends since sediment source is the underlying control on sediment characteristics.

All 1544 samples collected were analysed by laser diffraction using a Coulter™ LS230 instrument to determine the particle size distribution. The accuracy and precision obtainable using this instrument are discussed in Blott *et al.* (2004). For each sample, measurement was made using a subsample of *c.* 0.5–1g, taken with a spatula from a well-homogenized 30 g subsample. A range of statistical parameters to characterize the particle size distribution was obtained, including the mean, median, mode, standard deviation, skewness (degree of symmetry or distortion to one side of the distribution average) and kurtosis (degree of peakedness). In addition, cumulative percentile values (at which a specified percentage of the particles are finer) D_{10}, D_{25}, D_{50}, D_{75} and D_{90} were obtained. Sediment sorting has been described using the $D_{90}-D_{10}$ range, which provides a measure of the dispersion of the sample distribution, with lower values indicating a better-sorted material. The statistical parameters were obtained directly from the Coulter™ LS230 software, which uses the mathematical 'method of moments' to calculate statistics. The geometrical scale has been used in this study as it places equal emphasis on small differences in fine particles and larger differences in coarse particles and is suitable for well-sorted sediments. The relative merits of using different statistical models and summary parameters have previously been discussed by Blott and Pye (2001).

A subset of 582 samples was analysed by inductively coupled plasma atomic emission spectrometry (ICP-AES) using a Perkin Elmer Optima 3300RL instrument to determine the geochemical composition. The number of samples selected for geochemical analysis at each site was dependent upon system size, with a minimum of three samples from smaller systems and up to eleven samples at larger ones. In addition, even larger sample numbers have been analysed for selected dune systems as a separate exercise (e.g. at Sefton).

A standardized particle size fraction of 63–250 μm, obtained by dry sieving, was used to minimize variation in the geochemical properties due to particle size. A subsample of approximately 2–3 g was selected with a spatula from a well-homogenized sieved sample and ground using a stainless steel disc mill of the Tema® type. Each sample was then prepared by two different methods

significance of such variation for present purposes, the vertical particle size variation was investigated by analysing samples at 0.5 m intervals to a depth of 6 m in an auger hole drilled through a foredune crest at Formby, Merseyside (Fig. 2). No significant depth variation in particle size parameters was found (Fig. 3). The results of other studies which have analysed samples from different depths in this (Neal 1993; Pye & Neal 1993) and other coastal dune systems also suggest that, in most cases, vertical variation within individual dunes is small compared with spatial variations within dune systems and between different dune systems.

Generally, samples were taken from the surface (0–5 cm) but in some instances were obtained at depths of up to 50 cm to avoid zones that had undergone pedogenesis. Variation in particle size characteristics has been shown to occur at the laminae scale as a result of processes such as ripple migration, particularly in beach environments (Emery 1978; Grace *et al.* 1978). Samples were obtained from 5 cm depth increments in order to obtain characteristics averaged over several laminae. The sediment sample size required to represent the whole population is dependent upon the particle size and heterogeneity of the sediment. Samples weighing 200–400 g were taken, which is more than sufficient to incorporate the natural laminae scale variation. However, larger samples would be necessary if the sediment was very heterogeneous or a high percentage of very coarse sand and gravel-sized particles were present (some coastal dune deposits contain foreign particles of non-aeolian origin).

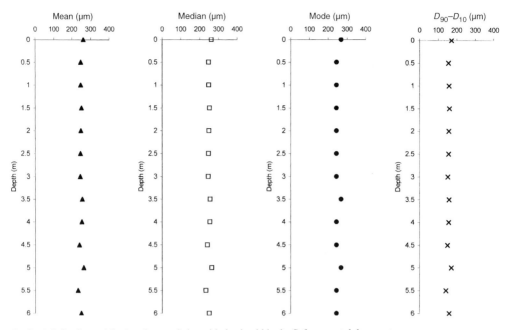

Fig. 3. Variation in particle size characteristics with depth within the Sefton coastal dune system.

using subsamples of the powdered material weighing 0.2 g ±0.001 g each. Two sample preparation methods were used: (1) sodium hydroxide fusions, to determine silica, aluminium oxide and zirconium; and (2) hydrofluoric and perchloric acids, to determine other oxides and trace elements. These solutions were prepared according to standard methods detailed in Thompson and Walsh (1989). Ten major oxides were determined (SiO_2, Al_2O_3, Fe_2O_3, MgO, CaO, Na_2O, K_2O, TiO_2, P_2O_5 and MnO) and 20 trace elements, including rare-earth elements (Ba, Co, Cr, Cu, Li, Ni, Sc, Sr, V, Zn, Zr, Pb, Y, La, Ce, Nd, Sm, Eu, Dy and Yb).

Methods of data comparison

Comparisons of the particle size distributions of samples are made using the range of statistical parameters and cumulative percentile values calculated to characterize the distributions. The most robust measures for discriminating between samples have been found to be the modal particle size, the median and the measure of sediment sorting ($D_{90}-D_{10}$) (Blott *et al.* 2004). The number of modes, their position and relative importance are useful discriminatory features of particle size distributions, particularly in the case of polymodal samples when certain statistical measures, such as the mean, skewness and kurtosis, do not adequately describe the particle size distribution. Visual comparison of the shape

of particle size distribution curves is useful and best made by overlaying the curves, although this approach is non-quantitative. To ascertain the degree of closeness between samples, those with D-values and mean values of within ±10% of each other are identified, as well as samples within ± one modal class. Further discussion of different particle size distribution attributes can be found in Blott *et al.* (2004).

In the comparison of geochemical composition data, the raw data are initially assessed visually to establish degree of similarity/difference between groups. To ascertain the degree of closeness between samples in terms of each element and oxide, values of within ±5% to ±20% of each other were identified. Samples that have a number of elements/oxides that have been identified as mathematically close can be considered to be 'similar'. A Pearson's correlation matrix can be calculated which involves the simultaneous measurement of the similarity between numerous samples, taking into account the results of all analysed elements/oxides. A correlation value of between 0.950 and 0.990 indicates a high degree of correlation, values greater than 0.990 indicate a very high correlation, while values of 1.000 represent a perfect correlation ('exact' similarity). This type of correlation must be used in conjunction with detailed examination of the dataset, as it is possible for a poor correlation to arise due to a large difference in just one element or oxide. For further discussion of geochemical data comparison methods see Pye and Blott (2004).

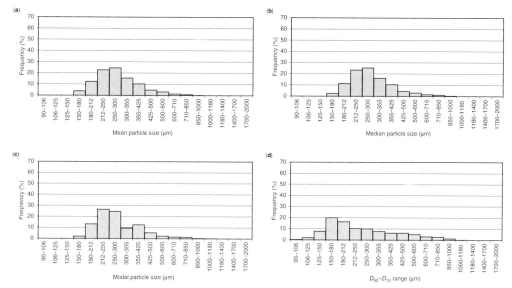

Fig. 4. Frequency histograms of: (**a**) mean particle size, (**b**) median particle size, (**c**) modal particle size; and (**d**) $D_{90}-D_{10}$ range of all samples from dune sites in England and Wales.

Particle size distribution and multi-element composition are useful characteristics for establishing the existence or otherwise of similarities between samples although, to ascertain exact matches, other characteristics in common are also required, including mineralogical composition (determined by microscopy, X-ray diffraction and/or energy dispersive X-ray microanalysis), and particle shape and surface texture (determined by optical microscopy, scanning electron microscopy and image analysis; Pye 2004).

Particle size analysis results

The coastal dune sediments in England and Wales were generally found to be unimodal with <10% of the distribution present in a coarser subpopulation or fine tail. A wide range of mean size values, from very fine sand to very coarse sand (117 μm–1066 μm) occurs, while the median and mode have higher lower and upper limits (153 μm–1114 μm and 154 μm–1091 μm, respectively). The most frequent mean and median particle size is 250–300 μm, while the most frequent modal size class is 212–250 μm. The sediment sorting ($D_{90}-D_{10}$ range) varies between 94 μm and 1232 μm, with the 150–180 μm size interval being the most frequent (Fig. 4).

Spatial variation was observed, with dunes in Kent, Sussex, north Devon, parts of Wales, and the northwestern coast of England being finer and better sorted, while those in Cornwall are coarsest and least

well sorted. The average mean size for each dune site varies cyclically along each coast, reflecting variation in local geology and transport processes (Fig. 5). The notably coarser and more poorly sorted nature of the dune sediments in Cornwall reflects high wind and wave energy and localized sediment sources, including shell debris.

Variation in particle size characteristics within each dune site surveyed is generally low, with a relative standard deviation (the SD expressed as a percentage of the mean) of between 5% and 15%. Some dune sites were found to be exceptionally homogeneous. For example, the Sefton coastal dune system is predominantly composed of fine to medium, well-sorted sands. A narrow range of mean, median and modal particle size values occurs, with the 212–250 μm size interval being the most dominant, with frequency values in excess of 70% (Fig. 6). The median and modal particle size values only fall within two size intervals. A greater range of $D_{90}-D_{10}$ values occurs on the Sefton coast, although this is largely due to two isolated samples that are more poorly sorted, and most samples (70%) are well sorted (150–180 μm) (Fig. 6). The uniformity within the Sefton coastal dune system is a reflection of the homogeneity of the sediment source, which is reworked glacial deposits from the eastern Irish Sea, and the efficiency of marine processes, largely wave action and tidal currents (Pye & Neal 1994). Other dune systems have a high degree of homogeneity in terms of particle size characteristics including many in Wales (e.g. Sites 68b, 69b, 72a, 74, 91b and 104a & b).

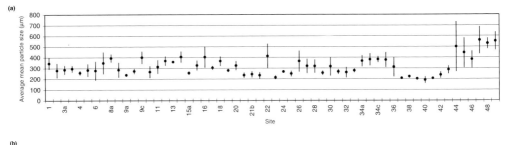

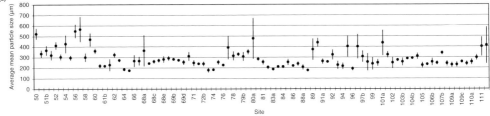

Fig. 5. Average mean particle size plotted with one standard deviation for dune sites: (**a**) on the east and south coasts of England; and (**b**) on the west coast of England and in Wales.

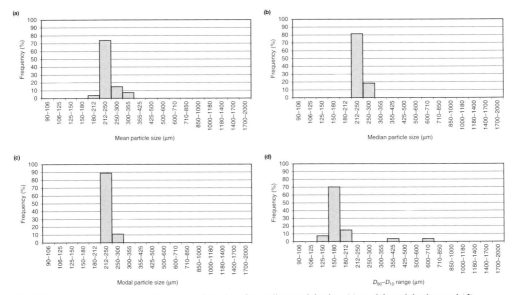

Fig. 6. Frequency histograms of: (**a**) mean particle size, (**b**) median particle size, (**c**) modal particle size; and (**d**) $D_{90}-D_{10}$ range of samples from the Sefton dune system.

By contrast, dune systems such as those at Bigbury Bay, south Devon, have a wide range of mean, median and modal particle size values (212–250 μm to 850–1000 μm) and sediment sorting (180–212 μm to 1000–1180 μm), with no dominant size interval, although two populations can be identified from the histograms (Fig. 7). Bigbury Bay is composed of numerous small sites that lie within different morphological settings.

There are examples of embayment (Thurlestone Sands) and estuarine settings (Cockleridge) as well as an embayment system positioned near to an estuary mouth (Meadowsfoot Beach). Differences in particle size characteristics correlate with different morphological settings, reflecting differences in the process regime and possibly sediment source. Dunes within embayment settings are coarser and less well sorted, while those in estuarine settings are finer and

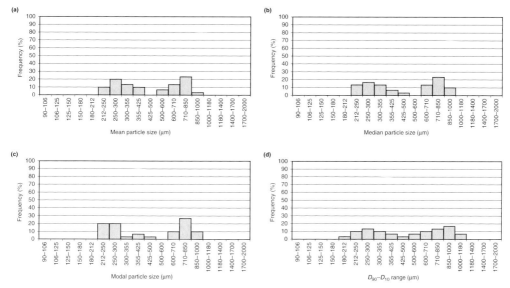

Fig. 7. Frequency histograms of: (**a**) mean particle size, (**b**) median particle size, (**c**) modal particle size and (**d**) $D_{90}-D_{10}$ range of samples from Bigbury Bay dune system.

better sorted, and those that occur within an estuary/embayment setting are intermediate. Although within-site variation is greatest at Bigbury Bay, comparable degrees of variation occur at the dune systems Kenfig (68a), Frainslake Sands (80a) and Maryport to Grune Point (112). The degree of within-site variation can usually be predicted from knowledge of the process regime and sediment sources supplying the system.

Geochemical analysis results

Geochemical composition can be used as a proxy for the mineralogical content of the dune sediments. Silica, which is the most abundant element, is related mainly to the presence of the resistant mineral quartz. The oxides of potassium, sodium and aluminium mainly reflect the presence of potassium and sodium feldspars. Calcium oxide provides a useful proxy for the calcium carbonate content, although some calcium is also present in feldspars. The contribution of heavy minerals is represented by the presence of iron, titanium, zirconium and other trace elements.

Examination of ratios of certain oxides/elements can provide additional useful information. The SiO_2/Al_2O_3 ratio provides an estimate of the quartz to feldspar ratio. The Al_2O_3/Fe_2O_3 ratio can be used as a proxy of the feldspar to iron ratio, while the Al_2O_3/K_2O ratio provides information about the ratio of all feldspars to potassium rich feldspars

(orthoclase) and micas. The SiO_2/CaO is a proxy for the quartz to calcium carbonate ratio.

Most of the dune sediments are very siliceous, with low levels of CaO and trace elements, thus suggesting common or similar sources. However, SiO_2 values range from 19.2% to 99.8%, illustrating the presence of additional localized sources. In particular, dunes in Cornwall have lower SiO_2 contents, diluted by CaO contents in excess of 15% on the north coast, while on the south coast heavy metal enrichment occurs in granitic areas (Fig. 8).

The dune systems of England and Wales have a wide range of SiO_2/Al_2O_3 and Al_2O_3/K_2O ratio values, but the majority of dune samples analysed fall within a narrow range of values, with 85% of samples having a SiO_2/Al_2O_3 ratio value of less than 80, and 90% having an Al_2O_3/K_2O ratio value of less than 6 (Fig. 9a, & b). The frequency distribution of Al_2O_3/Fe_2O_3 ratio values differs as, although the classes 1.50–1.75 and 1.75–2.00 are the most frequent, these do not represent the majority of dune samples and several other classes are almost as frequent (Fig. 9c).

A wide range of SiO_2/CaO ratios was found, although the majority of samples have a value of less than 100 (Fig. 9d). The very siliceous nature of some samples relative to other geochemical components has led to SiO_2/CaO ratio values in excess of 500, although this was found for only 6% of the dataset and largely spatially confined to dune sediment samples from Norfolk (Sites 32, 33, 34b, 34c and 35) and the southern coast of England (Sites 40–43).

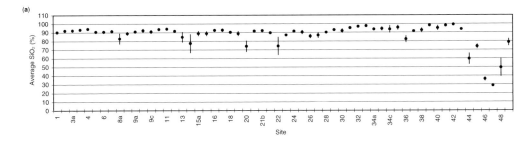

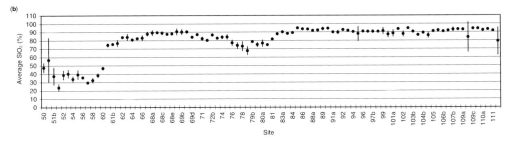

Fig. 8. Average SiO_2 content plotted with one standard deviation for the 63–250 μm fraction of dune sediment samples from dune sites: (**a**) on the east and south coasts of England; and (**b**) on the west coast of England and in Wales.

Ratios of oxides/elements can be used to differentiate between dune areas (Fig. 10). Most notably, sand dunes in Studland Bay were found to have exceptionally high values for the ratios of SiO_2/Al_2O_3 and Al_2O_3/K_2O, indicating provenance from a localized sediment source rich in SiO_2 with relatively low concentrations of feldspars and micas (mainly from Tertiary Bagshot Sands and Bracklesham Beds). However, the ratios of these two oxides were not found to be very diagnostic in differentiating between other dune systems.

An indication of the geochemical variation within sites can be obtained by calculating the relative standard deviation (RSD) for the oxides and elements. Higher RSD values indicate the existence of greater variation. This calculation identifies sites with a numerically large range of values for a given variable as having great variation. However, sites with numerically small differences in a given variable can also have large RSD values where the concentrations are very low. Thus high RSD values identify sites with high relative variation but the actual variation in values may not be important geologically. For example, at the dune site Caister to Great Yarmouth, a high RSD value of 115% was calculated for the CaO content. Values at this site have a narrow range of 0.02–0.78%; thus all values can be considered as low. At Lelant in St Ives Bay an RSD value of 70% was obtained, which is still relatively high, but the values have a much wider range (3.69–27.14%).

Six oxides/elements were studied to assess the degree of within-site variation: SiO_2, Fe_2O_3, CaO,

Cu, Ba and La. SiO_2 varies the least, having RSD values of up to 47%, while greatest variation occurs in lanthanum, with values of up to 200%.

Some dune sites have consistently low geochemical within-site variation in terms of most elements/oxides with RSD values of less than 20% (e.g. Northam Burrows, Freshwater West and Haverigg Haws), while others have consistently high RSD values (e.g. Caister to Great Yarmouth, and Maryport to Grune Point). Some sites have great variation in one or a few elements/oxides but low variation in others. However, the majority of sites have an intermediate degree of variation that is neither exceptionally high nor low in comparison to the other sites. There appears to be no spatial pattern in those sites with greater geochemical variation.

Variation within an individual dune system can take either of two forms: (1) all samples analysed may have slightly different geochemical composition, giving a relatively large range, or (2) there may be only a small number of samples (outliers), which are distinctly different to the others. Variation of the first type can arise due to the presence of localized sources of sediment, such as cliff erosion or fluvial input, in addition to the main marine beach source. Variation of the second type may also reflect the existence of a very localized secondary sand source, or the effects of anthropogenic contamination. The geochemical signature from localized inputs is likely to be strongest close to the source and become diluted in the direction of sediment transport due to deposition or mixing with other source material downdrift.

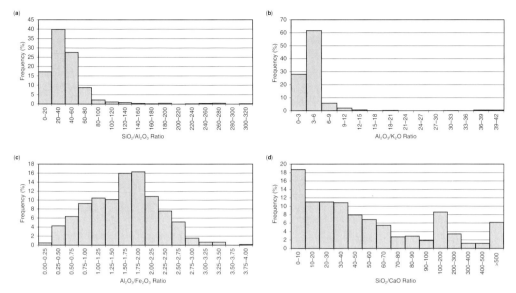

Fig. 9. Frequency histograms of: (**a**) SiO_2/Al_2O_3 ratio; (**b**) Al_2O_3/K_2O ratio; (**c**) Al_2O_3/Fe_2O_3 ratio; and (**d**) SiO_2/CaO ratio for the 63–250 μm particle size fraction of all samples from dune sites in England and Wales.

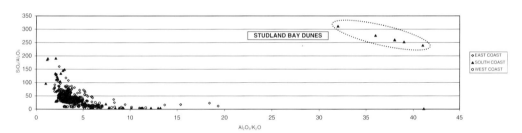

Fig. 10. Bivariate plot of the Al_2O_3/K_2O ratio v. the SiO_2/Al_2O_3 ratio for the 63–250 μm particle size fraction of dune sediment samples from the different coasts of England and Wales.

Samples from older parts of a dune system, such as hind dunes or relict perched or clifftop frontal dunes, may have different geochemistry to the younger parts due to: (1) former existence of different sources to those currently available, and/or (2) the effects of post-depositional alteration of the sediment by weathering and diagenesis. Weathering leads to the breakdown of many silicate minerals and the leaching of minerals such as calcium carbonate, leaving a residual of the more resistant quartz and heavy minerals such as zircon, rutile and tourmaline (Pye 1983). In addition, anthropogenic interference can affect dune sediment geochemical variation within sites, either directly through land-use activities such as golf courses and military operations or indirectly through beach nourishment and dumping of industrial waste onto beaches.

Forensic case examples

Murder investigation, Lincolnshire coast, UK

The dune sediment database has been used to determine the origin of sand on a glove implicated in body parts disposal (Lincolnshire Police, Operation Goldfinch). Evidence of temporary burial of a murder victim's limbs was discovered in the coastal sand dunes at Chapel Six Marshes, Lincolnshire (Figs. 11 & 12). Following police enquires, a blood-stained shirt, identified as belonging to the victim, was recovered buried in dunes close to a beach access point at the southern end of Chapel Six Marshes car park. No human remains were found, however, although the recovery of some blood-stained clothing suggested that some body parts may originally have been present and later

Fig. 11. Temporary burial site in sand dunes at Chapel Six Marshes, Lincolnshire.

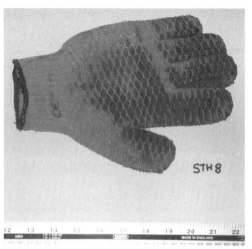

Fig. 13. Work glove with traces of decomposition fluids and sand grains trapped in the webbing.

Fig. 12. Profile of temporary burial site in sand dunes.

removed. During searches of the surrounding area a glove was also found discarded at the roadside, several kilometres inland near Orby, with traces of decomposition fluids and also sand grains trapped in the glove webbing (Figs. 13 & 14).

Initial optical examination of the sediment on the glove illustrated the presence of clean, well-sorted sand grains with an absence of clay and iron oxide coatings or associated cement/plaster particles that would have indicated a likelihood of the grains being derived from building sand. The presence of marine shell fragments within the sediment indicated that the grains had instead probably been derived from a beach or dune environment. Some fines were also present

which probably represent pre-existing dirt on the glove.

The particle size characteristics of the sand on the glove were determined, and it was found to be similar to, but significantly different from, sand at the temporary burial site in the dunes where the shirt was discovered at Chapel Six Marshes car park (Table 1; Fig. 15). The sand on the glove was also found to be significantly different from sand in a second pit, located within 100 m of the temporary burial pit, which was investigated because ground appeared to be disturbed (Pit 2 in Table 1). The particle size characteristics of the glove sand were then compared with data in the England and Wales dune database. Several areas were identified as possessing dune sediments with similar particle size characteristics to those on the glove (i.e. modal values within ± one class and D_{25}, D_{75} and median values within ±10% of the glove values), including parts of Northumberland, Lincolnshire, Pembrokeshire, Anglesey, north Wales and the sites Dawlish Warren (43), Penhale (53) and Drigg (110b). The sand on the glove therefore could not be sourced exclusively to Lincolnshire on the basis of particle size characteristics alone. However, given the location where the glove was found, it was considered most likely that the sand originated somewhere on the Lincolnshire coast between Mablethorpe and Gibraltar Point (Fig. 14).

Dune database samples from the Lincolnshire coast that were similar to the glove sediment in terms of particle size occur to the north and south of the temporary burial pit (CLEO 5B at Wolla Bank and CLEO 3B at Chapel St Leonards), and between Skegness and Gibraltar Point (GIB 5B, GIB 4G, GIB 3B and GIB 2B) (Table 2). Wolla Bank is both near to

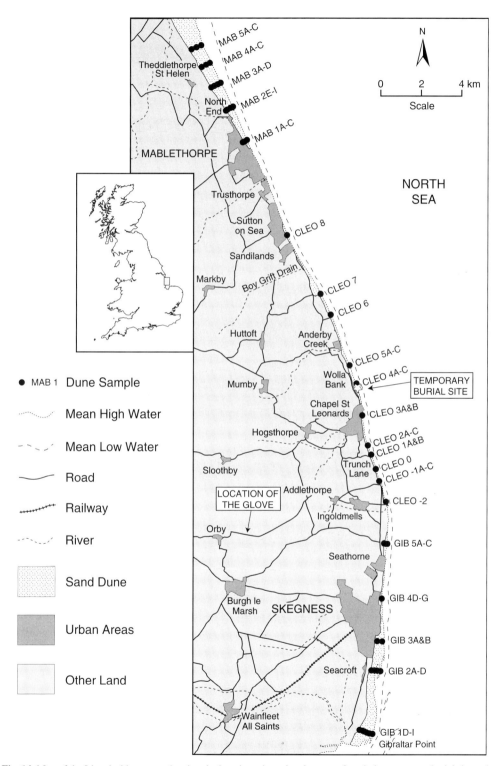

Fig. 14. Map of the Lincolnshire coast, showing the location where the glove was found, the temporary burial site and dune-sampling locations for which data were already available in the database.

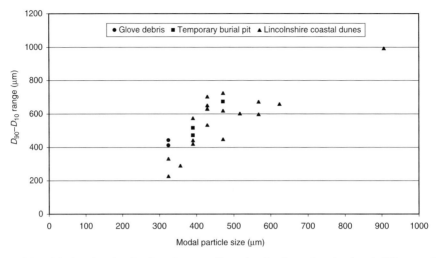

Fig. 15. Modal particle size plotted against $D_{90}-D_{10}$ range illustrating that the sand on the glove is different to that in the temporary burial site but closely resembles that in some of the sand dunes in other areas nearby

Table 1. *Particle size characteristics of sand on the suspect's glove and samples from the temporary burial pit in the dunes (Pit 1) and disturbed ground near to the temporary burial pit (Pit 2)*

Sample ID	Mean (μm)	Median (μm)	Mode (μm)	D_{10} (μm)	D_{25} (μm)	D_{75} (μm)	D_{90} (μm)	$D_{90}-D_{10}$ (μm)	Clay (%)	Silt (%)	Sand (%)
Glove 1	263.9	324.7	324.3	118.6	249.5	408.9	529.8	411.2	0.550	7.27	92.18
Glove 2	252.8	324.4	324.3	84.5	246.4	409.5	527.4	442.9	0.720	8.42	90.86
Burial pit 1A	437.3	424.2	390.9	267.1	329.1	573.2	783.6	516.5	0.000	0.14	**99.86**
Burial pit 1B	518.7	514.4	471.1	291.4	374.7	725.2	964.2	672.8	0.000	0.09	**99.91**
Burial pit 1C	414.9	408.8	390.9	258.2	321.1	533.1	729.3	**471.1**	0.000	0.42	**99.58**
Pit 2A	511.2	510.5	471.1	302.0	383.7	692.9	899.2	597.2	0.000	0.13	**99.87**
Pit 2B	472.7	461.6	429.2	284.3	352.9	633.0	851.8	567.5	0.000	0.12	**99.88**

Values in bold are within ± one modal class or within ±10% of the values obtained for sand from the glove

Table 2. *Particle size characteristics of dune sediments collected from the Lincolnshire coast (dune sites 24–27 on Fig. 1) as part of the coastal dune sediment database*

Sample ID	Mean (μm)	Median (μm)	Mode (μm)	D_{10} (μm)	D_{25} (μm)	D_{75} (μm)	D_{90} (μm)	$D_{90}-D_{10}$ (μm)	Clay (%)	Silt (%)	Sand (%)
HORSE 1	**263.6**	265.4	269.2	191.5	**222.9**	314.8	363.4	171.9	0.000	0.00	**100.00**
HORSE 2	**269.9**	269.8	269.2	195.2	**226.2**	322.2	373.5	178.3	0.000	0.00	**100.00**
MAB 11A	**241.3**	255.7	269.2	178.1	212.6	306.8	359.9	181.8	0.160	1.71	**98.13**
MAB 11B	210.5	230.3	245.2	156.2	190.7	275.3	320.6	164.4	0.350	2.45	**97.20**
MAB 10A	352.0	359.1	269.2	184.5	217.2	308.1	358.4	173.9	0.014	0.83	**99.16**
MAB 10B	**230.0**	254.0	269.2	168.4	208.1	306.3	358.7	190.3	0.400	2.39	**97.21**
MAB 9	**256.6**	264.7	269.2	186.5	220.3	317.1	368.6	182.1	0.085	0.73	**99.19**
MAB 8	**242.3**	253.4	245.2	179.2	211.8	302.6	352.2	173.0	0.150	1.07	**98.78**
MAB7A	**256.7**	267.5	269.2	184.7	221.4	321.0	371.8	187.1	0.066	0.96	**98.97**
MAB 7B	**232.1**	240.2	245.2	165.2	198.3	288.5	335.2	170.0	0.010	0.96	**99.03**
MAB 6A	**232.8**	261.6	269.2	162.9	212.9	315.8	368.1	205.2	0.180	3.51	**96.31**
MAB 6B	218.8	225.8	223.4	152.9	184.5	274.3	323.2	170.3	0.019	0.80	**99.01**
MAB 5A	**257.0**	266.2	269.2	189.0	**222.7**	316.4	365.5	176.5	0.050	0.81	**99.14**

Table 2. continued

Sample ID	Mean (μm)	Median (μm)	Mode (μm)	D_{10} (μm)	D_{25} (μm)	D_{72} (μm)	D_{90} (μm)	$D_{90}-D_{10}$ (μm)	Clay (%)	Silt (%)	Sand (%)
MAB 5B	**249.2**	258.5	269.2	179.5	214.8	309.0	360.5	181.0	0.009	0.99	**99.00**
MAB 5C	**247.5**	260.4	269.2	175.8	214.7	312.7	365.0	189.2	0.002	1.41	**98.59**
MAB 4A	**260.0**	269.0	269.2	191.7	**225.0**	320.1	369.0	177.3	0.045	0.78	**99.18**
MAB 4B	**252.1**	270.9	269.2	187.0	**224.3**	325.9	381.0	194.0	0.190	1.98	**97.83**
MAB 4C	**263.1**	263.7	269.2	193.3	**222.8**	311.2	359.3	166.0	0.000	0.00	**100.00**
MAB 3A	**290.0**	287.5	269.2	196.7	**235.1**	355.9	442.3	245.6	0.089	0.70	**99.21**
MAB 3B	**272.2**	274.4	269.2	199.9	**232.1**	323.9	369.9	170.0	0.000	0.00	**100.00**
MAB 3C	213.4	242.7	245.2	151.5	198.5	292.8	344.0	192.5	0.240	4.30	**95.46**
MAB 3D	**250.6**	258.9	269.2	184.0	217.1	307.3	356.2	172.2	0.014	0.89	**99.10**
MAB 2E	**234.2**	241.0	245.2	172.5	202.0	286.4	331.8	159.3	0.023	0.87	**99.11**
MAB 2F	**235.2**	246.7	245.2	174.0	205.8	294.9	341.0	167.0	0.110	1.37	**98.52**
MAB 2G	**237.0**	243.4	245.2	173.8	203.8	289.6	334.8	161.0	0.004	0.80	**99.20**
MAB 2H	**233.8**	248.1	245.2	172.9	206.4	296.7	343.1	170.2	0.110	1.76	**98.13**
MAB 1A	**265.7**	267.3	269.2	191.6	**223.5**	318.8	368.4	176.8	0.000	0.00	**100.00**
MAB 1B	**241.7**	242.2	245.2	178.1	204.9	285.7	329.3	151.2	0.000	0.00	**100.00**
MAB 1C	**236.9**	248.9	245.2	167.5	204.8	299.5	349.1	181.6	0.002	1.41	**98.59**
CLEO 8	**232.4**	233.2	223.4	152.3	189.3	288.2	362.6	210.3	0.000	1.68	**98.32**
CLEO 7	501.7	504.5	517.2	313.2	390.8	651.2	823.3	510.1	0.000	0.15	**99.85**
CLEO 6	538.3	573.6	684.2	304.9	401.8	803.9	1020.0	715.1	0.093	1.23	**98.68**
CLEO 5A	374.0	370.5	**356.1**	256.5	305.0	452.5	**555.2**	298.7	0.000	0.37	**99.63**
CLEO 5B	328.6	**342.4**	**356.1**	216.1	**273.1**	**431.5**	**554.9**	338.8	0.130	1.99	**97.88**
CLEO 5C	335.4	359.3	**356.1**	221.3	286.5	**450.1**	**582.2**	360.9	0.140	2.77	**97.09**
CLEO 4A	366.0	371.3	**356.1**	248.3	302.5	455.8	**561.2**	312.9	0.089	0.75	**99.16**
CLEO 4B	421.6	424.8	429.2	272.4	337.5	537.7	688.9	**416.5**	0.000	0.59	**99.41**
CLEO 3A	323.8	**337.3**	**356.1**	234.7	279.8	**404.9**	**476.0**	241.3	0.120	1.05	**98.83**
CLEO 3B	300.5	**306.4**	**295.5**	188.8	**242.4**	391.0	**517.7**	328.9	0.072	1.15	**98.13**
CLEO 2A	365.7	371.0	**356.1**	245.8	300.1	461.3	**577.1**	331.3	0.090	0.85	**99.06**
CLEO 2B	217.5	264.6	269.2	156.4	214.0	321.8	379.6	223.2	0.083	4.95	**94.22**
CLEO 2C	**262.9**	273.8	269.2	157.9	215.0	345.4	438.9	281.0	0.042	2.27	**97.69**
CLEO 1A	393.8	394.2	390.9	257.4	316.0	490.7	615.5	358.1	0.000	0.38	**99.62**
CLEO 1B	353.1	360.4	**356.1**	236.2	290.3	**446.2**	**551.2**	315.0	0.090	0.92	**98.99**
CLEO 0	388.9	457.1	471.1	224.2	333.2	623.9	855.7	631.5	0.400	4.70	**94.90**
CLEO −1A	471.1	467.8	471.1	296.3	367.5	601.9	777.9	**481.6**	0.000	0.13	**99.87**
CLEO −1B	506.0	540.6	567.8	296.9	400.5	717.8	921.2	624.3	0.090	1.36	**98.55**
CLEO −1C	171.5	262.9	**295.5**	20.53	105.7	**397.1**	644.4	623.9	1.370	17.63	81.00
CLEO −2	428.4	435.3	429.2	256.7	331.7	580.8	785.8	529.1	0.001	0.91	**99.09**
GIB 5A	353.3	**354.0**	**324.3**	220.2	275.0	469.4	649.3	**429.1**	0.055	1.35	**98.60**
GIB 5B	328.1	**340.1**	**324.3**	209.0	**264.6**	448.4	616.0	**407.0**	0.160	2.33	**97.51**
GIB 5C	294.4	290.5	269.2	187.7	**231.0**	370.9	**487.1**	299.4	0.003	0.97	**99.03**
GIB 4D	492.9	499.8	517.2	286.0	373.1	666.1	855.1	569.1	0.000	0.17	**99.83**
GIB 4E	348.7	**348.4**	**324.3**	219.0	**272.2**	457.5	627.4	**408.4**	0.083	1.09	**98.83**
GIB 4F	**285.2**	**297.7**	**295.5**	195.3	**240.9**	367.6	450.8	255.5	0.130	2.27	**97.60**
GIB 4G	**265.2**	292.5	**295.5**	120.6	219.6	**382.7**	524.9	404.3	0.130	4.79	95.08
GIB 3A	425.1	427.2	429.2	270.5	338.7	536.4	672.9	**402.4**	0.000	0.14	**99.86**
GIB 3B	300.9	**298.0**	**295.5**	199.1	**240.6**	375.1	**489.4**	290.3	0.050	0.97	**98.98**
GIB 2A	421.0	414.9	**356.1**	238.0	303.7	594.2	818.6	580.6	0.000	0.45	**99.55**
GIB 2B	324.4	**321.4**	**324.3**	217.5	**261.1**	397.5	**484.5**	267.0	0.000	0.41	**99.59**
GIB 2C	**264.7**	272.2	269.2	186.3	**223.3**	332.0	395.5	209.2	0.010	0.91	**99.08**
GIB 2D	**271.2**	275.2	269.2	174.4	218.9	354.1	472.3	297.9	0.120	2.07	**97.81**
GIB 1D	312.1	292.3	269.2	194.5	**233.3**	392.2	655.9	**461.4**	0.064	0.71	**99.23**
GIB 1E	**280.9**	269.9	245.2	180.9	217.8	349.2	**539.7**	358.8	0.160	1.15	**98.69**
GIB 1F	**254.3**	255.5	245.2	171.7	207.8	323.6	438.1	266.4	0.310	1.80	**97.89**
GIB 1G	**248.6**	246.0	245.2	163.4	198.9	309.3	407.2	243.8	0.044	1.84	**98.12**
GIB 1H	**249.2**	248.4	245.2	162.5	199.1	318.9	438.8	276.3	0.150	1.99	**97.89**
GIB 1I	350.9	**341.4**	**324.3**	205.0	**260.3**	476.3	745.4	540.4	0.092	1.03	**98.88**

Data are ordered north to south between Horse Shoe Point and Gibraltar Point. Values in bold are within ± one modal class or within ±10% of the values obtained for sand from the glove

Table 3. Geochemical composition of sand on the suspect's glove and dune sediments collected from the Lincolnshire coast (dune sites 24–27 on Fig. 1) as part of the coastal dune sediment database for England and Wales

Major oxide/element	Glove	Mini-mum	Maxi-mum	HORSE 1	HORSE 2	MAB 11A	MAB 11B	MAB 10A	MAB 10B	MAB 5A	MAB 5C	MAB 2E	MAB 2G	CLEO 6	CLEO 4A	CLEO 4B	CLEO 1B	GIB 4D	GIB 4F	GIB 2B	GIB 1D	GIB 1E	GIB 1G	GIB 11
SiO_2 (%)	80.62	82.51	95.23	**90.97**	**91.88**	**90.86**	**86.99**	**90.23**	**91.40**	**90.41**	**88.82**	**95.23**	**88.26**	**87.50**	**84.72**	**82.51**	**87.61**	**83.00**	**84.90**	**88.80**	**89.57**	**85.78**	**90.98**	**83.09**
Al_2O_3 (%)	2.35	1.79	2.96	**2.07**	**2.02**	**2.04**	**2.36**	**2.06**	**2.18**	**2.05**	**2.13**	1.79	**2.15**	**2.53**	**2.48**	2.96	**2.51**	**2.68**	**2.62**	**2.18**	**2.30**	**2.29**	**2.28**	2.89
Fe_2O_3 (%)	4.27	1.17	5.80	1.64	1.35	2.03	3.10	1.75	1.91	2.11	2.23	1.17	2.21	**4.04**	**3.56**	5.65	3.26	5.80	**4.79**	3.08	2.49	2.44	1.48	5.74
MgO (%)	0.93	0.25	0.81	0.54	0.46	0.44	0.61	0.46	0.44	0.55	0.55	0.25	0.61	**0.81**	0.60	0.67	0.60	0.67	0.65	0.60	0.68	0.68	0.43	0.69
CaO (%)	3.46	1.01	3.34	2.72	2.33	1.79	2.56	2.22	1.92	2.53	2.43	1.01	**2.92**	2.75	2.08	2.09	1.90	2.60	2.35	2.44	**3.53**	**3.34**	2.24	2.23
Na_2O (%)	1.49	0.23	0.32	0.28	0.29	0.26	0.29	0.27	0.27	0.27	0.27	0.23	0.28	0.32	0.28	0.28	0.31	0.23	0.26	0.28	0.30	0.30	0.30	0.27
K_2O (%)	0.71	0.49	0.72	0.56	**0.57**	0.52	0.56	0.54	**0.65**	0.53	0.51	0.56	**0.57**	0.51	0.52	0.50	**0.62**	0.49	0.49	0.50	**0.58**	**0.61**	**0.72**	0.55
TiO_2 (%)	0.65	0.09	0.68	0.10	0.09	0.14	0.29	0.11	0.13	0.15	0.14	0.09	0.16	0.35	0.40	**0.68**	0.36	**0.59**	0.46	0.23	0.16	0.19	0.10	**0.63**
P_2O_5 (%)	0.09	0.03	0.09	0.04	0.04	0.04	0.05	0.04	0.05	0.04	0.05	0.03	0.06	0.07	0.06	0.06	0.06	0.06	0.07	0.06	0.06	0.06	0.04	**0.09**
MnO (%)	0.08	0.02	0.13	0.03	0.02	0.03	0.05	0.03	0.03	0.03	0.03	0.02	0.04	**0.07**	**0.09**	0.13	**0.07**	0.13	**0.08**	0.05	0.04	0.04	0.02	0.11
Ba (ppm)	239	134	533	144	142	134	**224**	134	149	142	140	143	**223**	**219**	310	533	**268**	444	308	175	172	174	188	151
Co (ppm)	5	1	8	2	1	2	**4**	1	1	2	3	1	3	**5**	**5**	7	**4**	8	**6**	**4**	3	3	1	7
Cr (ppm)	112	2	38	4	2	5	12	4	5	7	6	2	6	26	21	38	22	32	20	12	7	9	5	32
Cu (ppm)	24	3	7	3	3	5	5	4	4	4	4	3	4	7	6	6	5	6	6	5	5	4	4	6
Ni (ppm)	10	7	14	**8**	**8**	**8**	**10**	**8**	**8**	**9**	**10**	7	**9**	14	**12**	13	**11**	13	**11**	**10**	**10**	**9**	**8**	**12**
Sc (ppm)	4	1	7	1	1	2	3	2	2	2	2	1	2	5	5	7	**4**	7	5	3	2	2	1	6
Sr (ppm)	81	48	92	**79**	**71**	60	**74**	**69**	**67**	**75**	**72**	48	**86**	**90**	**79**	**76**	**74**	**89**	**78**	**76**	99	**92**	**77**	**75**
V (ppm)	58	8	78	14	11	18	34	14	17	19	20	8	20	**51**	**48**	78	46	72	**51**	30	22	24	13	76
Zn (ppm)	70	19	55	34	32	34	38	27	21	33	37	19	38	36	36	46	33	55	48	32	32	33	34	44
Zr (ppm)	748	67	995	119	67	111	481	120	145	199	146	157	281	296	220	**878**	296	**886**	**776**	333	202	176	104	995
Pb (ppm)	144	5	16	10	7	12	16	11	6	10	16	5	11	11	11	14	12	15	12	8	12	11	10	13
Y (ppm)	18	5	27	6	7	11	14	8	8	8	**15**	6	9	**15**	**16**	27	13	23	**20**	10	11	6	5	22
La (ppm)	15	4	13	5	4	4	10	4	7	7	7	4	9	8	6	**12**	8	**13**	**13**	6	6	6	5	11
Ce (ppm)	34	8	30	8	8	9	20	8	14	14	13	8	19	19	16	**30**	15	24	**30**	14	14	13	12	27
Nd (ppm)	16	0	15	0	0	0	11	0	0	0	0	0	10	10	0	30	0	**15**	**15**	0	0	0	0	**13**
Sm (ppm)	3.0	0.7	2.3	1.1	1.0	0.8	1.3	1.1	1.7	1.2	1.2	0.8	1.3	1.1	1.3	1.7	1.8	2.0	1.8	1.5	1.7	1.6	0.7	2.3
Eu (ppm)	0.5	0.0	0.7	0.0	0.0	0.0	0.0	0.0	0.0	0.0	0.0	0.0	0.0	**0.6**	**0.5**	0.0	0.0	**0.6**	**0.6**	0.0	0.0	0.0	0.0	0.7
Dy (ppm)	2.4	0.6	3.4	1.1	0.8	1.3	**2.0**	1.1	1.2	1.3	1.3	0.6	1.6	**2.3**	**2.3**	3.4	**2.1**	3.4	**2.5**	1.5	1.3	1.3	0.8	3.0
Yb (ppm)	2.1	0.0	3.0	0.6	0.5	0.8	1.2	0.8	0.7	0.7	0.8	0.0	0.8	**1.7**	**2.2**	3.0	**1.7**	2.8	**1.9**	1.1	0.8	0.8	0.5	**2.4**

Data are ordered north to south between Horse Shoe Point and Gibraltar Point. Values in bold are within ±20 % of the values obtained for the sand from the suspect glove.

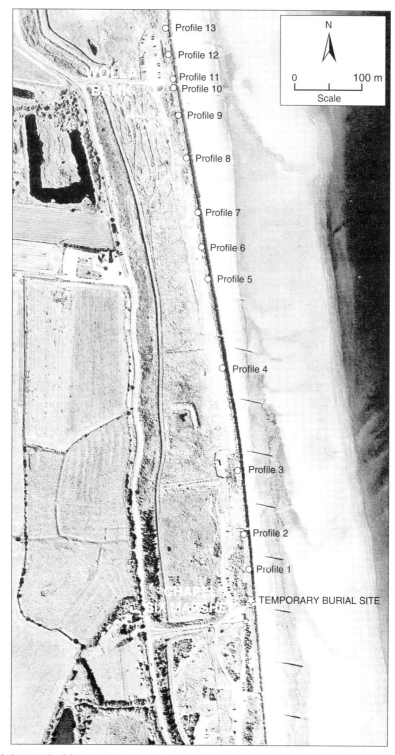

Fig. 16. Aerial photograph of the coastline between Chapel Six Marshes and Wolla Bank, Lincolnshire, showing the temporary burial site and the secondary dune-sampling locations (Profiles 1 to 13) on the frontal dunes.

Table 4. *Particle size characteristics of dune sediments from Profiles 1–13 between Chapel Six Marshes and Wolla Bank on the Lincolnshire coast (see Fig. 16 for position of profiles)*

Sample ID	Mean (μm)	Median (μm)	Mode (μm)	D_{10} (μm)	D_{25} (μm)	D_{75} (μm)	D_{90} (μm)	$D_{90}-D_{10}$ (μm)	Clay (%)	Silt (%)	Sand (%)
Profile 1A	689.9	771.6	905.1	308.9	510.1	1036.0	1303.0	994.1	0.055	0.405	**99.54**
Profile 1B	542.7	547.9	567.8	322.4	415.5	722.1	921.6	599.2	0.000	0.05	**99.95**
Profile 1C	575.4	586.8	623.3	332.5	436.2	780.7	991.9	659.4	0.000	0.10	**99.90**
Profile 2	501.4	502.3	471.1	269.5	358.8	731.1	995.7	726.2	0.000	0.43	**99.57**
Profile 3	482.3	481.8	471.1	315.6	386.9	603.2	764.4	**448.8**	0.000	0.19	**99.81**
Profile 4	463.6	463.9	429.2	283.4	355.7	617.2	817.9	534.5	0.000	0.49	**99.51**
Profile 5	409.7	404.2	390.9	259.0	318.8	525.6	701.7	**442.7**	0.000	0.40	**99.60**
Profile 6	483.2	483.9	429.2	270.8	351.9	691.9	923.4	652.6	0.000	0.47	**99.53**
Profile 7	396.1	392.2	390.9	250.9	310.3	505.4	671.6	**420.7**	0.000	0.50	**99.50**
Profile 8	328.3	**333.9**	**324.3**	235.1	277.9	**399.2**	463.2	228.1	0.000	0.14	**99.86**
Profile 9A	556.7	567.1	567.8	314.1	413.2	770.4	986.9	672.8	0.000	0.13	**99.87**
Profile 9B	515.2	521.5	517.2	299.0	388.2	701.0	901.3	602.3	0.000	0.15	**99.85**
Profile 9C	319.0	**321.4**	**324.3**	208.0	**255.6**	**410.7**	**539.5**	331.5	0.096	1.29	**98.61**
Profile 10	478.3	475.6	429.2	273.6	350.5	673.2	905.2	631.6	0.000	0.47	**99.53**
Profile 11A	356.2	**353.8**	**356.1**	240.4	288.8	**434.9**	**531.7**	291.3	0.074	0.43	**99.66**
Profile 11B	519.1	515.9	471.1	305.5	388.2	702.9	925.3	619.8	0.000	0.14	**99.86**
Profile 11C	452.1	437.6	390.9	268.7	335.5	599.5	844.5	575.8	0.000	0.13	**99.87**
Profile 12	519.7	510.7	429.2	288.5	370.2	744.8	993.1	704.6	0.000	0.12	**99.88**
Profile 13	485.2	479.6	429.2	284.7	359.7	675.3	914.9	630.2	0.003	0.47	**99.53**

Bold values are within ± one modal class or within ±10% of the values obtained for the glove

Chapel Six Marshes and remote, and was thus a more likely location than the town of Chapel St Leonards or the coast from Skegness to Gibraltar Point.

Comparison with the national dune sand geochemical database indicated that very few dune sand samples from outside Lincolnshire had many element concentration values within ±10% or ±20% of those obtained from the sand on the glove. When dune sand samples taken only from Lincolnshire coast were considered, the highest level of similarity observed was with control samples taken from the frontal dunes between Anderby Creek and Skegness (Fig. 14; Table 3). An exact match between the control dune sand samples and the sand on the glove would be unlikely due to the presence of body fluids and pre-existing material; as a consequence the sand on the glove has higher MgO, Na_2O, P_2O_5, Zn, Cr, Cu and Pb concentrations than the control samples from this area.

The geochemical composition of the sand on the glove was found to be most similar to that of the dunes in the vicinity of Chapel Six Marshes (CLEO 4A) and Anderby Creek (CLEO 6), to the north of Wolla Bank. Further control samples were therefore collected at approximately 50 m intervals between the temporary burial site at Chapel Six Marshes and just north of Wolla Bank (Fig. 16). The particle size characteristics of the sand on the glove closely resembled those of sand dunes in other areas nearby the burial site (Table 4). In particular, the glove was found to have the same modal characteristic as sand

dunes at Profile 9C (back dune at Wolla Bank), a few hundred metres north of Chapel Six Marshes (Fig. 17). A slightly larger tail of fine particles was present in the sand taken from the glove, due to the existence of pre-existing dirt particles.

The similarity of the sand on the glove to control samples taken from a short stretch of the dunes about 500 m from the known disposal site was confirmed by optical microscopy and geochemical analysis. However, police searches of this area revealed no further body parts, suggesting that, if any had originally been deposited, all had subsequently been removed. Some weeks later, parts of two legs were recovered from a location in north Norfolk close to the Lincolnshire border, and a partially decomposed torso was recovered adjacent to the A1 dual carriageway near Sawtry in Cambridgeshire. The hands, arms and head of the victim were never recovered.

Illegal release of degus into a nature conservation area, Sefton coast, UK

In a case of eco-vandalism, approximately 100 degus (exotic South American rodents) were released on the Queen's Silver Jubilee Trail, near Southport on the Sefton coast, Merseyside (Figs 18 & 19). The rodents released posed a serious threat to other wildlife within the sand dunes, including several endangered species, such as the red squirrel,

(a)   Glove debris

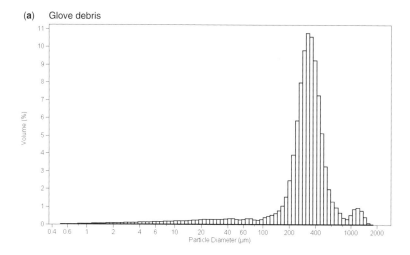

(b)   Dune back, 50 m south of beach access, Wolla Bank, Lincolnshire

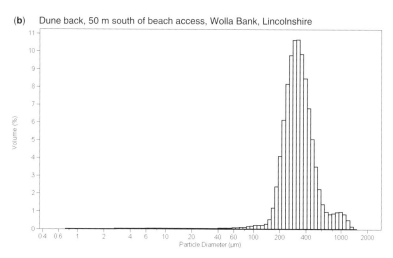

Fig. 17. Particle size distributions of: (**a**) sand on the suspect glove and (**b**) sand dune control location at Wolla Bank, Lincolnshire (Profile 9C). The secondary coarse mode in both cases represents comminuted shell debris. Solid lines represent cumulative frequency curves.

natterjack toad and sand lizard. A sand-covered lid from a dustbin, suspected to have been used in transportation of the degus to the release site, was seized from the home of the principal suspect (Fig. 20). This sand was compared with sediment from the release site of the degus as well as with database dune sediments from numerous locations along the Merseyside and Lancashire coasts. Comparison of particle size data immediately indicated that the sand on the bin lid could not possibly have been derived from the crime scene, or indeed anywhere else on the Sefton coast, being both coarser and more poorly sorted (Table 5; Figs 21 & 22). The bin lid sand was also compared with sediment samples in the dune database taken from dune systems further north

(Sites 107a, Lytham; 107b, Fleetwood), but was again found to be both coarser and more poorly sorted. However, the bin lid sand, although not identical in terms of particle size characteristics, was found to be more comparable with imported sand artificially placed in bunkers on the golf course at Fleetwood.

The sand from the dustbin lid was also found to be geochemically very different from the control sample from the degus release site and dune sediments from the Sefton coast recorded in the database (Table 6). The dustbin lid sand was predominantly composed of CaO, while the Sefton coastal dune sediments are predominantly siliceous. In addition, the dustbin lid sand had low abundances

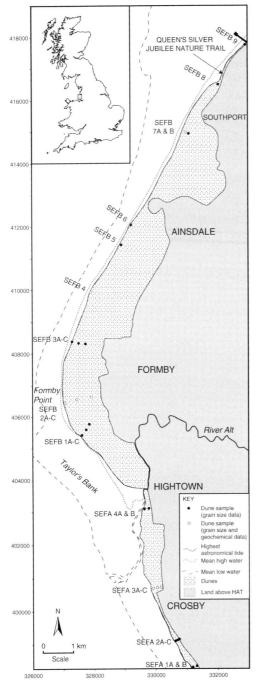

Fig. 18. Location map of the Sefton coast, showing the Queen's Silver Jubilee Nature Trail (the location of the degus release), the extent of the dune system and dune sampling locations.

of Al_2O_3, Fe_2O_3, K_2O, Ba and Zr, but greater concentrations of strontium, relative to the control dune sediments.

Optical microscopy also showed the dustbin lid sand to be very different in terms of mineralogical composition. It was concluded that the sand on the dustbin lid was calci-sand, widely sold in pet stores for use in reptile cages and fish tanks. Reference to the national sand dune database indicated that no dune sites exist in the UK which could have provided a source of similar calcareous sand to that present on the bin lid. Consequently, it was not possible to link the suspect to the scene of the crime based on the combined sedimentological and mineralogical evidence obtained for the dustbin lid sand.

Conclusions

The sedimentological properties of coastal sand dune sediments in England and Wales vary at both regional and local scales, although it cannot be said that every individual dunefield is 'unique'. Dune systems in Cornwall are notably different from others elsewhere, being coarser, more poorly sorted and possessing relatively low silica contents. More subtle particle size and chemical compositional differences are evident between other groups of sites.

The degree of within-site variation in terms of particle size characteristics and geochemistry is generally low. However, some sites are exceptionally homogeneous in terms of particle size characteristics (e.g. Sefton), while others are more diverse (e.g. Bigbury Bay); these differences can be explained by the nature of the prevailing process regime and differences in sediment provenance. High geochemical variation within a site is usually due to sediment being supplied from localized sediment sources and/or anthropogenic contamination.

The two case studies discussed in this paper demonstrate how database information can be used in forensic investigations. The existing dune sediment database provides a useful tool for screening samples and identifying locations in which more detailed sampling and analysis should be undertaken. However, further work is required to extend the database, both in terms of additional samples and the types of data included. Particle size characteristics and geochemical data need to be used in combination with mineralogical and biological data (e.g. pollen, seeds, diatoms) to provide better environmental discrimination.

Fig. 19. Part of the Queen's Silver Jubilee Nature Trail, within the Sefton coastal dune system near Southport, where the degus were released.

Fig. 20. Suspect dustbin lid with adhering sand thought to have been used in the transportation of the degus.

Table 5. *Particle size characteristics of sand from the suspect dustbin lid, the control sample from the degus release site and database dune sediments from various locations on the Sefton coast (see Fig. 18 for sampling locations)*

Sample ID	Mean (μm)	Median (μm)	Mode (μm)	D_{10} (μm)	D_{25} (μm)	D_{75} (μm)	D_{90} (μm)	$D_{90}-D_{10}$ (μm)	Clay (%)	Silt (%)	Sand (%)
Dustbin lid	478.9	558.3	567.8	356.7	451.6	688.1	827.3	470.6	0.690	2.69	96.62
Release site	196.7	229.2	223.4	111.1	181.2	287.0	360.0	248.9	0.470	7.14	**92.39**
SEFA 1A	229.4	230.2	223.4	166.1	193.3	273.9	316.3	150.2	0.000	0.00	**100.00**
SEFA 1B	212.7	232.1	245.2	155.0	191.0	278.8	326.2	171.2	0.170	2.95	**96.88**
SEFA 2A	223.1	223.9	223.4	162.1	187.7	266.3	306.6	144.5	0.000	0.00	**100.00**
SEFA 2B	235.9	236.7	245.2	172.3	199.4	279.9	322.7	150.4	0.000	0.00	**100.00**
sefa 2C	222.8	223.6	223.4	162.7	188.0	265.0	304.7	142.0	0.000	0.00	**100.00**
SEFA 3A	238.4	238.9	245.2	171.7	199.9	284.9	331.0	159.3	0.000	0.00	**100.00**
SEFA 3B	247.4	248.0	245.2	177.2	206.5	297.6	345.4	168.2	0.000	0.00	**100.00**
SEFA 3C	216.5	228.5	223.4	158.3	189.2	273.8	318.2	159.9	0.150	1.43	**98.42**
SEFA 4A	240.1	240.9	245.2	171.0	200.4	289.0	335.7	164.7	0.000	0.00	**100.00**
SEFA 4B	256.9	257.4	269.2	180.6	213.0	301.9	365.0	184.4	0.000	0.00	**100.00**
SEFB 1A	233.3	233.6	223.4	169.2	196.5	277.8	321.7	152.5	0.000	0.00	**100.00**
SEFB 1B	245.6	246.4	245.2	177.5	206.1	294.1	338.6	161.1	0.000	0.00	**100.00**
SEFB 1C	236.7	237.8	245.2	170.3	199.0	282.7	327.9	157.6	0.000	0.00	**100.00**
SEFB 2A	245.8	246.6	245.2	176.4	205.6	295.3	341.8	165.4	0.000	0.00	**100.00**
SEFB 2B	237.9	249.3	245.2	172.5	205.8	300.1	349.6	177.1	0.100	1.26	**98.64**
SEFB 2C	351.9	244.7	245.2	176.2	204.7	293.2	343.3	167.1	0.000	0.00	**100.00**
SEFB 3A	247.4	248.0	245.2	180.7	209.0	294.1	337.1	156.4	0.000	0.00	**100.00**
SEFB 3B	198.3	233.5	245.2	122.1	188.0	280.6	326.8	204.7	0.250	4.62	**94.13**
SEFB 3C	222.9	240.4	245.2	166.7	199.6	286.6	331.5	164.8	0.170	2.27	**97.56**
SEFB 4	245.2	246.6	245.2	179.4	208.0	291.6	334.0	154.6	0.000	0.00	**100.00**
SEFB 5	254.1	255.1	269.2	187.7	216.7	300.1	241.0	153.3	0.000	0.00	**100.00**
SEFB 6	265.5	267.5	269.2	189.2	221.4	320.6	370.9	181.7	0.000	0.00	**100.00**
SEFB 7A	245.5	246.7	245.2	180.4	208.5	291.2	333.2	152.8	0.000	0.00	**100.00**
SEFB 7B	243.2	251.1	245.2	175.8	208.1	301.5	350.8	175.0	0.047	0.83	**99.12**
SEFB 8	280.9	248.9	223.4	171.1	202.0	322.2	559.9	388.8	0.000	0.00	**100.00**
SEFB 9	244.5	229.9	223.4	162.3	189.6	283.2	359.8	197.5	0.000	0.00	**100.00**
SEFB 10	325.1	280.3	245.2	183.5	221.0	389.3	**889.4**	705.9	0.000	0.00	**100.00**

Values in bold are within ± one modal class or within ± 10% of the values obtained for the sand from the dustbin lid.

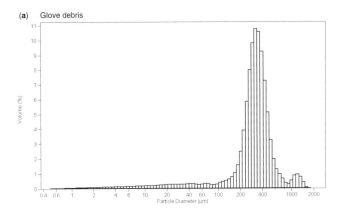

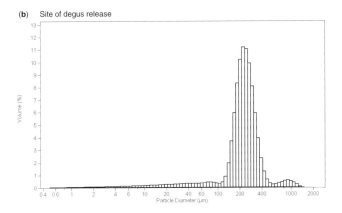

Fig. 21. Particle size distributions of: (**a**) sand from suspect dustbin lid, and (**b**) sand from the degus release site.

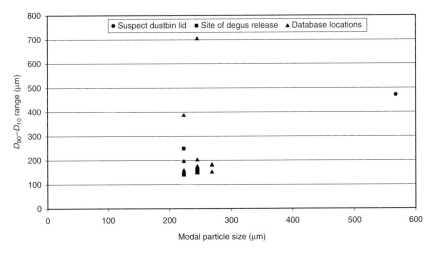

Fig. 22. Modal particle size plotted against $D_{90}-D_{10}$ range illustrating the coarser and more poorly sorted nature of the sediment from the suspect's dustbin lid

Table 6. *Geochemical composition of sand from the suspect dustbin lid, the control sample from the degus release site and database dune sediments from various locations on the Sefton coast (see Fig. 18 for sampling locations)*

Major oxide/ element		Dustbin lid	Release site	SEFA 3A	SEFA 3B	SEFA 3C	SEFB 2A	SEFB 2B	SEFB 2C	SEFB 4	SEFB 7A	SEFB 10
SiO_2	(%)	0.74	90.60	92.11	91.82	91.43	90.05	90.55	91.84	88.33	93.03	90.99
Al_2O_3	(%)	0.36	2.54	2.14	2.23	2.41	2.37	2.38	2.19	2.23	1.89	2.44
Fe_2O_3	(%)	0.12	1.42	1.17	0.94	1.06	1.33	1.38	0.81	1.53	0.82	0.86
MgO	(%)	0.23	**0.24**	0.33	0.28	0.32	0.40	0.35	0.30	0.51	0.29	**0.25**
CaO	(%)	49.38	1.38	1.39	1.45	1.25	2.11	1.55	1.44	2.41	1.36	1.91
Na_2O	(%)	0.01	0.31	0.28	0.27	0.31	0.30	0.36	0.26	0.28	0.25	0.31
K_2O	(%)	0.00	0.89	0.79	0.81	0.89	0.82	0.83	0.76	0.71	0.67	1.03
TiO_2	(%)	0.01	0.08	0.12	0.10	0.19	0.16	0.16	0.10	0.21	0.09	0.07
P_2O_5	(%)	0.01	0.05	0.05	0.03	0.04	0.04	0.04	0.04	0.05	0.03	0.04
MnO	(%)	0.02	0.03	**0.02**	**0.02**	**0.02**	0.04	0.03	**0.02**	0.04	**0.02**	**0.02**
Ba	(ppm)	62	172	160	168	171	165	165	154	143	137	195
Co	(ppm)	0	1	1	1	2	2	2	1	2	1	1
Cr	(ppm)	1	13	15	9	21	20	19	11	29	9	8
Cu	(ppm)	2	5	**2**	**2**	3	**2**	3	1	**2**	1	**2**
Li	(ppm)	0	4	12	12	12	12	8	12	12	12	13
Ni	(ppm)	6	10	**9**	7	**9**	10	**9**	**8**	10	**8**	**8**
Sc	(ppm)	1	**1**	2	**1**	2	2	2	**1**	3	**1**	**1**
Sr	(ppm)	115	64	55	57	56	71	61	55	76	55	84
V	(ppm)	0	11	12	11	17	18	17	12	23	10	10
Zn	(ppm)	21	37	27	33	14	15	16	13	**23**	**18**	30
Zr	(ppm)	8	45	97	107	197	221	157	76	184	73	16
Pb	(ppm)	6	15	11	12	10	9	8	**7**	11	8	13
Y	(ppm)	4	6	5	5	6	6	6	**4**	7	**4**	**4**
La	(ppm)	1	5	6	6	12	6	9	4	10	3	2
Ce	(ppm)	3	11	11	11	19	12	24	10	19	9	7
Nd	(ppm)	2.0	5.7	6.6	4.5	12.5	6.6	9.5	4.3	10.4	3.5	2.6
Sm	(ppm)	0.3	1.1	1.4	0.7	1.6	1.2	2.3	0.7	1.7	0.2	**0.3**
Eu	(ppm)	0.1	0.3	0.3	0.3	0.3	0.4	0.4	0.3	0.5	0.2	**0.1**
Dy	(ppm)	0.4	0.9	1.0	0.8	1.3	1.0	1.1	0.6	1.1	0.7	0.8
Yb	(ppm)	0.2	0.5	0.4	0.4	0.6	0.6	0.4	0.6	0.7	0.4	0.4

Values in bold are within ±20% of the values obtained for the sand from the dustbin lid

References

BLOTT, S. J., CROFT, D. J., PYE, K., SAYE, S. E. & WILSON, H. E. 2004. Particle size analysis by laser diffraction. *In*: PYE, K. & CROFT, D. J. (eds) *Forensic Geoscience: Principles, Techniques and Applications*. Geological Society, London, Special Publications, **232**, 63–73.

BLOTT, S. J. & PYE, K. 2001. GRADISTAT: a particle size distribution and statistics package for the analysis of unconsolidated sediments. *Earth Surface Processes and Landforms*, **26**, 1237–1248.

EMERY, K .O. 1978. Grain size in laminae of beach sand. *Journal of Sedimentary Petrology*, **48**, 1203–1212.

GAO, S. & COLLINS, M. B. 1992. Net sediment transport patterns inferred from particle-size trends, based upon definition of 'transport vectors'. *Sedimentary Geology*, **80**, 47–60.

GAO, S. & COLLINS, M. B. 1994. Analysis of particle size trends, for defining sediment transport pathways in marine environments. *Journal of Coastal Research*, **10**, 70–78.

GRACE, J. T., GROTHAUS, B. T. & EHRLICH, R. 1978. Size fre-

quency distributions taken from within sand laminae. *Journal of Sedimentary Petrology*, **48**, 1193–1202.

GREENWOOD, B. 1978. Spatial variability of texture over a beach-dune complex, North Devon, England. *Sedimentary Geology*, **21**, 27–44.

JAY, H. 1998. *Beach-dune sediment exchange and morphodynamic responses: implications for shoreline management, the Sefton coast, NW England*. PhD thesis, University of Reading.

MCLAREN, P. 1981. An interpretation of trends in particle size measures. *Journal of Sedimentary Petrology*, **51**, 611–624.

MCLAREN, P. & BOWLES, D. 1985. The effects of sediment transport on particle-size distributions. *Journal of Sedimentary Petrology*, **55**, 457–470.

MOTYKA, J. M. & BRAMPTON, A. H. 1993. *Coastal Management. Mapping of Littoral Cells*. HR Wallingford, Report SR 328.

NEAL, A. 1993. *Holocene development of a coastal barrier-dune system, northwest England*. PhD thesis, University of Reading.

PYE, K. 1982. Negatively skewed aeolian sands from a

humid tropical coastal dunefield, northern Australia. *Sedimentary Geology*, **31**, 249–266.

PYE, K. 1983. Coastal dunes. *Progress in Physical Geography*, **7**, 531–557.

PYE, K. 1991. Beach deflation and backshore dune formation following erosion under storm surge conditions: an example from Northwest England. *Acta Mechanica* Supplement, **2**, 171–181.

PYE, K. 1994. Properties of sediment particles. *In*: PYE, K. (ed.) *Sediment Transport and Depositional Processes*. Blackwell, Oxford, 1–24.

PYE, K. 2004. Forensic examination of rocks, sediments, soils and dust using scanning electron microscopy and X-ray chemical microanalysis. *In*: PYE, K. & CROFT, D. J. (eds) *Forensic Geoscience: Principles, Techniques and Applications*. Geological Society, London, Special Publications, **232**, 103–122.

PYE, K. & BLOTT, S. J. 2004. Comparison of soils and sediments using major and trace element data. *In*: PYE, K. & CROFT, D. J. (eds) *Forensic Geoscience: Principles, Techniques and Applications*. Geological Society, London, Special Publications, **232**, 183–196.

PYE, K. & NEAL, A. 1993. Late Holocene dune formation on the Sefton coast, northwest England. In: PYE, K.

(ed.) *The Dynamics and Environmental Context of Aeolian Sedimentary Systems*. Geological Society Special Publications **72**, London, 201–217.

PYE, K. & NEAL, A. 1994. Coastal dune erosion at Formby Point, north Merseyside, England: causes and mechanisms. *Marine Geology*, **119**, 39–56.

SALISBURY, E. J. 1952. *Downs and Dunes: Their Plant Life and its Environment*. G. Bell, London.

SAYE, S. E. 2003. *Morphology and sedimentology of coastal sand dune systems in England and Wales*. PhD Thesis. University of London.

THOMPSON, M. & WALSH, J. N. 1989. *A Handbook of Inductively Coupled Plasma Spectrometry*. 2nd edition. Blackie, Glasgow.

WILSON, K. 1960. The time factor in the development of dune soils at South Haven Peninsula, Dorset. *Journal of Ecology*, **48**, 341–359.

WINSPEAR, N. R. & PYE, K. 1995. Sand supply to the Algodones dunefield, south-eastern California, USA. *Sedimentology*, **42**, 875–891.

WINSPEAR, N. R. & PYE, K. 1996. Textural, geochemical and mineralogical evidence for the sources of aeolian sand in central and southwestern Nebraska, U.S.A. *Sedimentary Geology*, **101**, 85–98.

'Unique' particles in soil evidence

R. SUGITA & Y. MARUMO

National Research Institute of Police Science, 6–3-1 Kashiwanoha, Kashiwa, Chiba 277–0882,
Japan (e-mail: sugita@nrips.go.jp)

Abstract: Soil is presented as evidence as a mixture of both natural and artificial materials.
'Unique' particles found in soil may provide useful information to indicate the origin of the soil and,
when found in soil evidence, may provide rapid discrimination between samples. Evidence in poor
condition may also be examined for these diagnostic particles. This paper describes some case
examples and research, involving geological particles, plant fragments and algae, in which microscopy played an important role.

Soil is found as evidence in various forms, for
example as a cake from the sole of suspect's shoe, on
tapings from car carpets, as a clod of earth from a
spade and/or from a grave where a body was
exhumed. Soil is often the subject of forensic laboratories' attempts to prove the relationship between a
suspect and the scene of a crime, due to the adventi-
tious transfer of soil from one to the other. The complexity of soil makes forensic examination
potentially difficult because it contains many components, such as minerals, rock fragments, organic
matter, plant fragments and biogenic inorganic/
organic 'shells' (Murray & Tedrow 1992; Marumo
2003). The combination of time-consuming conventional methods such as colour comparison (Dudley
1975; Janssen *et al.* 1983; Antoci & Petraco 1993;
Sugita & Marumo 1996), particle size analysis
(Dudley 1977; Dudley & Smalldon 1978; Robertson
et al. 1984; Sugita & Marumo 2001) and mineralogical tests (Graves 1979; Ugolini *et al.* 1996) is necessary for detailed examination. These methods
require a certain amount of evidentiary and control
material for analysis and comparison to be achieved.
However, when 'unique' particles are found in soil
evidence, more precise and rapid discrimination can
be achieved, even if the amount of evidence recovered is small. Microscopy is the most useful technique for the detection of such characteristic
particles. Examples of these found in soil, and other
related evidence, are presented in this paper.

The uniqueness of Japanese soil

Minerals and rock fragments are the major constituents of soil, and the geological background influences the composition of particles present. Japan is
an arc of islands located in the temperate zone, with
relatively heavy rainfall. The prevailing mid-latitude
westerly winds blow throughout the year, bringing
with them the *kosa*, the aeolian dust from China.
Although there are various rocks types present,

only limited primary minerals are found in most of
soils where human activities occur. Surface soil is
immature because there is a continuous supply of
volcanic ash and/or the *kosa* on the plains, and relatively flat land, where relatively rapid leaching by
heavy rainfall occurs. Artificial alteration of surface
soil is also very significant due to construction and
agriculture.

Volcanic glasses and rock fragments

A wide area of the country is covered by layers of volcanic ash, as well as alluvial deposits with minerals
and glass assemblages derived from volcanoes and
volcanic events (Machida 1999). Volcanoes have supplied volcanic ash around the mountains and, on
occasions, to a distance of hundreds of kilometres
away when blown by the westerly winds. Rock fragments and lavas derived from volcanic events sometimes have unique colours and shapes. Shapes of
volcanic glasses other than pumice and scoria are
roughly classified into four types: bubble-wall,
fibrous, porous and massive. Their transparency
differs according to their thickness, colour and inclusions. The unique appearance of volcanic glasses may
afford rapid discrimination of soil evidence, and it is
very helpful to screen for them in the control samples
in many cases. The refractive indices (RI) of volcanic
glasses, as studied by Quaternary geologists, are a
useful property when characterizing key beds in stratigraphy (Machida & Arai 1992). The potential to discriminate by using RI's was studied and the results
from eleven representative Japanese tephra are shown
in Figure 1. Measurements were carried out with
a glass refracture index measurement system
(GRIM2), using the conventional method for industrial glasses. Two silicone oils, which are usually used
separately for measuring the RIs of soda lime and
borosilicate glasses, were blended to cover the range
of RIs of volcanic glasses. The range of RI. was considered as a useful indicator for distinguishing

From: PYE, K. & CROFT, D. J. (eds) 2004. *Forensic Geoscience: Principles, Techniques and Applications*. Geological
Society, London, Special Publications, **232**, 97–102. © The Geological Society of London, 2004.

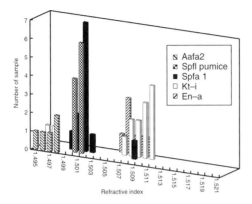

Fig. 2. Photograph of diagnostic red spherical particles in soil samples from a criminal case.

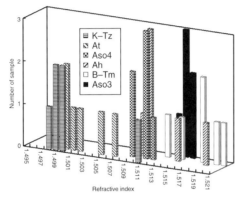

Fig. 1. The results of volcanic glass refractive indices (RIs) measurement by GRIM2 for eleven samples of Japanese tephra.

between ash types, and the results showed that this technique was applicable to forensic analysis.

In one case study, red spherical particles of several millimetres in diameter were found in sand and gravel fractions (Fig. 2). Such colour is sometimes observed in volcanic rocks and sediments that have been exposed to high temperature and highly oxidizing conditions. The quantity of these unique particles present was used as a comparison method because significant differences were observed between control samples.

In another case study, a body was found on basaltic lava at the foot of Mt Fuji. The existence of fragments of basalt from the suspect's car indicated that he had been in the area, because the soil where he lived did not contain any such fragments.

Inclusions

Inclusions, which have been trapped during crystallization, are not uncommon in primary minerals. They are often observed in minerals derived from volcanic sources, but are also present in igneous, metamorphosed and metasomatic rocks. The formation of inclusions has long been studied by petrologists, mineralogists and economic geologists, but they have not been used in forensic science. The shape and content of inclusions indicate the origin of the minerals, and should be the same in both, recovered and control samples if they are derived from the same soil.

In another murder case, samples of taped debris from the suspect's car and sand from a beach where it was suspected that the crime had been committed were submitted for examination. There were a limited number of particles on the tapes and they were typical volcanic minerals, namely quartz, plagioclase, pyroxene and magnetite. On first inspection they seemed to contain no characteristic particles to distinguish between them. However, pyroxenes, including large opaque minerals, were unique to both the taped samples and to sand from the beach.

Zeolites

Zeolites are utilized as soil improvement agents in farmland (Ming & Allen 2001). The majorities of zeolites used for this purpose in Japan are clinoptilolite and mordenite (Minato 1994a, b). In a survey of soil samples from ricefields in Nirasaki, an area about 100 km west of Tokyo, zeolites were detected by X-ray diffraction (Fig. 3) in only a limited area. These zeolites did not occur naturally, but had been added as improvement agents and therefore could be used as 'unique', or diagnostic, particles.

Plant fragments

Plant fragments and related matter are also very commonly observed in soil evidence. Comparison of the structures of this material in recovered and

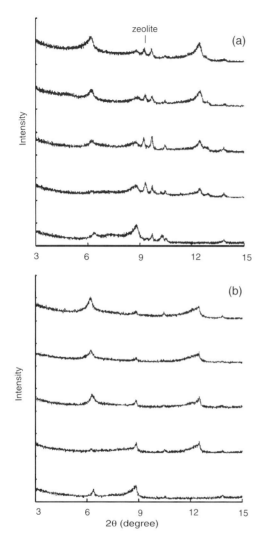

Fig. 3. X-ray diffractograms of the less-than 0.05 mm in diameter fraction from soil: (**a**) with zeolites; and (**b**) without zeolites.

Fig. 4. Secondary electron image of a plant fragment recovered from soil in a murder case, with a 'tick hole' (arrow).

Fig. 5. Photograph of a leaf from a control camphor tree with 'tick holes' (arrows).

Epidermis of leaves and seed coats from beans

The epidermis is the cover or 'skin' of the plant body and has many structural features. The seed coat protects the enclosed embryo (Esau 1977). Both the epidermis and the seed coat may be resistant to decomposition or digestion, and the structures may persist for long periods. Stomata on the epidermis of leaves have characteristic patterns depending on the species, which is useful for discrimination. Observation of the stomata of the epidermis and the structure of the seed coat by the transmitted light microscope and/or the scanning electron microscope are powerful methods for discrimination.

In a kidnap and murder case that occurred in the summer, the body of the victim was found in a river about a week after she was kidnapped. The epidermis of vegetable leaves was found in the victim's stomach contents, and subjected to examination in order to determine the type of vegetable. By researching the vegetables commonly sold in the area, the leaf was found to be that of a cabbage, which the victim's friend testified she had eaten before the kidnapping. This indicated that she was

control samples, using optical microscopes and the scanning electron microscope, have been successfully carried out, although little systematic research has been done, as indicated by Bock and Norris (1997).

In yet another murder case, fragments of leaves with 'tick holes' (Fig. 4) were found in a cake of soil from a suspect's shoe. The leaves had holes produced by ticks at the branching points of veins. Such holes are characteristic of the camphor tree (Fig. 5) (Kitayama & Murata 1979). This information revealed the crime scene, as there was only one camphor tree present in the vicinity.

Table 1. *Thicknesses of palisade parenchyma (PAL) and hourglass cells of hypodermis (HYP) and the rate between them (HYP/PAL)*

		Thickness of PAL (μm)		Thickness of HYP (μm)		Ratio of thickness (HYP/PAL)	
Japanese name	English name	Min.	Max.	Min.	Max.	Average	SD
morokko	kidney bean	50	50	–	–	–	–
kintoki		37.5	50	–	–	–	–
kinusaya	pea	40	50	–	5	0.10	–
soramame	broad bean	110	120	15	15	0.13	±0.01
azuki	adzuki bean	50	50	2	3	0.05	±0.01
kariwamame	soybean	20	50	10	40	1.01	±0.50
aomame		35	38	25	38	0.90	±0.15
enrei		35	45	30	50	0.99	±0.14
tambaguro		40	60	20	25	0.47	±0.05

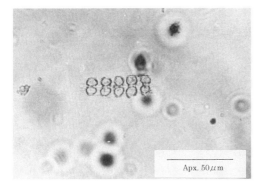

Fig. 6. Photograph of plant opals recovered from a cake of mud found on a screwdriver.

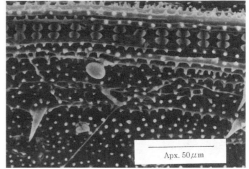

Fig. 7. Secondary electron image of plant opal from a control sample of *Oryza sativa* L.

killed just after she was kidnapped, prior to full digestion of the cabbage.

In a case of illegally dumped industrial waste, it was necessary to identify the type of waste, and information provided suggested that it might be from a type of bean. A classification of beans has been established, based on the identification of starches using the polarized light microscope; however, in this case, the morphology of the starch had been altered by decomposition, although fragments of the seed coats had survived. A study of nine types of bean showed that the thickness and shapes of the palisade parenchyma and hourglass cells of the hypodermis were characteristic as shown in Table 1. The seed coats found in the waste were then compared with bean types often used in industry and matched to one of them.

Plant opals

Plant opal, or phytolith, has been studied from an archaeological viewpoint (Pearsall 1989; Piperno

1988), and a forensic study was carried out by Marumo and Yanai (1986). Plant opal is an amorphous silicate formed in cells or in the structures of plant bodies. Its shape and structure vary depending upon species and cell types. Assemblages of plant opals in soil reflect the flora of the soil sample site.

In a case of arson, a hole had been made with a screwdriver in a traditional clay wall, consisting of a mixture of clay and rice grass (*Oryza sativa* L.). A small cake of mud on the screwdriver was found to contain plant opals with a shape and structure characteristic of *Oryza sativa* L. (Figs. 6 & 7), thereby matching it to the clay wall.

Microalgae

Algae exist almost everywhere, not only in the hydrosphere, but also in the pedosphere, the atmosphere, and even the lithosphere (Barns & Neirzwicki-Bauer 1997). Some microalgae, for example diatoms and flagellates, have organic or inorganic shells that may remain after the decomposition of the other organic components. Diatoms, known by their unique siliceous shells, have been well studied by diatomologists and palaeontologists, and the environment where they live, or have lived, may be deduced. Diatoms are also utilized in forensic science, especially to determine whether drowning has occurred (McLaughlin 1989–1995; Pollamen 1998; Krstic *et al.* 2002). Comparison of diatomaceous flora in recovered and control soil samples is also used in a forensic context.

In a large number of ricefields in Japan, fresh water, which provides a suitable habitat for particular types of diatoms, is artificially introduced during spring and summer. Microalgae in soil, collected in relation to an arson case, were cultured on agar media and examined with an optical microscope. Control samples collected from three farms were cultured for comparison. Suspensions of the control soils were spread on agar culture media, and the resulting colonies were mounted on glass slides for observation. As a result, several kinds of blue-green algae (Cyanophyceae) and green algae (Chlorophyceae) were observed in samples from the farms, but the flora were all different. Comparison was not successful in this case but it demonstrated the potential use of microalgae in forensic investigation.

Conclusion

Most of the examples in this paper involved the use of optical microscopy, which is one of the oldest methods for particle identification but still has high discriminating power. The examples describe particles familiar to forensic soil examiners, although opportunities to work with each type do not occur often for a single scientist. These examples are strongly dependent on cumulative information gathered over time and from techniques from every field related to forensic geoscience.

References

ANTOCI, P. R. & PETRACO, N. 1993. A technique for comparing soil colors in the forensic laboratory. *Journal of Forensic Science*, **38**, 437–441.

BARNS, S. M. & NIERZWICKI-BAUER, S. A. 1997. Microbial diversity in ocean, surface and subsurface environments. *In*: BANFIELDS, J. F. & NEALSON, K. H. (eds) *Geomicrobiology: Interactions Between Microbes and Minerals*. Reviews in Mineralogy, **35**, 35–79.

BOCK, J. H. & NORRIS, D. O. 1997. Forensic botany: an under-utilized resource. *Journal of Forensic Sciences*, **42**, 364–367.

DUDLEY, R. J. 1975. The use of colour in the discrimination between soils. *Journal of the Forensic Science Society*, **15**, 209–218.

DUDLEY, R. J. 1977. The particle size analysis of soils and its use in forensic science: The determination of particle size distributions within the silt and sand fractions. *Journal of Forensic Sciences*, **16**, 219–229.

DUDLEY, R. J. & SMALLDON, K. W. 1978. The objective comparison of the particle size distribution in soils with particular reference to the sand fraction. *Medicine, Science and the Law*, **18**, 278–281.

ESAU, K. 1977. *Anatomy of Seed Plants*. John Wiley & Sons, New York.

GRAVES, W. J. 1979. A Mineralogical soil classification technique for the forensic scientist. *Journal of Forensic Sciences*, **24**, 323–338.

JANSSEN, D. W., RUHF, W. A. & PRICHARD, W. W. 1983. The use of clay for soil color comparisons. *Journal of Forensic Sciences*, **28**, 773–776.

KITAYAMA, S. & MURATA, G. 1979. *Coloured Illustrations of Woody Plants of Japan*. Vol. II. Hoikusha Publishing Co. Ltd, Osaka.

KRSTIC, S., DUMA, A., JANEVSKA, B., LEVKOV, Z., NIKOLOVA, K. & NOVESKA, M. 2002. Diatoms in forensic expertise of drowning: a Macedonian experience. *Forensic Science International*, **127**, 198–203.

MACHIDA, H. 1999. Quaternary widespread tephra catalog in and around Japan: recent progress. *Daiyonki-Kenkyu*, **38**, 194–201.

MACHIDA, H. & ARAI, F. 1992. *Atlas of Tephra in and around Japan*. University of Tokyo.

MCLAUGHLIN, R. B. 1989–1995. Diatoms: diatom microscopy. *Microscope*, **37–43**.

MARUMO, Y. 2003. Forensic examination of soil evidence. *Japanese Journal of Science and Technology for Identification*, **7**, 95–111.

MARUMO, Y. & YANAI, H. 1986. Morphological analysis of opal phytoliths for soil discrimination in forensic science investigation. *Journal of Forensic Sciences*, **31**, 1039–1049.

MINATO, H. 1994a. Mineralogy of natural zeolite resources. *In*: No. 111 Committee (ed.) *Natural Zeolite and Its Utilization. Development of New Utilization of Minerals*. Japan Society for the Promotion of Science, Tokyo, 8–23.

MINATO, H. 1994b. Introduction for utilization of natural zeolite in Japan. *In*: No. 111 Committee (ed.) *Natural Zeolite and Its Utilization. Development of New Utilization of Minerals*. Japan Society for the Promotion of Science, Tokyo, 232–233.

MING, D. W. & ALLEN, E. R. 2001. Use of natural zeolites in agronomy, horticulture, and environmental soil remediation. *In*: BISH, D. L. & MING, D. W. (eds) *Natural Zeolites: Occurrence, Properties, Applications*. Reviews in Mineralogy and Geochemistry, **45**, 619–654.

MURRAY, R. C. & TEDROW, J. C. F. 1992. *Forensic Geology*, Prentice Hall, New Jersey.

PEARSALL, D. M. 1989. *Paleoethnobotany. A Handbook of Procedures*. Academic Press, San Diego.

PIPERNO, D. R. 1988. *Phytolith Analysis. An Archaeological and Geological Perspective*. Academic Press, San Diego.

POLLAMEN, M. S. 1998. Diatoms and homicide. *Forensic Science International*, **91**, 29–34.

ROBERTSON, J., THOMAS, C. J., CADDY, B. & LEWIS, A. J. M. 1984. Particle size analysis of soils: a comparison of dry and wet sieving techniques. *Forensic Science International*, **24**, 209–217.

SUGITA, R. & MARUMO, Y. 1996. Validity of color - examination for forensic soil identification. *Forensic Science Intentional*, **83**, 201–210.

SUGITA , R. & MARUMO, Y. 2001. Screening of soil evidence by a particle combination of simple techniques: validity of particle size distribution. *Forensic Science International*, **122**, 155–158.

UGOLINI, F. C., CORTI, G., AGENELLI, A. & PICCARDI, F. 1996. Mineralogical, physical and chemical properties of rock fragments in soil. *Soil Science*, **161**, 521–542.

WANOGHO, S., GETTINBY, G. & CADDY, B. 1987. Particle size distribution analysis of soils using laser diffraction. *Forensic Science International*, **33**, 117–128.

WANOGHO , S., GETTINBY, G., CADDY, B. & ROBERTSON, J. 1989. Determination of particle size distribution of soils in forensic science using classical and modern instrumental methods. *Journal of Forensic Sciences*, **34**, 823–835.

Forensic examination of rocks, sediments, soils and dusts using scanning electron microscopy and X-ray chemical microanalysis

KENNETH PYE[1,2]

[1]*Kenneth Pye Associates Ltd, Crowthorne Enterprise Centre, Crowthorne Business Estate, Old Wokingham Road, Crowthorne RG45 6AW, UK (e-mail: k.pye@kpal.co.uk)*
[2]*Department of Geology, Royal Holloway, University of London, Egham Hill, Egham TW20 0EX, UK*

Abstract: Scanning electron microcopy (SEM) and energy-dispersive X-ray (EDX) microanalysis are powerful techniques for forensic and wider environmental analysis of a range of materials, including rocks, sediments, soils and dusts. Methods of analysis have evolved rapidly over the past 40 years and computer-controlled, variable pressure SEMs with integrated EDX now provide the opportunity for rapid, automated analysis of large numbers of samples and particulates within individual samples. However, interpretation of the data requires care and experience on the part of the operator, and samples should always be checked by visual inspection. Early SEM work on rocks and sediments mainly used the secondary electron (SE) mode to produce topographical contrast on rough surfaces, but more recent studies have utilized the capacity of backscattered electron (BSE) imaging to image both topographical and atomic number contrast. BSE microscopy, combined with X-ray mapping, provides a rapid means of locating unusual particles and grain coatings, and of mapping their distribution, which may be of diagnostic or discriminatory importance. In the past, much attention has been given to grain surface textural features (mainly of quartz) but many such studies have suffered from a high degree of subjectivity, poor reproducibility, lack of discriminatory power, and high cost both in terms of time and money. The application of digital imaging and statistical data-processing techniques can to some extent reduce these problems but, in general, chemical characterization of particles offers a more powerful approach. This paper provides an overview of these techniques, discusses their limitations and illustrates some of the forensic and wider environmental applications.

Scanning electron microscopy (SEM) and X-ray chemical microanalysis are techniques now routinely-used in the examination and analysis of a wide range of materials, including those of interest for forensic applications. In most investigations, preliminary assessment of forensic exhibits and samples of interest is undertaken visually, using a magnifying lens and high-intensity light source, followed by low-power binocular microscope examination. These procedures allow the identification of areas and materials of interest which can then be removed, or sometimes examined *in situ*, at higher magnifications using SEM and microanalytical techniques.

There is a very large number of published papers dealing with examination of rocks, sediments, soils, dusts and related materials, including industrial products such as brick and concrete, using SEM and various forms of X-ray chemical microanalysis. However, there is no existing overview of the application of these techniques in the context of forensic investigations. Previous forensic studies of sand, silt and dust particles (e.g. Fitzpatrick & Thornton 1974; Graves 1979; McCrone 1992; Petraco 1994*a, b*; Demmelmeyer & Adam 1995; Palenik 1998; Marumo & Sugita 2001) have frequently relied heavily on optical microscopy and have not taken advantage of the full body of geological knowledge or the full armoury of investigative approaches available. This paper provides an overview of SEM and related chemical investigative methods which have been employed in environmental geoscience studies. The relative value and limitations of these techniques for forensic applications is also discussed.

Principles of scanning electron microscopy and X-ray microanalysis

The initial development of the SEM is widely attributed to Ardenne (1938) who first added scan coils to a transmission electron microscope (TEM) in order to create what in effect was the world's first scanning transmission electron microscope (STEM) (Goldstein *et al.* 1981). The first true SEM used to examine thick specimens was described by Zworykin *et al.* (1942), who recognized that secondary electron (SE) emission would produce topographical contrast and allow examination of fine surface detail. However, after the Second World War, further significant development of the SEM was

From: PYE, K. & CROFT, D. J. (eds) 2004. *Forensic Geoscience: Principles, Techniques and Applications.* Geological Society, London, Special Publications, **232**, 103–122. © The Geological Society of London, 2004.

undertaken at Cambridge University (Oatley & Everhart 1957; Everhart & Thornley 1960; Oatley *et al.* 1965; Oatley 1966, 1972). The first commercial instruments were made in 1965 by the Cambridge Scientific Instrument Company. The advantages of these instruments over conventional optical micro-scopes and TEMs, namely greater depth of field and the ability to obtain relatively high-resolution images at magnifications ranging from $\times 20$ to *c.* $\times 80000$, were quickly recognized, and the tech-nique had already become widely used in scientific research by the early 1970s.

Since that time there have been many further developments, including the development of the field emission SEM, which provides ultra-high reso-lution on account of its high brightness electron source, the development of other contrast mecha-nism detectors, including those for backscattered electrons and atomic number contrast, crystallo-graphic (electron channelling and magnetic) con-trast and cathodo-luminescence (CL) (Wolf & Everhart 1969; Krinsley & Hyde 1971; Joy 1974; Grant 1978; Krinsley & Tovey 1978; Robinson 1980). Other more recent developments include the variable pressure SEM (VP-SEM), which allows examination of specimens without the need for coating with gold or carbon, and the environmental SEM (ESEM), in which sensitive and even 'wet' samples can be examined (Robinson & Nickel 1979). Most recently, the computer-controlled SEM (CC-SEM), which allows computerized integration of the imaging and microanalysis processes, includ-ing digital image processing and automated X-ray microanalysis, has greatly increased the potential for characterization/classification of large numbers of particles as a routine procedure.

Elemental analysis of objects within the SEM first became possible with the addition of a coupled X-ray detector in 1968. Development of the dedicated X-ray analytical electron microprobe had in fact been taking place in parallel with that of the TEMs and SEMs since the late 1940s. Important pioneer-ing work carried out by Castaing at the University of Paris was followed by further development in Cambridge and elsewhere during the 1960s (Long 1962; McKinley *et al.* 1966). By the early 1970s electron microprobe analysis (EMPA) had become a routine tool for the quantitative chemical analysis of many materials, including rocks (mainly in thin section). Several reviews of these developments have been published (e.g. Oatley 1972; Anderson 1973; Holt *et al.* 1974; Reed 1975, 1993, 1996; Goldstein & Yakowitz 1975; Goldstein *et al.* 1981, 2002; Newbury *et al.* 1986; Joy *et al.* 1986; Heinrich & Newbery 1991; Potts *et al.* 1995).

The past 30 years have also seen the development of a variety of other microscopes and analytical microprobe instruments which complement that of the basic SEM and EMPA. These include X-ray microscopes, electron-tunnelling microscopes, acoustic microscopes, scanning auger microscopes (SAM) and confocal optical microscopes for imaging, and techniques such as electron energy loss spectroscopy (EELS), proton-induced X-ray emis-sion (PIXE), X-ray fluorescence (XRF) and ion microprobe analysis (secondary ion mass spectrom-etry, SIMS) and laser microprobe analysis (LIMA) or laser ablation inductively coupled plasma mass spectrometry (LA-ICP-MS) for chemical characteri-zation (Joy *et al.* 1986; Meeks 1990; Hinton 1995; Kelley 1995). However, most of these techniques are used for specialist rather than routine applications, and the most widely used instrument remains the SEM with an integrated energy dispersive X-ray (EDX) and/or wavelength dispersive X-ray (WDX) microanalysis system. In general, EDX systems are used to obtain rapid qualitative or semi-quantitative analysis of elements present in concentrations gen-erally of a percent or more, while WDX systems provide more precise quantitative analyses, particu-larly for certain light elements (atomic number: $Z < 11$) and trace elements. Modem EDX detectors are routinely capable of detecting elements in the range from boron ($Z = 5$) to uranium ($Z = 92$). By careful control of instrumental operating conditions, meaningful compositional information can be obtained from particles as small as $3 \mu m$ in diameter, and qualitative compositional indications can be obtained from particles down to submicron in size.

Analysis of rocks, sediments, soils and similar materials using SEM and EDX analysis

Prior to the early 1980s most examinations of rocks, minerals and related materials, such as brick, were undertaken using the secondary electron (SE; topo-graphical contrast) imaging mode, and natural frac-ture surfaces or artificially etched surfaces to highlight fine structural detail (e.g. O'Brien 1968, 1981). Such techniques have remained a useful tool in studies of the internal fabric and microstructure of fine-grained rocks (e.g. O'Brien & Slatt 1990; Bennett *et al.* 1991), and of porosity and diagenetic mineralogy in oil reservoir and similar rocks (e.g. Waugh 1978; Welton 1984). However, in the early 1980s, it was recognized that additional information could be gained by examined polished sections and blocks using the backscattered scanning electron (BSE) mode (Krinsley *et al.* 1983; Pye & Krinsley 1983, 1984), or using combined BSE, SE and even cathodo-luminescence (CL) signals (Pye & Windsor-Martin 1983). BSE images are heavily dependent on the mean atomic number of the target (Fig 1 & 2), with the result that different mineral and

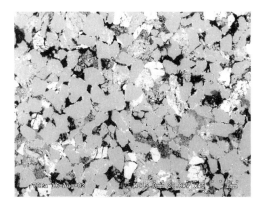

Fig. 1. BSE image of a polished thick section of a quartz sandstone showing quartz grains (mid-grey), resin-filled porosity (black) and other minerals with higher average atomic number than quartz (shades of light grey to white).

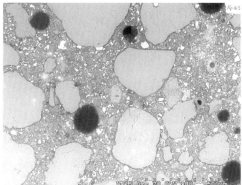

Fig. 2. BSE image of a polished section of concrete, recovered from the suspension of a murder suspect's car.

other phases can easily be distinguished, mapped and their relative abundances quantified (Pye 1984*a*). Since that time BSE examination of rocks and similar materials has become routine (Krinsley *et al.* 1998).

Early work in the 1970s and 1980s on bulk samples of sediments and soils also focused largely on microstructure and fabric as indicators of earth surface processes, environments of deposition, and their relationship to engineering properties (e.g. Derbyshire 1978; Osipov & Sokolov 1978; Smart & Tovey 1981). As in the case of rocks, examination was initially undertaken almost exclusively using the SE imaging mode, with attempts made to quantify the fabric from the images obtained using simple digital computer techniques (e.g. Tovey 1980). However, from the later 1990s onwards, studies made increasing use of BSE images, which are more amenable to computer quantification and image processing (e.g. Tovey & Krinsley 1992; Tovey & Hounslow 1995). Sophisticated methods are now available which allow high reproducibility measurements of fabric to be made rapidly, although careful sample preparation remains a key requirement, and it is a relatively straightforward process to undertake quantitative comparison of digital images obtained from two or more samples (Krinsley *et al.* 1998).

In forensic work, it is a frequent requirement that samples are compared using several different techniques, which preferably should be non-destructive, or at least minimally destructive. Using the BSE-imaging mode in a variable pressure computer-controlled scanning electron microscope (VP-CC-SEM), both topographical and atomic number information can be obtained, often without the need for polishing, impregnating or coating the samples. If images are obtained at standard magnifications and with stan-

dardized orientations, comparison of microfabric features and mineral distributions can be conducted either on-line or off-line.

In our work we use a Hitachi S–3000-N VP-CC-SEM which allows examination of non-conductive samples in their natural state. The system has integrated Oxford Instruments ISIS Series 300 EDX hardware (Li–Si detector and ATW2 thin window for light element detection down to $Z = 5$) and spectral acquisition and processing software which allows X-ray elemental mapping in conjunction with SE and BSE image acquisition. Using the SEM data manager, scanned images and the operating conditions under which they were obtained are automatically stored. Images and X-ray maps or spectra can be selected for immediate printing, stored on the system computer hard drive, or transferred electronically to an outside PC for further off-line processing. An optional Particle Characterization and Imaging (PCI) system is available that allows advanced digital image acquisition and management with the networking and database capabilities.

Examples of a simple forensic application of the SEM system are shown in Figure 3. In this instance the objective of the investigation was to compare the internal fabric (microstructure) of chalk rock particles found on a car owned by a murder suspect with chalk particles collected from a trackway along which the murder victims' bodies had been transported to a deposition site. Comparisons were also made with control samples of chalk from numerous other known locations. Internal microfabric provided one of the parameters used for comparison, others being major and trace element composition (determined by inductively coupled plasma spectrometry), microfaunal assemblages and quantitative spectroscopic colour determination. Each block was scanned to identify possible variations in texture, heavy mineral content and elemental concentrations due to diagenesis or weathering. None were detected.

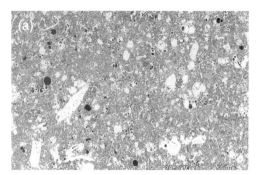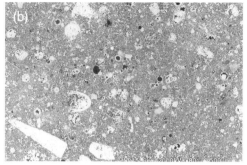

Fig. 3. BSE images showing: (**a**) the internal microstructure of a piece of Chalk rock recovered from the underside of a murder suspect's car compared with (**b**) a control sample of Chalk taken from the surface of a trackway leading to the body deposition site.

Examination of individual sediment particles

Characterization of particle morphology

The morphology of individual particles can also provide extremely valuable evidence in a forensic context and may be especially helpful where samples taken from a suspect's clothing or footwear are known to be mixtures from more than location (and hence where bulk methods of comparison are inappropriate). Initial examination is usually undertaken by low-power binocular microscopy of the 'as received' sample, but subsequent quantitative particle typology work is frequently carried out on selected particle size fractions, sometimes after cleaning of the grains to make characterization easier and more accurate. The analysis can be performed either on entire, unmounted grains or after the grains have been impregnated in resin, sectioned and polished.

A wide range of methods can be applied to characterize different properties of the particles, including size, shape, surface texture, particle elemental composition, mineral composition and nature of surface coatings or adhering particles. In some forensic contexts it may be sufficient to identify and characterize a small number of unusual grain types, but if these are not present there is little alternative but to make comparisons based on quantitative or semi-quantitative comparisons of the frequencies of different grain types. In the latter situation, issues of representative sampling, subsampling, sample pre-treatment and handling, and the accuracy and precision of the characterization method employed are of critical importance.

Quartz grain surface textures

Quartz is the most commonly found mineral on the surface of the earth and is a major constituent of most clastic sediments. Scientific interest in quartz is

long-standing, and it has been studied from a variety of petrographical and chemical points of view. The usefulness of variations in quartz grain shape and gross surface texture as indicators of geological source and transport history has long been recognized, but until the 1960s relatively few studies had been performed and those used only optical microscopy (e.g. Pettijohn 1949; Milner 1962).

Biederman (1962), Porter (1962) and Krinsley and Takahashi (1962a, b, 1964) were the first to recognize the potential of electron microsocopy in studies of quartz sand grain surface features. The first applications studies were carried out using replicas of the surface texture examined by TEM (e.g. Krinsley & Funnell 1965), but by the mid-1960s the benefits of direct observation of quartz grain surfaces in the SEM had been recognized (Krinsley & Donahue 1968a). From the late 1960s, until the early 1980s numerous quartz grain surface textural studies were undertaken. Many, although not all (see Krinsley & Trusty 1986 for an overview), were concerned with environmental discrimination and depositional history, that is identifying the environment of origin and process history of the grains based on their gross shape and fine surface textural detail (Margolis & Krinsley 1971; Krinsley & Doornkamp 1973; LeRibault, 1977, 1978). Reviews of this early work, including available sample preparation techniques, were provided by Smart & Tovey (1981, 1982), Bull (1981, 1986) and Bull et al. (1986). However, after the early 1980s, interest declined as many scientists became disillusioned with the imprecision of the technique and the low effort-to-results ratio. It became evident that no individual characteristics can be regarded as diagnostic of any given process or depositional environment, although combinations of features, particularly when treated statistically, may allow environmental discrimination (Culver et al. 1983; Elzenger et al. 1987). However, even this may not always be possible because many sediments are complex and have polygenetic histories. The interpretation of the ultimate sedimentary process is frequently complicated by the preceding depositional history of the particles (Elzenger *et al.* 1987), or by post-depositional weathering and diagenetic changes. Although SEM analysis of quartz surface textures continues to be undertaken (e.g. Mahaney 2002; Mahaney *et al.* 2004), its role in modern investigations is mainly supportive.

The decline in popularity of SEM surface textural studies was partly related to recognition that the technique, at least in its early form, is highly subjective and the results obtained, except in the simplest of environmental situations, are often equivocal and have low reproducibity. As noted by Tovey *et al.* (1978, p. 393): 'The identification of microtextural features is a largely subjective, pattern recognition exercise and on occasions some controversy exists over the identification of particular features. Agreement is often lacking when grains with complex environmental histories are examined and the relative importance of the individual features is assessed'.

Bias in results may also creep in due to grain selection, which is required at several stages during sample preparation. Many early studies referred simply to the 'selection of grains at random', but the procedures by which this was undertaken were often ill-defined. Some authors hand-picked what they took to be monocrystalline quartz grains using tweezers under a binocular microscope. Others scattered grains by hand onto a layer of adhesive or sticky tape mounted on an SEM stub. Tovey *et al.* (1978) emphasised the need to minimize possible bias in grain selection and proposed a scattering method through a fluid to mount more than 200 grains on a single 1 cm stub. Subsequent subsamples from the mounted population were selected by two methods: (1) by defining an area, and (2) by selecting individual grains at random. Using the latter techniques, statistically significant bias was observed among several different operators. Using the first technique, bias was found to be less, provided that sufficient grains were examined. However, few subsequent workers have adopted this method of sample preparation.

A further area of uncertainty and bias can arise from the use of non-standardized grain size fractions. The shapes, as well as the mineral and chemical composition, of sediment particles, are well known to vary as a function of particle size, and therefore it is important to examine features on several relatively narrowly defined size fractions (Tovey *et al.* 1978). Many workers have selected grains from one or more sieved fractions, for example the 1000–500, 250–500, 125–250 and 63–125 μm fractions, but some have simply used the bulk material.

There has also been disagreement regarding the minimum number of grains which should be analysed in order to obtain statistically representative results. Some authors have considered as few as 10 (Blackwelder & Pilkey 1972), 15 (Krinsley & Doornkamp 1973) or 20 grains (Baker 1976; Bull 1978, 1981) to be acceptable. Tovey *et al.* (1978) recommended a minimum of 30 grains per sample, Whalley & Langway (1984) used 40 grains, and Tovey & Wong (1978) recommended 50 grains, concluding that (p. 197) 'a sample size of 20 grains is certainly too small'. Culver *et al.* (1983) used 30–50 grains per sample, while Margolis & Krinsley (1973) recommended 50–100 grains and Vincent (1976) employed 200 grains per sample. Even this number is small by comparison with the minimum number considered necessary for statistically valid pollen results (e.g. Birks & Birks 1980; Moore *et al.* 1991) or the minimum number of grains used in conventional modal mineralogy studies undertaken by point counting (Galehouse 1971).

From a theoretical point of view, the minimum number of grains required to accurately characterize a sample will clearly depend on the degree of diversity of the grain population. A population consisting only of one grain type (which is virtually never found in nature) would theoretically only require a single grain, but in order to establish the probability that only a single grain type is present at the 95% confidence level at least 95 grains would have to be examined (cf. Dryden 1931). In reality, most natural sediment and soil samples contain several recognizable quartz grain types, at least three or four and sometimes as many as 15. Some of these, particularly the more unusual and potentially more discriminatory types, may be present only in 1% or 2% abundance, or even less. Consequently, if only 10–50 grains are examined, there is a significant chance that some of the rarer types will not be recorded. When dealing with natural soils or sediments, at least 300 grains should be examined and classified in order to make meaningful comparisons between samples and, if meaningful quantitative comparisons are to be made, at least 1000 grains should be analysed. Given the time-consuming nature of manual grain-surface textural analysis, this may not be feasible if more than a few samples are involved in the investigation. In such cases, automated particle-grain form analysis methods offer a more practical alternative.

The number and nature of identifiable features recognizable on quartz sand grains has also been a matter of debate and confusion. Different authors have recognized and recorded different numbers of features, ranging from 22 (Margolis & Kennett 1971; Bull 1978, 1981; Cater 1984), 29 (Elzenga *et al.* 1987), 30 (Higgs 1979), 32 (Culver *et al.* 1983), 34 (Goudie & Bull 1984) to more than 100

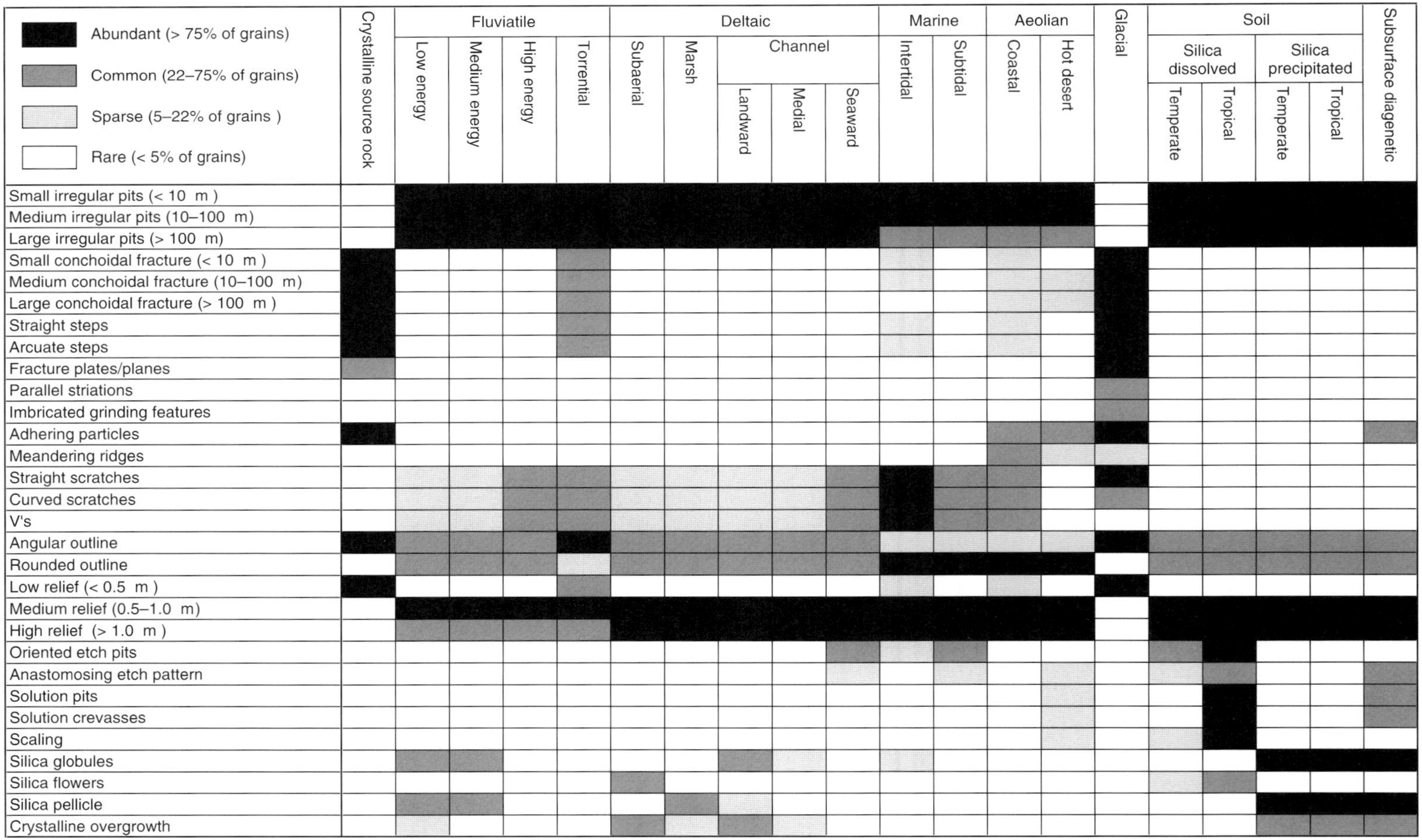

Fig. 4. Some quartz grain surface textural features and their suggested sedimentary environmental associations. Modified after Higgs 1979.

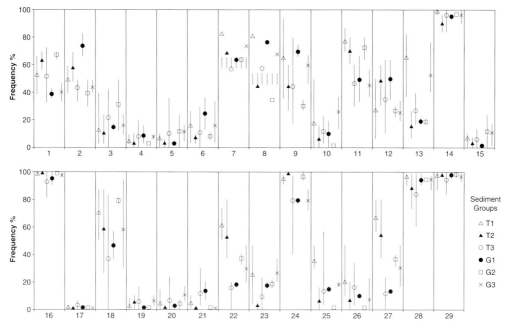

Fig. 5. Mean and range of the frequencies of 29 surface textural characteristics on quartz sand grains for different sediment groups from the Twente–Emsland (T) area of the eastern Netherlands and Goirle in the southern Netherlands. Modified after Elzenga *et al.* 1987. T1, periglacial aeolian coversand with homogeneous sandy texture; T2, periglacial aeolian coversand with regular alternation of coarser and finer beds; T3, originally aeolian coversand reworked by fluvial current flow; G1, periglacial aeolian coversand with fine, homogeneous sandy texture; G2, periglacial aeolian deposit grading from coversand to sandy loam (silt); G3, former periglacial aeolian deposits reworked by small-scale fluviatile action.

(LeRibault 1977, 1978). The meaning of some of the descriptive terms is obscure, and their relationship to specific environments or environmental processes is very tenuous in some cases. Nonetheless, several authors have suggested that certain groups of features are characteristic of particular environments (Krinsley & Doornkamp 1973; Higgs 1979; Fig. 4).

Some studies have merely recorded the presence or absence of particular features on each grain, while others have recorded relative abundance, either in numerical or descriptive terms (e.g. abundant [>75% of grains], common [22–75% of grains], sparse [5–22% of grains] and rare [<5% of grains]; Higgs 1979). Some authors have also tried to combine relative frequency of occurrence on grains with extent of coverage on individual grains (e.g. Setlow 1978). In the study by Culver *et al.* (1983), counts of grains displaying each feature were initially made and the percentage occurrence of each feature in each sample calculated. However, in this case the data were then 'degraded' by converting the percentage count data to binary (presence or absence) form. This experimental investigation, using five experienced scanning electron microscopists and eight coded samples from different environments, showed that there was considerable

operator variance in terms of recognition of individual surface features, although 39 out of 40 overall interpretations of the environment of origin were correct. However, the environments selected for study were quite distinctive: aeolian (hot), Millstone Grit regolith and some fluvial action, beach, glacio-fluvial, sandstone source rock and fluvio-glacial, aeolian (temperate), glacial and grus (weathered granitic regolith). Moreover, the number of variables (textural features) considered in the analysis was limited to 20 after combination of some of the original 32 features searched for, making this something of an 'unfair' test.

Several authors have noted that the quality of results and soundness of environmental interpretations can only be as good as the original data (quality of observations and recording), which is highly operator-dependent. Acquisition of good data requires experience and patience. Statistical analysis of the data is only worthwhile when sufficient grains and samples have been examined and sufficient counts made of each of the features of interest. A first stage is to calculate the mean, standard deviations and range of the frequencies of each surface characteristic for each of the different sediment groups under consideration (e.g. Fig. 5). The data can then

Fig. 6. BSE image of a relatively unabraded but fractured euhedral quartz grain (uncoated sample).

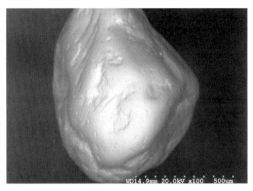

Fig. 7. BSE image of a well-rounded, near-spherical quartz grain with predominant chemical etching and precipitation features on its surface (uncoated sample).

be analysed further using techniques such as principal components analysis (Elzenga *et al.* 1987) or canonical analysis (Culver *et al.* 1983).

Work of the type described above is very time-consuming and expensive. For this reason, some workers have sought to develop simplified forms of analysis and sample comparison which are more practical for reconnaissance study purposes. Using such an approach, a number of different 'grain types' in a sample are identified on the basis of combinations of their dominant features. These types can be denoted simply as Type I, Type II, Type III etc., with accompanying notes of the principal features present: e.g. Type I – euhedral quartz overgrowths dominant (Fig. 6); Type II – well-rounded, low-relief grains with chemical etching and precipitation features (Fig. 7), Type III – angular grains with fresh conchoidal breakage features (Fig. 8). Counts can be made of each grain type in different samples and the results summarized either in tabular or graphical matrix form (e.g. Fig. 9).

While this approach has the advantage of being relatively rapid, it is highly subjective and likely to be subject to a high degree of operator variation. The grain types identified will vary from one geological situation to another and the individual types are simply operationally defined to meet individual circumstances. From a forensic perspective, it is virtually impossible to reproduce the results using different operators except in the simplest of situations, for example where a sample is obviously dominated by one or two easily recognizable grain types, such as the Types I, II and III referred to above. In many situations, 15 or more grain types may be identified by one operator, but ambiguity in grain/feature descriptions and classification may mean that fewer than half that number can be classified by another, equally experienced, examiner. In view of this, the approach should be considered as unsuitable for forensic casework applications.

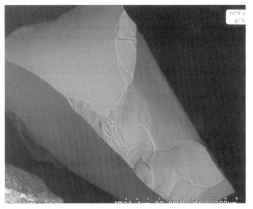

Fig. 8. BSE image of an angular, elongate quartz grain showing fresh conchoidal and other fracture features (uncoated sample).

Since the late 1970s the desirability of objective, quantitative analysis involving large numbers of grains has been recognized. Several authors have advocated the use of techniques such as Fourier analysis (Dowdeswell 1982; Ehrlich *et al.* 1987; Mazzullo & Ritter 1991; Pye & Mazzullo 1994) and fractal analysis (Orford & Whalley 1983, 1987) to quantify the forms of quartz and other sand grains. Data can be obtained either by digitizing individual photographic images obtained in the SEM or collectively using some form of optical microscope/image analysis system. Such techniques have the advantages of providing objective, high-resolution, mathematical descriptions of grain form, often for hundreds or thousands of grains, and acquisition of data in a form which is readily amenable to statistical analysis. An example is shown in Figures 10 and 11, where Fourier grain-shape data for two populations of Quaternary dune sand in the same geographical

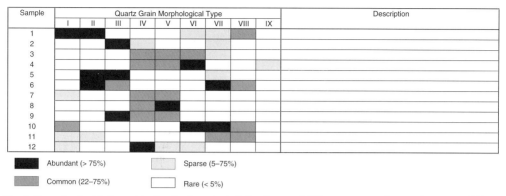

Fig. 9. Example of a tabular matrix showing hypothetical abundances of different quartz grain types in a group of 12 sediment samples.

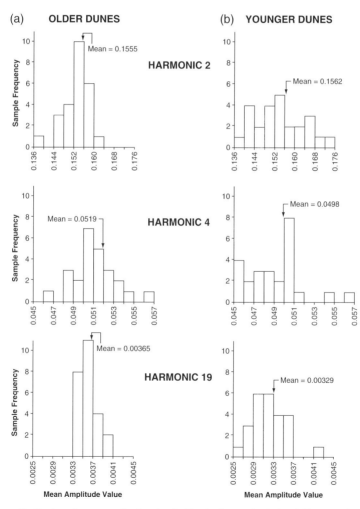

Fig. 10. Frequency distribution of mean amplitude value for Fourier harmonics 2, 4 and 19, comparing sediment samples from: (**a**) 'older' (Early Holocene), and (**b**) 'younger' (Late Holocene) coastal dunes at Ramsay Bay, Hinchinbrook Island, North Queensland. Modified after Pye & Mazzullo 1994.

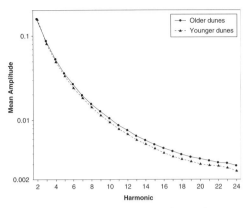

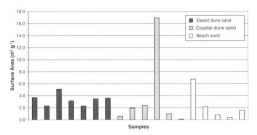

Fig. 12. Mean surface area for the 250–355 μm size fraction of beach, coastal dune and desert dune sands from several different parts of the world, determined by a nitrogen gas adsorption technique. Modified from Blott *et al.* 2004.

Fig. 11. Overall mean amplitude values for Fourier harmonics 2–24, showing systematic differences between the 'older' and 'younger' dune sample sets from Ramsay Bay. Modified after Pye & Mazzullo 1994.

area are compared. The results show that, while both populations have broadly similar characteristics in terms of the low order harmonics (which reflect gross grain shape), there are significant differences in terms of the higher order harmonics (which reflect the finer surface textural detail).

A disadvantage of the type of data shown in Figures 10 and 11 is that it relates only to 2-D form data (i.e. the grain outline morphology is digitized in silhouette). This can result in loss of information relating to fine surface textural detail. However, other methods can provide complementary 3-D information. For example, Blott *et al.* (2004) describe a nitrogen-gas absorption technique which can provide sensitive measurements of the surface area of a range of particle types, including sand grains, the information being used as a proxy quantitative measure for surface textural detail. Initial tests have shown significant differences between sands from different depositional environments and different geographical origins (e.g. Fig. 12).

Analysis of particle coatings

Most studies of quartz sand grain surface textures have recommended cleaning of the grains with stannous chloride or other agent before examination, in order to make the surface features more visible (Krinsley & Doornkamp 1973; Smart & Tovey 1982). However, from a forensic point of view this amounts to throwing away potentially valuable evidence, since the composition and texture of any coatings and adhering particles may provide important discriminatory criteria. Quartz grains in the environment commonly have adhering particles and coatings

which are sensitive indicators of their environment of origin. For example, marine, brackish and freshwater diatoms, together with the presence or absence of salt crystals, provide indicators of environmental salinity (Fig. 13). Grain coatings, which may consist of clay minerals, iron and/or manganese oxides, calcium carbonate or organic matter, are often present and may give indications of type of soil or sedimentary environment. By first examining quartz (and other sand) grains without cleaning, and using the BSE mode, the distribution of coatings can be determined and their elemental composition identified by EDX (Fig. 14 & 15). Standardless semi-quantitative chemical analysis can be performed by raster analysis scans of the coating material, and comparisons made with control material. Such coatings, both natural and experimental, can provide useful environmental tracers (Evans and Tokar 2000).

In the case of pebbles and small rock particles, coatings may include fragments of lichen, algae, diagenetic cement, desert varnish or pollution patina which may provide useful clues to origin (e.g. Wilson 1978; Jones *et al.* 1981; Dorn & Oberlander 1981; Schiavon 2002).

Non-quartz minerals and other particles

Although quartz has been of primary interest in SEM surface textural studies, other minerals have been studied to varying degrees. These include heavy minerals (Setlow & Karpovich 1972), micas (Katz & Pilkey 1987), flint, chert and obsidian (Linde 1986), volcanic ash particles (Marshall 1987), and clay minerals and clay aggregates (McHardy & Birnie 1987). Studies have been carried out on silt-sized particles (Krinsley & McCoy 1978; Pye 1984*b*) as well as sand, but only a few studies have also considered gravel-sized particles (e.g. Krinsley & Donahue 1968*b*). From a forensic comparison point of view, less common

Fig. 13. BSE image showing a freshwater diatom within the clay coating on a sand grain recovered from the lungs of a murder victim (uncoated sample).

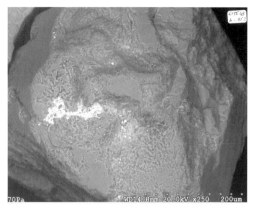

Fig. 15. BSE image showing partial coating of Ti–Cr oxide (white) and patchy euhedral crystal face development (uncoated sample).

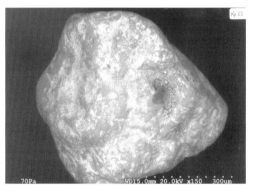

Fig. 14. BSE image of quartz grain with extensive Fe, Mn and Ti oxide coatings. Note clay in large pit on mid-right (uncoated sample).

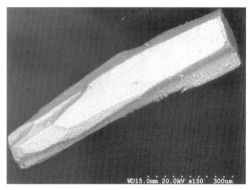

Fig. 16. BSE image of part of a calcite inoceramid prism, derived from the shell of a Cretaceous bivalve (uncoated sample).

constituents may be more important than the quartz. For example, Figure 16 illustrates part of a calcite inoceramid prism, which once formed part of the shell of a Cretaceous bivalve), found in mud taken from a murder suspect's car. Such particles were also found to be relatively abundant on a track near the body deposition site where Chalk rock had recently been imported to improve trafficability. Other particles of possible forensic interest include charcoal, wood fragments of varying types and pollen (Scott & Collinson 1978; Moore *et al.* 1991).

Search and identification of 'unusual' particles

As noted above, sediments and soil often contain unusual particles which result from human activities such as mining, construction, industrial manufacture and waste disposal. These often have a fairly restricted geographical distribution, and individual assemblages may be 'unique' to quite small areas. For example, many alluvial, estuarine and some coastal deposits are contaminated with metallic grains, alloys and compounds of Pb, Zn, Cu, Fe, Cr, Sb and Au, (e.g. Parkman *et al.* 1996; Large *et al.* 2001; Chapman *et al.* 2002; Langhmi & Watt 2003; Moles *et al.* 2003; Pirrie *et al.* 2003). Many sediments and soils also contain deposited airborne contaminants, notably industrial fly ash, which may take one of several forms: siliceous glassy spheres, metallic spherules and carbonaceous particles (Watt 1998). These particles are highly suited to examination and analysis by SEM/EDX and EMPA. BSE imaging, X-ray mapping, EDX spot analyses and automatic particle classification provide a useful routine method of examination (Pirrie *et al.* 2004), although in most circumstances other techniques

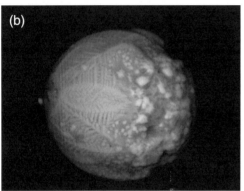

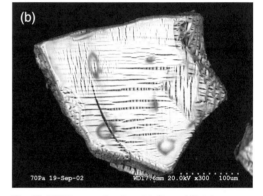

Fig. 17. BSE images of: (**a**) a Ce–La-rich spherical particle on the surface of a taping sample, and (**b**) the same particle at higher magnification showing fine surface textural detail (uncoated sample).

Fig. 18. BSE images showing: (**a**) Ca- and P-rich particles (white) and vegetation matter fragments (dark) recovered from the intestine of a child murder victim, and (**b**) a higher magnification image of one of the Ca–P particles showing apparent 'plywood' texture (uncoated sample).

such as X-ray diffraction (XRD), inductively coupled plasma atomic and optical emission spectrometry (ICP-AES and ICP-OES) optical microscopy and laser Raman spectroscopy will also be required to achieve full petrographic and chemical characterization.

BSE imaging of uncoated samples in a VP-SEM provides a rapid method of scanning dust samples and tapings samples taken from clothing, furniture, vehicles etc. for unusual (or 'exotic') particles which may be of forensic interest. For example, Figure 17 shows images of a spherical particle, composed mainly of the rare-earth elements Ce and La, recovered from a taping of clothing in a murder investigation. Such particles are generated in considerable numbers by heavy smokers who use certain types of disposable cigarette lighter. A number of unusual particles, composed largely of the elements Ca and P, recovered from the intestinal contents of a murdered child suspected of having been fed some form of 'potion' prior to death, are shown in Figure 18. The texture of the particles shows similarities to the 'twisted plywood' texture seen in some samples of

bone, ivory and teleost scales (Giraud-Guille 1988), although their exist origin remains undetermined. Figure 19 shows a small Pb-rich particle on the collar of a blood-stained shirt recovered from a location where human remains are believed to have been buried and later removed. The victim, some of whose dismembered remains were eventually recovered elsewhere, was thought to have been shot in the head, although the head was never recovered. DNA analysis of blood on the shirt confirmed that it belonged to the victim.

The sample of clothing shown in Figure 20 was taken from a crime suspect who was thought to have created a fire to destroy incriminating evidence. EDX analysis of the particles on the clothing fibres showed that they consisted overwhelmingly of carbon with traces of sulphur and chlorine, consistent with having been in proximity to a fire.

Further examples are a concentration of bluebell pollen grains within part of an anther recovered from the shoe of a person suspected of walking to a crime

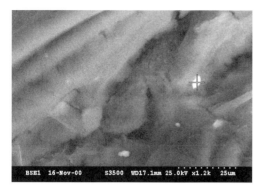

Fig. 19. BSE image showing Pb-rich particle (white) on part of a shirt from a suspected shooting victim (uncoated sample).

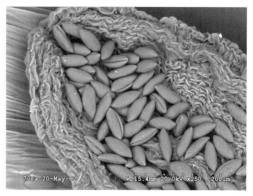

Fig. 21. BSE image of bluebell pollen grains and part of an anther recovered from footwear (uncoated sample).

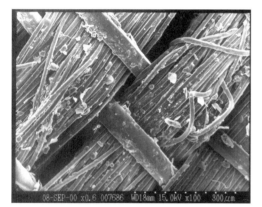

Fig. 20. BSE image showing smoke/fire debris particles on clothing fibres (uncoated sample).

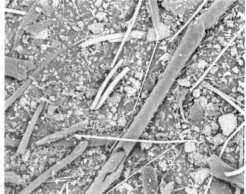

Fig. 22. BSE image of dust sample collected within a contaminated industrial building (uncoated sample).

scene through an area of woodland containing numerous bluebells (Fig. 21) and a tape-lift of a dust deposit in industrial premises where serious damage to sensitive equipment had occurred (Fig. 22). Analysis of the debris revealed the presence of large numbers of calcium silicate particles indistinguishable in texture and composition from those found in dust associated with concrete paving slab cutting outside the building.

Determination of sample mineralogical composition

A frequently used method in sediment and soil comparison studies is determination of the modal mineralogy. This has traditionally been accomplished by point counting of grains under a binocular or transmitted light microscope (Milner 1962; Tickell 1965; Galehouse 1971), although in the case of silts and clays it may also sometimes be undertaken using

semi-quantitative or quantitative X-ray diffraction (Wilson 1987; Moore & Reynolds 1997). A sediment or soil is normally composed of several different individual minerals, lithic (rock) fragments, biological material including microfossils, and sometimes non-crystalline natural materials, such as glass, or particles of human origin such as paint, brick, concrete, tile etc. Studies may focus on one or more of these groups of particles, such as the heavy mineral or light mineral fraction after heavy liquid separation, or they may consider the entire particle population. This type of study has been carried out on numerous occasions in relation to forensic geological investigations (e.g. Graves 1979; Demmelmeyer & Adam 1995). However, in the past 10 years increasing use has been made, both in geological and forensic studies, of automated SEM-EDX and microprobe analysis (Hunt *et al.* 1992; Watt 1990, 1998; Sitzmann *et al.* 1999; Pirrie *et al.* 2004).

An example of a small automated particle analysis

Table 1. *Example data file for automated particle analysis (stub 4, particles 1–14). Data from Croft (2003)*

ID	Stage x	Stage y	Stage z	Wt% O	Wt% Al	Wt% Fe	Wt% Si	Wt% S	Wt% Zr	Wt% Ti	Wt% Ca	Wt% Ni	Class	Size	Shape
1	18.30	11.26	13.44	8.55	3.45	85.4	2.56	0	0	0	0	0	Fe*	1.17	1.11
2	17.10	11.71	13.44	30.8	2.36	47.8	3.14	0.77	1.92	13.1	0	0	FeSiTi	1.17	1.11
3	17.10	11.71	13.44	32.1	2.41	46.1	3.01	0.72	2.88	12.6	0	0	FeSiTi	1.17	0.93
4	17.10	11.71	13.44	31.3	2.24	45.9	2.62	0.63	2.05	15.5	0	0	FeSiTi	0.68	0.57
5	17.10	11.71	13.44	30.1	2.48	49.5	2.82	0.49	1.95	12.6	0	0	FeSiTi	0.96	0.83
6	17.10	11.71	13.44	31.1	3.03	56.9	4.03	0.64	3.28	0.94	0	0	FeSiTi	0.96	0.83
7	17.10	11.71	13.44	28.8	2.71	49.0	3.43	0.59	1.78	13.5	0	0	FeSiTi	1.17	0.93
8	17.10	11.71	13.44	31.0	2.54	46.1	3.06	0.84	2.8	13.6	0	0	FeSiTi	1.17	1.11
9	17.10	11.71	13.44	29.3	2.27	48.8	2.72	0.65	2.28	13.8	0	0	FeSiTi	1.79	0.95
10	17.10	11.71	13.44	28.7	2.33	50.3	2.91	0.58	2.76	11.9	0.45	0	FeSiTi	0.96	0.92
11	18.89	11.71	13.44	31.7	1.29	49.7	3.07	0	0	14.1	0	0	FeSiTi	1.52	0.97
12	18.89	11.71	13.44	9.72	2.02	9.06	5.15	0	0	0	0.63	73.4	NiFeSi	6.10	1.60
13	16.51	12.15	13.44	24.9	0.68	40.7	2.33	0	0	14.5	16.6	0	FeSiTi	0.68	0.57
14	16.51	12.15	13.44	19.5	1.21	43.7	2.2	0	0	9.86	23.4	0	FeSiTi	0.96	0.83

Table 2. *Size and shape parameters of the three SEM stubs – bulk source soil, pink carpet material and beige carpet material. Data from Croft (2003)*

Sample	Bulk source soil Day 0 Size	Shape	Pink carpet soil sample (bP–BB-10) Size	Shape	Beige carpet soil sample (bB–BB-10) Size	Shape
Minimum	3.01	0.57	3.01	0.57	3.01	0.57
Maximum	31.32	5.75	34.89	5.61	54.25	12.80
Mean	6.15	1.03	5.32	0.96	6.21	1.04
SD	4.86	0.70	4.16	0.66	5.50	0.89

data file created using the Hitachi VP-SEM and Oxford Instruments ISIS system at the Royal Holloway Electron Microscopy Laboratory is shown in Table 1. After initial reconnaissance and appraisal, the computer software was 'programmed' to record the concentrations of selected elements and to classify the particles containing the elements of interest according to their relative concentrations. In this instance the particles of interest were mainly Fe-, Ti- and Ni-bearing metallic particles and silicates. Based on re-examination of the raw data, it is possible to reclassify particles off-line on completion of the analysis, if necessary. The size and shape (in terms of a shape factor) of each particle of interest was also recorded. Table 2 shows summary size and shape data for another simple example involving a comparison of soil deposits on two carpets with a soil source. The frequency distribution or particles in terms of recorded size and shape classes for the three samples showed close similarity when compared as histograms (Fig. 23).

Conclusions

SEM and EDX analysis have an important part to play in forensic investigations which involve soils, sediments, rocks and dust deposits. Several attributes of these materials can be examined and quantified using these techniques, including fabric (internal microstructure), particle and elemental distribution, modal mineral composition inferred from elemental data, and individual particle size, shape and surface texture. To date, however, forensic investigators have not made full use of the potential of such information.

Very small amounts of particulate material can be located and analysed using BSE imaging and EDX, often *in situ* and with a minimal effect on the integrity of an exhibit. However, the information obtained is only of use if it can be put into context. A number of comparative particulate databases already exist and others are under further development.

Integrated CC-SEM-EDX systems have made automated analysis a sensible proposition for routine

A   **B**

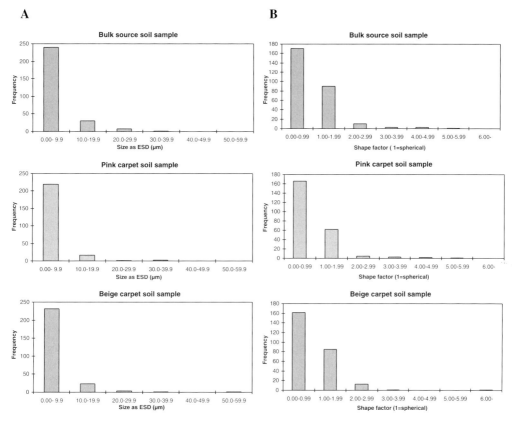

Fig. 23. Frequency histograms showing distribution of: (**a**) particle sizes, and (**b**) particle shapes in soil deposits from two carpets compared with a source soil, determined by automated analysis in a computer-controlled variable pressure SEM (Hitachi S3000-N SEM, equipped with Oxford Instruments Series 300 ISIS X-ray analysis software). Data from Croft (2003).

examination of large numbers of samples, but results should always be checked by manual examination of selected specimens.

Many earlier sedimentological studies of particle shape and surface texture have suffered from biased or otherwise unrepresentative results due to inappropriate grain selection procedures, two few particles being analysed, the subjective nature of the procedures used to characterize and classify the grains, and over-reliance on qualitative evaluation of the results. Methods which are overly subjective and prone to high operator variation are unsuitable for use in a forensic context. Techniques are now available which permit analysis of large numbers of grains and quantitative evaluation of the results, but further development and testing is required. However, large numbers of particles are not always required to provide significant forensic evidence, notably when dealing with the presence of unusual ('exotic') particle types which have very restricted distribution in the environment, either singly or as part of a characteristic assemblage. Such particles often provide the strongest evidence of a sample 'match'.

Technical support provided by M. Faulkner and P. Goggin at the Royal Holloway Electron Microscope Unit is gratefully acknowledged. Additional technical assistance and data were provided by D. Croft and S. Blott.

References

ANDERSON, C. A. (ed.) 1973. *Microprobe Analysis*. Wiley, Chichester.

ARDENNE, M. VON 1938. Das Elektronen-Rastermikroskop – Theoretische Grundlagen. *Zeitschrift fur Physik*, **109**, 553–572.

BAKER, H. W. 1976. Environmental sensitivity of submicroscopic surface textures on quartz sand grains – a statistical evaluation. *Journal of Sedimentary Petrology*, **46**, 871–880.

BENNETT, R. H., BRYANT, W. R. & HULBERT, M. H. (eds)

1991. *Microstructure of Fine-Grained Sediments: From Mud to Shale*. Springer Verlag, Berlin.

BIEDERMAN, E. W. 1962. Destruction of shoreline environments in New Jersey. *Journal of Sedimentary Petrology*, **32**, 181–200.

BIRKS, H. J. B. & BIRKS, H. H. 1980. *Quaternary Palaeoecology*. Arnold, London.

BLACKWELDER, P. & PILKEY, O. 1972. Electron microscopy of quartz grain surface textures: the US eastern Atlantic coastal margin. *Journal of Sedimentary Petrology*, **42**, 520–526.

BLOTT, S. J., AL-DOUSARI, A., PYE, K. & SAYE, K. 2004. Three-dimensional characterization of sand grain shape and surface texture using a nitrogen gas adsorption technique. *Journal of Sedimentary Research*, **74**, 156–159.

BULL, P. A. 1978. A quantitative approach to scanning electron microscope analysis of cave sediments. *In*: WHALLEY, W.B. (ed.) *Scanning Electron Microscopy in the Study of Sediments – A Symposium*. Geo Abstracts, Norwich, 201–206.

BULL, P. A. 1981. Environmental reconstruction by electron microscopy. *Progress in Physical Geography*, **5**, 368–397.

BULL, P. A. 1986. Procedures in environmental reconstruction by SEM analysis. *In*: SIEVEKING, G. DE G. & HART, M. B. (eds) *The Scientific Study of Flint and Chert*. Cambridge University Press, Cambridge, 221–226.

BULL, P. A., WHALLEY, W. B. & MAGEE, A. W. 1986. *An Annotated Bibliography of Environmental Reconstruction by SEM 1962–1985*. British Geomorphological Research Group Technical Bulletin, **35**. Geo Abstracts, Norwich.

CATER, J. M. L. 1984. An application of scanning electron microscopy of quartz sand surface textures to the environmental diagnosis of Neogene carbonate sediments, Finestrat Basin, south-east Spain. *Sedimentology*, **31**, 717–731.

CHAPMAN, R., LEAKE, B. & STYLES, M. 2002. Microchemical characterization of alluvial gold grains as an exploration tool. *Gold Bulletin*, **35**, 53–65.

CROFT, D. 2003. *Forensic geoscience: Development of techniques for soil analysis*. PhD thesis, Royal Holloway, University of London.

CULVER, S. J., BULL, P. A., CAMPBELL, S., SHAKESBY, R. A. & WHALLEY, W. B. 1983. Environmental discrimination based on quartz grain surface textures: a statistical investigation. *Sedimentology*, **30**, 129–136.

DEMMELMEYER, H. & ADAM, J. 1995. Forensic investigation of soil and vegetable materials. *Forensic Science Review*, **7**, 120–136.

DERBYSHIRE, E. 1978. A pilot study of till microfabrics using the scanning electron microscope. *In*: WHALLEY, W. B. (ed.) *Scanning Electron Microscopy in the Study of Sediments – A Symposium*. Geo Abstracts, Norwich, 41–59.

DORN, R. I. & OBERLANDER, T. M. 1981. Rock varnish origin, characteristics, and usage. *Zeitschrift fur Geomorphologie*, **25**, 420–436.

DOWDESWELL, J. A. 1982. Scanning electron micrographs of quartz sand grains from cold environments examined using Fourier shape analysis. *Journal of Sedimentary Petrology*, **52**, 1315–323.

DRYDEN, A. L. 1931. Accuracy in percentage representation of heavy mineral frequencies. *Proceedings of the National Academy of the United States*, **17**, 233–238.

EHRLICH, R., KENNEDY, S. K. & BROTHERHOOD, C. D. 1987. Respective roles of Fourier and SEM techniques in analyzing sedimentary quartz. *In*: MARSHALL, J. R. (ed.) *Clastic Particles: Scanning Electron Microscopy and Shape Analysis of Sedimentary and Volcanic Clasts*. Van Nostrand Reinhold, New York, 292–301.

ELZENGA, W., SCHWAN, J, BAUMFALK, Y. A., VANDENBERGH, J. & KROOK, L. 1987. Grain surface characteristics of periglacial aeolian and fluvial sands. *Geologie en Mijnbouw*, **65**, 273–286.

EVANS, J. E. & TOKAR, F. J., Jr. 2000. Use of SEM/EDS and X-ray diffraction analyses for sand transport studies, Lake Erie, Ohio. *Journal of Coastal Research*, **16**, 926–933.

EVERHART, T. E. & THORNLEY, R. F. M. 1960. Wide-band detector for microampere low-energy electron currents. *Journal of Scientific Instruments*, **37**, 246–248.

FITZPATRICK, F. & THORNTON, J. I. 1974. Forensic science characterization of sand. *Journal of Forensic Sciences*, **4**, 460–475.

GALEHOUSE, J. S. 1971. Point counting. *In*: CARVER, R. E. (ed) *Procedures in Sedimentary Petrology*. Wiley Interscience, New York, 385–407.

GIRAUD-GUILLE, M. M. (1988) Twisted plywood architecture of collagen fibrils in human compact bone osteons. *Calcified Tissue International*, **42**, 167–180.

GOLDSTEIN, J. I. & YAKOWITZ, H. (eds) 1975. *Practical Scanning Electron Microscopy*. Plenum Press, New York.

GOLDSTEIN, J. I., NEWBURY, D. E., ECHLIN, P., JOY, D. C., FIORI, C. & LIFSHIN, E. 1981. *Scanning Electron Microscopy and X-ray Microanalysis*. 1st edition. Plenum Press, New York.

GOLDSTEIN, J. I., NEWBURY, D. E., ECHLIN, P., JOY, D. C., FIORI, C. & LIFSHIN, E. 2002. *Scanning Electron Microscopy and X-ray Microanalysis*. 3rd edition. Plenum Press, New York.

GOUDIE, A. S. & BULL, P. A. 1984. Slope process change and colluvial deposition in Swaziland: an SEM analysis. *Earth Surface Processes and Landforms*, **9**, 289–299.

GRANT, P. 1978. The role of the scanning electron microscope in cathodoluminescence petrology. *In*: WHALLEY, W. B. (ed.) *Scanning Electron Microscopy in the Study of Sediments – A Symposium*. Geo Abstracts, Norwich, 1–9.

GRAVES, W. J. 1979. A mineralogical soil classification technique for the forensic scientist. *Journal of Forensic Sciences*, **24**, 323–337.

HEINRICH, K. F. J. & NEWBURY, D. E. 1991. *Electron Probe Quantitation*. Plenum Press, New York.

HIGGS, R. 1979. Quartz grain surface features of Mesozoic–Cenozoic sands from the Labrador and western Greenland continental margins. *Journal of Sedimentary Petrology*, **49**, 599–610.

HINTON, R. W. 1995. Ion microprobe analysis in geology. *In*: POTTS, P. J., BOWLES, J. F. W., REED, S. J. B. & CAVE, M. R. (eds) *Microprobe Techniques in the Earth Sciences*. Chapman & Hall, London, 235–289.

HOLT, D. B., MUIR, M. D. P. R., GRANT, P. R. & BOSARVA, I.

M. (eds) 1974. *Quantitative Scanning Electron Microscopy*. Academic Press, London.

HUNT, A., JOHNSON, D. L., WATT, J. M. & THORNTON, I. (1992). Characterising the sources of lead in house dust by automated scanning electron microscopy. *Environmental Science and Technology*, **26**, 1513–1523.

JONES, D., WILSON, M. J. & McHARDY, W. J. 1981. Lichen weathering of rock forming minerals: application of scanning electron microscopy and microprobe analysis. *Journal of Microscopy*, **124**, 95–104.

JOY, D. C. 1974. Electron channeling patterns in the SEM. *In*: HOLT, D. B., MUIR, P. R., GRANT, P. R. & BOSARVA, I. M. (eds) *Quantitative Scanning Electron Microscopy*. Academic Press, London, 131–182.

JOY, D. B., ROMIG, A. D. & GOLDSTEIN, J. I. (eds) 1986. *Principles of Analytical Electron Microscopy*. Plenum Press, New York.

KATZ, S. D. & PILKEY, O. H. 1987. An analysis of detrital mica grain morphology in two North Carolina fluvial networks. *In*: MARSHALL, J. R. (ed.) *Clastic Particles: Scanning Electron Microscopy and Shape Analysis of Sedimentary and Volcanic Clasts*. Van Nostrand Reinhold, New York, 328–339.

KELLEY, S. P. 1995. Ar–Ar dating by laser microprobe. *In*: POTTS, P. J., BOWLES, J. F. W., REED, S. J. B. & CAVE, M. R. (eds) *Microprobe Techniques in the Earth Sciences*. Chapman & Hall, London, 327–358.

KRINSLEY, D. H. & DONAHUE, J. 1968*a*. Environmental interpretation of sand grain surface textures by electron microscopy. *Bulletin of the Geological Society of America*, **79**, 743–748.

KRINSLEY, D. H. & DONAHUE, J. 1968*b*. Pebble surface textures. *Geological Magazine*, **105**, 521–525.

KRINSLEY, D. H. & DOORNKAMP, J. C. 1973. *Atlas of Quartz Sand Surface Textures*. Cambridge University Press, Cambridge.

KRINSLEY, D. H. & FUNNELL, B. W. 1965. Environmental history of quartz sand grains from the Lower Middle Pleistocene of Norfolk, England. *Quarterly Journal of the Geological Society of London*, **121**, 435–461.

KRINSLEY, D. H. & HYDE, P. 1971. Cathodoluminescence studies of sediments. *In*: JOHARI, O. (ed.) Proceedings of the 4th Annual SEM Symposium, *IHT Research Institute, Chicago*. Scanning Electron Microscopy, SEM Inc., AMF O'Hare, 409–416.

KRINSLEY, D. H. & McCOY, F. 1978. Aeolian quartz sand and silt. *In*: WHALLEY, W. B. (ed.) *Scanning Electron Microscopy in the Study of Sediments – A Symposium*. Geo Abstracts, Norwich, 249–260.

KRINSLEY, D. H. & TAKAHASHI, T. 1962*a*. The surface texture of sand grains, an application of electron microscopy. *Science*, **135**, 923–925.

KRINSLEY, D. H. & TAKAHASHI, T. 1962*b*. Application of electron microscopy to geology. *Transactions of the New York Academy of Sciences*, **25**, 3–22.

KRINSLEY, D. H. & TAKAHASHI, T. 1964. A technique for the study of surface textures of sand grains with electron microscopy. *Journal of Sedimentary Petrology*, **34**, 423–436.

KRINSLEY, D. H. & TOVEY, N. K. 1978. Cathodoluminescence in quartz sand grains. *In*: JOHARI, O. (ed) *Scanning Electron Microscopy*. SEM Inc., AMF O'Hare, Illinois, **1**, 887–894.

KRINSLEY, D. H. & TRUSTY, P. 1986. Sand grain surface textures. *In*: SIEVEKING, G. DE G. & HART, M.B. (eds) *The Scientific Study of Flint & Chert*. Cambridge University Press, Cambridge, 201–207.

KRINSLEY, D. H., PYE, K., BOGGS, S., JR, & TOVEY, N. K. 1998. *Backscattered Scanning Electron Microscopy and Image Analysis of Sediments and Sedimentary Rocks*. Cambridge University Press, Cambridge.

KRINSLEY, D. H., PYE, K. & KEARSLEY, A. T. 1983. Application of backscattered electron microscopy in shale petrology. *Geological Magazine*, **120**, 109–208.

LANGHMI, H.W. & WATT, J. 2003. Evaluation of computer-controlled SEM in the study of metal-contaminated soils. *Mineralogical Magazine*, **67**, 219–231.

LARGE, D. J., FORTEY, N. J., MILODOWSKI, A. E., CHRISTY, A. G. & DODD, J. 2001. Petrographic observations of iron, copper and zinc sulphides in freshwater canal sediments. *Journal of Sedimentary Research*, **71**, 61–69.

LINDE, K. 1986. Scanning electron micrographs of quartz, flint and obsidian grains after experimental glacial, subaqueous or aeolian transportation. *In*: SIEVEKING, G. DE G. & HART, M. B. (eds) *The Scientific Study of Flint and Chert*. Cambridge University Press, Cambridge, 209–220.

LERIBAULT, L. 1977. *L'Exoscopie des Quartz*. Masson, Paris.

LERIBAULT, L. 1978. The exoscopy of quartz sand grains. *In*: WHALLEY, W. B. (ed.) *Scanning Electron Microscopy in the Study of Sediments – A Symposium*. Geo Abstracts, Norwich, 319–328.

LONG, J. V. P. 1962. Recent advances in electron-probe analysis. *Advances in X-ray Analysis*, **6**, 276–290.

McCRONE, W. C. 1992. Forensic soil examination. *Microscope*, **40**, 109–121.

McCRONE, W. C., DELLY, J. G. & PALENIK, S. 1973. *The Particle Atlas*. Vol. III. *The Electron Microscopy Atlas*. Ann Arbor Science Publishers Inc., Ann Arbor, 575–794.

McHARDY, W. J. & BIRNIE, A. C. 1987. Scanning electron microscopy. *In*: WILSON, M. J. (ed.) *A Handbook of Determinative Methods in Clay Mineralogy*. Blackie, Glasgow, 174–208.

MAHANEY, W. C. 2002. *Atlas of Sand Grain Surface Textures and Applications*. Oxford University Press, Oxford.

MAHANEY, W. C., DIRSZOWKSY, R. W., MILNER, M. W., MENZIES, J., STEWART, A., KALM, V. & BEZADA, M. 2004. Quartz microtextures and microstructures owing to deformation of glaciolacustrine sediments in the northern Venezuelan Andes. *Journal of Quaternary Science*, **19**, 23–33.

MARGOLIS, S. V. & KENNETT, J. P. 1971. Cenozoic palaeoglacial history of Antarctica recorded in sub-Antarctic deep-sea cores. *American Journal of Science*, **270**, 1–36.

MARGOLIS, S. V. & KRINSLEY, D. H. 1971. Sub-microscopic frosting on eolian and subaqueous sand grains. *Geological Society of America Bulletin*, **82**, 3395–3406.

MARGOLIS, S. V. & KRINSLEY, D. H. 1973. Depositional histories of sand grains from surface textures. A comment. *Nature*, **245**, 30–31.

MARSHALL, J. R. (ed.) 1987. *Clastic Particles: Scanning*

Electron Microscopy and Shape Analysis of Sedimentary and Volcanic Clasts. Van Nostrand Reinhold Company, New York.

MARUMO, Y. & SUGITA, R. 2001. Forensic examination of soil evidence. *Proceedings of the 13th Interpol Forensic Science Symposium, 16–19 October 2001, Lyon, France*, 1–11.

MAZZULLO, J. & RITTER, C. 1991. Influence of sediment source on the shapes and surface textures of glacial quartz sand grains. *Geology*, **19**, 384–388.

MCKINLEY, T. D., HEINRICH, K. F. J & WITTRY, D. B. (eds) 1966. *The Electron Microprobe.* Wiley, New York.

MCVICAR, M. J. & GRAVES, W. J. 1997. The forensic comparison of soils by automated scanning electron microscopy. *Canadian Society of Forensic Scientists Journal*, **30**, 241–261.

MEEKS, N. D. 1990. Trace element detection in the SEM by X-ray induced fluorescence. *Microscopy and Analysis*, March, 23–27.

MILNER, H. B. 1962. *Sedimentary Petrography.* Vols 1 & 2. 4th edition. Macmillan, New York.

MOLES, N. R., BETZ, S. M., MCREADY, A. J. & MURPHY, P. J. 2003. Replacement and authigenic mineralogy of metal contaminants in stream and estuarine sediments at Newtownards, Northern Ireland. *Mineralogical Magazine*, **67**, 305–324.

MOORE, D. M. & REYNOLDS, R. C., JR. 1997. *X-ray Diffraction and the Identification and Analysis of Clay Minerals.* 2nd edition. Oxford University Press, Oxford.

MOORE, P. D., WEBB, J. A. & COLLINSON, M. E. 1991. *Pollen Analysis.* 2nd edition. Blackwell, Science, Oxford.

NEWBURY, D. E., JOY, D. C., ECHLIN, P., FIORI, C. E. & GOLDSTEIN, J. I. 1986. *Advanced Scanning Electron Microscopy and X-ray Microanalysis.* Plenum Press, New York.

OATLEY, C. W. 1966. The scanning electron microscope. *Science Progress, London*, **54**, 483–495.

OATLEY, C. W. 1972. *The Scanning Electron Microscope.* Cambridge University Press, Cambridge.

OATLEY, C. W. & EVERHART, T. E. 1957. The examination of p-n junctions with the scanning electron microscope. *Journal of Electronics*, **2**, 568–570.

OATLEY, C. W., NIXON, W. C. & PEASE, R. F. W. 1965. Scanning electron microscopy. *Advances in Electronics and Electron Physics*, **21**, 181–247.

O'BRIEN, N. R. 1968. Electron microscope of black shale fabric. *Naturwissenschaften*, **10**, 490–491.

O'BRIEN, N. R. 1981. SEM study of shale fabric – a review. *In*: Scanning Electron Microsocopy. SEM Inc., AMF O'Hare, **1**, 569–575.

O'BRIEN, N. R. & SLATT, R. M. 1990. *Argillaceous Rock Atlas.* Springer Verlag, Berlin.

ORFORD, J. D. & WHALLEY, W. B. 1983. The use of fractal dimension to quantify the morphology of irregular-shaped particles. *Sedimentology*, **30**, 655–668.

ORFORD, J. D. & WHALLEY, W. B. 1987. The quantitative description of highly irregular sedimentary particles: the use of the fractal dimension. *In*: MARSHALL, J. R. (ed.) *Clastic Particles: Scanning Electron Microscopy and Shape Analysis of Sedimentary and Volcanic Clasts.* Van Nostrand Reinhold, New York, 267–280.

OSIPOV, V. I. & SOKOLOV, V. N. 1978. Microstructure of Recent clay sediments examined by scanning electron microscopy. *In*: WHALLEY, W. B. (ed.) *Scanning Electron Microscopy in the Study of Sediments – A Symposium.* Geo Abstracts, Norwich, 29–40.

PALENIK, S. J. 1998. Microscopy and microchemistry of physical evidence. *In*: SAFERSTEIN, R. (ed.) *Forensic Science Handbook.* Vol. II. Prentice Hall, New Jersey, 161–208.

PARKMAN, R. H., CURTIS, C. D., VAUGHAN, D. J. & CHARNOCK, J. M. 1996. Metal fixation and mobilization in the sediments of the Afon Goch estuary, Anglesey – Dulas Bay, Anglesey. *Applied Geochemistry*, **11**, 203–210.

PETRACO, N. 1994a. Microscopic examination of mineral grains in forensic soil analysis – Part 1. *American Laboratory*, **26** (6), 35–40.

PETRACO, N. 1994b. Microscopic examination of mineral grains in forensic soil analysis – Part 2. *American Laboratory*, **26** (14), 33–39.

PETTIJOHN, F. J. 1949. *Sedimentary Rocks.* Harper & Bros, New York.

PIRRIE, D., POWER, M. R., ROLLINSON, G., CAMM, G. S., HUGHES, S. H., BUTCHER, A. R. & HUGHES, P. 2003. The spatial distribution and source of arsenic, copper, tin and zinc within surface sediments of the Fal Estuary, Cornwall. *Sedimentology*, **50**, 579–595.

PIRRIE, D., BUTCHER, A. R., POWER, M. R., GOTTLIEB, P. & MILLER, G. L. 2004. Rapid quantitative mineral and phase analysis using automated scanning electron microscopy (QuemSCAN): potential applications in forensic geoscience. *In*: PYE, K. & CROFT, D. J. (eds) *Forensic Geoscience: Principles, Techniques and Applications.* Geological Society, London, Special Publications, **232**, 123–136.

PORTER, J. J. 1962. Electron microscopy of sand surface textures. *Journal of Sedimentary Petrology*, **32**, 124–135.

POTTS, P. J., BOWLES, J. F. W., REED, S. J. B. & CAVE, M. R. (eds) 1995. *Microprobe Techniques in the Earth Sciences.* Chapman & Hall, London.

PYE, K. 1984a. Rapid estimation of porosity and mineral abundance in backscattered electron images using a simple SEM image analyzer. *Geological Magazine*, **121**, 81–84.

PYE, K. 1984b. SEM investigations of quartz silt micro-textures in relation to the source of loess. *In*: PECSI, M. (ed.) *Lithology and Stratigraphy of Loess and Paleosols.* Hungarian Academy of Sciences, Budapest, 139–151.

PYE, K. & KRINSLEY, D. H. 1983. Mudrocks examined by backscattered electron microscopy. *Nature*, **301**, 412–413.

PYE, K. & KRINSLEY, D. H. 1984. Petrographic examination of sedimentary rocks in the SEM using backscattered electron detectors. *Journal of Sedimentary Research*, **54**, 877–888.

PYE, K. & MAZZULLO, J. 1994. Effects of tropical weathering on quartz grain shape: an example from northeastern Australia. *Journal of Sedimentary Research* (A), **64**, 500–507.

PYE, K. & WINDSOR-MARTIN, J. 1983. SEM analysis of shales and other geological materials using the Philips Multi-Function Detector System. *The Edax Editor*, **13**, 406.

REED, S. J. B. 1975. *Electron Microprobe Analysis*. 1st edition. Cambridge University Press, Cambridge.

REED, S. J. B. 1993. *Electron Microprobe Analysis*. 2nd edition. Cambridge University Press, Cambridge.

REED, S. J. B. 1996. *Electron Microprobe Analysis and Scanning Electron Microscopy in Geology.* Cambridge University Press, Cambridge.

ROBINSON, B. W. & NICKEL, E. 1979. A useful new technique for mineralogy: rhe BSE / low vacuum mode of SEM operation. *American Mineralogist*, **64**, 1322–1328.

ROBINSON, V. N. E. 1980. Imaging with backscattered electrons in a scanning electron microscope. *Scanning*, **3**, 15–26.

SCHIAVON, N. 2002. Biodeterioration of calcareous and granitic building stones in urban environments. *In*: SIEGESMUND, S., WEISS, T. & VOLLBRECHT, A. (eds) *Natural Stone, Weathering Phenomena, Conservation Strategies and Case Studies*. Geological Society, London, Special Publications, **205**, 195–205.

SCOTT, A. C. & COLLINSON, M. E. 1978. Organic sedimentary particles: results from scanning electron microscope studies of fragmentary plant material. *In*: WHALLEY, W. B. (ed.) *Scanning Electron Microscopy in the Study of Sediments – A Symposium*. Geo Abstracts, Norwich, 137–167.

SETLOW, L. W. 1978. Age determination of reddening in coastal dunes in northwest Florida, USA, by use of scanning electron microscopy. *In*: WHALLEY, W. B. (ed.), *Scanning Electron Microscopy in the Study of Sediments – A Symposium*. Geo Abstracts, Norwich, 283–305.

SETLOW, L. W. & KARPOVICH, R. P. 1972. 'Glacial' microtextures on quartz and heavy mineral sand grains from the littoral environment. *Journal of Sedimentary Petrology*, **42**, 864–875.

SITZMANN, B., KENDALL, M., WATT, J. & WILLIAMS, I. 1999. Characterisation of airborne particles in London by computer-controlled scanning electron microscopy. *Science of the Total Environment*, **241**, 63–73 .

SMART, P. & TOVEY, N. K. 1981. *Electron Microscopy of Soils and Sediments: Examples*. Clarendon Press, Oxford.

SMART, P. & TOVEY, N. K. 1982. *Electron Microscopy of Soils and Sediments: Techniques*. Clarendon Press, Oxford.

TICKELL, F. G. 1965. *The Techniques of Sedimentary Mineralogy: Developments in Sedimentology*. Vol. 4. Elsevier, Amsterdam.

TOVEY, N. K. 1980. A digital computer technique for orientation analysis of micrographs of soil fabric. *Journal of Microscopy*, **120**, 303–315.

TOVEY, N. K. & HOUNSLOW, M. W. 1995. Quantitative microporosity and orientation analysis in soils and sediments. *Journal of the Geological Society, London*, **152**, 119–129.

TOVEY, N. K. & KRINSLEY, D. H. 1992. Mapping the orientation of fine-grained materials in soils and sediments.

Bulletin of the International Association of Engineering Geologists, **46**, 93–101.

TOVEY, N. K. & WONG, K. Y. 1978. Preparation, selection and interpretation problems in SEM studies of sediments. *In*: WHALLEY, W. B. (ed.) *Scanning Electron Microscopy in the Study of Sediments – A Symposium*. Geo Abstracts, Norwich, 181–199.

TOVEY, N. K., EYLES, N. & TURNER, R. 1978. Sand grain selection procedures for observation in the SEM. *In*: JOHARI, O. (ed.) Scanning Electron Microscopy, 1978, volume 1. SEM Inc., AMF O'Hare, **1**, 393–400.

VINCENT, P. J. 1976. Some periglacial deposits near Aberystywth. *Biultyn Periglacjalny*, **25**, 59–65.

WATT, J. 1990. Automated feature analysis in the scanning electron microscope. *Microscopy and Analysis*, **15**, 25–28.

WATT, J. 1998. Automated characterization of individual carbonaceous fly ash particles by computer controlled scanning electron microscopy: analytical methods and critical review of alternative techniques. *Water, Air and Soil Pollution*, **106**, 309–327.

WATT, J. M. & JOHNSON, D. L. 1987. Characterization of dust by scanning electron microscopy with energy-dispersive spectroscopy. *In*: THORNTON, I. & CULBARD, E. (eds) *Lead in the Home Environment*. Science Reviews Ltd, London, 85–96.

WAUGH, B. 1978. Diagenesis in continental red beds as revealed by scanning electron microscopy. *In*: WHALLEY, W. B. (ed.) *Scanning Electron Microscopy in the Study of Sediments – A Symposium*. Geo Abstracts, Norwich, 329–346.

WELTON, J. E. 1984. *SEM Petrology Atlas*. The American Association of Petroleum Geologists, Tulsa.

WHALLEY, W. B. (ed.) 1978. *Scanning Electron Microscopy in the Study of Sediments – A Symposium*. Geo Abstracts, Norwich.

WHALLEY, W. B. & LANGWAY, C. C., JR. 1984. A scanning electron microscope examination of subglacial quartz grains from Camp Century Core, Greenland – a preliminary study. *Journal of Glaciology*, **25**, 125–131.

WILSON, M. J. 1987. X-ray powder diffraction methods. *In*: WILSON, M. J. (ed.) *A Handbook of Determinative Methods in Clay Mineralogy*. Blackie, Glasgow and London, 26–98.

WILSON, P. 1978. Quartz overgrowths from the Millstone Grit sandstones (Namurian) of the southern Pennines as revealed by scanning electron microscopy. *Proceedings of the Yorkshire Geological Society*, **42**, 289–295.

WOLF, E. D. & EVERHART, T. E. 1969. Annular diode detector for high angular resolution pseudo-kikuchi pattern. *In*: Scanning Electron Microscopy, SEM Inc., AMF O'Hare, 43–44.

ZWORYKIN, V. K., HILLIER, J. & SNYDER, R. L. 1942. A scanning electron microscope. *Bulletin of the American Society for Testing of Materials*, **117**, 15–23.

Rapid quantitative mineral and phase analysis using automated scanning electron microscopy (QemSCAN); potential applications in forensic geoscience

DUNCAN PIRRIE[1], ALAN R. BUTCHER[2,3], MATTHEW R. POWER[1], PAUL GOTTLIEB[2,3] & GAVIN L. MILLER[2]

[1]*Camborne School of Mines, School of Geography, Archaeology and Earth Resources, University of Exeter, Redruth, Cornwall TR15 3SE, UK (e-mail: dpirrie@csm.ex.ac.uk)*
[2]*CSIRO Minerals, Queensland Centre for Advanced Technologies, Technology Court, Pullenvale, QLD 4069, Australia*
[3]*Present address: Intellection Pty Ltd, Milton, Brisbane, Queensland, QLD 4064, Australia*

Abstract: QemSCAN is a scanning electron microscope (SEM) system, initially designed to support the mining industry by providing rapid automated quantitative mineral analyses. The system is based upon Carl Zeiss SEMs fitted with up to four light-element energy dispersive X-ray spectrometers. Representative subsamples are mounted into either resin or wax blocks and polished prior to analysis, or can be mounted onto carbon tape. During analysis, X-ray spectra are collected at a user-defined pixel spacing and are acquired very rapidly (c. 10 ms per pixel). The measured spectra are automatically compared against a database of known spectra and a mineral or phase name is assigned to each measurement point by the QemSCAN computer software programs. In this way the near-surface qualitative elemental composition of each particle is systematically mapped, assigned to a mineral name or chemical compound/species, and digital pixel maps of each particle are created. Depending upon a range of parameters, including the particle size and the user-defined pixel spacing (which can vary between 0.20 μm and 25 μm), approximately 1000 particles, each 1–10 μm in size, can be measured per hour using a 1 μm pixel spacing. In addition to providing a qualitative elemental analysis and mineralogical or phase assignment for each particle, data relating to particle size, shape and calculated specific density are also generated. In this study, the potential application of this automated SEM system in forensic geoscience was evaluated by the analysis of: (1) a series of soil samples, and (2) a series of dust samples from an industrial complex. In both case studies, the mineralogy/phase composition of each sample analysed was found to be distinctive. In addition, textural data for the soil samples and particle shape data for the dust samples show that they can be clearly distinguished. Automated SEM using QemSCAN has clear potential application in the analysis of soil or other trace evidence in forensic case work.

The analysis of the mineralogy or phase composition of soil and other trace evidence, such as paint flakes or glass, can provide the key evidential data in some serious crime investigations (e.g. Murray 2000). However, although it is widely recognized that soil samples can provide key forensic evidence, soil analysis is generally regarded as complicated (e.g. Sugita & Marumo 1996), and the large number of samples required for soil analysis can result in the need for rapid screening methods (Sugita & Marumo 1996, 2001). Consequently, some soil analyses have focused on parameters such as soil colour or particle size distribution. However, the mineralogy and/or phase composition of soil or other trace evidence can also be of major significance in forensic geoscience (e.g. Brown *et al.* 2002). Traditionally, mineralogical analysis of soil samples has utilized optical light microscopy or X-ray diffraction, and detailed protocols for sample examination have been developed (e.g. Graves 1979; Petraco & DeForest 1993; Brown *et al.* 2002). However, soil mineralogy is not cur-

rently widely used in forensic investigations, in part due to the limitations of data generated using optical microscopy including: (1) operator dependence for sample analysis, and (2) the time, and therefore cost, required for detailed quantitative analysis.

In other areas of forensic geoscience where particulate phase analysis is required, scanning electron microscopy (SEM) linked with either energy dispersive or wavelength dispersive spectrometers (EDS/WDS) has been widely adopted. In particular, SEM-EDS analysis is the most important and routinely used method for the analysis of gunshot residues, which can be characterized on the basis of particle elemental composition linked with particle shape (e.g. Garofano *et al.* 1999; Brozek-Mucha & Jankowicz 2001; Romolo & Margot 2001; ASTM 2002). Manual SEM analysis of samples for gunshot residues is both time-consuming and operator dependent (Smith 1991). Consequently a number of automated SEM-based analysis systems have been developed, as reviewed by Romolo & Margot

From: Pye, K. & Croft, D. J. (eds) 2004. *Forensic Geoscience: Principles, Techniques and Applications*. Geological Society, London, Special Publications, **232**, 123–136. © The Geological Society of London, 2004.

(2001). SEM-EDS analysis is also directly applicable to the mineral/phase analysis of soil or other trace evidence, but, manual SEM examination is still time-consuming and operator dependent. In other disciplines, automated SEM analysis, also referred to as computer-controlled scanning electron microscopy (CC-SEM) has been used, for example in analysis of mineral matter in coal (Galbreath *et al.* 1996; Wigley *et al.* 1997; Zhang *et al.* 2002), characterization of airborne particulate material (Sitzmann *et al.* 1999) and contaminated soil analysis (Kennedy *et al.* 2002). However, automated SEM-EDS has not been routinely adopted in forensic mineralogy/phase analysis studies to date.

In this paper we examine the potential of utilizing an existing SEM-based automated mineral analysis system (QemSCAN) in forensic geoscience, using the analysis of both soil and dust samples as examples.

Mineralogy/phase analysis in forensic geoscience

Applications

Mineralogical analysis has been widely adopted in two main areas of forensic science: (1) in the characterization of soil samples, and (2) in the analysis of trace material evidence including dust samples (e.g. Demmelmeyer & Adam 1995). Soil forensic analysis may also include the examination of soil colour, density gradient distribution, particle size analysis, mineralogy, plant and other biogenic component analysis, organic matter, soil chemistry, thermoluminescence and pH (e.g. Fitzpatrick & Thornton 1974; Demmelmeyer & Adam 1995; Sugita & Marumo 1996, 2001; McVicar & Graves 1997; Marumo *et al.* 1999; Brown *et al.* 2002). Palynological studies also have considerable application in soil analysis, potentially providing high-resolution ($\sim$10 m scale) spatial fingerprinting (e.g. Bruce & Dettmann 1996; Horrocks & Walsh 1999; Horrocks *et al.* 1999). Mineralogical analysis of both soil and dust samples have significant potential in providing evidentiary data in serious crime investigations (e.g. Murray 2000) while, particle size distributions, particle shape and overall texture may also be important.

Existing methods

Commonly applied methods of mineral analysis in forensic geoscience include polarizing light microscopy (e.g. Graves 1979; McCrone 1992; Petraco & DeForest 1993), X-ray diffraction (e.g. Marumo *et al.* 1986; Brown *et al.* 2002) and SEM linked with EDS or WDS analysis (e.g. McVicar & Graves 1997; Ward 2000; Cengiz *et al.* 2003). While well-

established protocols have been developed for light microscope analysis of the mineralogy of dust and soil samples (e.g. Graves 1979; Petraco & DeForest 1993), such methods are partially operator-specific and time-consuming if a statistically reliable dataset is to be generated. SEMs linked to either EDSs or WDSs can be used to identify minerals and other chemical phases based upon quantitative and reproducible analysis. However, manual SEM-based techniques are still operator dependent and sample analysis is time-consuming (Smith 1991). For example the manual examination of samples for gunshot residues under SEM involves screening of the sample under backscatter electron imaging (BEI) (Smith 1991). Consequently, in gunshot residue analysis automated SEM-EDS based systems have been developed and are widely utilized (Romolo & Margot 2001). In addition, there is also interest in automated SEM mineral analysis for soil samples, and McVicar and Graves (1997) reported on the development and validation of an automated SEM system whereby mineral identification was based upon the acquisition of EDS spectra and the matching of the spectra with known mineral phases. This method provided routine mineral analysis in 6 s, based upon the measurement of one spectrum per particle. However, an existing automated SEM-based mineral analysis system (QemSCAN), in which particles are mapped by the acquisition of EDS spectra on a pixel by pixel basis (Fig. 1), may have significant potential in forensic geoscience.

Analysis using the QemSCAN system

QemSCAN is an automated SEM system, initially designed to support mineral-processing applications in the mining industry (e.g. Sutherland & Gottlieb 1991; Khosa *et al.* 2003) by providing rapid quantitative mineral analyses. The latest generation of QemSCAN is based upon a Carl Zeiss EVO-50 SEM fitted with up to four energy dispersive X-ray spectrometers. Routine analysis of particulate material, such as processed mineral products, involves disaggregating and microriffling the samples to obtain a representative subsample. This subsample is then mixed with resin, allowed to set and then polished. In addition to particulate material, small (up to 3 cm) consolidated samples, such as rock chips or ceramics (e.g. brick fragments), can also be prepared as polished blocks and examined. In some cases, for example the analysis of coal, samples are embedded into blocks constructed using carnauba wax, and then polished (cf. Matsuoka *et al.* 2002). The carnauba wax provides a sufficient contrast in backscatter coefficient between the mounting medium and the coal samples to allow the coal to be measured. Furthermore, a series of trials have been

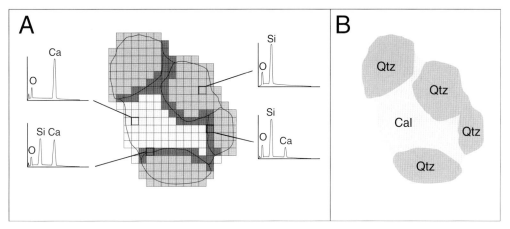

Fig. 1. Schematic diagram illustrating the EDS-based mineral analysis system. (**a**) Systematic mapping of a particle based on the user-defined pixel spacing. (**b**) The final particle image shows that, in this example, the particle has been interpreted as composed of four grains of quartz (Qtz) and a single grain of calcite (Cal).

carried out where dust samples were mounted onto double-sided carbon tape and analysed. This procedure negates the need for resin or wax embedding and polishing, and is more consistent with current sampling practice in forensic SEM-EDS investigations (e.g. McVicar & Graves 1997; Ward 1999).

Particles are automatically located within the sample using the contrast in backscatter coefficient between the mounting medium (resin or wax) and the particles. The use of backscattered electrons in SEM to allow the recognition of particles based on compositional contrast is a widely utilized tool in the examination of sediments and sedimentary rocks (e.g. Pye & Krinsley 1983, 1984; Krinsley *et al.* 1998). Once particles are located, the electron beam is rastered over the surface at a user-defined pixel spacing (typically between $0.2 \mu m$ and $25 \mu m$) and X-ray energy spectra are very rapidly acquired (Figure 1). As shown in Fig. 1, the whole particle is mapped at a pixel spacing of $1 \mu m$, resulting in the acquisition of 244 discrete spectra. The measured spectra are compared against a database of known spectra (the species identification programme) developed by CSIRO Minerals over the last 10–15 years and a mineral or phase name is assigned to each pixel (Fig. 1). A database of known X-ray spectra based on EDS analyses has also recently been developed by the Federal Bureau of Investigation (Ward 2000). Using QemSCAN each individual pixel is identified and assigned to a mineral or phase name in about 10 ms and, taking into account the time taken to move the stage, in excess of 100 000 pixels can be quantified per hour. Therefore, depending upon the pre-defined pixel spacing and the particle size, the system can routinely quantify and map the composition of approxi-

mately 1000 particles, each $1–10 \mu m$ in size per hour. Where a measured pixel overlaps between the particle and the surrounding resin, or occurs along a boundary between two or more discrete mineral phases within the particle, then the image analysis software will assign the pixel systematically to either of the two phases or to a separate new discrete phase based upon predefined rules (Fig. 1). In addition, the QemSCAN system can measure and identify many non-crystalline phases which have a distinct elemental composition. For example, the characterization of smelter waste, fly ash or ceramic products can be carried out by combining the elemental composition of the particle with data on particle size or shape. If a spectrum is measured which cannot be identified using the species identification programme then it is characterized as 'other'. The coordinates of each particle are automatically recorded, allowing the operator manually to re-examine the composition of phases reported as 'other' during routine analysis.

In addition to providing mineralogical or phase assignment for each particle based upon the elemental composition, the system also automatically provides data on particle size (area), particle shape and calculated particle density. However, it should be noted that, with samples prepared as polished blocks, the particle size analysis algorithm takes into account various stereological issues and is based upon the assumption that area equals volume and that the grains are spherical.

The QemSCAN system can be operated in four modes: (1) particle mineral analysis, (2) bulk mineral analysis, (3) trace mineral search, and (4) field image scan (Fig. 2). Particle mineral analysis systematically maps the composition of each discrete particle,

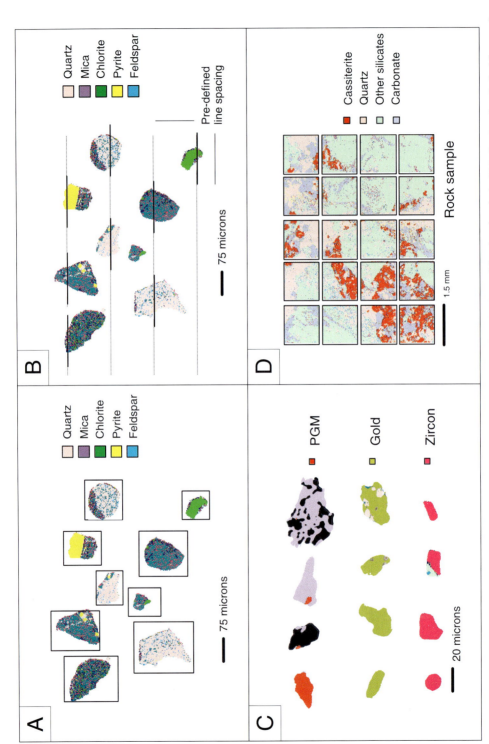

Fig. 2. Automated mineral analysis can be carried out using QemSCAN in four different operational modes: (**a**) particle mineral analysis; (**b**) bulk mineral analysis; (**c**) trace mineral analysis; and (**d**) field area analysis.

A

Quartz
Mica
Chlorite
Pyrite
Feldspar

75 microns

B

Quartz
Mica
Chlorite
Pyrite
Feldspar

Pre-defined
line spacing

75 microns

C

PGM

Gold

Zircon

20 microns

D

Cassiterite
Quartz
Other silicates
Carbonate

Rock sample

1.5 mm

or a pre-defined number of particles, within the prepared sample. This provides quantitative data on mineralogy or phase composition along with particle grain size and shape. Bulk mineral analysis is based on the analysis of particles along scan lines with a pre-defined spacing, and provides quantitative data on the bulk sample mineralogy or phases present. The advantage of a bulk mineral analysis is greater analytical speed when compared with particle mineral analysis. The trace-mineral search programme enables the operator to measure only those particles which have a backscatter coefficient above a pre-defined threshold set by the operator. For example, a backscatter threshold could be set so that only particles with a brightness greater than the common rock-forming silicates will be measured. In this mode of operation common phases can be excluded allowing the very rapid identification of phases present at potentially only trace levels; this mode of examination is for example, comparable with existing automated SEM analysis systems for gunshot residues (Romolo & Margot 2001). Field image scanning is based upon the mapping of a non-particulate sample, such as a rock fragment or a ceramic block. The sample is divided analytically into a series of 'fields'; each field is measured and the pixels are assigned to a mineral species or phase and then the adjacent fields are 'stitched' together to provide a mineralogical or phase compositional map of the scanned area.

The analytical speed of acquisition of each X-ray spectrum means that the QemSCAN system can collect a very large number of individual analyses in a short period of time (in excess of 100 000 analyses per hour). Thus routine measurement rapidly provides quantitative modal mineralogy or phase composition data. In addition, as a large number of grains are measured (in some cases all grains in the prepared sample), rare mineral phases cannot be overlooked during sample analysis. This, coupled with the inherent operator independence, means that the data generated by this system are statistically reliable and reproducible. During data processing the system operator needs to define mineral groupings for the acquired dataset, but as long as the same groupings are retained the data output will remain directly comparable between different samples – effectively all samples can be examined under the same set of operator-defined rules during both data acquisition and processing.

Examples of mineral and phase analysis using QemSCAN

In this paper two studies investigating the potential application of QemSCAN analysis in forensic geoscience are described. These investigations were initially carried out as part of a larger environmental mineralogy feasibility study, and consequently the sampling carried out is not comparable with most real-life forensic investigations in terms of either sample size or sampling methodology. These two case studies do, however, demonstrate the general application of QemSCAN analysis of soil and dust samples. Samples were analysed at the CSIRO Minerals QemSCAN Laboratory at the Queensland Centre for Advanced Technology, Brisbane, Australia. The instrument used in this study was a QemSCAN automated mineral analyser, based on a LEO 440 electron microscope, with three Oxford light element EDS detectors. The samples were measured under fully automated conditions, using a 3 nA beam current, at 25 keV with a working distance of 25 mm.

Case study 1: characterizing soil mineralogy

Seven soil samples were collected from three different localities around the Brisbane area, each of which is characterized by different underlying bedrock geology (Table 1; Fig. 3). Approximately 100 gm of soil was collected at each site, both from the surface and, at Moggill State Forest, from two different depths within the soil profile (Table 1). The dried samples were gently riffled and screened at 5 mm; aggregated soil particles <5 mm in diameter were not disaggregated. The <5 mm size fraction was mounted in resin blocks and measured using the particle mineral analysis mode with a pixel spacing of 1.7 μm. The total number of pixels measured in each sample was typically >500 000, allowing between 551 and 1106 particles to be characterized in each sample (Table 2).

The bulk mineralogy of all of the samples is comparable, being dominated by quartz, chlorite and kaolinite (Fig. 4; Table 2). Minor phases include alkali feldspar, plagioclase, muscovite, biotite, hornblende, rutile, ilmenite, siderite and Fe oxides. The category 'other silicates' comprises Fe–Mn silicates and tourmaline, while 'other' includes zircon, barite, sulphides and carbonates other than siderite. However, based upon both the overall modal mineralogy, and in particular, the trace mineralogy (Fig. 4; Table 2), each sample can be distinguished.

Two samples (MSF1 and MSF2) were collected at depths of 40 cm and 20 cm respectively within a soil profile developed above Devonian–Carboniferous metasediments of the Neranleigh–Fernvale beds (Cameron 1999). The overall mineralogy of these two samples is very similar, being dominated by quartz, chlorite and kaolinite, although there is an apparent increase in chlorite and decrease in kaolinite with depth in the soil profile; quartz modal abundance is very similar. In contrast, sample HR1 was a

Table 1. *Soil samples collected for mineralogical analysis in the Brisbane area*

Sample number	Location	Soil depth	Soil colour	Bedrock geology
MSF1	Moggill State Forest	40 cm deep in roadcut	Red	Devonian–Carboniferous metasediments
MSF2	Moggill State Forest	20 cm deep in roadcut	Brown	Triassic basalts
HR1	Hawkesbury Road	Surface sample	Grey	Pleistocene fluvial terraces overlying Triassic sediments
MPF1	Moggill Pineapple Farm	Surface sample	White-grey marly	Pleistocene fluvial terraces overlying Triassic sediments
MPF2	Moggill Pineapple Farm	Surface sample	Grey	Pleistocene fluvial terraces overlying Triassic sediments
MPF3	Moggill Pineapple Farm	Surface sample	Red-brown	Pleistocene fluvial terraces overlying Triassic sediments
MPF4	Moggill Pineapple Farm	Surface sample	Red	Pleistocene fluvial terraces overlying Triassic sedimernts

surface soil sample from an area underlain by altered/weathered Triassic Sugars basalt. This sample has significantly less quartz (26.9%), the highest abundance of plagioclase (albeit only 2.2%) and is also extremely kaolinite-rich (59.2%).

Four samples were collected from the Moggill Pineapple Farm (samples MPF1–4) from soils developed upon Pleistocene fluvial terraces overlying Triassic sediments of the Tivoli Formation. The four soil samples are markedly different in colour, ranging from white-grey marly to red soils. Each soil is mineralogically distinctive in terms of both the major and minor mineral phases (Fig. 4; Table 2). There is a significant variation between the samples in the modal mineralogy as shown in Figure 4. Note also that there is an apparent increase in Fe oxide

content within the red-brown (MPF3) and red soils (MPF4) (cf. Sugita & Marumo 1996).

Importantly, the samples are not only mineralogically distinctive, but also texturally distinctive (Fig. 5). Primary soil textures have been retained within the larger soil particles which were not disaggregated during sample preparation, as shown most clearly in the red soils from Moggill State Forest (Fig. 5). In addition, Figure 5 shows that all of the soils appear distinctly visually different as a consequence of primary variations in both modal mineralogy and texture.

Case study 2: mineralogical analysis of atmospheric particulate matter

Atmospheric particulate matter can be produced as a result of natural processes, such as weathering and erosion, with possibly long-distance sediment transport, or may be derived from anthropogenic activity, which may also either be widely dispersed or local in terms of source area. In this case study a series of samples of atmospheric particulate matter was collected to see whether their mineralogy/phase composition could be distinguished and interpreted in terms of sediment source areas. Samples were collected by brushing small sediment samples (c. 1–2 gm) into clean sample vials at four sites around a laboratory and office complex in Queensland (Table 3; Fig. 6). The site is surrounded by bushland and residential areas in a semi-rural part of Brisbane's outer suburbs. There is a coal-fired power station approximately 20 km to the southwest, but there is no other major industrial activity close to the site. The progressive construction of a housing estate to the west of the site has been ongoing over the last

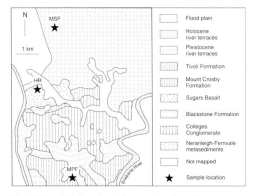

Fig. 3. Simplified geological map, based on Cameron (1999), showing the locations for the sampled soils. The three sample sites were chosen because of their contrasting bedrock geology. MSF, Moggill State Forest; HR, Hawkesbury Road; MPF; Moggill Pineapple Farm.

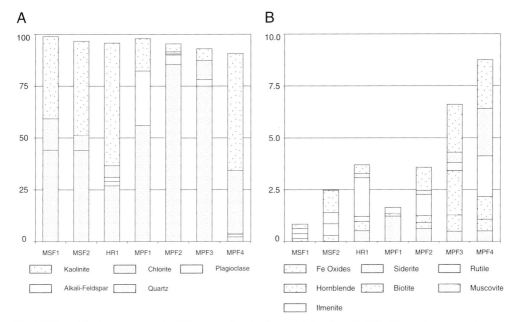

Fig. 4. Soil modal mineralogy data. (**a**) Histograms showing the modal abundance (wt%) of the major mineral phases present within the soil samples. Phases <0.5% have been excluded for clarity. Although each sample is dominated by quartz, chlorite, kaolinite, plagioclase and alkali feldspar there are clear differences between each sample. (**b**) Histogram showing the modal abundance (wt%) of trace mineral phases present within the soil samples. Phases <0.1% have been excluded for clarity. Each sample is distinctive based upon both the relative abundance of trace mineral phases and the relative abundance of each phase.

Table 2. *Soil modal mineralogy data. The modal data are based on the measurement of between 551 and 1106 particles in each sample and their automated assignment to discrete phases*

Sample number	MSF1	MSF2	HR1	MPF1	MPF2	MPF3	MPF4
No. of particles measured	551	1027	1106	1065	1090	774	1002
Modal mineralogy							
Quartz	44.1	43.9	26.9	56.0	85.5	74.9	2.1
Alkali-feldspar	0.1	0.2	2.2	26.3	4.6	3.2	1.4
Plagioclase	0.0	0.2	1.9	0.0	0.5	0.1	0.0
Muscovite	0.0	0.1	0.5	1.2	0.6	0.5	0.5
Biotite	0.0	0.0	0.1	0.1	0.3	0.8	0.6
Chlorite	15.2	7.3	5.6	0.1	1.0	9.1	30.7
Hornblende	0.1	0.3	0.5	0.0	0.3	2.1	1.1
Kaolinite	39.7	45.5	59.2	15.7	3.9	5.7	56.5
*Other silicates	0.0	0.1	0.2	0.0	0.0	0.1	0.1
Ilmenite	0.2	0.6	0.2	0.0	0.1	0.0	0.0
Rutile	0.2	0.5	1.9	0.3	1.0	0.4	2.0
Siderite	0.0	0.0	0.2	0.0	0.2	0.5	2.3
Fe oxides	0.2	1.1	0.4	0.0	1.1	2.3	2.4
†Others	0.0	0.2	0.2	0.2	0.9	0.2	0.4
Total	100.0	100.0	100.0	100.0	100.0	100.0	100.0

*Includes Fe–Mn silicates and tourmaline.
†Includes zircon, barite, sulphides and carbonates other than siderite.

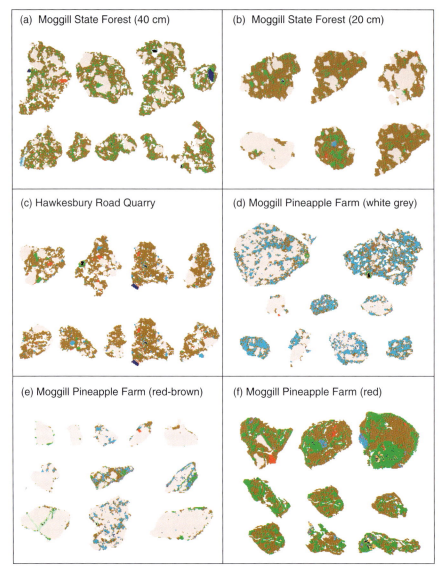

Fig. 5. Representative QemSCAN images of soil particles. Sample numbers: (**a**) MSF1, (**b**) MSF2, (**c**) HR1, (**d**) MPF1, (**e**) MPF3, (**f**) MPF4. Note that the samples are both mineralogically and texturally distinct. Dominant mineral phases are: brown, kaolinite; pink, quartz; green, chlorite; red, rutile; light blue, plagioclase; medium blue, alkali feldspar.

couple of years and is therefore a potential dust source. Samples were collected from: (1) a canteen roof, (2) a mineral loading bay tarpaulin, (3) a workshop window-sill, and (4) an office window-sill (Fig. 6). The samples were then prepared as resin blocks without size sorting and analysed using QemSCAN with a pixel spacing of 0.2 μm. Only the data for the <15 μm–>0.5 μm particle size fraction are reported here. The total number of particles measured per sample ranged between 2006 and 2050, based upon

the acquisition of between 398 000 and 634 000 discrete spectra per sample. Samples were analysed using the particle mineral analysis mode. The very small pixel stepping interval selected (0.2 μm) resulted in relatively long analytical run times (between 4 h 30 min and 7 h per sample).

The mineralogy/phase compositions are shown in Figure 7 and Table 3. As can be seen, each sample is distinctive based upon the overall modal mineralogy/phase analysis, although quartz, kaolinite, K–Al

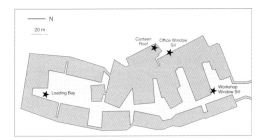

Fig. 6. Schematic diagram of the laboratory and office-building complex near Brisbane, from which the dust samples were collected. The four sampling sites are indicated. Dark grey, buildings; light grey, access roadway which leads to the mineral loading bay area. Note that material will be transported past the workshop window-sill sampling point en route to and from the mineral loading bay.

and Ca–Al-silicates (feldspars) and Mg–Fe–Al silicates are present in all of the samples (Fig. 7). Fe oxides are only present as trace phases in both the canteen roof and office window-sill samples, but are common in both the workshop window-sill and the loading bay samples, with 46.2 and 27.4 wt%, respectively. If the Fe oxide particles are excluded from the dataset and the modal mineralogy is recal-

culated, then all of the samples are much more comparable in overall composition. Minor phases recognized include: calcite, Fe sulphides, rutile, ilmenite, and 'others' including gypsum, magnesium compounds and Al–Fe oxides. In addition, a number of platy particles composed of titanium, copper and zinc are interpreted as paint flakes (cf. SWGMAT 2002).

When the Fe oxide particles are examined in more detail, it is apparent that there is a distinct difference in Fe oxide particle shape between the loading bay and the workshop window-sill samples. The workshop window-sill sample contains a greater proportion of spherical Fe oxide particles when compared with the loading bay tarpaulin sample (Fig. 8). To quantify this further, using the sample analysis software, all particles containing greater than 10% by weight Fe oxide were selected and a shape filter was applied to these grains. The shape filter allows the proportion of the measured grains with a very high sphericity to be quantified. Based on this analysis, 27.1% of the Fe oxide grains in the workshop window-sill sample have very high sphericity in comparison with only 4.6% from the loading bay tarpaulin.

While the overall bulk composition of the samples, excluding the Fe oxide grains, are comparable, and possibly indicative of the provenance areas

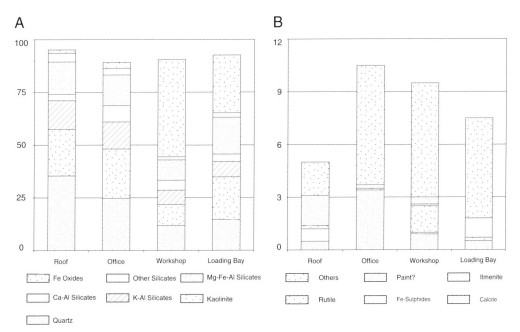

Fig. 7. Dust modal mineralogy data. (**a**) Histograms showing the modal abundance (wt%) of the major mineral phases present within the dust samples. Note the abundance of Fe oxides in both the mineral loading bay and workshop window-sill sample sites, when compared with the canteen roof and office window-sill sample sites. (**b**) Histogram showing the modal abundance (wt%) of trace mineral phases present within the dust samples. Note that paint flakes are only abundant in the canteen roof sample, while calcite is most abundant in the samples from the office window-sill.

Table 3. *Modal mineralogy/phase composition data for the dust samples*

Sample site	Canteen roof	Office window-sill	Workshop window-sill	Mineral-loading bay tarpaulin
No. of particles measured	2050	2006	2008	2039
Modal mineralogy (%)				
Quartz	35.4	24.8	11.8	14.6
Kaolinite	22.1	23.3	10.0	20.2
K–Al silicates	13.6	12.9	6.8	7.3
Ca–Al silicates	3.0	7.7	4.7	3.5
Mg–Fe–Al silicates	15.4	14.5	9.6	17.5
Other silicates	4.1	3.1	1.5	2.1
Fe oxides	1.6	3.0	46.2	27.4
Calcite	0.5	3.4	0.9	0.5
Fe sulphides	0.7	0.0	0.1	0.0
Rutile	0.2	0.1	1.5	0.2
Ilmenite	0.0	0.2	0.1	1.1
Paint?	1.7	0.0	0.4	0.0
*Others	1.9	6.8	6.5	5.7
Total	100.0	100.0	100.0	100.0

*Includes gypsum, Mg compounds and Al–Fe oxides.

for the dominant sediment types (e.g. both local and remote sources, which may include large-scale dust storms prevalent in this area), the other compositional differences identifiable in the samples can best be interpreted as reflecting specific, and probably very proximal, particulate sources. The Fe oxide particles on the loading-bay tarpaulin are interpreted as being dominated by fugitive dusts from iron-ore processing operations in that area. In contrast, the common spherical particles classified as Fe oxides from the workshop window-sill (Fig. 8) are interpreted as representing solidified droplets of molten metal fume derived from the workshop area. Thus, although both these sites contain abundant Fe oxides, the two sites can be distinguished on the basis of the shape analysis function of the Fe oxide grains and interpreted in terms of localized anthropogenic activity. The other samples are also distinctive, based on the presence of trace phases. The canteen roof samples contain an increased abundance of particles interpreted to be paint flakes, while the office window-sill sample shows an increase in the abundance of calcite. Both of these observations are best interpreted in terms of localized sediment sources, most obviously with the paint flakes being derived directly from the painted canteen roof.

These data suggest that, although there is a background compositional signature, the mineralogy/phase composition of each sample is distinctive and, at least in part, reflects localized proximal anthropogenic activity. Thus, from a forensic examination viewpoint, particulate trace evidence from these different locations would indeed be site-specific.

Discussion

While the evidential value of soil samples in forensic examinations has been demonstrated based on a variety of techniques, there is still considerable scope for the increased utilization of this type of evidence in forensic geoscience. Recent work has shown that palynological studies have the ability to identify discrete soil samples with a high degree of spatial resolution and thus have considerable application in forensic studies (e.g. Bruce & Dettmann 1996; Horrocks & Walsh 1999; Horrocks *et al.* 1999). Soil mineralogy and texture should be of equal evidentiary value but few studies have been published (Brown *et al.* 2002). There are a number of problems inherent in current mineral analysis methods (e.g. Marumo *et al.* 1999) when applied to forensic studies; these are largely a consequence of the typically very small sample sizes available for examination. Existing light microscopy methods in forensic geoscience commonly only examine a limited particle size fraction, and opaque phases and non-mineral phases are not identified (e.g. Graves 1979). Generating quantitative modal mineralogy data, which typically requires approximately 500 grains to be determined per sample, is generally considered to be very time-consuming. Graves (1979) established a procedure for point counting forensic samples, which grouped mineral species into a more restricted classification scheme that enabled individual samples to be counted in about 2 h. However, by utilizing this system, there is the potential for valuable species specific mineralogical data to be overlooked. If a smaller number of grains is examined, then there is a risk that the data presented are not quantitative. In addition, in many soil samples there will be a considerable clay and silt-sized grain size fraction, which is difficult to identify mineralogically using light microscopy and, commonly, the available sample size is too small for semi-quantitative X-ray diffraction analysis.

Many of the limitations of light microscopy can be overcome by using SEM-EDS analysis where mineral or phase composition is inferred based upon the elemental composition of the sample, and details of particle size, shape and texture can also be obtained. Sample analysis can be carried out on very small samples, and the whole size fraction can be examined. However, the main limitation on manual SEM analysis is that it is operator dependent and generating quantitative modal data is difficult and time-consuming (cf. McVicar & Graves 1997); conversely the skilled operator can provide high-level interpretation of textural/mineralogical data (Pye 2004). Automated SEM analysis reduces operator-dependence and allows quantitative data to be collected, as is well illustrated by existing literature relating to the analysis of gunshot residue, (e.g. Romolo & Margot 2001). While a range of automated SEM systems have been developed, this study suggests that the existing QemSCAN technology, already widely adopted in the mining industry (e.g. Sutherland & Gottlieb 1991) and used in a range of other applications (e.g. Pirrie *et al.* 2003) has considerable potential in forensic geoscience investigations. Potential advantages of the system include:

(1) Analytical speed – the ability to routinely measure >100000 X-ray spectra per hour means that a very large number of particles can be quantified.

(2) Operator independence – once the operator has selected the mode of operation, number of particles to be measured and pixel spacing, then data acquisition is fully automated and operator independent. Subsequent data processing is required, but again pre-defined rules can be applied ensuring that the data generated are both fully quantitative and reproducible.

(3) The data output not only includes mineral or phase analysis, but also provides data on particle shape, particle size, particle by particle or bulk elemental analysis, and calculated particle density. Once a sample has been measured, any of these parameters can be used in the interpretation and description of the sample.

(4) The visual data output, although false colour images, allows direct visual comparison of samples.

However, with the current configuration of the QemSCAN system there are also a number of clear limitations:

(1) Samples have previously been routinely prepared as polished blocks. Consequently, surface textural data are not measured. Future test work is required to develop measurement of unpolished samples mounted on double-sided tape further, thus still permitting surface textural data to be collected.

(2) Mineral or phase names are assigned based upon the comparison of the unknown measured spectra against a database of known spectra – the species identification programme. The existing species identification programmes have been developed for general mineralogical and metallurgical applications. While these are broadly applicable for forensic samples, the development of a forensic-specific species identification programme is required.

(3) Although particles such as coal grains can be imaged, other organic components within the samples cannot be identified. In addition, mineral polymorphs cannot be distinguished on the basis of elemental analysis alone.

Advanced automated mineral/phase analysis of trace evidence, such as soil, when combined with the examination of the soil palynology provides the possibility of identifying a very distinctive trace evidence 'fingerprint'. This trace evidence fingerprint may link a suspect or a victim to a crime scene. In other cases it may help to narrow the search area for a missing person. The potential significance of soil or other particulate forensic evidence in serious crime investigations should not be overlooked.

We acknowledge support for part of this study from the Southwest Regional Development Agency, Cornwall Enterprise and Prosper. DP acknowledges the support of a CSIRO Visiting Scientist award in 2002. MRP is funded through European Social Fund grant number 011019SW1. We are grateful to D. Croft for assistance in sourcing background references on forensic geoscience and to S. Atkinson (Camborne School of Mines) for his help in tracing literature. A. Wolfe (Centre for Forensic Science, Toronto) also kindly provided relevant reprints. The assistance of S. Peck (CSIRO) during the soil sampling is much appreciated. We are also grateful to P. Wiltshire for helping us understand the complexity of casework. Editorial comments by K. Pye which assisted us in sharpening up this paper are gratefully acknowledged.

References

ASTM. 2002. Standard guide for gunshot residue analysis by scanning electron microscopy/energy dispersive spectroscopy. *Annual Book of ASTM Standards 2002*. American Society for Testing Materials West Conshohocken, **14.02**, 528–530.

(a) Workshop   (b) Loading Bay

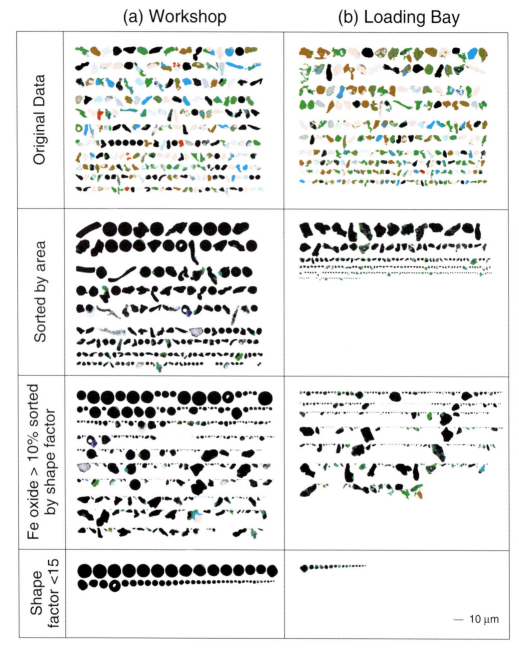

Fig. 8. QemSCAN particle images for grains from: (**a**) the workshop window-sill and (**b**) the loading-bay tarpaulin. The original data image shows that a wide range of mineral particles are present in both samples. However, these two samples can be distinguished from the canteen roof and office window-sill samples on the basis of the abundance of particles of Fe oxide (the black grains in the images). By applying a filter to the data, particles with >10% Fe oxide can be examined further. These grains are shown sorted by area, where it is clear that there are more larger Fe oxide grains in the workshop window-sill sample when compared with the loading-bay sample. The grains were then sorted by shape, and it is clear that the workshop window-sill sample contains a large number of more spherical Fe oxide grains than the loading-bay sample. Finally, the data was sorted by the application of a shape factor filter which enables the quantification of the number of Fe oxide particles that have high sphericity. This analysis confirms the abundance of highly spherical particles in the sample from the workshop window-sill when compared with the loading-bay tarpaulin.

BROWN, A. G., SMITH, A. & ELMHURST, O. 2002. The combined use of pollen and soil analyses in a search and subsequent murder investigation. *Journal of Forensic Science*, **47**, 614–618.

BROZEK-MUCHA, Z. & JANKOWICZ, A. 2001. Evaluation of the possibility of differentiation between various types of ammunition by means of GSR examination with SEM-EDX method. *Forensic Science International*, **123**, 39–47.

BRUCE, R. G. & DETTMANN, M. E. 1996. Palynological analyses of Australian surface soils and their potential in forensic science. *Forensic Science International*, **81**, 77–94.

CAMERON, I. 1999. *A Green and Pleasant Land*. Ian Cameron, Queensland.

CENGIZ, S., KARACA, A. & CAKIR, I. 2003. SEM-EDS analysis and discrimination of forensic soil. *Proceedings of the American Academy of Forensic Sciences, Annual Meeting, Chicago, 2003*, 36.

DEMMELMEYER, H. & ADAM, J. 1995. Forensic investigation of soil and vegetable materials. *Forensic Science Review*, **7**, 120–137.

FITZPATRICK, F. & THORNTON, J.I. 1974. Forensic science characterisation of sand. *Journal of Forensic Science*, **4**, 460–475.

GALBREATH, K., ZYGARLICKE, C. *ET AL.* 1996. Collaborative study of quantitative, coal mineral analysis using computer-controlled scanning electron microscopy. *Fuel*, **75**, 424–430.

GAROFANO, L., CAPRA, M., FERRARI, F., BIZZARO, G. P., DI TULLIO, D., DELL'OLIO, M. & GHITTI, A. 1999. Gunshot residue. Further studies on particles of environmental and occupational origin. *Forensic Science International*, **103**, 1–21.

GRAVES, W. J. 1979. A mineralogical soil classification technique for the forensic scientist. *Journal of Forensic Science*, **24**, 323–338.

HORROCKS, M. & WALSH, K. A. J. 1999. Fine resolution of pollen patterns in limited space: Differentiating a crime scene and alibi scene seven meters apart. *Journal of Forensic Science*, **44**, 417–420.

HORROCKS, M., COULSON, S. A. & WALSH, K. A. J. 1999. Forensic palynology: variation in the pollen content of soil on shoes and in shoeprints in soil. *Journal of Forensic Science*, **44**, 119–122.

KENNEDY, S.K., WALKER, W. & FORSLUND, B. 2002. Speciation and characterisation of heavy-metal contaminated soils using computer-controlled scanning electron microscopy. *Environmental Forensics*, **3**, 131–143.

KHOSA, J., MANUEL, J. & TRUDU, A. 2003. Results from a preliminary investigation of particulate emission during the sintering of iron ore. *Transactions of the Institution of Mining and Metallurgy*, **112**, C25–C32.

KRINSLEY, D. H., PYE, K., BOGGS, S. & TOVEY, N. K. 1998. *Backscattered Scanning Electron Microscopy and Image Analysis of Sediments and Sedimentary Rocks*. Cambridge University Press, Cambridge.

MARUMO, Y., NAGATSUKA, S. & OBA, Y. 1986. Clay mineralogical analysis using the <0.05 mm fraction for forensic science investigation – its application to volcanic ash soils and yellow-brown forest soils. *Journal of Forensic Sciences*, **31**, 92–105.

MARUMO, Y., SUGITA, R. & SETA, S. 1999. Soil as evidence in crime investigation. *ICPR*, **474–475**, 75–84.

MATSUOKA, K., ROSYADI, E. & TOMITA, A. 2002. Mode of occurrence of calcium in various coals. *Fuel*, **81**, 1433–1438.

MCCRONE, W. C. 1992. Forensic soil examination. *Microscope*, **40**, 109–121.

MCVICAR, M.J. & GRAVES, W.J. 1997. The forensic comparison of soils by automated scanning electron microscopy. *Journal of the Canadian Society of Forensic Scientists*, **30**, 241–261.

MURRAY, R. 2000. Devil in the details – the science of forensic geology. *Geotimes*, 2000, 14–17.

PETRACO, N. & DEFOREST, P. R. 1993. A guide to the analysis of forensic dust specimens. *In*: SAFERSTEIN, R. (ed.) *Forensic Handbook*. vol. III. Prentice Hall, New Jersey, 24–69.

PIRRIE, D., POWER, M. R., ROLLINSON, G., CAMM, G.S., HUGHES, S.H., BUTCHER, A. R. & HUGHES, P. 2003. The spatial distribution and source of arsenic, copper, tin and zinc within the surface sediments of the Fal Estuary, Cornwall, UK. *Sedimentology*, **50**, 579–595.

PYE, K. 2004. Forensic examination of sediments, soils, dusts and rocks using scanning electron microscopy and X-ray chemical microanalysis. In: PYE, K & CROFT, D. J. (eds) *Forensic Geoscience: Principles, Techniques and Applications*. Geological Society, London, Special Publications, **232**, 103–122.

PYE, K. & KRINSLEY, D.H. 1983. Mudrocks examined by backscattered electron microscopy, *Nature*, **301**, 412–413.

PYE, K. & KRINSLEY, D. H. 1984. Petrographic examination of sedimentary rocks in the SEM using backscattered electron detectors. *Journal of Sedimentary Petrology*, **54**, 877–888.

ROMOLO, F. S. & MARGOT, P. 2001. Identification of gunshot residue: a critical review. *Forensic Science International*, **119**, 195–211.

SITZMANN, B., KENDALL, M., WATT, J. & WILLIAMS, I. 1999. Characterisation of airborne particles in London by computer-controlled scanning electron microscopy. *Science of the Total Environment*, **241**, 63–73.

SMITH, M. A. 1991. Particle analysis by the scanning electron microscope. *International Symposium on the Forensic Aspects of Trace Evidence*. FBI Academy, Quantico, US Department of Justice, 31–39.

SUGITA, R. & MARUMO, Y. 1996. Validity of color examination for forensic soil identification. *Forensic Science International*, **83**, 201–210.

SUGITA, R. & MARUMO, Y. 2001. Screening of soil evidence by a combination of simple techniques: validity of particle size distribution. *Forensic Science International*, **122**, 155–158.

SUTHERLAND, D. N. & GOTTLIEB, P. 1991. Application of automated quantitative mineralogy in mineral processing, *Minerals Engineering*, **4**, 753–762.

SWGMAT (Scientific Working Group on Materials Analysis). 2002. Standard guide for using scanning electron microscopy/X-ray spectrometry in forensic paint examinations. *Forensic Science Communications*, **4**, [On-line] www.fbi.gov/hq/lab/fsc

WARD, D. C. 1999. A small sample mounting technique for scanning electron microscopy and X-ray analysis. *Forensic Science Communications*, **1**. [On-line]

WARD, D. C. 2000. Use of an X-Ray spectral database in forensic science. *Forensic Science Communications*, **2**. [On-line] www.fbi.gov/hq/lab/fsc

WIGLEY, F., WILLIAMSON, J. & GIBB, W. H. 1997. The distribution of mineral matter in pulverised coal particles in relation to burnout behaviour. *Fuel*, **76**, 1283–1288.

ZHANG, L., SATO, A. & NINOMIYA, Y. 2002. CCSEM analysis of ash from combustion of coal added with limestone. *Fuel*, **81**, 1499–1508.

Mineralogy and microanalysis in the determination of cause of impact damage to spacecraft surfaces

G. A. GRAHAM[1,4*], A. T. KEARSLEY[2,**], G. DROLSHAGEN[3], J. A. M. McDONNELL[1], I. P. WRIGHT[1] & M. M. GRADY[4]

[1]Planetary & Space Sciences Research Institute, The Open University, Walton Hall, Milton Keynes MK7 6AA, UK

[2]School of Biological & Molecular Sciences, Oxford Brookes University, Headington, Oxford OX3 0BP, UK

[3]TOS-EMA, European Space Research Technology Centre, The European Space Agency, Keplerlaan 1, 2201 AZ Noordwijk, The Netherlands

[4]Department of Mineralogy, The Natural History Museum, Cromwell Road, London, SW7 5BD, UK

[*]Institute for Geophysics & Planetary Physics, Lawrence Livermore National Laboratory, Livermore, CA 94551, USA (e-mail: graham42@llnl.gov)

[**]Present address: Electron Microscopy & Mineral Analysis, Department of Mineralogy, The Natural History Museum, Cromwell Road, London SW7 5BD, UK

Abstract: Cosmic dust grains are the abundant, fine-grained end-member of a range of extraterrestrial materials travelling through space. These particles can impact orbiting space vehicles (e.g. satellites and the International Space Station) at velocities ranging from 10 to 72 kms^{-1}. Impact damage resulting from such a collision could potentially disable or limit the operational use of a spacecraft. There is great commercial interest from the satellite companies and space agencies to understand the nature and proportion of impacts that are caused by cosmic dust particles to assist in risk management studies and for protective shielding optimization. The successful recovery of any surface that has been exposed to the near-Earth environment offers an excellent opportunity to search for micrometre-scaled impact features and the associated projectile residues using scanning electron microscopy and X-ray microanalysis.

Cosmic dust is the finer grain size portion of the natural particulate flux of extraterrestrial materials orbiting within the solar system, and entering it from interstellar space (Fig. 1). The annual flux of cosmic dust accreted by the Earth has been estimated as $0.39-1.2 \times 10^{-4}$ per year, from the analysis of impact craters preserved on surfaces returned from the long duration exposure facility (LDEF) (Love & Brownlee 1993). These particles are a potential hazard to orbiting space vehicles (e.g. satellites, the International Space Station and dedicated scientific experiments such as the Hubble space telescope, HST). When dust particles collide with spacecraft, the impact speed will be greater than $2-3 kms^{-1}$ and is therefore termed as hypervelocity in nature (Burchell et al. 1999). The impact velocities for cosmic dust or, as they are termed, micrometeoroids when captured in the near-Earth environment, can range from $11 kms^{-1}$ (defined by the escape velocity of the Earth) to approximately $72 kms^{-1}$ (Taylor 1995), with a mean meteoroid encounter velocity of $20 kms^{-1}$ (Mandeville 1993). There is a significant

need to understand and study these natural cosmic bullets considering the physical damage and effects, e.g. plasma discharge (Foschini 1998) that result from an impact event. To date there have been no confirmed failures of an operational spacecrafts as the result of a micrometeoroid collision. However a number of papers have discussed the potential hazard during meteoroid storms such as the Perseid or the Leonids (e.g. Beech & Brown 1993; McBride & McDonnell 1999). It was speculated that the failure of the Olympus telecommunications satellite during the Perseid meteors shower may have been caused by a dust impact (Caswell et al. 1995). McBride & Taylor (1996) used flux models to determine the risk to satellite tethers from the small-sized micrometeoroids and orbital debris. More recently it has been postulated that an anomalous response from pixels on the pn-charge coupled device (CCD) camera of X-ray astronomy observatory XMM-Newton was also the result of a micrometeoroid impact event (Struder et al. 2001).

In addition to micrometeoroid collisions there is a

From: PYE, K. & CROFT, D. J. (eds) 2004. *Forensic Geoscience: Principles, Techniques and Applications.* Geological Society, London, Special Publications, **232**, 137–146. © The Geological Society of London, 2004.

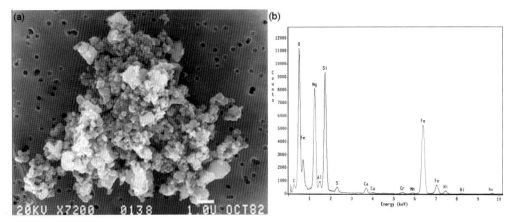

Fig. 1. (**a**) Secondary electron micrograph of a typical interplanetary dust particle (approximately 20 μm in diameter) that has been collected in the stratosphere. (**b**) The typical bulk energy dispersive X-ray spectrum acquired for an interplanetary dust particle (although the S peak is a little low). Source: J.P. Bradley and Z. Dai, Lawrence Livermore National Laboratory, USA.

population of artificial space debris that has been generated by the growth in the use of Earth orbits for military, scientific, and commercial telecommunication satellites. Space debris is very diverse in size, ranging from large, metre-scale spent rocket bodies (e.g. upper stages) and satellites to micrometre-scale fragments of material (e.g. Crowther 2002). For this paper it is the ability to distinguish between the micrometre-sized space debris that is of interest. This size regime of space debris is generated from the degradation of spacecrafts over time, so can include a number of different compositions for example droplets of leaked coolant (liquid Na and K) from nuclear power plants of some satellites (Zolensky *et al.* 1993) and solid rocket fuel exhaust products consisting of Al_2O_3 (Hörz *et al.* 2002). The typical impact encounter velocities for these space debris microparticles are up to $15 \, \mathrm{km \, s^{-1}}$ (Mandeville 1993).

Micrometeoroid and space debris impacts have been investigated using a number of different techniques, such as real-time particle impact detectors fitted to spacecraft (as described by Iglseder *et al.* 1993; Kuitunen *et al.* 2001). Dedicated spacecraft with numerous detection experiments, for example LDEF (e.g. Zolensky *et al.* 1995) and the European Space Agency's European Retrievable Carrier (e.g. Drolshagen *et al.* 1995; McDonnell *et al.* 1995) or individual impact experiments (e.g. McDonnell *et al.* 1984; Hörz *et al.* 2000) enable the analysis of the preserved impact damage in the laboratory as spacecrafts and the experiments were retrieved after extensive periods of exposure in low-Earth orbit (LEO). Results from these studies allow particle flux models to be developed that significantly enhance our understanding of the near Earth environment

(e.g. McDonnell *et al.* 1997; McBride *et al.* 1999). The availability of these exposed surfaces has enabled detailed chemical residues studies using scanning electron microscopy (SEM) and X-ray microanalysis to ascertain the origin of the projectile in terms of cosmic dust and space debris (e.g. Bernhard *et al.* 1993; Yano *et al.* 1994; Zolensky *et al.* 1995). In addition to these dedicated experiments, it is also possible to carry out similar investigations on any surface from a spacecraft that is recovered (e.g. Warren *et al.* 1989; Drolshagen *et al.* 1995; Graham *et al.* 1997). In this paper we discuss the interpretation of impact residues preserved on materials returned from the HST, with particular focus on the interpretation of micrometeoroid remnants.

Opportunities to study surfaces exposed in the near-Earth environment

The types of substrate available for micrometeoroid and orbital debris damage assessment from a particular mission will depend on a diverse range of factors, including their expendability from an operational spacecraft and the ease of recovery from LEO. Shuttle Orbiter missions have been particularly successful in retrieval of material from American, Russian, European and Japanese spacecraft. Examples have included several samples of thermal insulation blankets (e.g. Warren *et al.* 1989), but very diverse materials can be returned from a single space mission, such as the solar cells, polymer stiffeners and metal support struts from the third HST service mission in March 2002. The retrieval of one of the solar array assembly panels from the 1993 HST

Fig. 2. (**a**) An optical photograph of one of the solar array panel wings recovered during the 2002-service mission completely extended during the preliminary post-flight investigation at the European Space Research Technology Centre of the European Space Agency. (**b**) An optical photograph of a selected area showing the individual solar cells. Some impact damage sustained by the individual solar cells is clearly visible in the image. (**c**) An optical photograph of the primary deployment mechanism arm that also shows evidence of a hypervelocity impact encounter (see insert – the diameter of the crater is 2mm).

service mission and both of the assembly panels in the 2002 mission has provided comparable surfaces from two intervals of LEO exposure (Fig. 2). In this paper we discuss: (1) solar cells from the upper blanket of the solar array assembly of HST, returned to Earth in 1993, after exposure for over 3.5 years (Drolshagen *et al.* 1995) and (2) stiffener material exposed for over 8 years and returned by the 2002 service mission. The solar cells are a laminate of bor-osilicate glass on silicone resin, silicon photovoltaic layer and metallic conductor, all supported upon a

stiff woven glass-fibre blanket impregnated with resin (Fig. 3). The stiffener is a thicker glass-fibre blanket, impregnated with polysulphone resin and capped with a layer of special epoxy resin (Fig. 3).

The 2002 post-flight investigation of the retrieved HST solar arrays initially focused on a global optical photographic survey in the clean-room facility at the European Space Research Technology Centre of the European Space Agency (CESA). During this stage of the survey it was possible to extract several impact features from the bulk material, including the impact

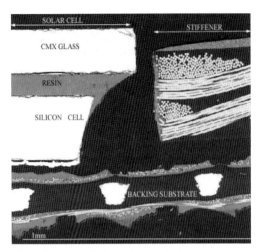

Fig. 3. Secondary electron micrograph of a cross-section made of an individual solar cell and supporting resin stiffener. Source: Graham *et al.* 1999*a*.

on the stiffener that is discussed in this paper. The typical impact features generated by micrometeoroids and space debris on most surfaces have diameters ranging from micrometres to millimetres. Although they can be examined using optical microscopy, the use of an SEM enables detailed examination at much higher magnifications that reveals fine discrete features. Furthermore, since impact-derived residue materials are often only preserved as nanometre- and micrometre-sized remnants, it is only really possible to contemplate their study using electron or ion microscopy. The attachment of an energy dispersive X-ray spectrometer (EDS) allows the determination of the elemental compositions of both the target material and the remnants of the projectile. For the materials discussed herein the analyses were carried out using the JEOL JSM 840 SEM fitted with an Oxford Instruments e-XL EDS system at Oxford Brookes University. The analytical work was carried out at an accelerating voltage of 20 kV with a beam current of 2 nA and a working distance of 32 mm. Further detailed high-resolution secondary electron imaging was carried out on a Philips XL30 field emission scanning electron microscope (FE-SEM) at the Natural History Museum, London, with a range of accelerating voltages from 5–15 kV with a beam current of 1 nA and a working distance of 10 mm.

The interpretation of preserved micrometeoroid chemistries

While identifying the chemistry of impacts generated by space debris particles is clearly important in

terms of the overall information that detailed chemical residue studies can achieve (e.g. Bernhard *et al.* 1993; Graham *et al.* 1999*a*), they will not be discussed at length within the context of this paper. The analytical protocols employed for the recognition of orbital debris particles are essentially the same as those for natural particles, and the same techniques are employed on each crater, as the nature of the projectile will rarely be known prior to SEM examination. A broad knowledge of materials science, especially metallurgy, is helpful in the recognition of space debris residues which are often found to be aerospace alloys, degraded paint particles or burnt solid rocket motor fuel (Graham *et al.* 1999*a*).

To be able to recognize and interpret possible remnants of micrometeoroids, it is important to have detailed knowledge of the distinctive composition of pristine particles. Many of the natural dust particles that are collected in the stratosphere (e.g. the interplanetary dust particles (IDPs) of Mackinnon *et al.* 1982; Schramm *et al.* 1989) and micrometeorites in polar ice have been described as 'chondritic', in reference to their bulk chemistry, which is believed to approximate to the unequilibrated composition seen in the most primitive types of meteorites, the chondrites. This probably reflects the average composition of volatile-free solar system materials, and they are thus regarded as very primitive in their retention of unfractionated mineral compositions. Most workers believe chondritic materials to be derived from asteroids of C and S types, as well as possible cometary sources. The most abundant components of chondritic particles are enriched in Mg, Si, S, Fe and Ni (Fig. 1), but these may be present within differing minerals. The detailed studies of intact IDPs collected from the stratosphere suggest that there are broadly two groups, based on the major mineral components: (1) anhydrous mafic silicates, and (2) hydrous phyllosilicates (Klöck & Stadermann 1994). The anhydrous mafic silicates are dominated by either olivines or pyroxenes (Thomas *et al.* 1993). One subdivision of anhydrous IDPs is dominated by fine-grained microcrystalline aggregates, composed of 5–50 nm grains of Mg–Fe silicates, Fe–Ni sulphides and Fe–Ni metal embedded in a fine-grained carbonaceous matrix (Klöck & Stadermann 1994). The hydrous IDPs are dominated by phyllosilicate mineralogies (smectites such as saponite, and serpentine) although minor anhydrous mineral phases, e.g. Fe–Ni sulfides, Mg–Fe silicates, Mg–Fe carbonates and magnetite may also be identified, particularly in smectite-rich IDPs (Klöck & Stadermann 1994).

However, the micrometeoroid remnants preserved on a spacecraft surface after a hypervelocity collision at $20 \, \mathrm{km \, s^{-1}}$ have experienced dramatic conversion of kinetic energy into shock heating and decompression during impact (e.g. Bernhard *et al.*

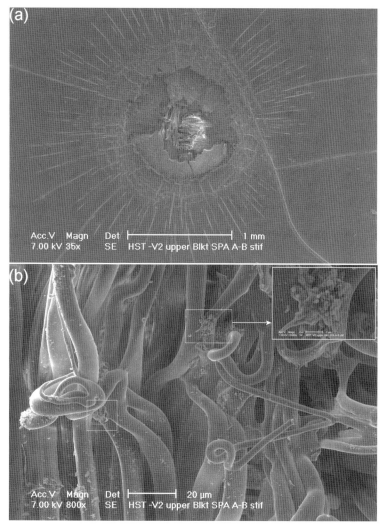

Fig. 4. (**a**) Secondary electron micrograph of an impact crater preserved in the stiffener support material that is part of the solar blanket assembly. (**b**) Secondary electron micrograph of the inner crater, the individual glass-fibres of the stiffener have be melted and deformed during the hypervelocity encounter with the projectile. Remnants of the projectile are highlighted in the square boxes and the insert (scale bar is 2 μm) shows a higher magnification micrograph of one of the remnants.

1993). Conventional physical models suggest that such events should make profound alteration to the appearance and composition of remnants retained within an impact feature (Fig. 4). In our experience of over 200 small impact craters we observe that little evidence remains of the original particle outline shape or internal construction, although a diverse assemblage of different residue compositions may be found together. This is particularly apparent in highly asymmetrical impact craters, where the particle trajectory was at a low angle to the target surface, resulting in a lower effective impact velocity and consequent less damage to the projectile components. Experimental studies by Wallis *et al.* (2002) have demonstrated that there is some dependence of impact crater shape upon the mineral species responsible, but it appears that this alone will not be a reliable indicator of primary projectile composition. Fortunately, discrete residue particles may yield X-ray spectra that are very close to those from meteoritic minerals. A 'near-intact' olivine particle was identified during the analysis of impact features preserved on the thermal control blanket from the Solar Maximum satellite (Rietmeijer & Blandford

1988). However, significant changes to mineral properties have also been reported, and Zolensky *et al.* (1993) observed that remnant mineral components in impact features from LDEF showed evidence of intense shock metamorphism, planar deformation to crystal structure and subsequent recrystallization, creating 120° grain intersections on remnant orthopyroxene material. In extreme cases the particle can be completely vaporized during impact (e.g. Bernhard *et al.* 1993).

Some authors have suggested that the chemistry of the micrometeoroid remnants will also be highly modified; Amari *et al.* (1993) identified enrichments in the refractory elements Al, Ca and Ti, compared them to IDPs collected in the stratosphere, and suggested that this was the result of evaporation and loss of more volatile elements during impact. It should however be noted that discrete Ca- and Al-rich inclusions (CAIs) are not uncommon in several types of carbonaceous chondrite meteorites, and could thus be the primary impacting material in this case. Our experiments with laboratory impacts using a buckshot of well-characterized mineral projectiles (e.g. nepheline and jadeite) fired at velocities of up to $6 \mathrm{km s^{-1}}$ in a light gas gun (Graham *et al.* 1999*b*) do suggest that impact onto metal substrates may be so damaging that volatile elements such as the alkali metals (Na and K) are indeed lost from the residues. However, where the impact was upon solar cell glass substrates, the projectile remnants may preserve original stoichiometric composition, including the alkali metals. Yano *et al.* (1994) observed that remnants of micrometeoroid impactors identified in aluminium clamps from LDEF, while similar in composition to chondritic IDPs in terms of Mg, Si, Fe and S elemental abundance, were depleted in Ni and Ca. We have observed that the heterogeneity of mineral abundance on a scale of tens of micrometres within most groups of chondrite meteorites means that it is possible to find a wide range of apparently representative elemental compositions when sampling impacts by very small (micrometre-scale) particles.

Although a characteristic chemical composition may be retained, the crystallographic structure of the original parent mineral will almost certainly be lost, thereby removing one of the most diagnostic features and preventing firm identification where a chemical compound may exist as more than one polytype. For some extraterrestrial materials, the suite of elements present may be so distinctive that a positive mineral identification can be made (e.g. the co-occurrence of P, Ni and Fe is a reliable indicator of the meteoritic phosphide mineral schreibersite, distinct from nickel-poor cementite). Where there are no distinctive associations, an added difficulty may arise if the residue has mixed with the target substrate chemistries. The elemental 'fingerprints'

that could be used as indicators of micrometeoroid origin (e.g. Si, Mg and Ca) may also be present as fundamental elemental components of the substrate (e.g. magnesium fluoride coated calcium-bearing borosilicate glass, as employed in the solar cells). Impact residues are often a complex blend of both the projectile and the impacted material (Figs 5 & 6), and mixing can occur on a very fine submicrometre scale (Graham *et al.* 2001). Remarkably, the fine structure and apparent retention of stoichiometry of composition in many residues suggests that the entire process of shock melting, residue emplacement and solidification must occur so rapidly that there is little opportunity for migration of elements within melt or equilibration of composition.

Before any meaningful analysis of residues can be performed, it is important to characterize the composition of the impact substrate. Energy dispersive X-ray microanalysis (EDS) is well suited to this task as spectra can be obtained from small points even within impact craters of complex topography, and the technique is effectively non-destructive. The X-ray spectra obtained may contain substantial artefacts from the complex shape of the material, and it is very important to acquire comparative 'background' spectra from comparable locations. Recognition of extraneous material is limited to levels that exceed detection limits for more subtle techniques, such as wavelength dispersive spectrometry or secondary ion mass spectrometry, but is less constrained by the awkward shape of the substrate. Modern EDS detectors allow recognition of light elements such as C and O within spectra, which is particularly important for space debris studies. When the presence of an extraneous chemical component has been recognized, it is still important to identify its origin. We have employed a classification scheme extended from those developed by earlier researchers such as Zolensky *et al.* (1993). In impact craters on the HST solar cells a micrometeoroid origin is indicated by Mg-, S-, Fe- and Ni-rich residues. If all of these elements are intimately associated and occur mixed with the solar cell glass, this may indicate a hydrous silicate (serpentine or smectite) origin. A residue rich in Fe and Mg, but lacking in S and Ni, is likely to be derived from an anhydrous silicate such as olivine or pyroxene; such residue is often localized as micrometre-scale patches (Fig. 5). Discrete metallic droplets of Fe and Ni (Fig. 6), which lack significant Mn or Cr content (as might be found in a steel alloy) have compositions typical of those found in many types of meteorite. Iron and nickel sulphides are also common residues from micrometeoroid impacts. Although other micrometeoroid residues might be expected to occur, albeit more rarely, their compositions (e.g. Ca carbonate) are more difficult to distinguish from the possible surface contamination that

(a)

(b)

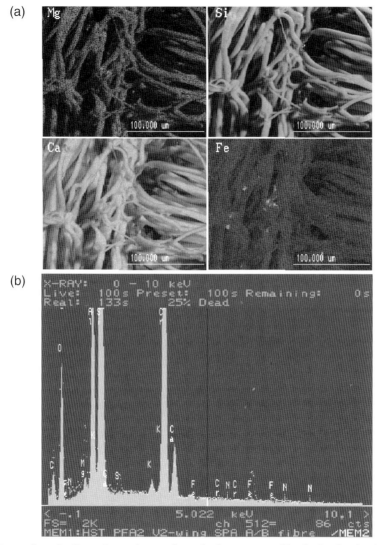

Fig. 5. (**a**) Energy-dispersive X-ray elemental maps of a patch of extraneous residue that has fused to the deformed glass-fibres. The remnant is enriched in Fe and Mg, a combination that would strongly suggest a micrometeoroid origin. (**b**) An ED spectrum shows Mg, Fe, Ni and S (dotted line above background levels), confirming the impact as generated by a micrometeoroid. Discrete areas enriched in Fe and S indicates that the impacting particle contained a metallic sulfide as well as mafic silicates.

may occur during retrieval and post-flight handling of spacecraft materials. As the original micrometeoroids were probably each composed of several different minerals, it is not surprising to find that a single impact event may retain a suite of distinctive residues (Fig. 6).

The classification of some impact residues in terms of either micrometeoroid or orbital debris origin is difficult, and it may not be possible to give a totally unambiguous answer where interpretation can only be based on 'major element' fingerprints

of the original intact material (i.e. at wt % oxide fraction levels). For example, although it is highly likely that a residue composed of Al and O is the remnant of solid rocket motor debris, it could be from the natural mineral corundum (Al_2O_3), which has been identified in primitive meteorites, although this is extremely rare. Were trace element and isotopic microanalyses feasible on such small grains (as may become possible with new generations of ion microprobes), it should be possible to resolve their origin.

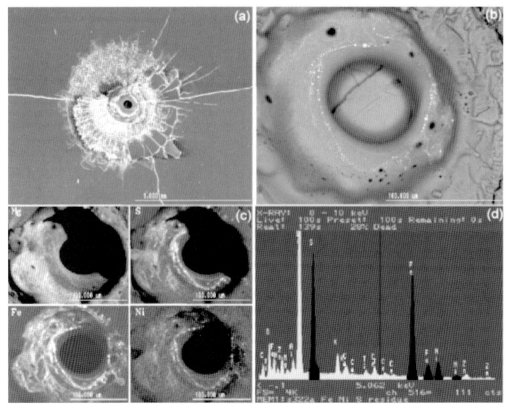

Fig. 6. (a) Secondary electron micrograph of a typical impact crater preserved in a solar cell that has been damaged by a micrometeoroid. **(b)** Backscattered electron micrograph of the inner crater melt pit, some remnants of the original projectile are visible as bright droplets in the image and the glass melt is vesicular. **(c)** The EDS X-ray maps for Mg, Fe, S and Ni suggest that remnants of two different mineral phases have been retained within the impact crater. Areas of high iron enrichment with less Mg,S and Ni are likely to be derived from an hydrous mafic silicate (smectite or serpentine), higher Ni and S with Fe shows as discrete droplets of Fe–Ni sulphide. **(d)** EDS X-ray spectrum identifying the Fe–Ni sulphide component (peaks marked in black) of the remnant micrometeoroid.

Application of data acquired from impact residue studies

The previous section has showed that X-ray micro-analysis can lead to detailed interpretation of the material associated with a specific impact feature preserved on a surface that has been exposed in LEO. If this kind of study is carried out on a statistically significant number of impacts, then the data is of value to the wider research community, who are interested in the ratio of micrometeoroid to space debris impacts. For example our previous study of over 177 impact craters preserved in solar cells retrieved during the 1993 HST service mission suggested that in the small size range of impact damage (50 μm to 1 mm craters) the predominant projectiles were micrometeoroids (Graham *et al.* 2001). This chemical data was used to compared predicated ratios of micrometeoroid to space debris impacts

from the particle flux models thereby acting as a validation tool (McBride 2002). However, due to the infrequency of either a dedicated experiment (such as LDEF) or individual surfaces/structures from operational spacecraft being retrieved from the LEO environment, detailed chemical residue studies have only been performed on a limited number of surfaces (e.g. Warren *et al.* 1989; Bernhard *et al.* 1993; Graham *et al.* 1999a). It is only possible to give 'time-averaged' snapshots of the LEO microparticle environment because it is not possible to assign a particular time or date to a specific impact event. This is because the retrieved surfaces are rarely equipped with dedicated particle impact sensors and data recorders. It has been suggested that a new generation microparticle detector could be constructed that contains both particle collection (e.g. low-density silica aerogel, Hörz *et al.* 2000) and particle sensor experiments (McDonnell *et al.* 2000), and that this could be exposed to LEO environment on the International Space Station.

Conclusion

Orbiting spacecraft (e.g. satellites) are prone to hyper-velocity collisions with both orbital debris (micrometre to metre in diametre) and micrometeoroids. The damage that results from such a collision can be investigated in the laboratory if and when an exposed surface is retrieved. As the impact-derived residue will be a complex mixture of the impacted substrate and the hyper-velocity particle, it is important to have a good understanding of the likely composition of the original projectiles. While a materials scientist can interpret the elemental remnants of paint fragment or metallic fragmentation debris, the interpretation of the micrometeoroid remnants in terms of mineralogy is clearly within the scope of the geoscientist. The classification of the small-scale impact threat in terms of micrometeoroid versus orbital debris is an important aspect of monitoring the particulate population of material in the near Earth environment. It is, however, highly unlikely that the kind of detailed chemical residue studies described in this paper would ever be used for litigation reasons. Currently the legal aspects of the human utilization of space are focused on reducing the generation of orbital debris (e.g. Perek 2002).

L. Gerlach is thanked for selecting and extracting the material from the HST Solar Array Wing used in this study. C. Jones from the Natural History Museum is thanked for assisting with the scanning electron microscopy work. UniSpace Kent is thanked for financial support under ESA contract no.16283/02/NL/LvH, which allowed G. A. Graham to attend the preliminary post-flight investigation of the HST Array. ESA contract nos. 13308/98/NL/MV and 16349/02/NL/VD have, in part, supported this work. The recent work carried by G. A. Graham was performed under the auspices of the US Department of Energy, National Nuclear Security Administration by the University of California, Lawrence Livermore National Laboratory under contract No. W-7405–Eng-48. Finally the comments of the reviewers greatly improved this paper.

References

AMARI, S., FOOTE, J., SWAN, P., WALKER, R. M., ZINNER, E. & LANGE, G. 1993. SIMS chemical analysis of extented impacts on the leading and trailing edges of LDEF experiment AO187–2. *In*: LEVINE, A. S. (ed.) *LDEF – 69 Months in Space. 2nd Post-Retrieval Symposium, Washington DC,* NASA CP-3194, 513–528.

BEECH, M. & BROWN, P. 1993. Impact probabilities on artificial satellites for the 1993 Perseid meteoroid stream. *Monthly Notices of the Royal Astronomical Society,* **262**, L35–L36.

BERNHARD, R. P., DURIN, C. & ZOLENSKY, M. E. 1993. Scanning electron microscope / energy dispersive X-ray analysis of impact residues in LDEF Tray clamps. *In*: LEVINE, A. S. (ed.) *LDEF – 69 Months in Space. 2nd Post-Retrieval Symposium,* Washington DC, NASA CP-3194, 541–550.

BURCHELL, M. J., COLE, M. J., MCDONNELL, J. A. M. & ZARNECKI, J.C. 1999. Hypervelocity impact studies using the 2 MV Van de Graaff accelerator and two-stage light gas gun of the University of Kent at Canterbury. *Measurement Science and Technology,* **10**, 41–50.

CASWELL, R. D., MCBRIDE, N. & TAYLOR, A. 1995. Olympus end of life: a Perseid meteoroid impact event? *International Journal of Impact Engineering,* **17**, 139–150.

CROWTHER, R. 2002. Space junk: protecting space for future generations. *Science,* **296**, 1241–1242.

DROLSHAGEN, G., MCDONNELL, J. A. M., STEVENSON, T., ACETI, R & GERLACH, L. 1995. Post-flight measurements of meteoroid/debris impact features on EURECA and the Hubble Solar Array. *Advances in Space Research,* **16**, 85–89.

FOSCHINI, L. 1998. Electromagnetic interference from plasmas generated in meteoroid impacts. *Europhysics Letters,* **43**, 226–229.

GRAHAM, G. A., KEARSLEY, A.T., GRADY, M.M., WRIGHT, I.P., GRIFFITHS, A.D. & MCDONNELL, J.A.M. 1999*a*. Hypervelocity impacts in low-Earth orbit: cosmic dust versus space debris. *Advances in Space Research,* **23**, 95–100.

GRAHAM, G. A., KEARSLEY, A. T., GRADY, M. M., WRIGHT, I. P., HERBERT, M. K. & MCDONNELL, J. A. M. 1999*b*. Natural and simulated hypervelocity impacts into solar cells. International *Journal of Impact Engineering,* **23**, 319–330.

GRAHAM, G. A., MCBRIDE, N. *ET AL.* 2001. The chemistry of micrometeoroid and space debris remnants captured on the Hubble Space Telescope solar cells. *International Journal of Impact Engineering,* **26**, 263–274.

GRAHAM, G. A., SEXTON, A., GRADY, M. M. & WRIGHT, I. P. 1997. Further attempts to constrain the nature of impact residues in the HST solar array panels. *Advances in Space Research,* **20**, 1461–1465.

HÖRZ, F., ZOLENSKY, M. E., BERNHARD, R. P., SEE, T. H. & WARREN, J. L. 2000. Impact features and projectile residues in aerogel exposed on *Mir. Icarus,* **147**, 559–579.

HÖRZ, F., BERNHARD, R. P., SEE, T. H. & KESSLER, D. J. 2002. Metallic and oxidized aluminum debris impacting the trailing edge of the Long Duration Exposure Facility (LDEF). *Space Debris,* **2**, 51–66.

IGLSEDER, H., MUNZENMAYER, R., SVEDHEM, H. & GRUN, E. 1993. Cosmic dust and space debris measurements with the Munich dust counter on board the satellites *Hiten* and *Brem-Sat. Advances in Space Research,* **13**, 129–132.

KLÖCK, W. & STADERMANN, F.J. 1994. Mineralogical and chemical relationships of interplanetary dust particles, micrometeorites and meteorites. *In*: ZOLENSKY, M. E., WILSON, T. L., RIETMEIJER, F. J. M. & FLYNN, G.

(eds) *Analysis of Interplanetary Dust Particles.* American Institute of Physics Conference Proceedings, **310**, 51–88.

KUITUNEN, J., DROLSHAGEN, G. *ET AL.* 2001. DEBIE – First standard in-situ debris monitoring instrument. *Proceedings of the 3rd European Conference on Space Debris.* ESA, SP-473, 185–190.

LOVE, S. G. & BROWNLEE, D. E. 1993. A direct measurement of the terrestrial mass accretion rate. *Science*, **265**, 550–553.

MACKINNON, I. D. R., MCKAY, D. S., NACE, G. & ISAACS, A. M. 1982. Classification of the Johnson Space Center Stratospheric Dust Collection. *In*: Proceedings of the 13th Lunar Planetary Science Conference, *Journal of Geophysical Research*, A413–421.

MANDEVILLE, J. C. 1993. Orbital debris and meteoroids: Results from retrieved spacecraft surfaces. *Advances in Space Research*, **13**, 123–127.

MCBRIDE, N. 2002. Dust characterization in the near Earth environment. *In*: GREEN, S. F., WILLIAMS, I. P., MCDONNELL, J. A. M. & MCBRIDE, N. (eds) *Dust in the Solar System and Other Planetary Systems.* Committee on Space Research (COSPAR), Colloquia Series, **15**, 343–358.

MCBRIDE, N. & MCDONNELL, J. A. M. 1999. Meteoroid impacts on spacecraft: sporadics, streams and the 1999 Leonids. *Planetary & Space Science*, **47**, 1005–1013.

MCBRIDE, N. & TAYLOR, E. A. 1996. The risk to satellite tethers from meteoroid and debris impacts. *In*: *Proceedings of the 2nd European Conference on Space Debris.* ESA, SP-392, 643–647.

MCBRIDE, N., GREEN, S. F. & MCDONNELL, J. A. M. 1999. Meteoroids and small sized debris in the low earth orbit and at 1 Au: results of recent modeling. *Advances in Space Research*, **23**, 73–82.

MCDONNELL, J. A. M., BURCHELL, M. J. *ET AL.* 2000. APSIS – Aerogel Position-Sensitive Impact Sensor: Capabilities for in-situ collection and sample return. *Advances in Space Research*, **25**, 315–322.

MCDONNELL, J. A. M., CAREY, W. C. & DIXON, D. G. 1984. Cosmic dust collection by the capture cell technique on the space shuttle. *Nature*, **309**, 237–240.

MCDONNELL, J. A. M., DROLSHAGEN, G., GARDNER, D.J., ACETI, R. & COLLIER, I. 1995. EURECA's exposure in the near Earth space environment. Hypervelocity impact cratering distributions at a time of space debris growth. *Advances in Space Research*, **16**, 73–83.

MCDONNELL, J. A. M., RATCLIFF, P. R., GREEN, S. F., MCBRIDE, N. & COLLIER, I. 1997. Micro-particle populations at LEO altitudes: recent spacecraft measurements. *Icarus*, **127**, 55–64.

PEREK, L. 2002. Space debris at the United Nations. *Space Debris*, **2**, 123–136.

RIETMEIJMER, F. J. M. & BLANDFORD, G. E. 1988. Capture of an olivine micrometeoroid by spacecraft in low-Earth-orbit. *Journal of Geophysical Research*, **93**, 11943–11948.

SCHRAMM, L. S., BROWNLEE, D. E. & WHEELOCK, M. M. 1989. Major element composition of stratospheric micrometeorites. *Meteoritics*, **24**, 99–112.

STRUDER, L., ASCHENBACH, B. *ET AL.* 2001. Evidence for micrometeoroid damage in the pn-CCD camera system aboard XMM-Newton. *Astronomy and Astrophysics*, **375**, L5–L8.

TAYLOR, A. D. 1995. Earth encounter velocities for interplanetary meteoroids. *Advances in Space Research*, **17**, 205–209.

THOMAS, K. L, BLANDFORD, G. E., KELLER, L. P., KLÖCK, W. & MCKAY, D. S. 1993. Carbon abundance and silicate mineralogy of anhydrous interplanetary dust particles, *Geochimica et Cosmoschimica Acta*, **57**, 1551–1566.

WALLIS, D., SOLOMON, C. J., KEARSLEY, A. T., GRAHAM, G. & MCBRIDE, N. 2002. Modelling radially symmetric impact craters with Zernike polynomials. *International Journal of Impact Engineering*, **27**, 433–457.

WARREN, J. L., ZOOK, H. A. *ET AL.* 1989. The detection and observation of meteoroid and space debris impact features on the Solar Max satellite. *Proceedings of the 19th Lunar and Planetary Science Conference*, 641–657.

YANO, H., FITZGERALD, H. J. & TANNER, W. G. 1994. Chemical analysis of natural particulate impact residues on the long duration exposure facility, *Planetary and Space Research*, **9**, 793–802.

ZOLENSKY, M. E., ZOOK, H. A. *ET AL.* 1993. Interim report of the Meteoroid and Debris Special Investigation Group. *In*: LEVINE, A.S. (ed.) *LDEF 69 Months in Space, 2nd Post-Retrieval Symposium, Washington DC*, NASA CP-3194, 277–302.

ZOLENSKY, M. E., SEE, T. H. *ET AL.* 1995. Final activities and results of the Long Duration Exposure Facility meteoroid and debris special investigation group. *Advances in Space Research*, **16**, 53–65.

The archaeologist as a detective: scientific techniques and the investigation of past societies

JULIAN HENDERSON

Department of Archaeology, University of Nottingham, University Park, Nottingham NG7 2RD, UK (e-mail: julian.henderson@nottngham.ac.uk)

Abstract: The use of scientific techniques in the investigation of archaeological sites and artefacts has a long history. These days archaeological science as a discipline has matured to the extent that well-defined questions can be answered in increasingly refined ways. In this paper consideration of specific case studies highlights the kinds of investigations that have been carried out on archaeological materials. The research projects are described in ways that show parallel approaches to more recent types of research in police forensic work. The two case studies focused on are: (1) Islamic glass production – a cross-roads in technology? (eighth to twelfth centuries AD); (2) Ottoman Iznik pottery: the state of the art or the art of the State? (fifteenth to seventeenth centuries AD). A range of analytical techniques has been used, including electron microprobe analysis, inductively coupled plasma atomic emission spectrometry, scanning electron microscopy and mass spectrometry. Clearly these techniques provide different (and sometimes overlapping) information which help to answer research questions. The characterization of raw materials, production processes and distribution zones of the products all form part of a holistic approach. Ideally the results should be embedded in our knowledge of past societies, just as the interpretation of police forensic work should be.

The earliest archaeological science investigations go back as far as the eighteenth century: Klaproth delivered a lecture at the Royal Academy of Sciences and Belles-Lettres in Berlin on 9 July 1775, in which he discussed the gravimetric determinations of the elements making up Greek and Roman coins; Davy (1815) wrote about the pigments used by the ancients in paintings at Rome and Pompeii. Such were the early applications of science to archaeology; since then we have come a long way in creating connections between the two areas. Clearly focused research objectives which address a full integration of both areas have the widest potential to contribute both to science, with the possibility of stretching or developing existing techniques, and to archaeology, where a contribution to reconstructing past human behaviour becomes possible. Trying to create relationships between the numbers generated from chemical characterization of ancient materials and past human actions, whether 2 years or 2000 years ago, presents a range of challenges. Obviously the scientific techniques used for both the analysis of archaeological materials (Tite 1982; Pollard & Heron 1996, pp. 20–80; Henderson 2000, pp. 8–23) and the interpretation of past human behaviour or social processes (Lemonnier 1993, 3–4; Dobres & Hoffman 1994) based on those analyses (Henderson 2000, pp. 3–7) have potential contributions to make to forensic studies (Hunter *et al.* 1996).

The scientific techniques used for the characterization of ancient organic and inorganic materials are the same as those used by police forensic laboratories. These include micro-analytical techniques such as gas chromatography mass spectometry, (GC-MS) electron microprobe analysis (EMPA), scanning electron microscopy (SEM) and laser ablation inductively coupled plasma atomic emission spectrometry (LA-ICP-AES). The actual samples themselves may be very similar, such as microsamples of windscreen glass or food residues. The interpretation of the results may also involve the same and overlapping techniques: for example comparison of a new sample with an existing database of the analytical results of samples of known provenance, the use of statistical techniques to summarize data and to highlight differences and similarities between compositional groups (Fletcher & Lock 1991; Baxter 1994; Drennan 1996), and a broader assessment of probability, introducing other parameters, such as the cultural factors of date, associated artefacts and type of deposit. All of these techniques are germane to both archaeological science and police forensic investigations. Clearly the introduction of a new major component provides a fixed point in a comparison; equally the introduction of one or more trace elements as a result of a slight change in raw material source provides important clues for the interpretations of such investigations. As the size of databases of the chemical analyses of ancient (and modern) artefacts increases, it becomes possible to identify the use of modern raw materials in the production of fakes.

In order to exemplify these areas of overlap in the use of both analytical techniques and interpretation, this paper will focus on two case studies which deal with (Islamic) societies dating from the eighth–twelfth and fifteenth–seventeenth centuries

From: PYE, K. & CROFT, D. J. (eds) 2004. *Forensic Geoscience: Principles, Techniques and Applications.* Geological Society, London, Special Publications, **232**, 147–158. © The Geological Society of London, 2004.

AD. They deal with two rather different societies: in Syria and Turkey respectively. The first case study focuses on experimentation v. the *status quo* in glass production, the second on the emergence, development and decline of an Ottoman glazed ceramic technology. The interpretation of the data in both cases will be addressed in their respective socio-political contexts.

Islamic glass production in the eighth–twelfth centuries AD: a cross-roads in technology?

In AD 796 (AH 175) Harūn al-Rashid, the famous caliph of Baghdad, moved his court, army and treasury to the city of Raqqa in northern central Syria. As part of this move he built a second city next to the existing one, known as Rafika ('the companion'), a series of massive palace complexes, and founded a 2 km long extramural industrial complex. Harūn has come down in history as one of the most successful 'Abassid caliphs who was able to create an effective economic, political and military focus around his court. This allowed him to control the caliphate in a highly successful way. The 'Abbasid caliphate stretched from northern Africa to northern India, so this was a massive achievement. Harūn's reign represents one of the most successfully integrated Islamic caliphates. After he died in AD 809 political and economic structures became less tightly organized, and less successful, and by the eleventh century, the time of the Crusades, warring factions based on city states were common. One reflection of changing fortunes in Raqqa is the archaeological record of the industries there (Tonghini & Henderson 1999; Henderson *et al.* 2002).

The Raqqa ancient industry project was set up in order to investigate the evidence for glass and pottery production at Raqqa using archaeological, scientific (analytical) and remote sensing techniques. Six separate excavated sites have revealed comprehensive evidence for primary and secondary glass production in the eighth–ninth, eleventh and twelfth centuries as well as unglazed pottery production at these times and glazed pottery production in the eleventh and twelfth centuries. Evidence for the environmental impact of the industries (pollution, fuel use) has been examined, along with the characterization of the raw materials used and the artefacts made from them (Henderson 1999; Henderson *et al.* 2004). Remote sensing and the geoloical information system (GIS) have provided clear evidence for the distribution and siting of production sites in the urban landscape (Challis *et al.* in press).

The excavated evidence of primary glass production from raw materials found at Raqqa is frit – partially fused batch materials (Henderson 2000, pp.

Fig. 1. A three-chambered beehived-shaped glass furnace showing: (**a**) the firebox; (**b**) the middle chamber with 'glory' holes (**c**) the annealing chamber, with (**d**) its opening; (**e**) the crucibles; and (**f**) the floor of the annealing chamber. From Georgius Agricola, *De Re Metallica* (1556). This is one type of glass furnace that is thought to have been used in the ancient world.

88–89) – and raw glass of three plant ash compositions attached to remnants of single-chambered tank furnace fragments. The evidence for glass working consists of a series of moils (collars knocked off the blowing iron) of different diameters, rejected vessel fragments (cullet) and cast glass. Excavated 'structural' evidence for glass production consisted of a glass workshop with the remains of small beehive-shaped glass furnaces (such as in Figure 1) and massive dumps consisting of the destroyed remains of single-chambered tank furnaces (Henderson 1999, p. 230; Henderson 2000, p. 43, fig. 3.16). Glass was both made from primary raw materials and worked at Raqqa: this may be a typical model for glass production in the urban centres of the Islamic world (Henderson 2003); raw glass may also have been exported/ traded to other centres which specialized only in glass-working.

The scientific analysis of the evidence for glass

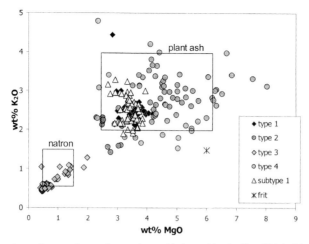

Fig. 2. Relative proportions of magnesium and potassium oxide impurities (wt% oxide) in Islamic eighth–ninth- and eleventh-century glasses from Raqqa, Syria, showing a gross distinction between glasses made from mineral and plant ash sources of alkali.

production has involved the use of EMPA, SEM and mass spectrometry (MS). EMPA was used for the determination of major, minor and some trace components (Henderson & McLoughlin 2002; Henderson et al. in press). An SEM has been used for the examination of crystalline materials in the glasses and MS for determining the stable isotopes in the glasses (Henderson et al. in press). The scientific analysis was carried out in order to determine the range of glass chemical compositions in use – and to detect any changes over time, to infer the raw materials used to make the glass, to investigate the possibility of glass recycling, to relate composition to vessel form, to relate glass technology to glaze technology and to evaluate whether the recipes used could be regarded as characteristic to production at the site or within a zone. This was the first time that an Islamic glass production site had been investigated in such comprehensive detail. The key to these investigations is the discovery of the evidence for the primary manufacture of glass in the shape of raw glass attached to furnace fragments. Without this evidence for primary glass production and its chemical characterization, it is difficult to stipulate where glass vessels might have been made; variations in vessel form, even if particular forms occur in geographical clusters, cannot provide such crucial evidence.

So far, 244 samples of glass have been analysed. The vast majority of the glasses analysed are of a soda-lime-silica composition. For the manufacture of soda glasses (type 3 in Figs 2–4) it is thought that the mineral natron or trona, a sodium sesquicarbonate, was combined with sand. Natron is a mixture of sodium carbonate and bicarbonate (Turner 1956, table IV). The sand is thought to contain shell parti-

cles which provided the lime component of the soda glass. For the manufacture of the second kind of soda-lime glass – the plant ash glass (types 1, 2 & 4 in Figs 2–4) – ashes of plants which grow on the margins of deserts were combined with quartz, a purer source of silica than sand. The plants themselves contain the calcium which provides the lime for the soda-lime glasses.

A variety of impurities are present in the natron (NaCl and $NaSO_4$) and plant ash (Mg, K, S, P and Cl), with others in the sand (e.g. Al, Fe and Ti) and quartz (Al). This is by no means an exhaustive list. Colorants such as cobalt and copper would probably have been added separately as part of colorant-rich materials, and would have introduced their own suites of impurities.

The glasses analysed derived from two periods of production (the eighth–ninth and the eleventh centuries). The period of glass production to be focused on here is the eighth–ninth centuries, the period when Harūn al-Rashid resided in Raqqa. A variety of vessel types and window glasses of a range of colours were investigated (Henderson 1999; Henderson & McLoughlin 2002; Henderson et al. 2004). The data were examined by plotting pairs of oxides against each other. This has the advantage of allowing the creation of relationships between chemical components and the inferred use of particular glass raw materials.

Figure 2 shows a plot of wt% magnesium oxide v. wt% potassium oxide in the glasses analysed. These oxides are thought to reflect the general type of alkali used in the glasses. A clear positive correlation is visible for most samples of natron glasses (type 3) plotted, a characteristic of many ancient glasses. The

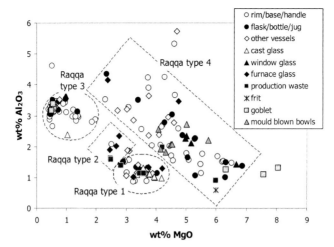

Fig. 3. Relative proportions of magnesium and aluminium oxide impurities (wt% oxide) in Islamic eighth–ninth- and eleventh-century glasses from Raqqa, Syria, showing the four different compositional types that have been detected.

inference is that the low magnesium oxide/potassium oxide glasses have been made with the mineral natron while those with higher levels have been made with plant ashes. The glasses which fall between the plant ash and natron glasses are probably a mixture of mineral and plant ash glass, although, as can be seen in Figure 4, the levels of calcium oxide show that glasses of types 1 (plant ash) and 3 (natron) found in Raqqa did not provide the base glass for this mixture. This already suggests that some recycling occurred at Raqqa, although at a relatively low level.

The range of plant species suitable for making glasses is quite wide. Three species of halophytic plants belonging to the Chenopodiceae family could have been used for glass production: *Salsola kali*, *Salsola soda* and *Hammada scoparia*. These plants grow on the margins of deserts and in maritime environments (Ashtor 1992, p. 494). Representatives of the Chenopodiceae family certainly grew in northern Syria (Ashtor 1992, p. 487–488) in the early Islamic period. Plants of the genus *Salsola* as well as *Salicornia*, grow near Raqqa today. The Arabic geographer al-Mukaddasī, writing in about 985, records that *Úshnān* (the literary Arabic word for the vernacular word *'Kali*) was exported from Aleppo some 150 kms upstream from Raqqa. (*Ushnān* is thought to be *Hammada scoparia* (Ashtor 1992, 482)). The purification of these plant ashes (known as *keli*, or *kali*) involved solution, filtration, concentration and crystallization, producing white salts of sodium carbonate. It is possible that variations in these procedures may have led to variations in the chemical composition of Islamic glass. From the last decades of the thirteenth century at the latest, Venetian glass-makers used imported plant ashes

from the Levant as a their soda source (Jacoby 1993, p. 68) and Italian glasses made from the ashes of *Salsola kali* have compositional similarities with those made in the Islamic Middle East (Verità 1985, p. 19–21, table I).

If we turn to the relative levels of wt% magnesium oxide v. wt% aluminium oxide (Fig. 3), we can observe compositional groupings which reflect the uses of differing silica sources. The presence of low levels of aluminium oxide in (ancient) glasses infers the use of a relatively pure source of silica, such as quartz pebbles from the Euphrates river valley; relatively high levels infer the use of aluminium oxide-rich impurities found in sand (Henderson 2000, p. 27). Type 1 glass, with low aluminium oxide and high magnesium and potassium oxides, is a plant ash-quartz glass; type 3 glass, with high aluminium oxide and low magnesium and potassium oxides, is a natron-sand glass; and, as mentioned above, type 2 appears to be a mixture of a plant ash and natron glass, but is not a mixture of the glasses so far analysed from Raqqa. At the high magnesium oxide end of the compositional range for type 4 we can suggest that quartz was melted with a different plant ash from that used to make type 1. A less likely interpretation is that a variation in the preparatory/ashing procedure was used. At the lower aluminium oxide end of the type 4 compositional range, it can be inferred that the same plant ash was used as for making type 1 but mixed with sand.

Four basic types of soda-lime-silica glass have therefore been detected, three made with plant ashes and one made with a mineral alkali (see Table 1). The existence of a wide range of type 4 compositions needs to be explained. The clear dilution line for type 4 shown in Figure 3 strongly suggests that different proportions of end member compositions have been mixed together to produce the compositions in between. Although this constitutes evidence for a form of glass recycling there is no apparent analytical evidence for the mixing of the other compositional types with each other, or with type 4. The inference is that experimentation with raw materials occurred at Raqqa in the late eighth to early ninth centuries. The correlation between glass colour and chemical composition (Henderson *et al.* 2004) would have provided glass-workers with a way of identifying glasses of those compositions with specific working properties. The use of a combination of sand and plant ash (high Al_2O_3, Raqqa type 4) to make ancient glass is unusual; another reported example, provided by the chemical analysis of the famous massive glass slab from Bet She'arim, Israel (Brill & Wosinsky 1965; Brill 1967; Freestone & Gorin-Rosen 1999) is probably contemporary with the Raqqa type 4 glass. This was originally interpreted as a failed experiment because of the excessive calcium oxide levels detected. However, it is now clear that successful experiments with similar raw materials were occurring elsewhere in the Islamic world and that the glass melt in the Bet She'arim can now be interpreted as a failed melt. Other supporting evidence for experimentation is the fact that a wide range of glass compositions were in use at this time, that type 4 (experimental) glasses were only in use for a relatively short time (*c.* 150 years), that the melting temperature of type 1 plant ash glass is *c.* 50°C lower than the natron glass found at Raqqa (McLoughin *et al.* 2000), and that this same plant ash glass became dominant after *c.* AD 1000. In addition there was a clear practical reason why this experimental phase occurred. The source of the mineral alkali, natron, at Wadi Natrūn, Egypt, which had been the primary alkali used for 1.6 Ka, was starting to run out, leading to lower total alkali levels in the glasses melted from it, ever-increasing glass melting temperatures, and an increase in the amount of the most expensive raw material used, the fuel (Henderson 2002, p. 601).

Although the alkaline fluxes used were a critical determinant in reducing the melting temperatures of the glass batch, the balance of other components, especially calcium oxide and silica, which would increase glass melting temperatures, were also significant. Figure 4 shows that another characteristic of type 4 glass is its lower calcium oxide levels, which would have helped to keep its melting temperatures low. Silica levels detected are comparable with those detected in type 3 and some type 1 glasses.

The discovery of this range of glass compositions and the evidence for experimentation with raw materials from a single production site is (so far) unique in the investigation of glass production in the ancient world. Until further archaeological and scientific

Table 1. *The inferred raw materials used for the production of the four types of soda-lime-silica glasses from eighth–ninth century Raqqa, Syria*

Glass type	Silica	Soda	Lime
Raqqa 1	Quartz	Plant ash	Plant ash
Raqqa 2	Sand/quartz	Plant ash	Plant ash
Raqqa 3	Sand	Natron	Shell fragments in sand
Raqqa 4 (Low MgO)	Sand	Plant ash	Plant ash Shell fragments
Raqqa 4 (High MgO)	Quartz	Plant ash	Plant ash

investigations of Islamic glass production sites occur, so as to place the discoveries from Raqqa in a broader context, it is difficult to be sure whether the plant ash glass made there was characteristic of the location or to the region. Contemporary glasses in use at Nishapur, Iran (some 1500 km from Raqqa) have been analysed by Brill (1995, 1999). These have a similar composition to those of Raqqa type 4 but can be distinguished on the basis of the relative wt% of calcium and sodium oxides (Henderson 2003, fig. 8). They can therefore be described as a compositional subtype of type 4 Raqqa glass. One interpretation of the existence of this compositional subtype is that, when plants of the same species, but growing in different areas, are ashed and then prepared in order to make glass, the glasses made from them are compositionally distinctive (Henderson 2003, p. 115). Given the large distances involved, this may not be deemed an especially important discovery, but it does show a compositional distinction that can be used to distinguish between contemporary glass vessels made in the two different zones. A glass production site which is geographically closer to Raqqa, and for which there are published analyses of *raw* plant ash glasses, is the eleventh–thirteenth-century site of Banias, Israel (Freestone *et al.* 2000, table 2). Even though the sites are geographically closer, the 'type 1' plant ash glasses from Banias, Israel (eleventh–thirteenth centuries) are compositionally distinguishable from contemporary ones found at Raqqa, especially in terms of relative levels of magnesium and potassium oxides (Fig. 5). The same reasons as given above can be suggested for the compositional distinction between Banias and Raqqa glasses. The tighter clustering of the Banias glass compositions may be explained by a number of possible factors. It may be a function of the smaller number of analyses of Banias glasses involved and/or it reflects the greater compositional variation of plants used to make the glass at Raqqa and/or that

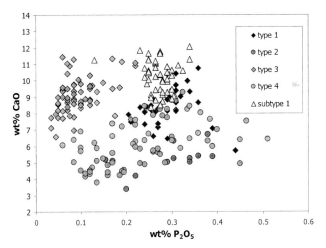

Fig. 4. Relative proportions of calcium oxide and phosphorus pentoxide (wt% oxide) in Islamic eighth–ninth- and eleventh-century glasses from Raqqa, Syria, showing characteristically low levels of calcium oxide in experimental type 4 glasses.

the ashes were prepared in a slightly different way to make glass at the two sites.

Although chemical analysis is a good way of characterizing materials, a far more powerful way is to combine these analyses with stable isotope determinations. For various reasons the two sets of results may not necessarily be correlated, and there is potential for mixing and recycling of metals (Rohl & Needham 1998, p. 36) and glasses which can blur the isotopic (and chemical) signatures of the raw materials used. Nevertheless, if isotopic and compositional data do correlate, confidence in the identification of production spheres is increased. If they don't correlate, the isotopic groupings are purely a reflection of the geological sources of raw material used. The compositional groupings result from a combination of using different types of primary raw materials and colorants together with variations in impurities associated with them and the potential addition of recycled glass to the original melt. Chemical analyses and isotope determinations have the potential to contribute to the difficult area of provenance (Wilson & Pollard 2001). Although obsidian has proved to be an ideal archaeological material for provenancing (Pollard & Heron 1996, p. 81; Henderson 2000, p. 305) glass has proved to be far trickier. In order to avoid circular arguments, initially it is of course crucial to determine stable isotope signatures of the primary raw materials (where identifiable) and the *raw glass* made from them found on production sites. It then becomes possible to compare the chemical analyses and isotope signatures for glass vessels of specific forms and decorations with the raw glass from production sites, potentially providing a provenance for the vessels.

Strontium isotope ratio determinations have provided distinctive signatures for natron glasses from four production sites: two in the Levant, one in Egypt, and one in the Eifel area in Germany (Freestone *et al.* 2003, pp. 27–29, fig. 3; Wedepohl & Baumann 2000). A number of parameters can affect the sources of strontium, producing variation in plants (Sillen & Sealy 1995; Price *et al.* 2002, p. 120). Moreover, the determination of strontium, lead and oxygen isotopes in plant ash glasses, when added to compositional evidence, could provide a reliable and critical means of distinguishing between glasses made in separate production zones, or even between glasses made on different production sites (Henderson *et al.* in press *b*). Indeed, for the first time, it has been possible to provide a provenance for plant ash glasses and should lead to the construction of trade networks for Islamic plant ash glass and glass vessels, and plant ash glasses made in other periods and areas. Its success depends on there being sufficient variation in the bedrock geology on which plants grow to provide distinctive isotopic signatures in the glasses made from them.

Ottoman Iznik pottery in the fifteenth–seventeenth centuries AD: the state of the art or the art of the state?

From *c*. 1480 until *c*.1620 the Ottoman world manufactured a highly decorated polychrome glazed ceramic with the widest range of colours found in any ancient pottery. Brightly coloured decorative designs were applied to a brilliant white background which set them off in a stunning way. Even though

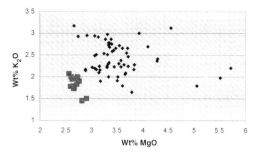

Fig. 5. Relative proportions of magnesium and potassium oxide impurities (wt% oxide) in Islamic glasses from eleventh-century Raqqa, Syria (lozenges) and eleventh-thirteenth-century Banias, Israel (squares) showing a compositional distinction between plant ash glasses made in two contemporary glass-making centres.

we have discovered minimal archaeological evidence for the production at Iznik (Atasoy & Raby 1989, fig. 44; Henderson 2000, pp. 184–186), it is clear from historical documents that the pottery was fired in kilns in Iznik (ancient Nicea) in Asia Minor.

In addition other centres also manufactured similar wares. John Carswell has noted that, in the sixteenth century, Kütahya, some 200 km south of Istanbul, produced pottery which is visually very similar to Iznik wares (Carswell & Dowsett 1972) and, on present analytical evidence (Tite 1989), appears to have been manufactured from the same raw materials. The archaeological evidence for its production there is also minimal (Şahin 1981). In eastern Turkey, Diyarbakir also produced provincial imitations of Iznik wares (Raby 1977–8; Atasoy & Raby 1989, p. 74). An illustrated reconstruction of the kind of updraught kiln in which Iznik pottery and tiles were fired has been published by Porter (1995, fig. 2). The glazed wares would have been separated by being located on kiln rods and plates.

The sixteenth century saw one of the most successful periods of the Turkish Ottoman Empire measured in terms of expansion and political influence, and included the period when Süleyman the Magnificent, who reigned from 1520 to 1566, held court in the Topkapi palace in Istanbul. This period saw a massive patronage of the arts, including the production of Iznik. The significance of pottery in court life is reflected in the fact that, even today, the Topkapi palace holds one of the largest collections of Chinese celadon pottery in the world (10 600 pieces), the remnants of what was imported in the sixteenth century. It is clear that such pottery had one of the highest social and economic values.

One of the advantages of working with material from a historical period is that, quite apart from Iznik being perceived as a high-quality product, it affords an independent means of assessing its social and

economic value. Atasoy & Raby (1989) in particular have reviewed the Ottoman documents that shed light on the production and use of the ware. An example of its use for a specific social occasion was the purchase in 1582 of 541 Iznik plates, dishes and bowls from the bazaar in Istanbul for the sumptuous banquets which were celebrated to commemorate the circumcision of Prince Mehmed, the son of Sultan Murad III (Atasoy & Raby 1989, p. 14). Other documents considered by Atasoy & Raby were schedules of fixed prices, inventories of the effects of deceased persons and palace registers. Ottoman society was a hierarchical one in which the military class was held in high esteem. They often practised farming and stock rearing, and owned slaves, and there is historical evidence that they owned Iznik pottery. A probate inventory dated 1548 for Hâce Ishak bin Abdürrezzak, a wealthy draper and clothier who owned houses in Edirne and farms in the country, mentions specifically that he owned Iznik pottery (Atasoy & Raby 1989, p. 26). Again these documents reinforce the link between high social status and the high value of Iznik pottery.

Given that Iznik had a high value we must investigate how the technology emerged and what characteristics led to the perfection of the technology. Iznik is what is known as a 'stonepaste' ceramic. The manufacture of stonepaste pottery (also known as fritware) is mentioned in Abūl Qa-sim's treatise, which describes the production of Iranian fritware, a ware which is allied to Iznik. It consists of two manuscripts dating to 1301 and 1583 respectively (Allan 1973). The body of Iznik pottery was made from ten parts of ground quartz, one part of fine white clay and one part of (glaze) frit. As in Islamic glass production (essentially the same thing), the quartz was provided by ground-up river pebble. The frit was made from a calcined halophytic plant (see above) and quartz. The pots are likely to have been biscuit fired (Henderson 2000, p. 124) and then decorated with slips and glazes. The brilliant white background to the glaze decoration was provided by a white slip; a red slip (Armenian bole) was also sometimes applied on top of the white slip. The glaze decoration was painted on the white slip layer often within 'black' outlines and the whole sealed in with a layer of colourless overglaze.

This was an existing technique which came into its own in fifteenth-century Egypt and Syria (Porter 1995, p. 17) where a mixture of Islamic- and Chinese-inspired designs can be seen in blue-and-white tiles (Porter 1995, fig. 85). As the pot was fired, a glassy network developed between the silica crystals (Fig. 6), and the use of this lead-rich glassy network improved the glaze fit. As the pot cooled down, the lead in the frit helped to match the coefficient of contraction of the lead in the glaze. All of these features can be seen in Figure 7, a backscattered SEM

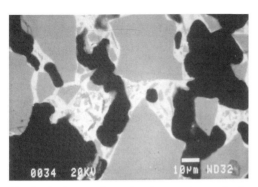

Fig. 6. A backscattered scanning electron micrograph of a polished section through the body of a typical Iznik body sherd showing the existence of a network of lead-rich glass (white) connecting angular silica crystals (grey). The black areas are voids.

Fig. 7. A backscattered scanning electron micrograph of a polished section through the glaze and body of a typical Iznik pottery sherd. The glaze layer (white) lies over a thin interaction layer between body and slip. The slip layer contains smaller silica crystals than the much thicker underlying body. The two layers are divided by a clear horizontal line. Vitreous areas are visible in the body appearing white.

Table 2. *A range of decorative styles used in Turkish pottery from 1430–c. 1650*

Date	Description
1430	'Masters of Tabriz' blue and white
*c.*1460	Miletus ware blue and green
1480	Blue-and-white from Iznik
1510	Blue-and-white from Iznik
1520	Blue-and-white from Iznik
1530	Tugrakes Spiral blue-and-white from Iznik
1540	Potters-style grape dish fragment blue-and-white with turquoise from Iznik
1550s	Damascus ware dark, blue, olive green, turquoise, 'black' outlines from Iznik
1560–70	Slip-painted. Red slip, blue-and-white from Iznik
1580	Rhodian ware. Bole red, blue and emerald green from Iznik
1580	Rhodian ware. Bole red, blue, turquoise, emerald green and 'black' from Iznik.
1580	Late blue-and-white from Iznik.
*c.*1650	Emerald green, blue, 'black' and red

image of a thick section through an Iznik sherd. Additional features are an interaction layer between glaze and slip and the use of smaller silica crystals in the slip layer rather than in the body of the pot.

Iznik first appeared in Ottoman Turkey *c.*1480. The range of colours used to make the glaze decoration gradually increased, from initial blue-and-white in 1480 to a complex polychrome decoration including cobalt blue, aubergine purple, emerald green, turquoise green and a tomato-coloured red slip. Dating of the different decorative styles has been established by using the secure construction dates of mosques and minarets, which themselves were decorated with Iznik tiles. Dating relies heavily on original research conducted by Lane (1957). Dates and stylistic phases of Iznik pottery (Carswell 1998), together with the types which preceded Iznik, are given in Table 2.

Having described some of the characteristics of the social and political context in which this Iznik pottery was made and used, we must consider the research questions which impinge on the first appearance of Iznik ware and its development. The questions are: (1) did Iznik technology develop from Chinese blue-and-white porcelain, local Miletus ware or 'Masters of Tabriz' ware; (2) does the body of Iznik pottery reflect changes in glaze compositions over time?

Given the great political strength of the Ottoman Turks, from the fifteenth century onwards the conditions lent themselves to the development of existing technologies and the introduction of new ones. Fourteenth–seventeenth century Yuan and Ming Dynasty Chinese blue-and-white porcelain clearly inspired the designs of Iznik blue-and-white wares (Carswell 1998, p. 12). The translucent nature of porcelain was also something to imitate. It might be argued that the very act of using the blue-and-white glaze designs found on porcelain generated an important part of the demand to produce Iznik. Another possible influence was the production of a local glazed earthenware called Miletus ware. This already displayed many of the decorative features of Iznik, such as bunches of grapes and spirals. Another family of pots is thought to have been made by the Masters of Tabriz, who arrived in Turkey from Iran in the 1410s and tiled a sequence of imperial build-

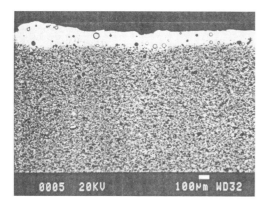

Fig. 8. A backscattered scanning electron micrograph of a polished section through the glaze and body of a Chinese porcelain sherd showing an entirely different structure from Iznik ware seen in Figure 7.

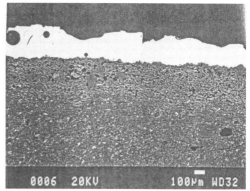

Fig. 9. A backscattered scanning electron micrograph of a polished section through the glaze and body of a sherd of Miletus ware. Cracks are visible in the earthenware body and there is a considerably lower incidence of silica crystals compared to Figures 7 and 8.

ings between 1420 and 1474 (Henderson & Raby 1989, pp. 117–118). Proof, if needed, that they were involved is provided by a Persian inscription on tiles decorating the Mehmed I complex at Bursa, which was built between 1419 and 1424: *amal-i-ustadan-i Tabriz* which means 'made by the Masters of Tabriz' (Porter 1995, p. 99). The products of the Masters of Tabriz may also have provided a contribution to the emergence of Iznik pottery and tile technology. However, there remains the question of whether the bodies of 'Masters of Tabriz', Chinese porcelain and Miletus ware have any similar characteristics.

One way of answering this is to compare the typical structure of the Iznik body (Fig. 7) with those of the other three types. It is clear that the typical structure of porcelain (Fig. 8) is distinct, with no slip layer and an entirely different mineralogical structure. The structure of Miletus ware is again distinctly different: although there is a slip layer, there is no frit in the body and the proportion of silica is clearly significantly lower: it is earthenware not stonepaste (Fig. 9). That leaves the 'Masters of Tabriz' ware. As can be seen in Figure 10, although there is a patchy development of a vitreous phase in some areas of the body, it is not distributed throughout the fabric in the same comprehensive way that it is in Iznik. Another feature of the 'Masters of Tabriz' fabrics which distinguishes them from Iznik is the much wider range of crystal sizes and types (Fig. 10). The development of a vitreous phase occurs in a range of pottery fabrics (Henderson 2000, p. 132), so it may not be regarded as an especially diagnostic feature of ceramic technology. However, a characteristic which sets the use of 'frit' in Iznik bodies apart from its predecessors (such as the 'Masters of Tabriz' ware) is the fact that Iznik frit is always lead-rich. Earlier, fifteenth-century tiles, for example, tended to contain either no detectable glass phases at all or a

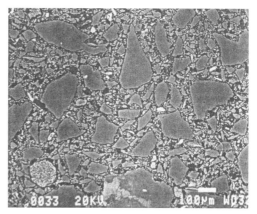

Fig. 10. A backscattered scanning electron micrograph of a polished section of a sample of the body of a tile ascribed to the Masters of Tabriz in which no visible glassy network is discernible.

sporadic distribution of lead- and tin-rich or alkaline glassy inclusions (Henderson & Raby 1989, pp. 122–123). The presence of lead in the glassy network in the body of Iznik pottery would help to match the relative coefficients of contraction of Iznik bodies and glazes, improving the all-important glaze fit (Henderson 2000, p. 124). The potter would have thrown scrap glass, or remnants of glaze scraped from the vessel in which it was prepared, into the clay and silica mix, a practice which was carried out in the manufacture of Delft wares (Caiger-Smith pers. comm.)

The use of lead brings us to a consideration of lead oxide in the glazes themselves (Tite *et al.* 1998). All Iznik glazes are, without exception, lead oxide-soda-lime-silica in composition (Henderson 1989; Tite *et*

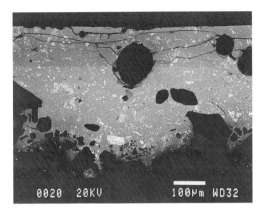

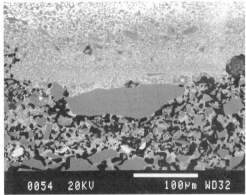

Fig. 11. A backscattered scanning electron micrograph through a polished section through opaque turquoise glaze and the surface of the body of a tile from the Edirne Yeşilce Cami dated 1440–1441. The agglomerated white crystals in the glaze are tin oxide. The upper, darker, layer of glaze contains a lower lead oxide level which means that it would have remained fluid for a shorter period of time than the lower layer and on cooling it would have contracted faster. This has led to a series of horizontal cracks in the outer glaze layer.

Fig. 12. A backscattered scanning electron micrograph through a polished section of red Armenian bole pigment which was used in the decoration of Iznik from a potsherd dating to *c*. 1580. The Fig. shows the deposit of very finely divided (pale grey) iron-rich pigment lying on the white slip layer which itself contains bright (white) chromite crystals. The chromite crystals which have become mixed in with the (white) slip layer were used as a dark pigment used for painting the outlines of glaze decoration on the (white) slip surface. At the bottom of the Fig. are several large silica crystals; these are the top surface of the pot body. There is a large silica crystal in the centre of the image lying in the top layer of the slip.

al. 1989). Another characteristic feature is the presence of tin oxide. Although from *c*. eighth-century tin oxide was used as a white opacifier in Islamic glazes, an Islamic technological innovation (Mason & Tite 1997), in Iznik glazes a white background is provided by a brilliant white slip instead, and although 3–6% tin oxide is present it is *dissolved* in the colourless overglaze. While lead oxide would already change the refractory properties of the glaze it is thought that tin oxide dissolved in the overglaze would, in addition, tend to increase its brilliance. This being the case, the presence of opacifying tin oxide crystals in Ottoman pre-Iznik glazes (Fig. 11) as late as 1474 (Henderson & Raby 1989, Plate 6), 6 years before the first Iznik pottery, indicates that the dissolution of tin oxide in Iznik glazes is a new technological development, which again sets it apart from its predecessors.

It would therefore appear that not only was the use of a lead-rich glass in the body of Iznik a new departure in Islamic pottery, but so was the solution of tin oxide in the glaze combined with a brilliant white slip layer. It can be suggested that, although stonepaste was not a new technology, the development of these specific features in Iznik pottery technology was. Neither the more refractory imported Chinese porcelain, the local earthenware Miletus ware nor the wares apparently made by the Masters of Tabriz displayed these technological features. Therefore several aspects of the production of Iznik pottery production involved technological innovation.

The second research question is whether glaze compositions and body structure are correlated. Although the range of glaze colours and decorative tricks increased between *c*. 1480 and 1600 (Table 2) the chemical composition of the glaze remained essentially the same. The colourless overglaze has a lead oxide-soda-lime-silica composition in all cases. The wide range of underglaze colours used involved the use of an ever-increasing range of mineral-rich compounds. The initial use of cobalt-rich minerals producing the blue colour in blue-and-white glazes was followed by the introduction of copper (II) (CuO) for the turquoise blue colour; manganese oxide for purple; iron oxide (ferrous and ferric) for emerald green; chromium oxide for the 'black' pigment used for painting the outlines of decoration; and a special painted-on slip, known as Armenian bole, coloured with iron-rich particles and reduced copper (I) (Cu_2O) which produced an opaque red and more yellow orange colours (Fig. 12). The use of chromium-rich minerals in Iznik pottery is the earliest in ancient glass and glaze technology. Chromic oxide was not introduced as a glass colorant until the nineteenth century.

Although the range of colorants increased over the 120-year production period, the principle characteristics of the fabric remained essentially the same. It was not until *c*. 1650 that the quality of Iznik glazes dropped, with the glaze rolling off the dry brownish bole. There was an attempt to use bone ash as a white opacifier in the glazes, which also resulted in the glaze shivering; the underglaze colours ran in the colourless glaze which was, nevertheless, of the same lead oxide-soda-lime-silica composition. Phosphorus pentoxide detected in the body may be a very early attempt to use it as a flux (Henderson 1989, p. 68).

The same very high level of skill in the manufacture of Iznik was retained for over some 120 years, so the decline in technology was in stark contrast. There are several possible explanations for this decline. One overwhelming reason was that the harsh economic environment in which the potters operated reduced the demand for fine Iznik dramatically (Atasoy & Raby 1989, p. 32) and formed part of the slow decay of the Ottoman Empire (Çarswell 1998, p. 106). Linked to this, the networks for importing raw materials must have collapsed. A Turkish traveller, Evliya Çelebi, commented that there were only nine pottery workshops left at Iznik in 1648, compared to more than 300 only 50 years earlier (Carswell 1998). Although difficult to ascertain, it appears that the actual knowledge involved in the production processes may also have been lost; even good modern copies of Iznik which can be bought in Istanbul still do not achieve the levels of perfection attained during the sixteenth century.

Conclusions

The two case studies presented here provide different levels of evidence for the production and use of glass and pottery. The example focusing on eighth–ninth-century glass discusses the complexities of interpreting the scientific analyses of the debris left by an industry. This forms the essential basis for further research in comparable production centres and zones, and, it is hoped, with the incorporation of isotope analyses, it will lead (for the first time) to the provenancing of Islamic glass in the region and beyond. The investigation of Iznik ware involved fewer samples, but a consistency in the technology suggests that the analysis of more samples would not necessarily clarify the situation. Although Kütahya was another production centre for so-called 'Iznik' ware, as yet there has been no way of distinguishing between the products of the two centres analytically. Perhaps a technique such as LA-ICP-AES or MS would provide the level of analytical sensitivity which could distinguish between the products from the two centres.

Clear parallels can be found in modern production, distribution and use of glass and pottery, including the basic raw materials used. Modern glass and clays do tend to be more refined than their ancient counterparts and, as such, would introduce

fewer diagnostic impurities. It is, however, these impurities which in the long term (when combined with isotope analyses) could lead to ancient glass provenance. For modern glass, however, minor differences in the purification of glass batch recipes or clay purification techniques could be a means of distinguishing between the products of factories or countries of origin. At least it could allow the possible sources to be narrowed down by a process of elimination. As with research into ancient glass and pottery it is necessary to assemble a good series of reference analyses thought to characterize the factory outputs at different times. The greater purity of the modern raw materials could reduce the potential for characterizing the products but, in spite of this, the chemical analysis of microsamples of window glass, for example, could provide vital information about the scene of a crime.

References

ALLAN, J. W. 1973. Abūl Qāsim's treatise on ceramics. *Iran*, **11**, 111–120.
ASHTOR, E. 1992. Levantine alkali ashes and European industries. *In*: KEDAR, B. Z. (ed.) *Technology, Industry and Trade*. Variorum, Vermont, 475–522.
ATASOY, N. & RABY, R. 1989. *Iznik, the Pottery of Ottoman Turkey*. Alexandria Press, London.
BAXTER, M. J. 1994. *Exploratory Multivariate Analysis in Archaeology*. Edinburgh University Press, Edinburgh.
BRILL, R. H. 1967. A great glass slab from Ancient Galilee. *Archaeology*, **20**, 88–95.
BRILL, R. H. 1995. Chemical analyses of some glass fragments from Nishapur in the Corning Museum of Glass. *In*: KRÖGER, J. (ed.) *Nishapur: Glass of the Early Islamic Period*. Metropolitan Museum of Art, New York, 211–233.
BRILL, R. H. 1999 *Chemical Analyses of Ancient Glasses. Vol. 2. The Tables*, The Corning Museum of Glass, New York.
BRILL, R. H. & WOSINSKI, J. F. 1965. A huge slab of glass in the ancient necropolis of Beth She'arim. *Comptes Rendus, VIIe Congrès International du Verre, Bruxelles, 28th Juin–3 Juillet 1965*. Section B. Paper 219.
CARSWELL, J. 1998. *Iznik Pottery*, British Museum Press, London.
CARSWELL, J. & DOWSETT, C. J. E. 1972. *Kütaha Tiles and Pottery from the Armenian Cathedral of St. James, Jerusalem*. Vols 1 & 2. Oxford University Press, Oxford.
CHALLIS, K., PRIESTNALL, G., GARDNER, A., HENDERSON, J. & O'HARA, S. (in press). The use of corona remotely-sensed imagery in dryland archaeology: an example from the Islamic city of al-Raqqa, Syria. *Journal of Field Archaeology*, **29** for 2002–4 (2004).
DAVY, H. 1815. Some experiments and observations on the colours used in painting by the Ancients. *Philosophical Transactions of the Royal Society of London*, **105**, 97–124.
DOBRES, M-A. & HOFFMAN, C. R. 1994. Social agency and

the dynamics of prehistoric technology, *Journal of Archaeological Method and Theory* **1**, 211–258.

DRENNAN, R. D. 1996 *Statistics for Archaeologists: A Commonsense Approach*. Plenum Press, New York.

FLETCHER, M. & LOCK, G. R. 1991. *Digging Numbers: Elementary Statistics for Archaeologists*, University Committee for Archaeology, Oxford.

FREESTONE, I. C. & GORIN-ROSEN, Y. 1999. The great slab at Bet She'arim, Israel. *Journal of Glass Studies*, **41**, 105–116.

FREESTONE, I. C., GORIN-ROSEN, Y. & HUGHES, M. J. 2000. Primary glass from Israel and the production of glass in late antiquity and the early Islamic period. *In*: NENNA, M. D. (ed.) *La Route du Verre*. Maison de l'Orient Méditerranéan-Jean Pouilloux, Lyon, 65–84.

FREESTONE, I. C., LESLIE, K. A., THIRLWELL, M. & GORIN-ROSEN, Y. 2003. Strontium isotopes in the investigation of early glass production: Byzantine and early Islamic glass from the Near East, *Archaeometry*, **45**, 19–32.

HENDERSON, J. 1989. Iznik ceramics: a technical examination. *In*: ATASOY, N. & RABY, J. (eds) *Iznik, The Pottery of Ottoman Turkey*. Alexandria Press, London, 65–70.

HENDERSON, J. 1999. Archaeological and scientific evidence for the production of early Islamic glass in al-Raqqa, Syria. *Levant*, **31**, 225–240.

HENDERSON, J. 2000. *The Science and Archaeology of Materials*. Routledge, London and New York.

HENDERSON, J. 2002. 'Tradition and experiment in first millennium A.D. glass production – the emergence of early Islamic glass technology in late antiquity. *Accounts of Chemical Research*, **35**, 594–602.

HENDERSON, J. 2003. Glass trade and chemical analysis: a possible model for Islamic glass production. *In*: FOY, D. & NENNA, M-D. (eds) *Échanges et commerce du Verre dans Le Monde Antique*. Monographies Instrumentum, Éditions Monique Mergoil, Montagnac, **24**, 109–123.

HENDERSON, J. & MCLOUGHLIN, S.D. 2002. Glass production in al-Raqqa: experimentation and technological changes. *Proceedings of the 15th Congress of the International Association for the History of Glass, New York*, 144–148.

HENDERSON, J. & RABY, J. 1989. The technology of fifteenth century Turkish tiles: an interim statement on the origins of the Iznik industry. *World Archaeology*, **21**, 115–132.

HENDERSON, J., CHALLIS, K., GARDNER, A., O'HARA, S. & PRIESTNALL, G. 2002. The Raqqa ancient industry project. *Antiquity*, **76**, (291), 33–34.

HENDERSON, J., MCLOUGHLIN, S. & MCPHAIL, D. (2004). Radical changes in Islamic glass technology: evidence for conservatism and experimentation with new glass recipes from early and middle Islamic Raqqa, Syria. *Archaeometry*, **46** (3).

HENDERSON, J., EVANS, J. A., SLOANE, H. J., LENG, M. J. & DOHERTY, C. (in press). The use of oxygen, strontium and lead isotopes to provenance ancient glasses in the Middle East. *Journal of Archaeological Science*.

HUNTER, J. R., ROBERTS, C. A. & MARTIN, A. (eds) 1996. *Studies in Crime: an Introduction to Forensic Archaeology*. Seaby/ Batsford, London.

JACOBY, D. 1993. Raw materials for the glass industries of Venice and the *Terraferma*, about 1370 – about 1460. *Journal of Glass Studies*, **35**, 65–90.

LANE, A. 1957. The Ottoman pottery of Isnik. *Ars Orientalis*, **2**, 247–281

LEMONNIER, P. (ed.) 1993. *Technological Choices Transformation in Material Culture Since the Neolithic*. Routledge, London and New York.

MASON, R. B. & TITE, M. S. 1997. The beginnings of tin-opacification of pottery glazes. *Archaeometry*, **39**, 1, 41–58.

MCLOUGHLIN, S., AL-SADIQ, N., HENDERSON, J. & MCPHAIL, D. S. 2000. On technological change in Islamic glass production in Raqqa, Syria. *Glass News*, **9**, 4–5.

POLLARD, A. M. & HERON, C. 1996. *Archaeological Chemistry*. The Royal Society of Chemistry, Cambridge.

PORTER, V. 1995 *Islamic Tiles*. British Museum Press, London.

PRICE, T. D., BURTON, J. H. & BENTLEY, R. A. 2002. The characterization of biologically available strontium isotope ratios for the study of prehistoric migration. *Archaeometry*, **44**, 117–136.

RABY, J. 1977–8. Diyarbakir: a rival to Iznik: a sixteenth century tile industry in Eastern Anatolia. *Istanbuler Mitteilungen*, **27/28**, 429–459.

ROHL, B. & NEEDHAM, S. 1998. *The Circulation of Metal in the British Bronze Age: The Application of Lead Isotope Analysis*. The British Museum, London, Occasional Papers, **102**.

ŞAHIN, E. 1981. Kütayha seramik teknolojisi ve çini firinlari hakkinda görüşler. *Sanat Tarihi Yilli ği*, **11**, 133–151.

SILLEN, A. & SEALEY, J. C. 1995 Diagenesis of strontium in fossil bone: a reconsideration of Nelson *et al.* (1986). *Journal of Archaeological Science*, **22**, 313–320.

TITE, M. S. 1982. *Methods of Physical Examination in Archaeology*. Seminar Press, London.

TITE, M. S. 1989. Iznik pottery: an investigation of the methods of production, *Archaeometry*, **31**, 115–132.

TITE, M. S., FREESTONE, I. C., MASON, R., MOLERA, J., VENDRELL-SAZ, M. & WOOD, N. 1998. Lead glazes in antiquity – methods of production and reasons for use. *Archaeometry*, **40**, 241–260.

TONGHINI, C. & HENDERSON, J. 1999. An eleventh century pottery production workshop at al-Raqqa, preliminary report. *Levant*, **30**, 113–127.

TURNER, W. E. S. 1956. Studies of ancient glass and glass-making processes. Part V. Raw materials and melting processes. *Journal of the Society of Glass Technology*, **40**, 277T–300T.

VERITÀ, M. 1985. L'invenzione del cristallo muranese: una verifica analitica della fonti storiche. *Rivista della Stazione Sperimentale del Vetro*, **1**, 17–36.

WEDEPOHL, K. A. & BAUMANN, A. 2000. The use of marine molluskan shells for Roman glass and local raw glass production in the Eifel area (Western Germany). *Naturwissenschaften*, **87**, 129–132.

WILSON, L. & POLLARD, A. M. 2001. The provenance hypothesis. *In*: BROTHWELL, D. R. & POLLARD, A. M. (eds) *Handbook of Archaeological Sciences*. John Wiley, Chichester, 507–517.

Forensic applications of Raman spectroscopy to the non-destructive analysis of biomaterials and their degradation

HOWELL G. M. EDWARDS

Department of Chemical and Forensic Sciences, University of Bradford, Bradford BD7 1DP, UK (e-mail: h.g.m.edwards@bradford.ac.uk)

Abstract: An initial survey of the advantages and disadvantages of Raman spectroscopic techniques for application to forensic crime scene analysis in a geoscience context is followed by some illustrative examples that demonstrate the potential information which can be forthcoming from Raman spectral data and molecular characterization. A range of specimens is reported, including Egyptian human mummies, ice-mummified bodies, resins and ivories; all of these can be related through a geoscience context and the potential for forensic application is indicated.

The analytical characterization of biomaterials and their degradation products is important for a wide range of specimens from forensic crime scene environments. The deposition of materials in burial environments starts a complex chain of events which results in loss or change of material through degradation and the acquisition of components from the surroundings. The recognition of the factors which affect specimen integrity and which may provide clues as to changes in molecular composition arising from the burial or storage environments is an essential part of the subsequent analytical portfolio.

Physical and biological attack on biomaterials often results in problems of specimen deterioration and fragility, with consequent changes in colour and the deposition of debris being observed. These effects, which are often particularly useful in the characterization of geological soil specimens in the neighbourhood of a crime scene, can hinder the application of optical spectrometric techniques for the evaluation of biomaterials associated with the scene. This can be especially relevant, for example, in the infrared examination of fabrics and artefacts from waterlogged burial sites, because the absorption by water of infrared radiation is significant; often, the desiccation of these specimens is not particularly achievable since specimen rupture or disintegration may result. Also, the loss of infrared radiation transmission at wavenumbers lower than $1000\,cm^{-1}$ in such cases, especially for cellulose materials with hydroxyl groups, means that access to an important region of the electromagnetic spectrum for the characterisation of mineral debris is denied. (Edwards 2001). Many of these problems can be avoided using Raman spectroscopy. In this paper, the nature of the method is described and a number of case illustrations are presented.

The Raman spectroscopic method

The viability of the application of Raman spectroscopic analytical techniques to forensic geoscience problems is receiving attention for several important reasons:

(1) The full vibrational spectroscopic wavenumber range from $c.\ 50-3500\,cm^{-1}$ is normally achievable; this means that structural information about human biological materials, such as skin, hair, nails, skeletal components (such as bones and teeth), fabrics, resins and waxes can be recorded along with associated inorganic materials from a scene, such as mineral pigments, geological fragments, metal oxides and paints.

(2) Specimen presentation to the spectrometer is generally very simple; no sample pre-treatment of a chemical or mechanical nature is required – the method is *non-destructive*, and this has some important ramifications for the preservation of evidential material or, when specimens are in only limited supply, the procurement of analytical information from several sequential techniques then becomes critically important for the maximizing of information.

(3) Instrumentation involving a dedicated Raman microscope facility and a remote probe sensor device are commercially available, so that very large specimens (e.g. statues or complete elephant tusks weighing 30 kg or more) or microscopic particles (e.g. dust or pigment grains on fabrics, of the order of only several ng or pg) can be examined. The sampling 'footprints' in these cases are typically about $100\,\mu m$ down to only $1-2\,\mu m$.

(4) Quantification, where achievable, depends on the linear dependence of the intensity of the scattered Raman radiation on the molecular species concentration (Long 2002); this has

From: PYE, K. & CROFT, D. J. (eds) 2004. *Forensic Geoscience: Principles, Techniques and Applications.* Geological Society, London, Special Publications, **232**, 159–170. © The Geological Society of London, 2004.

been used to particularly good effect in the Raman spectroscopic analysis of dry mixtures (Hendra *et al.*1991), and for pigment composition. Calibration in terms of a *molecular scattering coefficient* for each molecular species under study is necessary, since some materials have much stronger response to laser excitation than others, i.e. the sensitivity of the technique is not the same for all materials (Long 2002).

(5) Specialized Raman techniques such as SERS (surface-enhanced Raman spectroscopy), RRS (resonance Raman spectroscopy) and a combination of these, SERRS (surface enhanced resonance Raman spectroscopy) are now also finding applications in forensic analysis, in that they respond to special enhancement of sensitivity for material detection through the use of surface plasmon excitation or colour (laser excitation at or near the electronic absorption band maxima of a molecular species).

(6) The selection rules for vibrational Raman scattering, which dictate whether or not a particular band will be observable, depend on the bond polarizability changes during the vibration, unlike the infrared absorption technique which requires a dipole moment to be manifest for a band to appear in the spectrum (Long 2002). A consequence of this is that Raman bands for homopolar entities are often significantly stronger than can be observed in the infrared, if these species have an activity at all in the latter. For example, the ν (SS) mode in keratotic proteins, such as human skin, hair and nail, is observed near $500\,cm^{-1}$ in the Raman spectrum, and its wavenumber position is critically dependent on the C–S–S–C conformations present; this has been used to good effect to study the degradation of human hair in burial environments, degradation in the outermost *stratum corneum* layer of human skin and the results of the application of chemical oxidation treatment to human hair, such as bleaching and permanent waving. The –SS mode exhibits no dipole moment change during vibration and hence is not seen at all in the infrared absorption spectrum of human keratotics. In contrast, the Raman spectral scattering of bands which are highly polar is often significantly weaker than the infrared counterpart; hence, –OH and >C=O modes are very strong in the infrared but occur only weakly in the Raman spectrum. This means that the Raman spectra of cellulose fabrics such as cotton and linen are often clearly defined compared with their infrared counterparts, and that specimens such as human skin and nail, which can contain up to 150% water content, give good-quality Raman spectra (Edwards 2001).

Alongside these advantages, there are also several possible disadvantages which may be encountered in the application of Raman spectroscopy to forensic crime scene analysis: the most significant of these is the onset of fluorescence emission, which is several orders of magnitude more intense than the Raman effect, and which can swamp completely the weaker Raman spectral features in a broad, diffuse spectral 'background'. It is to be expected that materials excavated from some burial environments could have absorbed or leached out fluorescent chemical components which will affect the Raman spectral quality. Also, some degradative processes produce fluorescent products which result in inferior spectral data; a good example of this is provided by the thermal processing of some natural resins. Selection of a low-energy Raman excitation wavelength in the near infrared can provide a means of overcoming the fluorescence problem.

Hitherto, there has been a relatively large input of Raman spectroscopic data from materials in art collections and in museums, from which strategies for conservation have been determined based on the identification of degradative products and of original source materials. It is now realized that the molecular information provided from Raman spectroscopic studies of materials relevant to forensic crime scene analysis can be correlated with the elemental information from inductively coupled plasma atomic emission spectroscopy (ICP-AES) and scanning electron microscopy (SEM) techniques, often associated with additional species data from gas chromatography mass spectrometry (GC-MS) and x-ray diffraction (XRD) analyses, to provide a more complete analytical picture of the specimen.

Experimental procedure, equipment and materials

Fourier-transform Raman spectra (FT-RS) were excited using a Bruker IFS66/FRA106 instrument with a dedicated Raman microscope, operating with a Nd^{3+}/YAG laser at 1064 nm in the near infrared. The long wavenumber laser excitation facilitated the generation of spectra with minimal fluorescence emission from samples excavated from a variety of burial sites. Typically, the sample 'footprint' was $100\,\mu m$ in macroscopic mode and about $8\,\mu m$ in the microscopic mode using a $\times100$ microscope objective lens. With a $4\,cm^{-1}$ spectral resolution, some 2000–4000 spectral scans were accumulated over the wavenumber range $50–3500\,cm^{-1}$ at a rate of about 2 scans per s to provide a resultant spectrum with improved signal to noise ratio. Comparison with internal laser calibration facilitated the recording of spectral band wavenumber positions to a precision of $+1\,cm^{-1}$ or better for strong, sharp Raman bands.

Since the intensity of Raman scattering is dependent on λ^{-4}, where λ is the excitation wavelength, the selection of a laser operating in the ultraviolet or visible regions of the electromagnetic spectrum affords the greatest opportunity for the observation of relatively weak Raman spectra; for example the Raman spectrum of a specimen excited by 250 nm laser radiation in the ultraviolet will be some $\times 260$ stronger than that excited using 1064 nm radiation in the near infrared for identical incident laser powers and assuming that all other instrumental conditions are similar. However, high-energy ultraviolet radiation can also exceptionally cause specimen damage. Major advantages of visible and low-wavelength excitation in Raman spectroscopy are the rapidity of response and data acquisition afforded by charge-coupled diode array detectors, the μm^3 sample volume required for analysis and the potential for portable systems for in-field studies; the latter fortunately is now being realized and is relevant to forensic crime scene applications of the Raman technique. Currently, commercial diode array detectors are available in the near infrared for around 800 nm laser excitation and then only with limited Stokes wavenumber-shift range; this can restrict the observation of Raman spectra somewhat while off-setting the longer spectral accumulation times required by FT interferometric recording instruments operating at 1064 nm. The latter require several minutes for satisfactory data accumulation, which can be realized in seconds with lower wavelength excitation.

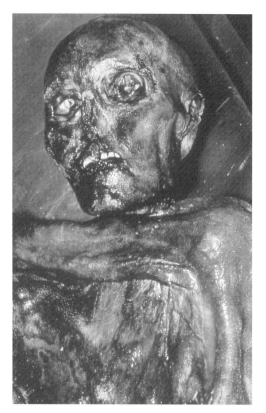

Fig. 1. The Alpine Iceman; a Neolithic ice-mummified body dating from 5.2 ka BP.

Specimens

The specimens selected for analysis here present some of the important facets on which novel information is available from Raman spectroscopy in a geoscience context, *sensu lato*:

1. Human mummified skin tissue from ice-mummies

The processes by which human or animal bodies are preserved are of historical and scientific interest since they reflect the cultural development and technologies of ancient peoples. Mummification can represent either a natural process of drying, which is often seen in mummies found in hot or cold deserts, or an artificial process whereby the body has been treated with different substances to promote preservation. The best-known examples of artificial mummification are provided by Middle Kingdom Egyptian burials recovered from pyramid or rock tombs. Surprisingly, it is not always clear whether the body has been mummified naturally or artificially, and this is an issue of

great importance in anthropological and archaeological research. It may be often assumed that mummies found in dry areas (deserts) are naturally mummified. However, some differences may be observed in the appearance of the skin from mummified bodies found in the same area which could suggest real differences in the process of mummification. Therefore, the novel application of a technique that can examine the constitution of mummified skin samples is needed; this is now provided by Raman spectroscopy. The advantages of skin as a material for Raman spectroscopic investigations are (1) it is easily available, and (2) the sampling does not require further destruction of the mummy.

In September 1991, the mummified body of a male now known as the Alpine Iceman ('Otzi') was found in thawing glacial deposits at a height of 3200 m in the Otzal Alps on the Austro–Italian border. Radiocarbon dating placed the age of the Iceman as 5.2 ka BP, and the artefacts and clothing found with the body have been the subject of detailed archaeological studies. The Iceman provides a unique example of a Neolithic man in generally well-preserved condition (Fig. 1). Raman

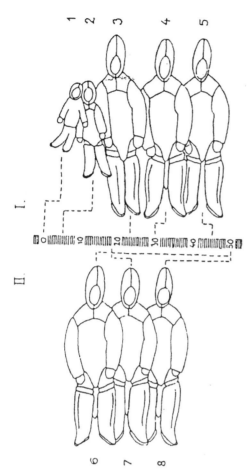

Fig. 2. Diagrammatic representation of the Greenland mummies in Graves I and II from an ice-bound rock tomb in Qilakitsoq; Grave I contained the mummified body of a 6-month-old baby studied here (denoted by 1 in Grave I). The other mummies (denoted 2–8) are of a 4-year-old boy and of women aged between 20 and 50 years. The mummies have been dated to 1475 ± 50. Courtesy of Hart Hansen *et al.* (1991).

Fig. 3. Mummified 6-month-old baby girl from the Qilakitsoq burial (Grave I, mummy 1 in Fig. 2).

spectroscopic studies were carried out non-destructively on a 16.2 mg sample removed from the hip of the Iceman, prior to the destruction of the specimen in the radiocarbon dating accelerator mass spectrometry (AMS) process. At that time, access to the mummy for *in situ* studies was not possible . Early optical microscopy studies revealed that the skin specimen appeared to have retained its integrity with regard to the stratum corneum and dermis. When recovered from the glacier, the Iceman weighed about 19 kg; clearly, extensive desiccation had occurred, and the Iceman provided a complete example of an ice-mummy, that is a human body whose tissues have been preserved through natural

mummification processes alone and without resort to exogenous, applied chemical treatment. There is still much conjecture as to how the Iceman died but, from artefacts associated with the body, it is thought that he was a wounded hunter-gatherer who was caught in bad weather and died from exposure (Spindler 1994).

The discovery of eight mummies in two graves (Fig. 2) located in a rock cleft near the abandoned settlement in Qilakitsoq in the Uummanaq district of northwestern Greenland in 1972 offered a second possibility for Raman spectroscopic study of ice-mummified human remains. The Qilakitsoq mummies are the oldest preserved human bodies in the Arctic and have been dated to about 1475 (±50), that is *c.* 0.5 ka BP. They were found fully dressed in equally well-preserved clothing. Extensive anthro-pological, radiological, odontological and dermato-logical studies have been undertaken (Hart Hansen *et al.* 1991). The archaeological excavation revealed that the mummies found in the first grave (Grave I in Fig.2) were a 6-month-old baby (Fig. 3), a 4-year-old boy and three adult women aged about 20–30 and 50 years. The mummies in Grave II were a woman aged about 30 years and two women aged about 50 years. Post-mortem examination failed to

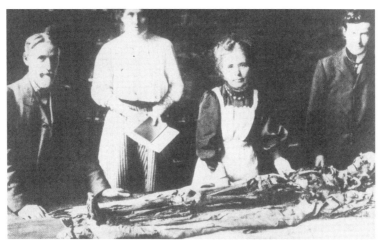

Fig. 4. The earliest recorded forensic scientific investigation of a mummy, Victoria University of Manchester, 1906; the archaeologist in the centre of the picture is Margaret Murray. The mummy of Nekht-Ankh, later re-investigated by Dr. Rosalie David, provided specimens for the Raman analyses cited here.

establish the cause of death; certainly, none of the mummies had any immediate life-threatening disease and all the women were healthy, with the exception of the 20-year-old, who had a kidney stone, and one of the 50-year-olds who exhibited skull bone damage ascribed to naso-pharyngeal carcinoma. The 4-year-old boy had signs of Legg-Calves-Perthes disease of the bone and Down's syndrome. Two skin punch specimens each were taken from the 4-year-old boy and the 20-, 30- and 50-year-old women (mummies no. 2, 3, 4 and 6, respectively); in addition, several finger nails from the 6-month-old baby (mummy no.1) were provided for analysis.

2. Mummified skin tissue from an Egyptian Twelfth-dynasty burial

The Egyptians believed that the spirit of a person could not continue to exist if the body disappeared, and therefore the body had to be preserved (David 1978). According to the Greek historian Herodotus, as a step in the embalming process the body was covered entirely with natron. Natron absorbs water and is also mildly antiseptic. Natron is a natural compound of sodium carbonate and bicarbonate that also contains sodium sulphate and sodium chloride (Andrews 1998). Such a material was found crystallized in considerable quantities along the edges of the lakes in the Wadi Natrūn, 65 km northwest of Cairo. In experiments designed to mimic an Egyptian-style mummification, more than 200 kg natron was used over 35 days to cover the body, and it was necessary to use a wide table such as those which have been excavated in ancient Egypt (Andrews 1998).

The 'Tomb of the Two Brothers', Khnum-Nakht and Nekht-Ankh, was discovered by the English archaeologist Sir William Flinders Petrie (1852–1942) in 1906. The contents of the burial site were passed to Manchester Museum, where they have since been studied, first by Margaret Murray (Fig. 4) and more recently by a team led by Rosalie David. The Two Brothers from the Twelfth dynasty, or Middle Kingdom (1991–1786 BC), came from Der Rifeh in Middle Egypt, and they are from the finest intact non-royal tomb ever found from that era (David 1978).

Four samples of the mummified skin studied here are from different areas on the embalmed mummy of Nekht-Ankh.

3. Identification of resin sources

Specimens of resins from different sources have been submitted for non-destructive analysis. These include particles adhering to ceramic shards from 1000-year-old native American Indian grave sites; a unique eye-bead from an Eighteenth-dynasty Egyptian cat mummy (c. 1400 BC), and ambers, copals and *dragon's blood* resins from documented geographical locations. Non-destructive Raman spectroscopic protocols have been established (Edwards *et al.* 2004) for the identification of different resins and for their attribution to botanical specimens for geographical sourcing. Possible future applications in a forensic arena are apparent (Edwards 2004).

4. Ivories and fragments associated with human skeletal remains

Ivory is the generic term for exoskeletal tooth (or 'tusk') formations associated with a limited group of terrestrial and marine mammalian species which include the African and Asian elephants, sperm whale, narwhal, hippopotamus, walrus, wart-hog, pig and mammoth. In common with teeth from other animal species, ivory consists of osteons comprising a matrix of hydroxyapatite with proteinaceous collagen. The composition of the inorganic and organic components of ivory varies with the enamel, dentine or pulp regions of the tusk. This is illustrated in Figure 5, where the enamel, dentine and pulpal cavity regions are clearly seen for a walrus tusk.

Identification of large ivory pieces, especially in sections, is generally fairly straightforward when an expert has some good surfaces or complete tusks for examination, and the characteristic pattern of tusk construction is usually definitive for the attribution of ivory to the mammalian species; the observation of Schreger line angles on polished sections is a normal forensic protocol for the discrimination between elephant ivories and mammoth ivory. However, the recognition of genuine ivory and its assignment to an individual species is rendered extremely difficult for small fragments or for highly carved artefacts. In this respect, the non-destructive evaluation of suspected ivory materials using Raman spectroscopy is likely to become a standard forensic application in the near future (Brody *et al.* 2001*b*).

To illustrate some of the major features of the Raman spectroscopic method applied to ivory identification, specimens of ivory from different mammalian species have been examined; the effect of burial environment or deposition on the chemical composition of ivories has also been discussed and its relevance for archaeological material highlighted. Specimens of archaeological ivory (Roman, *c.* AD 200) are compared with modern specimens obtained from HM Customs and Excise CITES seizures of contraband at Heathrow Airport, London (Edwards 2004; Edwards & Hassan 2004).

In several cases studied in our laboratory, ivory fragments have been associated with human skeletal remains; an example of this is discussed later in this paper.

Results and discussion

The Raman spectra of the inner and outer regions of the human skin specimens from the Alpine Iceman (Fig. 6 & 7) demonstrated that extensive protein changes have resulted from his burial in the glacier, with the broadening and diminution of the amide I band intensity near 1660cm^{-1} (Williams *et al.*

1995). The spectral shape of this feature is sensitive to keratotic changes in conformation of the –CONH peptide linkages such as the α- helical chains, β-sheet and β-turn. Degradation in burial environments affect the hydrogen bonding and protein integrity which become manifest in the spectral changes seen in skin, hair and nail specimens (Gniadecka *et al.* 1999).

The spectrum of the nails of the baby ice-mummy (Fig. 8) show that, in comparison with modern baby's nail clippings, little protein change has occurred during the 500+ years of ice entombment. In contrast, the skin of the baby ice-mummy has deteriorated in parts to an extent similar to that observed for the Iceman. The effect of fluorescence emission on the biomaterials from the burial environment is seen as a residual background on the spectra.

Some interesting phenomena could be characterized with the Raman spectroscopic examination of the ancient Egyptian embalmed mummy and comparison with the baby ice-mummy; Raman spectra obtained from different sites on skin samples from the Nekht-Ankh mummy embalmed by natron in the way used by ancient Egyptians showed various degrees of deterioration. However, in general both the lipids and the proteins were well preserved (Fig. 9). Here, the spectral stackplots comprise Raman spectra collected from eight different sites of preserved mummified skin tissue. The amide I and III protein bands can be used to monitor the presence of proteins and to give an idea about the secondary protein structure. In most of the spectral regions investigated, both helical and β-sheet secondary structures were observed. Sodium sulphate was the only artificial chemical found (Fig. 10). However, in all spectra where sodium sulphate was observed the degradation of biomaterial seemed to be strongest (note the broadening of the Raman bands). In general, the skin samples were very well preserved, taking into account the hot climate in the region where the mummy was found. The Nekht-Ankh mummy was as well preserved (4000a BP) as the most well preserved of the rock-cleft mummies (0.5ka BP) from Qilakitsoq, Greenland, the child mummy (Fig. 11).

The Raman spectrum of a material can be regarded as a 'fingerprint' of its composition and this can be used to good effect to study resins, waxes and gums. In illustration of this, a spectrum of Baltic amber is shown in Figure 12; here, the characteristic features are clearly identifiable and differentiate the fossilized resin from a particle found adhering to a ceramic shard in a grave site (Fig. 13) (Brody *et al.* 2001*a*). The latter is not amber, but matches that closely of a *Pinus* species. It should be noted that the analysis did not require the detachment of the resin particle from the shard. Recently, we have undertaken a comprehensive study of the important resin

Fig. 5. Tusk ivory, showing the distinctive enamel, dentine and pulpal cavity regions.

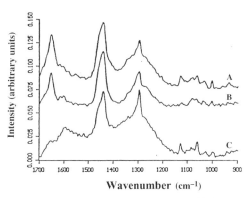

Fig. 6. FT-RS of the fingerprint region of: (**a**) dried contemporary stratum corneum; (**b**) freeze-dried contemporary stratum corneum; and (**c**) Iceman stratum corneum. The spectra demonstrate that the protein component of Iceman skin has degraded during its *c.* 5.2 ka. interment in an Alpine glacier, whereas the lipoidal moiety of the tissue is largely intact, although some oxidation of olefinic moieties is suggested. No evidence of modern external chemical contamination can be seen in the spectrum for Iceman skin.

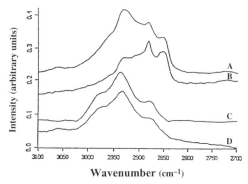

Fig. 7. FT-RS of the C-H stretching modes of: (**a**) contemporary human stratum corneum; (**b**) Iceman stratum corneum; (**c**) contemporary dermis, and (**d**) Iceman dermis. The spectra clearly show differences between the molecular structure of stratum corneum and dermal tissue, and that the Iceman has retained these distinct tissue layers.

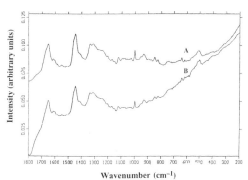

Fig. 8. Comparison of the FT-RS spectra of: (**a**) modern baby's nail clipping; and (**b**) nail from the 6-month-old Greenland baby (mummy no.1 in Fig. 2); 1064 nm excitation, 4 cm^{-1} spectral resolution, wavenumber range 200–1800 cm^{-1}.

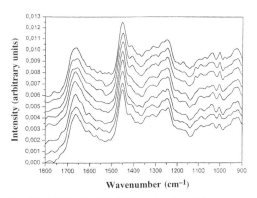

Fig. 9. FT-RS in the 900–1800 cm^{-1} region of the mummified skin of Nekht-Ankh, demonstrating the good state of preservation of the proteins compared with the ice-mummy in Fig. 6.

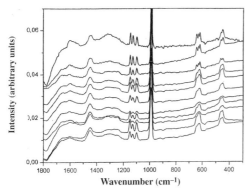

Fig. 10. FT-Raman spectra in the 300–1800 cm^{-1} region of the skin of the Nekht-Ankh mummy. Here, the presence of the exogenous chemical treatment used in the mummification process can be clearly seen; it is interesting that only Raman bands due to sodium sulphate can be observed as a triplet between 1100 and 1200 cm^{-1}, a strong band at 980 cm^{-1} and as features near 650 and 450 cm^{-1}, and none from the sodium carbonate present in the natron treatment.

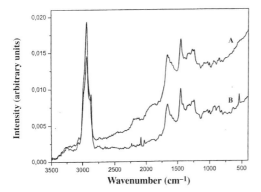

Fig. 11. A comparison between the FT-RS: (**a**) of skin from the Qilakitsoq child mummy (no.1) and (**b**) the most well-preserved skin from the Nekht-Ankh mummy.

family known as *dragon's blood*; these resins were known in antiquity and have been highly prized for their depth of colour and mystical association (Edwards *et al.* 2001). The original source of dragon's blood was *Dracaena cinnabari* from Socotra in the Indian Ocean; however, several 'alternatives' were often used and are passed off as dragon's blood today – these include *Daemonorops draco* and resins of the *Croton* genus.

Figure 14 shows the Raman spectrum of a genuine archival specimen of dragon's blood resin collected on Socotra, with that of a *Daemonorops draco* resin from the East Indies. Although the specimens are identical in appearance, these resins exhibit different spectral signatures, which therefore affords an opportunity to use the Raman spectroscopic technique as a non-destructive sourcing technology. This has some obvious application in forensic geoscience and will surely be utilized henceforth to label resin fingerprints with their geographical origins and possibly even identify suppliers (who tend to 'mix' resins and pigments to produce a desired colour palette). The spectrum in Figure 15 is that of a resin which has been degraded somewhat in its burial environment and some key biosignature bands have disappeared.

Several genuine ivories and fake ivories were analysed and key spectral markers were identified to assist in their attribution. Figure 16 shows three 'ivory' bangles dating, it was believed, from the early twentieth century and a necklace of indeterminate age; on the left is a highly coloured bangle having flower and insect motifs with residues of green and red applied decoration; in the middle is a large plain bangle; and at the right is a thin, small bangle – the last two items were devoid of surface decoration, but under the microscope evidence of subsurface structure could be seen. Raman spectra were obtained of each bangle and are shown in Figure 17; none of the spectra match that of genuine ivory, shown in Figure 18, in that the characteristic

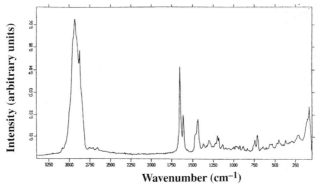

Fig. 12. FT-RS spectrum of Baltic amber, 1064 nm excitation, 100–3400 cm^{-1} range, 2000 scans at 4 cm^{-1} spectral resolution.

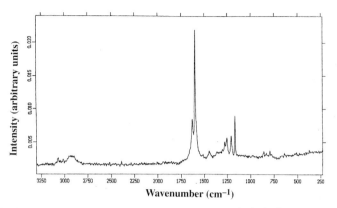

Fig. 13. FT-RS of resin particle adhering to a ceramic shard from a grave site in the Sonoran Desert, Arizona; although similar visually to amber, the Raman spectrum of the resin particle is clearly different and matches closely a *Pinus* resin, from several botanical possibilities in the neighbourhood.

amide protein vibrations in the $1200-1700\,cm^{-1}$ region are absent and the strong hydroxyapatite band at $960\,cm^{-1}$ is absent. In Figure 17, the upper spectrum matches that of cellulose nitrate in wavenumber positions of the key bands (Fig. 19), but the spectral intensity on the stackplot is lower. The small bangle, the second spectrum in the stackplot, corresponds to a mixture of polystyrene (ps) and polymethylmethacrylate (pmma); the third spectrum is polymethylmethacrylate. The bottom spectrum in the stackplot is that of the large bangle. Clearly, again, the large bangle is a fake ivory artefact in that the bands due to ps and pmma can be identified, so dating the artefact to the 1930s at the earliest and certainly not to the nineteenth century! An interesting feature in the Raman spectrum of the large bangle, however, are the bands at 1086, 712 and $280\,cm^{-1}$ which are characteristic of the mineral calcite. On further investigation, it was realized that powdered calcite (marble) had been added to the

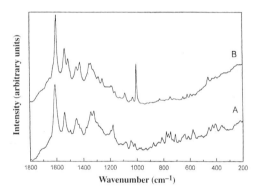

Fig. 14. FT-RS stackplot comparison of: (**a**) genuine dragon's blood resin from *Dracaena cinnabari* (source: Socotra, East Africa), and (**b**) *Daemonorops draco* (source: East Indies). Although the dark red resins are identical in colour and appearance, they have distinctive spectral signatures which afford a forensic application for potential geographical sourcing of materials.

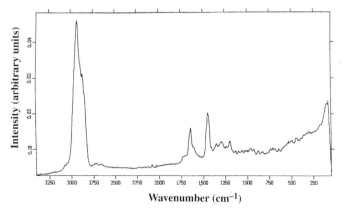

Fig. 15. FT-RS of partly degraded amber resin, showing the broadening of vibrational bands and disappearance of some key features.

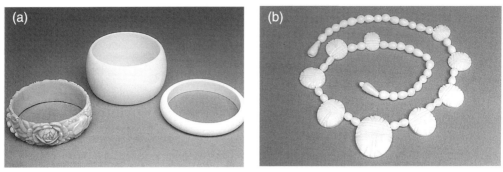

Fig. 16. (**a**) Three 'ivory' bangles dating from about 100 a BP and (**b**) an ivory necklace with scarab motifs of indeterminate age, but believed to be of an eighteenth-century origin.

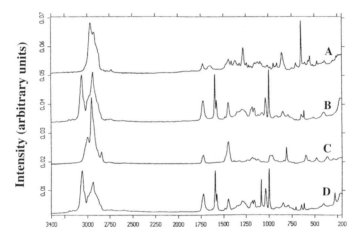

Fig. 17. FT-RS spectral stackplot of 'ivory' bangle specimens shown in Figure 16; (**a**) carved bangle (left); (**b**) small bangle (right); (**c**) polymethylmethacrylate sample; and (**d**) large bangle (centre). None of the bangles is ivory!

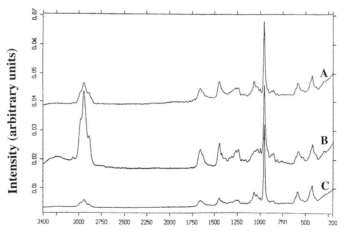

Fig. 18. FT-RS stackplot: (**a**) bone; (**b**) African elephant ivory and (**c**) mammoth ivory.

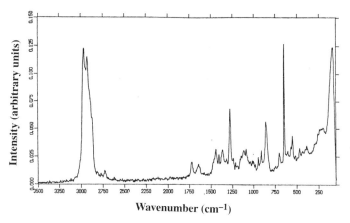

Fig. 19. FT-RS of cellulose nitrate.

polymer mixture (ps + pmma) to produce a composite of similar density to genuine ivory.

The forensic applications of non-destructive Raman spectroscopic analysis are clearly demonstrated.

Another development has been the attempt to characterize different species of mammalian ivory from their Raman spectra; this is of interest to law enforcement organizations such as HM Customs and Excise, who need to be able to identify not just genuine contraband ivory materials, but also to source these if possible, and attribution to the individual species is hence essential. Generically, ivory from different mammalian sources has a similar Raman spectrum, but it is possible using chemometric methods to identify the individual mammalian sources using up to 11 different combinations of bands in the Raman spectra (Brody *et al.* 2001*a*). A particularly intriguing problem relates to the appearance of mammoth ivory in contraband seizures. Figure 18 shows a stackplot of Raman spectra of, from the top, animal bone, African elephant ivory and mammoth ivory; for comparison purposes, the phosphate band at $960 \, cm^{-1}$ has been normalized between the three spectra. The spectra, although very similar superficially, exhibit some important differences, such as the ratio of the organic proteinaceous bands due to collagen (~ 3000, 1660, 1450, $1250 \, cm^{-1}$) to the phosphatic modes at 960, 620 and $450 \, cm^{-1}$. The protein bands are significantly depleted in intensity for the mammoth ivory specimen (Fig. 20), as expected from its age and deposition in its burial environment in which leaching out of the organic component from the inorganic matrix could have been expected. Key differences also occur between the animal bone and ivory specimens, especially in the intensity of the band near $3000 \, cm^{-1}$ and in the spectral shape of the complex features near 1100, 1250 and $900 \, cm^{-1}$.

Fig. 20. Specimen of mammoth ivory, in which the degradation of the collagen proteinaceous component and the leaching in of materials from the burial environment has produced a heterogeneous specimen.

A carved 'scarab' necklace, which is ivory-like in appearance but of uncertain attribution, is shown in Figure 16; the Raman spectrum in Figure 21 confirms the material as ivory.

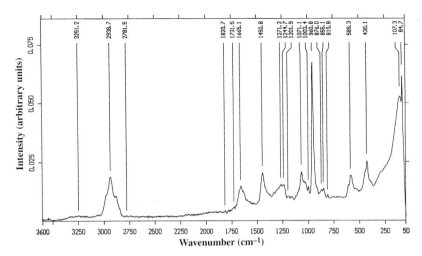

Fig. 21. The FT-RS of the necklace shown in Fig. 16 confirms its source as true ivory and most probably African elephant ivory. Here, the automatic spectral 'read-out' of wavenumbers has been retained to illustrate the ease of comparison of wavenumber data with standard specimens and for the incorporation of the latter into a database.

Conclusions

The non-destructive analytical capability of Raman spectroscopy for the characterization of biomaterials in geological environments is demonstrated with examples of resins, ivories and mummified tissue from a range of burial depositions. Although an increase in fluorescence emission background is expected with these samples, nevertheless suitably definitive spectra have, in most cases, been obtained and these have assisted in the provision of information relevant to sources, processing or degradation of the specimens in forensic geoscience scenarios. The potential of the technique for analytical advancement in the forensic area is highlighted.

References

ANDREWS, C. 1998. *Egyptian Mummies*. British Museum Press, London

BRODY, R. H., EDWARDS, H. G. M. & POLLARD, A. M. 2001*a*. Chemometrics methods applied to the differentiation of Fourier-transform Raman spectra of ivories. *Analytica Chimica Acta*, **427**, 223–232.

BRODY, R. H., EDWARDS, H. G. M. & POLLARD, A. M. 2001*b*. A study of amber and copal samples using FT-Raman spectroscopy. *Spectrochimica Acta (A)*, **57**, 1325–1338.

DAVID, A. R. 1978. *Mysteries of the Mummies: The Story of the Manchester University Investigation*. Book Club Associates, London.

EDWARDS, H. G. M. 2001. Raman spectroscopic applications to archaeological biomaterials. *In*: LEWIS, I. R. & EDWARDS, H. G. M. (eds) *Handbook of Raman Spectroscopy: From the Research Laboratory to the Process Line*. Marcel Dekker, New York, 1011–1044.

EDWARDS, H. G. M. (2004). Forensic applications of Raman spectroscopy in art. *In*: CLAYBOURN, M. (ed.) *Raman Spectroscopy in Forensic Science*. Humana Press.

EDWARDS, H. G. M. & HASSAN, N. F. N. 2004. Evaluation of Raman Spectroscopy for the non-destructive testing differentiation between elephant and mammoth ivories. *Asian Chemistry Letters*, **7**, 185–196.

EDWARDS, H. G. M., DE OLIVEIRA, L. F. C. & QUYE, A. 2001. Raman spectroscopy of coloured resins used in antiquity: dragon's blood and related substances. *Spectrochimica Acta (A)*, **57**, 2831–2842.

EDWARDS, H. G. M., DE OLIVIERA, L. F. C. & PRENDERGAST, H. V. 2004. Raman spectroscopic analysis of dragon's blood resins: a basis for distinguishing between *Dracaena* (Convallariaceae), *Daemonropos* (Palmae) and *Croton* (Euphorbiaceae). *Analyst* **129**, 134–138.

GNIADECKA, M., EDWARDS, H. G. M., HANSEN, J. P. H., NIELSEN, O. F., CHRISTENSEN, D. H., GUILLEN, S. E. & WULF, H. C. 1999. NIR-FT Raman spectroscopy of the mummified skin of the Alpine Iceman, Qilakitsoq Greenland mummies and the Chiribaya mummies from Peru. *Journal of Raman Spectroscopy*, **30**, 147–154.

HART HANSEN, J. P., MELDEGAARD, J. & NORDQUIST, J. 1991. *The Greenland Mummies*. British Museum Press, London.

HENDRA, P. J., JONES, C. A. & WARNES, G. 1991. *FT-Raman Spectroscopy*. Ellis-Horwood, London.

LONG, D. A. 2002. *Raman Spectroscopy*, John Wiley & Sons, Chichester.

SPINDLER, K. 1994. *The Man in the Ice*. Wiedenfeld and Nicholson, London.

WILLIAMS, A. C. EDWARDS, H. G. M. & BARRY, B. W. 1995. The iceman: molecular structure of 5200-year-old skin characterised by Raman spectroscopy and electron microscopy. *Biochimica et Biophysica Acta*, **1246**, 98–104.

Assessing element variability in small soil samples taken during forensic investigation

KYM E. JARVIS[1], H. ELIZABETH WILSON[2] & SARAH L. JAMES[2]

[1]*Department of Earth Sciences & Geography (NERC ICP Facility), Kingston University, Penrhyn Road, Kingston upon Thames KT1 2EE, UK (e-mail: kym.jarvis@kingston.ac.uk)*
[2]*Department of Geology, Royal Holloway, University of London, Egham Hill, Egham TW20 0EX, UK*

Abstract: Inductively coupled plasma analytical techniques are widely used in forensic geochemistry because they can provide concentration data for a wide range of major and trace elements relatively rapidly and at reasonable cost. A pilot study was undertaken to identify the relative importance of uncertainty resulting from instrumental measurement sources and that due to the procedures used to prepare the samples initially. Three soils with a range of major and trace element concentrations were collected to permit an evaluation of uncertainty. A reference sample of demonstrated homogeneity was also prepared and analysed. Samples were prepared in replicate (five preparations) of each, and assessment made of uncertainty in the instrumental measurement alone and for replicated preparations of the same material. Small sample sizes (0.05 g) were used to mimic the situation common in forensic investigation. Results show that, while instrumental variability may be an important factor during measurement, between-sample variation has a dominant effect on uncertainty in the final result. It is clear that, without replicated measurement and preparation, the uncertainty of the measured data is unknown. Thus, critical samples, on which a case might depend, must be analysed in a way that defines clearly that uncertainty.

Until recently, most police officers and, indeed, forensic scientists, were unaware of the potential role of using chemical variation in soils as a tool in criminal investigation. This approach is now becoming increasingly integrated into the armoury of forensic techniques used during investigation. The technique is one of comparative analysis in which soil from a crime scene is compared to soil from a suspect or their possessions. If the two can be shown to be sufficiently similar, the data may be used to support the proposition that the suspect is linked to the crime scene. The critical issues here are defining 'sufficiently similar' and assessing the level of support that can be attached to the preposition of the association. This involves consideration of several issues, including sample representativeness, measurement uncertainty (instrumental, preparation) and the extent of the variation found in natural samples.

The reliability of analytical techniques used in the academic and commercial world has been well studied and, by consensus, levels of accuracy and precision appropriate to the field are adopted. Geological research laboratories have for many years been faced with the problems of evaluating data quality from a wide range of instrumental analytical techniques (Thompson & Howarth 1976, 1978). In some cases cost has been the driving factor affecting protocols but it is understood that all elemental measurements are subject to error,

and for most geological studies an uncertainty of $\pm 10\%$ is accepted (geological processes usually result in elemental changes of $>10\%$). Unlike the case in many forensic investigations, the mass of sample available for analysis in geochemical studies is usually large (many hundreds of grams) and issues of homogeneity are not serious in most cases.

In addition to understanding data quality from a single instrument in a single laboratory, there has been a need to ensure consistency in results from one laboratory to another, often on a worldwide basis. To this end, the International Association of Geoanalysts proficiency testing scheme (GeoPT) was established in 1996 to allow participants to evaluate their own performance with respect to measurement of elemental concentration, in a range of geological sample types, against that of other laboratories and the standard of performance set by GeoPT (e.g. Thompson *et al.* 2000). Each participating laboratory is sent a subsample of a powdered material of demonstrable homogeneity (at a given mass). The results from each laboratory, from a range of analytical techniques, are compiled to show a comparison of results of each element from each laboratory. A consensus value is calculated. While the data for major element oxides often lies close to the consensus value, this is not usually the case for trace elements which often show a wide spread. This type of proficiency testing is not,

From: PYE, K. & CROFT, D. J. (eds) 2004. *Forensic Geoscience: Principles, Techniques and Applications*. Geological Society, London, Special Publications, **232**, 171–182. © The Geological Society of London, 2004.

however, designed to evaluate instrumental uncertainty. Sample homogeneity is evaluated prior to distribution of the sample by subsampling of the bulk material and analysis using precise analytical techniques such as X-ray fluorescence spectrometry. Thus via this scheme we have a way of evaluating uncertainty which can arise from different techniques or from different approaches to analysis in laboratories. The reasons for the lack of close agreement to a single consensus value for any particular element are numerous, and beyond the scope of this paper, but serve to highlight the need for a unified approach to preparation, analysis and data handling. For small sample masses, such as those encountered in forensic investigation, the general issues are exacerbated by sample size.

Given its impact on society, we must be aware that work carried out for the criminal courts must be of the very highest standard and that levels of uncertainty must be minimized: the data must be fit for purpose. In recent years there have been a number of studies which have utilized trace and major element data to characterize and source forensic samples using inductively coupled plasma (ICP) techniques. These have mainly concentrated on glasses (e.g. Suzuki *et al.* 2000, 2003; Duckworth *et al.* 2002), bullet lead (Keto 1999; Yourd *et al.* 2001) and document paper (Spence *et al.* 2000). However, there has been little attention to soil discrimination and the uncertainty attached to measurement using ICP techniques. Such a study appears overdue. To stimulate discussion on analytical reliability and the wider issue of levels of confidence, a pilot study was undertaken to evaluate the variability of replicate analyses of three soils of different chemical composition .

There are two aspects of total uncertainty in analytical measurement that together allow us to make an assessment of the reliability of data. Accuracy describes the closeness of the measured value to the true value and is controlled by both bias (a measure of systematic error) and/or precision (a measure of random error). To some extent accuracy can be assessed by analysing a 'control sample' or 'reference material' of accepted composition (within certain limits) and comparing measured concentrations for elements of interest with the certified or accepted values .

To quantify accuracy we can measure bias and precision. Bias is systematic and is assessed using a reference material where

$$\text{Absolute bias} = C_m - C_a \tag{1}$$

and

$$\text{Relative bias} = \frac{100\,(C_m - C_a)}{C_a} \tag{2}$$

where C_m is the measured concentration for an element and C_a is the reference or accepted value for the same element. Hence bias is calculated independently for each element. Uncertainty about the reference value is often not taken into account. Precision is a measure of random error and is often expressed using standard deviation (SD) or coefficient of variation (CV) (see Webster 2001).

An assessment of precision can be made for different parts of the analytical process. This is particularly important when a sample undergoes complex and lengthy preparation procedures prior to analysis, such as with ICP techniques. Preparation procedures can introduce selective contamination into a sample or samples, or may result in the selective loss of one or more elements. Weighing and handling errors can also be introduced. Moreover, the inherent heterogeneity of the sample may also be an important factor. It is therefore desirable to separate the reproducibility of the preparation process or initial heterogeneity of a sample (between-sample variation) from those caused by instrumental measurement variation. These can best be expressed by the use of variance (s^2), which is calculated as the square of the SD. Variance, unlike SD, is additive, which means that, if the error in an analysis originates from three independent (and random) sources, then the combined error is given by the sum of the variances, for example:

$$s^2 = s^2_A + s^2_B + s^2_C \tag{3}$$

Some statistical tests, such as the t-test, use variance as part of the test.

If data interpretation is to be made in the light of the above, then we need a way of making an assessment of the total uncertainty in a measurement. That is to say we need to be able to assess the size of the 'area' around the measured result in which the true value lies, and the probability of it doing so. It is only in this way that we can give an informed opinion as to the likelihood of two samples originating from the same population or to their similarity being a result of cumulative errors. We can consider standard uncertainty (u) as being equivalent to 1σ, and extended uncertainty (U) as equivalent to 2σ. The total uncertainty can be expressed as a sum of the random error or precision on the pre-analysis procedures (collection, preparation) plus the same parameter on the measurement plus the systematic error or bias on the pre-analysis procedures plus the same parameter on the measurement (Ramsey 1994, 1997).

The first objective of this study was to assess variation in measured data that results from changes in signal derived from the instrumentation itself. The second objective was to compare this variation with that obtained for a series of replicate preparations of the same sample.

Methodology

The soils selected for this study were: an allochthonous garden soil from Claygate, Surrey, a natural parkland soil overlying Hythe Beds, Dryhill, Kent; and a natural woodland soil overlying a Jurassic sandstone, Sheffield, South Yorkshire. All soils supported grasses, and the main rooting mat ramified through the upper 0.5 cm of soil. Soil profiles were taken from each site during November 2001 and refrigerated at 4 °C until they were prepared for analysis in December 2001. The soils were all taken from the A horizon, between the surface and 1.0 cm depth.

For each of the three soil samples, 2.0 g of bulk soil was sieved through 150 μm mesh using deionized water. The resulting filtrate was left to settle for 48 h and then dried at 40 °C. The dried sediment was then ground to a fine (<60 μm) using an agate pestle and mortar. Grinding of the sample aids mixing and improves homogeneity, while also producing a large surface area to speed sample digestion. This process produced approximately 0.6 g of prepared soil which was then stored in a glass vial. There has been much discussion in the published literature concerning the sample mass that should be prepared for analysis in order for it to be representative. For most routine geochemical studies this mass varies from 0.1 g to 0.5 g, depending on the particle size and the likely distribution of elements of interest. Where an element occurs in a relatively small number of discrete grains, e.g. gold, then it may be necessary to take 50–100 g or more of ground sample in order for the final analysis to be representative of the composition of the whole (Gy 1992). In many forensic investigations, an informed decision about sample mass required may not be possible where only a few grains (few mg) of soil evidence are available. Therefore, for this study we have gone some way to mimic this common situation and have taken ten subsamples of only 0.05 g of the <150 μm fraction of each of the three soil samples. The same weight of the reference material SARM1 (also known as granite NIM-G) was also prepared.

The samples were prepared for analysis using lithium metaborate alkali fusion. This is a well-established procedure for the preparation of soil and geochemical samples for measurement by ICP techniques (among others) (Totland et al. 1992), and involves mixing the sample powder with an alkali salt, lithium metaborate (LiBO2) in small crucibles. The sample mix is heated to >1000 °C, resulting in a low viscosity melt being formed. The melted material is dissolved in dilute nitric acid and diluted with high-purity water prior to analysis. This procedure has been shown to produce quantitative recovery for the elements considered below although it should be noted that losses for some other elements

may be significant (e.g. Pb, Tl) due to the high temperatures required to form the melt. Full details of the procedure used in this work are described by Thompson and Walsh (2003); reagent masses were adjusted for the reduced sample weights used in this study. Gallium was used as an internal standard for ICP atomic emission spectrometry (ICP-AES) measurement (Walsh 1992) to ensure optimum precision for the major elements. Procedural blanks (Jarvis et al. 2003) were also prepared.

Samples were analysed for ten major elements (Si, Al, Fe, Mg, Ca, Na, K, Ti, P, Mn) and eight trace elements (Ba, Co, Cr, Sc, Sr, V, Zn, Zr) using a Perkin Elmer Optima 3300RL ICP-AES and 20 trace (Ce, Cs, Dy, Er, Eu, Gd, Hf, Ho, La, Lu, Nb, Nd, Pr, Rb, Sm, Ta, Th, U, Y, Yb) elements using a Perkin Elmer Sciex Elan 5000 ICP-MS. Elements were chosen on the basis that they were not subject to volatilization during preparation nor to serious interference effects during analysis. Uncertainty from these sources is therefore minimized. Measured concentrations are reported in Appendices 1 and 2.

Detection limits

Detection limits were calculated for both ICP-AES and ICP mass spectrometry (ICP-MS) instruments by analysing a reagent blank solution ten times on each instrument. The lower limit of detection (LLD) is the limit of quantitative analysis and is calculated by multiplying the SD (s) of the blank measurements by three and then calculating the concentration equivalent. The resulting values are shown in Tables 1 and 2. For ICP-AES, major element LLDs are typically ~0.01 wt% oxide while trace elements vary from 0.33 to 2.00 p.p.m. calculated in the solid. For ICP-MS, LLDs are lower with typically <0.1 p.p.m. The LLD is an estimation of the lower limit of quantitative measurement. Above this value there is a 97% probability that the signal (equivalent to concentration) is a true signal from any particular element and not a result of random noise. In this study the lower limit of detection for cobalt (Co) by ICP-AES, for example, was 1.29 p.p.m. (Table 1). To achieve greater certainty that a measured signal is derived from an element of interest, the lower limit of measurement can be set at a value of six times the SD of the background signal (limit of determination, LoD) and for cobalt this would be equivalent to 2.59 p.p.m. The limit of quantitation LoQ ($10s$) may also be calculated and offers the analyst additional confidence that a measured signal does not derive from random background, and may be the preferred limit where interpretation of the data has some legal implication, (Potts, 2003). The LoQ calculated for cobalt was 4.31 p.p.m.

Table 1. *ICP-AES detection limits calculated using 10 replicate measurements of a reagent blank (concentrations for oxides in wt % and trace elements in p.p.m.)*

	LLD (3σ)	LoD (6σ)	LoQ (10σ)
SiO_2	0.04	0.09	0.14
Al_2O_3	0.01	0.02	0.03
Fe_2O_3	0.04	0.07	0.12
MgO	0.01	0.01	0.01
CaO	0.01	0.01	0.01
Na_2O	0.01	0.01	0.01
K_2O	0.01	0.02	0.03
TiO_2	0.01	0.01	0.01
P_2O_5	0.01	0.01	0.01
MnO	0.01	0.01	0.01
Ba	1.20	2.20	3.66
Sr	0.33	0.66	1.10
Y	0.74	1.48	2.47
Zr	1.99	3.98	6.64
Co	1.29	2.59	4.31
Cr	1.79	3.58	5.97
Sc	0.67	1.32	2.21
V	1.06	2.11	3.51
Zn	0.64	1.26	2.10

Table 2. *ICP-MS detection limits calculated using 10 replicate measurements of a reagent blank (concentrations in p.p.m.)*

	LLD (3σ)	LoD (6σ)	LoQ (10σ)
U	0.01	0.02	0.04
Th	0.07	0.14	0.24
Rb	0.13	0.27	0.44
Nb	0.71	1.42	2.37
Cs	0.03	0.06	0.10
Hf	0.76	1.52	2.53
Ta	0.03	0.05	0.08
Y	0.07	0.17	0.23
La	0.24	0.47	0.78
Ce	0.08	0.16	0.26
Pr	0.02	0.03	0.05
Nd	0.07	0.15	0.24
Sm	0.06	0.13	0.21
Eu	0.01	0.01	0.02
Gd	0.03	0.06	0.10
Dy	0.03	0.05	0.09
Ho	0.01	0.01	0.02
Er	0.01	0.02	0.03
Yb	0.03	0.06	0.09
Lu	0.03	0.05	0.09

Signal reproducibility

Instrumental variation was assessed by measuring one preparation of each sample five times consecutively without removal of the instrument sample probe from the solution. Sampling or inter-sample variation was assessed by measuring each preparation once.

Results & discussion

Instrumental variation

Tables 3 and 4 show the CV for the instrumental variation (five consecutive measurements) for each soil and the reference material SARM1 for both ICP-AES and ICP-MS. Major element oxides show values which are typically <4% with elements such as silicon and aluminium consistently showing very low values (<1%). Calcium in Soil 3 displays very poor reproducibility which is due to its low concentration (0.05% CaO). This is similarly the case for sodium in Soil 2. This dataset is typical for the determination of major elements by ICP-AES. The eight trace elements determined display a greater range of CV values, which are again largely controlled by concentration. For example, barium concentrations are between 100 and 360 p.p.m. with CV of <5%. Cobalt concentrations are low, between 10–30 p.p.m., only a few times higher than the LoQ, with CVs of 3–18%. For SARM 1, concentrations are <LLD. CVs calcu-

lated for trace elements measured by ICP-MS exhibit relatively similar values of between 1% and 8% for all elements.

If we now compare results from measurements of separate sample preparations of each material, we see a slightly different picture (Tables 5 & 6). Major element oxides show a greater range of values: from 0.01% to 67% with several values exceeding 5%. The iron content of SARM1, which is significantly above the LLD, has an anomalously high CV which may be due to inhomogeneity at the sample weight chosen. Iron can occur in discrete mineral grains which would not be uniformly sampled when taking only 0.05 g samples. This variation is often referred to as the 'nugget' effect. The trace elements similarly display a greater range of values, with almost half showing >5%. Values for trace elements by ICP-MS show variation from 2.29% to 23.6%, with figures typically between 5% and 10% with about one-third >10%. It is interesting to note that the values for the reference material of measurement by ICP-AES are similar to those for the three soil samples, while those for the trace elements by ICP-MS seem to give slightly better (lower) values than the soil samples. The reference material was included because it is likely to be more homogeneous (a commercially prepared material from a large mass of starting material) than the sample soils collected for this work. In a number of cases the instrumental precision exceeds the sampling or preparation precision (Table 7). For ICP-AES 29%

Table 3. *Instrument measurement precision (CV) calculated for major and trace elements by ICP-AES* (n = 5)

	Soil 1		Soil 2		Soil 3		SARM 1	
	Mean	CV	Mean	CV	Mean	CV	Mean	CV
SiO_2	71.9	0.182	64.9	0.269	51.6	0.135	77.0	0.44
Al_2O_3	7.36	0.201	5.43	0.335	12.2	0.124	12.4	0.55
Fe_2O_3	3.70	0.226	10.70	1.43	3.37	0.248	1.78	2.36
MgO	0.53	3.53	1.37	1.52	0.426	3.56	0.130	13.3
CaO	0.57	1.24	4.25	1.10	0.050	14.1	0.782	1.67
Na_2O	0.23	3.61	0.032	26.1	0.718	2.29	3.34	0.82
K_2O	1.81	3.24	3.00	0.67	1.26	2.28	5.20	0.64
TiO_2	0.61	3.54	0.250	0.000	0.786	2.64	0.090	0.000
P_2O_5	0.13	3.39	0.214	4.18	0.226	2.42	0.048	9.32
MnO	0.04	0.000	0.092	4.86	0.080	0.00	0.02	0.000
Ba	376	4.20	147	0.962	321	3.57	111	2.15
Sr	66.6	4.06	79.8	2.06	75.4	3.70	16.6	5.39
Zr	482	3.65	417	2.09	394	3.34	283	2.22
Co	9.20	16.1	27.8	3.01	10.8	17.8	1.40	81.4
Cr	120	4.26	160	0.279	122	3.73	16.0	7.65
Sc	7.20	6.21	23.6	13.3	9.20	4.86	9.20	27.1
V	78.4	4.65	162	2.03	74.6	3.98	4.00	46.8
Zn	160	3.96	69.8	3.98	64.8	5.92	42.4	3.16

Table 4. *Instrument measurement precision (CV) calculated for trace elements by ICP-MS* (n = 5)

	Soil 1		Soil 2		Soil 3		SARM 1	
	Mean	CV	Mean	CV	Mean	CV	Mean	CV
U	1.87	1.16	1.27	2.93	1.95	1.76	15.5	6.21
Th	6.59	1.56	7.54	1.96	7.44	3.00	52.0	5.81
Rb	77.5	3.08	110	2.48	70.0	3.32	305	7.29
Nb	13.7	5.00	12.8	3.90	13.1	3.49	51.7	7.26
Cs	3.49	4.20	3.50	4.20	2.42	5.66	0.878	6.99
Hf	11.9	1.47	11.8	1.26	6.88	4.78	13.2	7.53
Ta	0.87	3.28	0.66	4.05	0.82	3.69	4.65	4.92
Y	22.0	2.52	20.7	1.91	17.6	3.48	151	7.02
La	20.4	3.80	31.8	2.78	32.8	4.70	103	6.84
Ce	40.7	3.65	55.6	0.88	65.6	4.14	192	6.49
Pr	4.10	3.36	9.08	1.47	6.83	3.74	20.1	6.99
Nd	19.7	2.55	38.7	2.77	28.4	3.09	72.8	6.50
Sm	3.42	4.48	6.67	1.32	4.38	4.42	14.7	6.51
Eu	0.674	3.42	1.22	6.07	0.844	3.20	0.304	3.75
Gd	3.36	2.51	4.87	2.40	3.59	5.26	13.1	6.32
Dy	3.01	4.21	3.14	5.66	2.53	5.32	16.8	5.88
Ho	0.734	3.42	0.776	3.48	0.634	3.80	3.89	5.84
Er	1.85	2.82	1.78	3.90	1.57	4.44	11.9	7.02
Yb	1.98	2.21	1.65	2.99	1.52	3.10	14.5	6.23
Lu	0.352	5.08	0.388	4.96	0.28	4.13	2.19	7.87

of elements display instrument variation which is greater than preparation precision. This is the case for strontium, scandium and vanadium in all samples, and for selected major element oxides. For ICP-MS this is only the case for five measurements or 6% of the elements. The reason for these observations may be partially explained by the fact that signal reproducibility is intrinsically better by ICP-AES than ICP-MS. Measurements of concentration by ICP-AES are made via the emission of photons from the sample aerosol introduced into the ICP. For ICP-MS, although the excitation source is the same, measurements are made of the number of ions extracted from the ICP. The latter is a fundamentally less reproducible process than the measurement of light (Jarvis *et al.* 2003). To evaluate more rigorously

Table 5. *Preparation precision (CV) calculated for major and trace elements by ICP-AES* (n = 10)

| | Soil 1 | | Soil 2 | | Soil 3 | | SARM 1 | |
	Mean	CV	Mean	CV	Mean	CV	Mean	CV
SiO_2	72.6	2.33	64.6	1.41	52.1	2.05	76.3	1.11
Al_2O_3	7.34	3.80	5.42	1.54	12.4	3.71	12.2	1.01
Fe_2O_3	4.08	12.1	11.2	5.71	3.61	8.47	1.87	25.7
MgO	0.531	6.61	1.34	1.53	0.434	6.54	0.112	15.1
CaO	0.540	8.99	4.31	2.41	0.076	67.1	0.772	2.43
Na_2O	0.228	4.98	0.027	17.9	0.753	4.06	3.31	2.00
K_2O	1.88	2.51	2.98	1.69	1.30	2.87	5.17	2.66
TiO_2	0.638	3.20	0.286	7.77	0.809	2.64	0.087	5.55
P_2O_5	0.130	3.63	0.211	4.71	0.229	3.22	0.038	16.6
MnO	0.040	0.00	0.092	4.58	0.087	5.55	0.019	16.6
Ba	364	2.90	144	3.97	319	2.67	106	1.95
Sr	63.3	3.16	77.2	1.81	74.5	2.98	15.8	4.99
Zr	472	9.44	341	15.7	317	16.5	285	3.50
Co	10.7	24.9	28.2	10.0	11.1	20.1	0.800	165
Cr	107	11.2	169	8.33	101	9.48	11.4	24.9
Sc	7.00	0.00	17.8	7.40	9.00	0.000	6.60	20.5
V	74.6	3.41	157	1.90	73.7	2.79	2.50	50.8
Zn	157	15.9	72.9	20.6	74.0	20.6	47.7	20.6

the significance of the above observations, the data may be subject to a *t*-test.

The t-tests A t-test can be used to answer questions about the likely value of a population mean on the basis of the sample statistics. The probabilities of two populations having the same mean will be dependent on the standard deviation and sample (population) size. A two-sample t-test can be used in this case to answer the question 'Are the two datasets for each element drawn from the same population?' The answer will help us evaluate whether replicate preparations are needed or if replicate measurements of a single solution will give an answer with sufficient confidence to be used in forensic investigation.

From t-test tables we note that the critical value for acceptance of the null hypothesis is −2.160 to +2.160. The level of significance has been set to be 5%, that is there is a 1 in 20 chance of our conclusion being wrong. Thus if the calculated T value for an element exceeds the critical value then we can conclude that the means of the two populations are different. Results of the t-test are shown in Table 8. For almost one-third of the elements determined by ICP-AES the calculated t value exceeds the critical value, and for ICP-MS this is the case for 11% of measurements.

The interpretation of this information has far-reaching consequences. In a significant proportion of measurements by ICP-AES, instrument precision is the dominant factor affecting the reproducibility of the dataset. In the other two-thirds of cases, preparation precision is dominant in controlling data quality.

For ICP-MS measurements, preparation precision is the most important factor. In practice, measurements from both techniques are usually carried out on the same solution in order to give as wide an elemental coverage as possible. It is clear that, in order to be able to assess the data quality critically, and hence use that data in an appropriate and informed way, it is necessary to replicate both the instrument measurements and the sample preparation.

A graphical representation of measurement and sample preparation precision is shown in Figure 1 for one pair of trace elements, thorium and uranium, in Soil 1. The preparation variation is large and the area defined by these points illustrates where a measured value could lie. Although the mean values appear to lie close to one another, the spread of data is clearly very different. Figure 2 illustrates the areas defined by data for Soil 3, lying within one SD of the means. By definition, 67% of the data lie within this area. However, that still means there is a 33% chance of the value lying beyond this defined area – an unacceptable situation for a forensic investigation. The data on which Figures 1 and 2 are based allow an assessment of the degree of error that must be taken into account when interpreting the results for any single element or pair of elements. A single analysis of a single sample permits no assessment of uncertainty.

Conclusions

This study has shown that instrument measurement variation can play an important role in controlling

Table 6. *Preparation precision (CV) calculated for trace elements by ICP-MS* (n = 10)

	Soil 1		Soil 2		Soil 3		SARM 1	
	Mean	CV	Mean	CV	Mean	CV	Mean	CV
U	1.76	6.10	1.11	8.87	1.93	10.7	15.5	8.20
Th	7.24	19.6	7.26	4.44	8.11	23.6	52.0	7.21
Rb	72.0	4.44	109	2.29	65.7	3.23	312	8.84
Nb	12.8	5.09	13.9	9.26	12.72	5.01	51.1	7.81
Cs	3.30	6.10	3.41	4.97	2.28	8.27	0.89	12.0
Hf	11.6	7.75	9.24	15.6	7.53	15.5	12.9	7.35
Ta	0.86	13.1	0.732	8.12	0.837	9.48	4.79	7.61
Y	20.7	6.33	19.9	5.69	18.3	17.0	145	7.49
La	22.1	14.3	31.4	4.47	34.1	14.3	103	7.92
Ce	43.0	15.8	54.6	3.67	67.6	15.2	190	6.77
Pr	4.57	19.3	8.81	3.35	7.17	15.6	19.8	7.37
Nd	21.8	14.9	37.3	2.95	31.0	14.1	73.6	7.65
Sm	3.63	14.4	6.57	4.35	4.79	20.0	14.8	8.76
Eu	0.69	5.27	1.22	3.91	0.899	6.99	0.31	8.91
Gd	3.35	8.95	4.84	3.84	3.85	17.2	13.1	7.95
Dy	2.75	7.16	3.23	7.28	2.70	20.2	16.6	7.95
Ho	0.702	5.15	0.756	4.05	0.643	12.4	3.73	7.26
Er	1.74	6.54	1.68	6.90	1.57	13.7	11.1	8.46
Yb	1.87	6.30	1.55	11.0	1.60	15.0	13.8	9.13
Lu	0.328	6.55	0.37	7.42	0.269	12.4	2.01	8.92

ICP data quality. However, replicated sample preparation is perhaps more critical if a true estimate of uncertainty is to be made. Uncertainty increases as concentrations approach the LLD, and therefore each element in each sample must be individually assessed with respect to error. An opinion regarding whether or not a suspect sample belongs to a control population must take full account of both instrument limits of detection derived from reagent blanks and from intersample variation derived from replicate and reference material analyses. The occurrence of a large CV does not in itself preclude the use of the data for a particular element, but the magnitude of the uncertainty must be taken into account during data interpretation and presentation in a court of law.

While the results of this study serve to highlight the critical importance of evaluating instrumental reproducibility and sampling reproducibility, we are left with a practical issue concerned with investigation. That is, the availability of often small amounts of test sample and the fact that routinely employed sample preparation procedures, such as alkali fusion and acid digestion, result in permanent destruction of the original physical and chemical state of the sample. However, without replicated measurement, uncertainty cannot be ruled out. Single analyses might be appropriate for preliminary investigation of an area where gross characterisation is required, but critical samples on which a case might depend should be analysed in a way that defines clearly the context of the data. This aspect of

Table 7. *Elements which show measurement precision to be greater than preparation precision*

Technique	Sample	Elements
ICP-AES	Soil 1	K_2O, TiO_2, Ba, Sr, Sc, V
	Soil 2	MgO, Na_2O, MnO, Sr, Sc, V
	Soil 3	Ba, Sr, Sc, V
	SARM1	MgO, Ba, Sr, Sc, V
ICP-MS	Soil 1	Nb
	Soil 2	Rb, Eu
	Soil 3	Rb
	SARM1	Hf

replication of measurement, which forms the cornerstone of reliability, therefore leads us to recommend that the technique is used only when sufficient sample is available to allow an assessment of uncertainty.

Clearly the application of, in this example, soil chemistry can provide valuable information during forensic investigation and the use of ICP techniques can provide data for more than 40 elements. One way forward with this issue is the initial screening of samples using a non-destructive multi-element technique such as neutron activation analysis (NAA) or X-ray fluorescence (XRF). While there can be significant issues for very small samples by both these techniques, NAA in particular offers good accuracy and multi-element determination with sample chemistry and physical properties pre-

Table 8. *Two-tailed t-test comparing measurement and preparation precision. Values in italics exceed the critical value of −2.160 to +2.160*

ICP-AES	SiO$_2$	Al$_2$O$_3$	Fe$_2$O$_3$	MgO	CaO	Na$_2$O	K$_2$O	TiO$_2$	P$_2$O$_5$	MnO	Ba	Sr	Zr	Co	Cr	Sc	V	Zn
Soil 1	0.90	−0.16	1.72	0.06	1.35	−0.69	2.58	2.28	−0.79	0.00	1.77	2.69	0.47	1.16	−2.39	−1.47	−2.37	−0.31
Soil 2	−0.54	−0.29	1.57	−2.30	1.35	−1.49	−0.72	3.56	−0.57	0.00	−1.29	−3.21	−3.09	0.31	1.30	−5.16	−2.90	0.45
Soil 3	1.09	0.62	1.73	0.58	1.11	2.37	2.30	1.99	0.80	3.18	−0.35	−0.68	−3.17	0.26	−4.81	−1.47	−0.69	1.30
SRM	−1.71	−2.35	0.38	−1.93	−1.06	−0.86	−0.41	−1.36	−3.14	−0.69	−4.64	−1.78	0.36	<LLD*	−3.42	−2.67	−1.85	1.18

ICP-MS	U	Th	Rb	Nb	Cs	Hf	Ta	Y	La	Ce	Pr	Nd	Sm	Eu	Gd	Dy	Ho	Er	Yb	Lu
Soil 1	−2.20	1.00	−3.38	−2.48	−1.82	−0.66	−0.19	−2.11	1.15	1.15	0.74	1.43	0.86	1.11	−0.12	−2.60	−1.76	−1.93	−2.00	−2.14
Soil 2	−3.40	−1.85	−0.67	1.75	−1.01	−3.86	2.48	−1.47	−0.64	−1.03	−1.89	−2.37	−0.75	0.16	−0.34	0.74	−1.24	−1.72	−1.23	−1.61
Soil 3	−0.17	0.76	−3.62	−1.30	−1.49	1.21	0.40	0.48	0.59	0.41	0.66	1.31	0.91	1.85	0.85	0.65	0.24	−0.01	0.69	−0.45
SRM	−0.12	0.02	0.50	−0.25	0.31	−0.59	0.78	−1.06	0.00	−0.35	−0.39	0.27	0.26	0.61	−0.04	−0.24	−1.10	−1.53	−1.20	−1.86

* Concentration less than the lower limit of detection

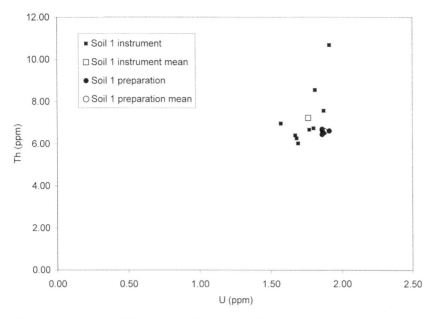

Fig. 1. Measured concentration of U and Th for Soil 1 showing difference between measurement and preparation precision.

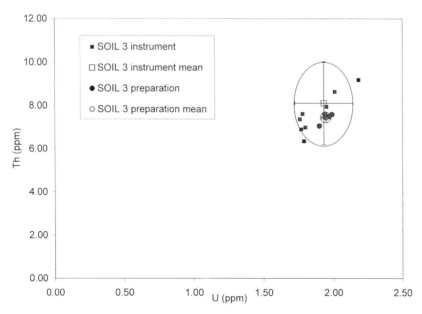

Fig. 2. Error bars at the 1σ level for U and Th determinations showing instrumental and preparation means shown for Soil 3.

served, and it is already used in some forensic investigations (Parry 2003). Initial screening of samples could permit a more informed decision about sample preparation approach and concentration levels present.

The authors would like to thank A. Moncrieff (Hawkins and Associates) for constructive criticism and support during the production of this manuscript. The NERC ICP-AES and ICP-MS Facilities are supported by the Natural Environment Research Council.

Appendix 1. Results of sample analysis by ICP-AES (concentrations in wt% for oxides and p.p.m. for trace elements

Table 1. *Intrasample analysis* (n = 5)

	Soil 1			Soil 2			Soil 3			SARM1		
	Mean	SD	CV	Mean	SD	CV	Mean	SD	CV	Mean	SD	CV
SiO_2	71.9	0.131	**0.182**	64.9	0.175	**0.269**	51.6	0.069	**0.135**	77.0	0.34	**0.44**
Al_2O_3	7.36	0.015	**0.201**	5.43	0.018	**0.335**	12.2	0.015	**0.124**	12.4	0.07	**0.55**
Fe_2O_3	3.70	0.008	**0.226**	10.70	0.153	**1.43**	3.37	0.008	**0.248**	1.78	0.04	**2.36**
MgO	0.53	0.019	**3.53**	1.37	0.021	**1.52**	0.426	0.015	**3.56**	0.130	0.017	**13.3**
CaO	0.57	0.007	**1.24**	4.25	0.047	**1.10**	0.050	0.007	**14.1**	0.782	0.013	**1.67**
Na_2O	0.23	0.008	**3.61**	0.032	0.008	**26.1**	0.718	0.016	**2.29**	3.34	0.03	**0.82**
K_2O	1.81	0.059	**3.24**	3.00	0.020	**0.67**	1.26	0.029	**2.28**	5.20	0.03	**0.64**
TiO_2	0.61	0.022	**3.54**	0.250	0.000	**0.000**	0.786	0.021	**2.64**	0.090	0.000	**0.000**
P_2O_5	0.13	0.004	**3.39**	0.214	0.009	**4.18**	0.226	0.005	**2.42**	0.048	0.004	**9.32**
MnO	0.04	0.000	**0.000**	0.092	0.004	**4.86**	0.080	0.000	**0.00**	0.02	0.000	**0.000**
Ba	376	15.76	**4.20**	147	1.41	**0.962**	321	11.4	**3.57**	111	2.39	**2.15**
Sr	66.6	2.70	**4.06**	79.8	1.64	**2.06**	75.4	2.79	**3.70**	16.6	0.894	**5.39**
Zr	482	17.59	**3.65**	417	8.71	**2.09**	394	13.2	**3.34**	283	6.30	**2.22**
Co	9.20	1.48	**16.1**	27.8	0.837	**3.01**	10.8	1.92	**17.8**	1.40	1.14	**81.4**
Cr	120	5.13	**4.26**	160	0.447	**0.279**	122	4.56	**3.73**	16.0	1.22	**7.65**
Sc	7.20	0.447	**6.21**	23.6	3.13	**13.3**	9.20	0.45	**4.86**	9.20	2.49	**27.1**
V	78.4	3.65	**4.65**	162	3.29	**2.03**	74.6	2.97	**3.98**	4.00	1.87	**46.8**
Zn	160	6.34	**3.96**	69.8	2.77	**3.98**	64.8	3.83	**5.92**	42.4	1.34	**3.16**

Table 2. *Inter sample analysis* (n = 10)

	Soil 1			Soil 2			Soil 3			SARM1		
	Mean	SD	CV	Mean	SD	CV	Mean	SD	CV	Mean	SD	CV
SiO_2	72.6	1.69	**2.33**	64.6	0.909	**1.41**	52.1	1.07	**2.05**	76.3	0.845	**1.11**
Al_2O_3	7.34	0.279	**3.80**	5.42	0.084	**1.54**	12.4	0.458	**3.71**	12.2	0.124	**1.01**
Fe_2O_3	4.08	0.492	**12.1**	11.2	0.638	**5.71**	3.61	0.306	**8.47**	1.87	0.481	**25.7**
MgO	0.531	0.035	**6.61**	1.34	0.021	**1.53**	0.434	0.028	**6.54**	0.112	0.017	**15.1**
CaO	0.540	0.049	**8.99**	4.31	0.104	**2.41**	0.076	0.051	**67.1**	0.772	0.019	**2.43**
Na_2O	0.228	0.011	**4.98**	0.027	0.005	**17.9**	0.753	0.031	**4.06**	3.31	0.066	**2.00**
K_2O	1.88	0.047	**2.51**	2.98	0.050	**1.69**	1.30	0.037	**2.87**	5.17	0.138	**2.66**
TiO_2	0.638	0.020	**3.20**	0.286	0.022	**7.77**	0.809	0.021	**2.64**	0.087	0.005	**5.55**
P_2O_5	0.130	0.005	**3.63**	0.211	0.010	**4.71**	0.229	0.007	**3.22**	0.038	0.006	**16.6**
MnO	0.040	0.000	**0.00**	0.092	0.004	**4.58**	0.087	0.005	**5.55**	0.019	0.003	**16.6**
Ba	364	10.6	**2.90**	144	5.70	**3.97**	319	8.52	**2.67**	106	2.058	**1.95**
Sr	63.3	2.00	**3.16**	77.2	1.40	**1.81**	74.5	2.22	**2.98**	15.8	0.789	**4.99**
Zr	472	44.5	**9.44**	341	53.6	**15.7**	317	52.5	**16.5**	285	9.989	**3.50**
Co	10.7	2.67	**24.9**	28.2	2.82	**10.0**	11.1	2.23	**20.1**	0.800	1.317	**165**
Cr	107	12.0	**11.2**	169	14.0	**8.33**	101	9.52	**9.48**	11.4	2.84	**24.9**
Sc	7.00	0.000	**0.00**	17.8	1.32	**7.40**	9.00	0.000	**0.000**	6.60	1.35	**20.5**
V	74.6	2.55	**3.41**	157	2.98	**1.90**	73.7	2.06	**2.79**	2.50	1.27	**50.8**
Zn	157	24.9	**15.9**	72.9	15.0	**20.6**	74.0	15.3	**20.6**	47.7	9.82	**20.6**

Appendix 2. *Results of sample analysis by ICP-MS (concentrations in p.p.m.)*

Table 1. *Intrasample analysis (n = 5)*

	Soil 1			Soil 2			Soil 3			SARM1		
	Mean	SD	CV	Mean	SD	CV	Mean	SD	CV	Mean	SD	CV
U	1.87	0.022	**1.16**	1.27	0.037	**2.93**	1.95	0.034	**1.76**	15.5	0.965	**6.21**
Th	6.59	0.103	**1.56**	7.54	0.148	**1.96**	7.44	0.223	**3.00**	52.0	3.02	**5.81**
Rb	77.5	2.38	**3.08**	110	2.72	**2.48**	70.0	2.32	**3.32**	305	22.2	**7.29**
Nb	13.7	0.684	**5.00**	12.8	0.499	**3.90**	13.1	0.459	**3.49**	51.7	3.75	**7.26**
Cs	3.49	0.146	**4.20**	3.50	0.147	**4.20**	2.42	0.137	**5.66**	0.878	0.061	**6.99**
Hf	11.9	0.176	**1.47**	11.8	0.148	**1.26**	6.88	0.329	**4.78**	13.2	0.992	**7.53**
Ta	0.87	0.029	**3.28**	0.66	0.027	**4.05**	0.82	0.030	**3.69**	4.65	0.23	**4.92**
Y	22.0	0.553	**2.52**	20.7	0.395	**1.91**	17.6	0.615	**3.48**	151	10.6	**7.02**
La	20.4	0.775	**3.80**	31.8	0.886	**2.78**	32.8	1.54	**4.70**	103	7.04	**6.84**
Ce	40.7	1.49	**3.65**	55.6	0.487	**0.88**	65.6	2.72	**4.14**	192	12.5	**6.49**
Pr	4.10	0.138	**3.36**	9.08	0.134	**1.47**	6.83	0.256	**3.74**	20.1	1.41	**6.99**
Nd	19.7	0.502	**2.55**	38.7	1.07	**2.77**	28.4	0.877	**3.09**	72.8	4.73	**6.50**
Sm	3.42	0.153	**4.48**	6.67	0.088	**1.32**	4.38	0.194	**4.42**	14.7	0.95	**6.51**
Eu	0.674	0.023	**3.42**	1.22	0.074	**6.07**	0.844	0.027	**3.20**	0.304	0.011	**3.75**
Gd	3.36	0.084	**2.51**	4.87	0.117	**2.40**	3.59	0.188	**5.26**	13.1	0.827	**6.32**
Dy	3.01	0.127	**4.21**	3.14	0.178	**5.66**	2.53	0.135	**5.32**	16.8	0.987	**5.88**
Ho	0.734	0.025	**3.42**	0.776	0.027	**3.48**	0.634	0.024	**3.80**	3.89	0.227	**5.84**
Er	1.85	0.052	**2.82**	1.78	0.069	**3.90**	1.57	0.069	**4.44**	11.9	0.833	**7.02**
Yb	1.98	0.044	**2.21**	1.65	0.049	**2.99**	1.52	0.047	**3.10**	14.5	0.906	**6.23**
Lu	0.352	0.018	**5.08**	0.388	0.019	**4.96**	0.28	0.01	**4.13**	2.19	0.172	**7.87**

Table 2. *Intersample analysis (n = 10)*

	Soil 1			Soil 2			Soil 3			SARM1		
	Mean	SD	CV	Mean	SD	CV	Mean	SD	CV	Mean	SD	CV
U	1.76	0.108	**6.10**	1.11	0.098	**8.87**	1.93	0.207	**10.7**	15.5	1.27	**8.20**
Th	7.24	1.42	**19.6**	7.26	0.322	**4.44**	8.11	1.92	**23.6**	52.0	3.75	**7.21**
Rb	72.0	3.19	**4.44**	109	2.49	**2.29**	65.7	2.12	**3.23**	312	27.6	**8.84**
Nb	12.8	0.650	**5.09**	13.9	1.28	**9.26**	12.72	0.637	**5.01**	51.1	3.99	**7.81**
Cs	3.30	0.202	**6.10**	3.41	0.169	**4.97**	2.28	0.189	**8.27**	0.89	0.11	**12.0**
Hf	11.6	0.903	**7.75**	9.24	1.44	**15.6**	7.53	1.16	**15.5**	12.9	0.95	**7.35**
Ta	0.86	0.113	**13.1**	0.732	0.059	**8.12**	0.837	0.079	**9.48**	4.79	0.36	**7.61**
Y	20.7	1.31	**6.33**	19.9	1.13	**5.69**	18.3	3.12	**17.0**	145	10.8	**7.49**
La	22.1	3.15	**14.3**	31.4	1.40	**4.47**	34.1	4.87	**14.3**	103	8.16	**7.92**
Ce	43.0	6.80	**15.8**	54.6	2.01	**3.67**	67.6	10.26	**15.2**	190	12.8	**6.77**
Pr	4.57	0.882	**19.3**	8.81	0.295	**3.35**	7.17	1.12	**15.6**	19.8	1.46	**7.37**
Nd	21.8	3.24	**14.9**	37.3	1.10	**2.95**	31.0	4.37	**14.1**	73.6	5.63	**7.65**
Sm	3.63	0.523	**14.4**	6.57	0.286	**4.35**	4.79	0.957	**20.0**	14.8	1.30	**8.76**
Eu	0.69	0.037	**5.27**	1.22	0.048	**3.91**	0.899	0.063	**6.99**	0.31	0.03	**8.91**
Gd	3.35	0.300	**8.95**	4.84	0.186	**3.84**	3.85	0.662	**17.2**	13.1	1.04	**7.95**
Dy	2.75	0.197	**7.16**	3.23	0.235	**7.28**	2.70	0.545	**20.2**	16.6	1.32	**7.95**
Ho	0.702	0.036	**5.15**	0.756	0.031	**4.05**	0.643	0.080	**12.4**	3.73	0.27	**7.26**
Er	1.74	0.114	**6.54**	1.68	0.116	**6.90**	1.57	0.215	**13.7**	11.1	0.94	**8.46**
Yb	1.87	0.118	**6.30**	1.55	0.171	**11.0**	1.60	0.240	**15.0**	13.8	1.26	**9.13**
Lu	0.328	0.021	**6.55**	0.37	0.03	**7.42**	0.269	0.033	**12.4**	2.01	0.18	**8.92**

References

DUCKWORTH, D. C., MORTON, S. J., BAYNE, C. K., KOONS, R. D., MONTERO, S. & ALMIRALL, J. R. 2002. Forensic glass analysis by ICP-MS: a multi-element assessment of discriminating power via analysis of variance and pairwise comparisons. *Journal of Analytical Atomic Spectrometry*, **17**, 662–668.

GY, P. M. 1992. *Sampling of heterogeneous and Dynamic Materials: Theories of Heterogeneity, Sampling and Homogenising.* Elsevier, Amsterdam.

JARVIS, K. E., GRAY, A. L. & HOUK, R. S. 2003. *Handbook of Inductively Coupled Plasma Mass Spectrometry,* Viridian Publishing, Woking.

KETO, R.O. 1999. Analysis and comparison of bullet leads by inductively coupled plasma mass spectrometry. *Journal of Forensic Science*, **44**, 1020–1026.

PARRY, S. J. 2003. *Handbook of Neutron Activation Analysis.* Viridian Publishing, Woking.

POTTS, P. J. 2003. *Handbook of Rock Analysis.* Viridian Publishing, Woking.

RAMSEY, M. H. 1994. Error estimation in environmental sampling and analysis. *In*: MARKERT, B. (ed.) *Sampling of Environmental Materials for Trace Analysis.* VCH, Weinheim, 93–108.

RAMSEY, M. H. 1997. Sampling and sample preparation. *In*: GILL, R. (ed.) Chapter 2 – *Modern Analytical Geochemistry*, Longman, Harlow, 12–28.

SPENCE, L. D., BAKER, A. T. & BYRNE, J. P 2000. Characterisation of document paper using elemental compositions determined by inductively coupled plasma mass spectrometry. *Journal of Analytical Atomic Spectrometry*, **15**, 813–819.

SUZUKI, Y., KASAMATSU, M., SUGITA, R., OHTA, H., SUZUKI, S., AND MARUMO, Y. 2003. Forensic discrimination of headlight glass by analysis of trace impurities with synchrotron radiation X-ray fluorescence spectrometry and ICP-MS. *Bunseki Kagaku*, **52**, 469–474.

SUZUKI, Y., SUGITA, R., SUZUKI, S. & MARUMO, Y. 2000. Forensic discrimination of bottle glass by refractive index measurement and analysis of trace elements with ICP-MS. *Analytical Sciences*, **16**, 1195–1198

THOMPSON, M. & HOWARTH, R. J. 1976. Duplicate analysis in geochemical practice. *Analyst*, **101**, 690–698.

THOMPSON, M. & HOWARTH, R. J. 1978. A new approach to the estimation of analytical precision. *Journal of Geochemical Exploration*, **9**, 23–30, 690–698.

THOMPSON, M. & WALSH, J. N. 2003. Handbook of inductively coupled plasma atomic emission spectrometry, 2nd edition. Viridian Publishing, Woking.

THOMPSON, M., POTTS, P. J., KANE, J.S. & WILSON, S. 2000. GeoPT5. International proficiency test for analytical geochemistry laboratories. Report on round 5. Geostandards Newsletter, *Journal of Geostandards and Geoanalysis*, **24**, E1–E28. [Electronic journal edition]

TOTLAND, M., JARVIS, I. AND JARVIS, K. E. 1992. An assessment of dissolution techniques for the analysis of geological samples by plasma spectrometry. *Chemical Geology*, **95**, 35–62

WALSH, J. N. 1992. Use of multiple internal standards for high precision, routine analysis of geological samples by inductively coupled plasma-atomic emission spectrometry. *Chemical Geology*, **95**, 113–121

WEBSTER, R. 2001. Statistics to support soil research and their presentation. *European Journal of Soil Science*, **52**, 331–340.

YOURD, E. R., TYSON, J.F. & KOONS, R. D. 2001. On-line matrix removal of lead for the determination of trace elements in forensic bullet samples by flow injection inductively coupled plasma mass spectrometry. *Spectrochimica Acta (B)*, **56**, 1731–1745.

Comparison of soils and sediments using major and trace element data

KENNETH PYE* & SIMON J. BLOTT

*Kenneth Pye Associates Ltd, Crowthorne Enterprise Centre, Crowthorne Business Estate, Old Wokingham Road, Crowthorne RG45 6AW, UK (*e-mail: k.pye@kpal.co.uk)*

Abstract: Analysis of geochemical data has become an important tool used in forensic comparisons of soils and sediments. The combined use of inductively coupled plasma optical emission spectrometry and mass spectrometry (ICP-OES and ICP-MS) instrumentation allows the abundance of up to 50 elements to be routinely determined in small samples (typically $>0.1\,g$). However, key issues concern the extent to which analysed subsamples are representative of the parent material from which they are taken, the best means of comparing datasets for different samples, and interpretation of the significance of the results obtained. ICP measurement precision for most elements is good, but it is important to understand the degree of variation that can arise due to subsampling procedures and selective transfer mechanisms relating to forensic exhibits, as well as the extent of spatial (and sometimes temporal) variability which exists in nature. Although analysis of several different size fractions is often helpful where sufficiently large samples are available, analysis of a standardized $<150\,\mu m$ fraction separated from a bulk sample in many cases provides adequate discrimination between samples and provides the most practical method for mass-screening of samples. Where possible, duplicate or triplicate sample preparations should be made, and several analytical determinations made on each prepared subsample. However, the additional time, cost and sample size requirements involved need to be weighed against the benefits of undertaking additional types of analysis on the samples. In order to obtain maximum information from multi-element geochemical data, the dataset should be evaluated using a variety of numerical, statistical and graphical procedures. This paper discusses a number of options for such data evaluation using a simple dataset example. Casework experience and experiments have shown that, even in complex situations, multi-element geochemical data can provide very sensitive environmental indicators. However, such data should normally be used in combination with the results of other analyses when making forensic comparisons of soils and sediments.

The comparison of soils and sediments for forensic purposes, and in wider environmental provenance studies, is normally undertaken on the basis of a number of independent criteria and using a number of different techniques. One powerful criterion for comparison is major and trace element composition, either in a bulk sample or in one or more separated grain size fractions. Concentrations of selected cations and anions in leachate samples, or certain isotopic ratios, may also usefully be determined. On the basis of these attributes, a chemical profile can be obtained from the sediment or soil in question. Some authors have used the term 'fingerprint' to define any characteristic, including a chemical profile, which is identifiable in both a sediment source and a sediment sink containing material from that source (e.g., Merefield *et al.* 2000; Rowan *et al.* 2000; Slattery *et al.* 2000). However, the term 'fingerprint' has connotations of singularity, if not uniqueness, which are rarely applicable to individual sediment or soil samples, and the present authors prefer the term *characteristic profile*. A characteristic profile is made up of several characteristics which may be useful individually as *tracers* or *identifiers* of association between soil/sediment sources and deposits. A

variety of characteristics may be useful in this way, including chemical and mineral composition, magnetic characteristics, grain size distributions, grain shape and surface textural properties. In this paper we restrict our discussion to major and trace element chemical characteristics.

The requirement for soil or sediment comparison may arise in a variety of circumstances of varying complexity. In the simplest case, a geoscientist is faced with two soil samples, A and B. It needs to be determined whether A and B are indistinguishable, broadly similar, or significantly different to each other. Where one sample is a 'questioned' material of unknown origin and the other is a 'control' (or 'exemplar') material from a known source, it is a frequent requirement to determine whether the questioned material could, or did, come from the control location, which may be a crime scene. If samples A and B are both from unknown locations, the key issues are: (1) whether they could, or did, come from the same source, and (2) the nature and location of that source.

There are instances in forensic geoscience where these questions can be answered with effective certainty, but such circumstances are very rare, and

From: PYE, K. & CROFT, D. J. (eds) 2004. *Forensic Geoscience: Principles, Techniques and Applications.* Geological Society, London, Special Publications, **232**, 183–196. © The Geological Society of London, 2004.

normally occur only where a perfect physical fit can be obtained, supported by chemical, mineralogical or textural data; for example between two pieces of broken rock or between a piece of rock and a cast formed in dried mud. However, in most situations, samples can only be compared on the basis of bulk sample or individual particle characteristics, and the results can only be interpreted as probabilistic. In general, the greater the number of characteristics which can be used for comparison, the greater the confidence which can be placed in any apparent similarity. The value of individual characteristics and techniques will vary with circumstances, some characteristics being more location-specific, and some techniques yielding results with greater inherent precision and resolving power. However, in many circumstances, major and trace elemental composition provides one of the most sensitive and precise criteria for comparison, and has a role to play in the vast majority of soil comparison studies.

Two soil samples will rarely, if ever, show perfect chemical similarity (i.e. in general parlance, provide a perfect chemical 'match'). This is true of samples taken only a few centimetres apart at the same time, and arises due to the inherent spatial variation which exists in soils and sediments, combined with the variability in results due to subsampling, both in the field and laboratory, and to errors associated with sample preparation and instrumental measurement. In general, the more chemically similar are two samples, the more likely it is that they come from either the same source or very similar, essentially chemically indistinguishable sources. On the other hand, if sample A is very different from sample B, then it is very unlikely that the bulk of soil A came from the same source, or even a similar source, as soil B. However, from a forensic investigation point of view this should not be seen necessarily as the end of the exercise, since soil A may contain a component, perhaps representing a few percent or less, of particles which originate from the same source as particles in soil B. This is one reason why forensic comparison of soils should never rely only on comparison of bulk properties such a multi-element geochemistry; microscopic examination and/or other tests should also be carried out to determine the possible presence of diagnostic tracers or identifiers. However, if there are significant bulk compositional similarities between soil samples A and B, then subsidiary questions need to be asked. For example, could sample A be derived from the same source as B, and is there only one source or several possible sources?

As noted above, a number of techniques usually need to be applied in combination to answer these questions. However, in some instances the information provided by a single technique may be sufficient. This most frequently occurs where one or

several highly unusual particles which do not occur naturally in an area, sometimes referred to as 'exotics', can be identified (usually using microscopic and chemical microanalysis techniques) in both the questioned sample and one or more control samples from the area of interest (e.g. the crime scene). Where no such particles can be identified, the analyst is forced to rely on comparison of bulk sediment properties, including multi-element chemical analysis. It is therefore important to understand the limitations of these methods and to employ a variety of methods for interrogation of the data obtained.

Development of analytical instrumentation over the past 25 years has meant that ever-smaller amounts of material can be analysed in terms of elemental composition with high precision and accuracy. A number of techniques are now available to determine the inorganic elemental composition of soils and sediments, including X-ray fluorescence (XRF), atomic absorption spectrophotometry (AAS), inductively coupled plasma (ICP) spectrometry, neutron activation analysis (NAA) and energy and wavelength dispersive X-ray (EDX and WDX) microanalysis (e.g. Butler 1986; Willard et al. 1988; Gill 1997; Parry 1997, 2003). Each technique has advantages and disadvantages in terms of ease of sample preparation, sample size requirements, cost, precision and accuracy of results for different elements, and extent to which the analysis is destructive or non-destructive. No technique is without its limitations, and selection of the most appropriate method of analysis is always dependent on the objectives and requirements of a particular investigation. Where sample size is not a limitation, as in mineral exploration and environmental contamination studies of soils and sediments, XRF is frequently used (e.g. Pirrie et al. 2003; Rawlins et al. 2003). However, in the past 15 to 20 years ICP spectrometry has become the preferred technique of many scientists who need to measure the abundance of a broad suite of elements (typically up to 50) in small samples. ICP spectrometry was initially developed in the 1960s and has since been developed for a range of specialist applications, including environmental and exploration geoscience (Walsh 1979, 1997; Walsh et al. 1981; Jarvis & Jarvis 1985; Thompson & Walsh 2003; Jarvis et al. 2003). In forensic science, ICP spectrometry has been used in the analysis of many types of trace evidence including glass (Skirda 1991), wood fragments (Miller 1991) and bullet fragments (Keto 1999). However, there are relatively few published studies involving its use in forensic soil investigations (e.g. Shinomiya et al. 1998).

There are two principal variants of the ICP instrumental analysis technique, namely ICP-OES (optical emission spectrometry, also termed ICP-AES, or atomic emission spectrometry) and ICP-MS (mass spectrometry). Each is ideally suited to a different range of elements, but used in combination can provide concentration data for around 50 elements in samples as small as 0.1 g. Even smaller samples (0.05 g) can be used, although smaller samples result in greater weighing errors, greater sample dilutions and poorer signal to noise ratios, and therefore poorer precision and accuracy (Thompson & Walsh 2003; Jarvis *et al.* 2003).

The main issue in ICP analysis, as with all methods used to analyse relatively small samples, is to obtain an analysed subsample that is representative of the whole. Subsampling invariably produces differences between analysed subsamples, and a major task is to keep these to a minimum through careful sample preparation protocols.

This paper describes a method of sample preparation for ICP analysis which has been found by the authors to work well in routine soil and sediment investigations, including forensic applications, and discusses a number of methods which can be used to interpret the results obtained. Issues of instrumental measurement and precision, subsampling variability, different scales of natural heterogeneity, and assessment of the statistical and evidential value of results are important but limitations of space preclude a full discussion here.

Sample preparation and handling procedures

Effects of particle size variation

The results obtained in any study of major and trace element composition will be affected by a number of factors, including: (1) the pattern of natural variability which exists in the field (background population); (2) the nature and efficiency of field sampling; (3) the nature and efficiency of laboratory handling procedures, including storage, subsampling, any pre-treatment and additions of dispersants, acids, water or flocculants; (4) the extent of operator errors in weighing, sample handling, notation etc.; (5) the reliability and stability of analytical instrumentation; (6) variations in any standards and reference materials used for instrument calibration and correction of results; (7) the nature of any additional post-collection data processing. A full discussion of each of these sources of variability is beyond the scope of this paper, but comments on the importance of grain size variation and sample handling are made below.

When analysing any property of a sediment or soil, it is important to consider the effects of particle size variations on the results. In all but the most mineralogically homogeneous materials, chemical composition is likely to vary significantly with grain size, silica, for instance, being relatively more abundant

in the coarser fractions, and trace metals being concentrated in the fine sand, silt and clay fractions (which contain most of the heavy minerals and clay minerals). Particle size distribution varies from sample to sample, but the extent of variation differs greatly from one geological material to another. At a small spatial scale (<1 m), some soils and sediments are relatively homogeneous in terms of particle size distribution and mineralogical/chemical composition, but others are very heterogeneous. Loess deposits and aqueous silts, for example, are often relatively homogeneous at such small spatial scales, while surface samples taken from a roadside location or waste tip often show a very high degree of particle size and compositional heterogeneity. It is also important to be aware that, during transfer of soil material onto forensic exhibits, such as clothing or footwear, grain size fractionation may take place. For example, wet mud or muddy water dripped onto a surface may not contain the full range of grain sizes present in the source material. Some minerals and other particle types are concentrated in certain grain size ranges so that a high degree of similarity between the source and transferred materials may not always be found, and there is a risk that a false conclusion may be reached regarding provenance unless the compositions of narrowly defined size fractions are compared.

Although useful information can often be gained from analysis of bulk soil and sediment samples (or at least the <2 mm fractions of samples from which gravel material has been removed), better comparisons may be made by comparing chemical data for several different size fractions, the size ranges chosen being related to the material in question. This is because, as noted above, chemical variations are often associated with particle size differences, and in some sediments and soils the particle size distribution is highly variable over quite short distances. This may make intersample comparison difficult and obscure larger scale concentration trends. Consequently, many studies, for example of heavy metal contaminants in sediments, have concentrated on the <60 um or <20 mm fractions, which show less small-scale intersample spatial variability than the coarser fractions and bulk samples (e.g. Ackermann *et al.* 1983).

In our experience, for a majority of circumstances, analysis of the <150 μm fraction allows adequate discrimination between samples and provides a relatively consistent indicative measure of the composition of a sample. It also provides the most practical and time/cost-effective method for mass screening of samples. However, in some circumstances, for example where the 'forensic' material in question is fine grained (e.g. muddy water stains on clothing), a more appropriate comparison may be made with the <20 μm and/or smaller size fractions taken from the control samples. More generally, where a sufficiently large amount of critical 'forensic sample' material is available, better discrimination may be obtained by separating a number of narrowly defined size fractions within the 0–150 μm range. However, with small samples this is often not possible, or the procedure may reduce the size of the separated subsamples to such an extent that additional errors are introduced as a result of inaccuracies in weighing, greater sample dilution and poorer signal to noise ratio during instrumental measurement. Since sample preparation for ICP analysis is destructive, the amount of sample material that is used for this purpose also reduces the options for other methods of analysis (e.g. pollen or diatom analysis). A judgement therefore has to be made about prioritization in the analytical strategy, taking into account the circumstances and objectives of each individual investigation.

All particle size separations are time-consuming, result in the creation of smaller subsamples for analysis, and increase the risk of additional sample contamination if dispersants and flocculants are used in the separation process. In forensic and environmental contamination studies, it is clearly desirable that all potential sources of analytical contamination are kept to a minimum, and that their effects on the results obtained are understood. Wherever possible, disposable, inert (e.g. plastic) sieve meshes should be used and the use of dispersants and flocculants avoided. 'Single-use' mesh is essential for forensic analysis where each size fraction must be retained for evidence purposes and the possibilities of cross-contamination from sample to sample need to be reduced to an absolute minimum. The use of traditional brass or steel sieves is generally unacceptable due to the obvious risk of metal contamination and also to the risk of cross-sample contamination; such sieves are also generally too expensive for one-time use.

It has, in fact, been long-standing practice in many geochemical laboratories to obtain different size fractions for analysis by 'washing' the sample through disposable, inert nylon mesh held in a circular polycarbonate frame, using distilled de-ionized water. While the polycarbonate holders themselves are not disposable, they can be easily cleaned with distilled de-ionized water or other cleansing agents between analyses. Separate sets are normally used for different groups of samples.

Several different mesh sizes are available commercially, the most commonly used being 150 μm, 60 μm and 20 μm. For routine forensic and environmental provenance work, we have found the 150 μm mesh to be most useful for a number of reasons, including the fact that it is relatively easy to disaggregate, disperse and wash soil and sediment samples through this grade of mesh using only dis-

tilled de-ionized water from a wash bottle and gentle pressure (e.g., from a rubber-tipped glass rod). Additionally, the <150 μm fraction has been found to provide good discrimination between most soil and sediment sample types since it contains most of the heavy minerals, micas and clays with which a majority of the trace elements are associated. The separated >150 μm fraction, on the other hand, can conveniently be examined by optical microscopy free from adhering silt, clay and organic matter.

The <150 μm fraction that passes through the mesh is transferred to a sterile, disposable, snap-top plastic beaker and allowed to settle for at least 24 h before being evaporated to dryness in $situ$ in an oven. Part of the dried powder (typically half) is then removed and ground in an agate pestle and mortar to a fineness of <10 μm. The remaining half of the sample is retained in $situ$ for other forms of analysis, or repeat analyses, as required. Adequate grinding of the material to be subjected to chemical analysis is an essential requirement to ensure homogeneity and minimization of the 'nugget' effect, which can arise if a sample contains an irregular distribution of a relatively small number of particles rich in certain elements (Ramsey 1997).

The procedure for dissolution of the ground fraction used to generate the results presented in this paper has been modified from Thompson and Walsh (2003), as described by Jarvis et $al.$ (2003) and undertaken routinely at the University of Greenwich. A quantity of the ground sample is combined with a lithium metaborate ($LiBO_2$) flux in the ratio 1:5 and fused at a nominal 1000 °C for 20 min. The molten bead is then poured into a vessel containing weak nitric acid and allowed to dissolve before being filtered and made up to volume. This method has been found to be suitable for the quantitative dissolution of major elements and most trace elements in rock samples. Sixteen elements are subsequently determined by ICP-OES: SiO_2, Al_2O_3, Fe_2O_3, MgO, CaO, Na_2O, K_2O, TiO_2, P_2O_5, MnO (in wt %), and Ba, Be, Ni, Sc, Y, Zn (in p.p.m.). A further 33 elements are then determined by ICP-MS: Co, Cr, Cu, Pb, Sr, V, Zr, Ga, Rb, Nb, Mo, Sn, Cs, Hf, Ta, W, Tl, U, Th, La, Ce, Pr, Nd, Sm, Eu, Gd, Tb, Dy, Ho, Er, Tm, Yb, Lu (in p.p.m.). This method is not suitable for the determination of volatile metals such as As, Cd, and Hg, and does not allow determination of Li due to the use of $LiBO_2$ flux during preparation.

The ICP-OES instrument (TJA Iris Advantage) used to generate the data presented here is operated using a series of quality control protocols developed over a number of years by David Wray at the University of Greenwich. The instrument is calibrated daily by the analysis of nine certified reference materials (CRMs), prepared in the same way as the unknown samples, and covering the range of values expected in the samples. Five CRMs are also analysed after every 20 samples and at the beginning and end of every batch. A monitor solution checks for drift after every five solutions, and results are automatically blank corrected. The ICP-MS instrument (Thermo Elemental X7 with a HPI interface) is calibrated after every ten samples using six synthetic standards prepared in $LiBO_2$ flux. A correction for drift is automatically extrapolated between sets of standards, and three blank corrections are performed. During runs on both instruments, five CRMs are analysed after every 20 samples and at the beginning and end of every batch. Results from six replicate analyses of the same sample (six sequential runs) indicate that the short-term precision of the instruments is good, with a coefficient of variation of less than 1% for most elements. Analysis of four replicate subsamples of the same sample (run within the same batch) also indicate that precision is usually reasonable, with coefficients of variation of 1–2% for ICP-OES elements and 2–3% for ICP-MS elements. Precision between batches of samples run over periods of several weeks or months is rather lower due to drift in the instrument and differences in the made-up solutions. Monitoring of standards over an 8-month period has demonstrated a precision of between 2% and 5% for most ICP-OES elements, and between 4% and 6% for most ICP-MS elements.

Procedures for interpretation of geochemical data

Even a relatively small batch of around 20 samples will yield approximately 1000 oxide and elemental concentration values, and interpretation can initially appear highly complex. Larger cases, which may involve hundreds of samples, generate proportionately more data. A method is therefore required to systematically inspect the data in order to extract trends between individual samples and different groups of samples. To illustrate this process we use here a simple real case example where five visually similar soil samples were taken from a suspect's footwear and compared with nine control samples taken by police at the scene of a crime. The objective was to determine whether the soil from the boots 'matched' any of the control samples and therefore might suggest that the suspect had been present at the scene of the crime.

A key step in this process is to understand the nature of the samples which are being compared; for example whether more than one type is likely to present on the footwear or from the control locations. Consequently it is an advantage if the samples themselves have been taken by the data analyst, and examined visually by optical microscopy prior to chemical analysis.

Although a wide range of statistical techniques is available for the analysis of geochemical data (e.g. Rollinson 1993; Swan & Sandilands 1995; Sapsford & Jupp 1996; Webster & Oliver 2001; Davis 2002), it is important that simple visual examination, tabular comparison and graphical plotting procedures should not be overlooked. An initial step is simply to visually examine the data in order to identify any apparent groupings and patterns of similarity between the samples. As well as inspecting the raw concentration data, it may also be helpful to 'normalize' the raw values to a common element, frequently Al_2O_3. Normalization removes the 'closure' problem with chemical data, which frequently arises due to the presence of unanalysed components such as organic matter, carbonate or sulphide. The presence of a large amount of organic matter in a sample which has not been ashed before weighing and acid digestion will produce apparent dilution. Moreover, quoted percentage (or p.p.m.) concentration values for measured elements are autocorrelated; that is the percentage of any one measured element is affected by the concentration of other measured elements. The major oxides SiO_2 and Al_2O_3, being the dominant constituents in most sediments and soils, will have the most influence, and taking ratios of other elements to either of these two oxides can be used to normalize the dataset.

Using either the raw or normalized values, the next stage is to identify broad-scale trends in the dataset. This can be done by highlighting those values in the questioned forensic samples which fall within the range of the control samples, plus or minus a 'degree of tolerance' to allow for measurement uncertainty and sampling error ($\pm10\%$ can be taken as a reasonable rule of thumb for most elements, although other threshold values, such as 3%, 5% or 20%, can be adopted for modelling purposes). The calculations can be performed automatically and highlighted using a simple macro in a computer spreadsheet package such as Microsoft® Excel, or by manual calculation and using a marker pen. The results of such a highlighting exercise are shown in Table 1. It is clear that the composition of target and control samples is similar, at the $\pm10\%$ level, for approximately half of the elements analysed. However, there are also significant differences for some elements, such as W, which is around an order of magnitude lower in the control samples, and Pb which is approximately half as abundant in the control samples. These elements are therefore indicative of potentially significant differences between samples. The magnitude of the differences, even for relatively few of the elements analysed, is sufficient at this stage to indicate that the questioned forensic samples do not closely 'match' the control samples. However, the fact that a small number of elements in the questioned samples fall outside the range of control samples, or vice versa, does not in itself

Table 1. *Major and trace element data for mud taken from a suspect's boots compared with control soil samples from the crime scene (<150 μm fraction, analysed by ICP-OES and ICP-MS). Boot mud values which fall within ±10% of the range of values for the control soils are highlighted*

		SiO_2	Al_2O_3	Fe_2O_3	MgO	CaO	Na_2O	K_2O	TiO_2	P_2O_5	MnO	Ba	Co	Cr	Cu	Ni	Pb	Sc	V	Sr	Zn	Zr	Ga	Rb	Nb
Boot mud	KP 1	**41.7**	**11.2**	5.08	**1.60**	**8.02**	**1.15**	**2.06**	**0.67**	**0.48**	0.07	**370**	**12.6**	116	137	**43.3**	123	**11.1**	102	**233**	331	177	**15.8**	**79.5**	31.0
	KP 2	**41.4**	9.83	4.76	1.81	9.76	0.77	1.72	0.59	0.90	0.09	365	**11.4**	**84.1**	86.8	**38.7**	44.1	**10.3**	**91.3**	189	372	184	**13.1**	69.9	20.7
	KP 3	**45.6**	**11.5**	5.05	1.72	**8.15**	0.86	**1.88**	**0.70**	0.49	**0.08**	356	**12.7**	101	113	**44.2**	43.9	**11.9**	100	204	345	208	**15.0**	**82.5**	19.9
	KP 4	37.7	**8.90**	4.04	1.76	9.52	1.49	2.61	**0.55**	1.54	0.08	424	14.5	76.4	**227**	**36.6**	103	7.59	76.2	**225**	507	166	**11.4**	**67.0**	**13.5**
	KP 5	**40.7**	**10.3**	4.60	1.66	**7.41**	**1.20**	**1.98**	**0.61**	0.77	0.07	**410**	**11.4**	91.4	119	**38.2**	59.1	9.73	**89.1**	197	342	159	**13.0**	**72.9**	17.4
Control soils	KP 17	55.8	12.7	7.19	1.41	4.14	1.07	2.35	0.76	0.28	0.09	376	16.7	102	599	52.6	27.7	14.5	109	243	453	282	15.7	90.1	13.7
	KP 18	51.9	11.9	6.74	1.40	6.08	1.47	2.26	0.71	0.37	0.10	467	16.3	94.7	309	46.6	21.5	13.2	101	314	658	229	14.5	84.9	12.2
	KP 19	48.5	11.1	5.99	1.48	8.16	2.11	2.25	0.67	0.39	0.08	337	14.5	87.5	282	40.0	12.1	12.4	92.9	428	202	234	13.1	83.3	11.5
	KP 20	50.4	10.6	5.83	1.31	7.19	1.56	2.21	0.69	0.46	0.08	422	13.1	86.7	295	38.8	20.3	11.9	88.8	352	196	226	12.8	86.1	11.9
	KP 21	48.4	11.1	6.78	1.41	8.01	1.47	2.23	0.63	0.34	0.08	335	14.4	91.4	310	47.1	20.3	12.9	94.7	333	197	228	13.3	82.5	11.4
	KP 22	51.7	11.8	6.63	1.31	5.42	1.30	2.27	0.72	0.39	0.09	351	15.7	95.0	248	43.2	17.7	13.3	100	282	171	230	14.3	88.1	13.2
	KP 23	52.6	12.3	6.74	1.35	5.46	1.20	2.27	0.72	0.30	0.09	389	16.0	98.3	626	51.3	26.8	13.3	102	278	451	259	14.7	84.5	13.2
	KP 24	43.5	9.68	5.83	1.40	9.76	1.70	2.01	0.56	0.38	0.08	337	12.2	81.0	927	44.2	18.8	11.1	83.0	413	638	202	11.6	68.4	9.97
	KP 25	52.1	12.5	6.96	1.43	5.56	1.47	2.31	0.73	0.32	0.09	359	16.1	97.9	209	49.3	18.3	13.8	105	293	157	239	15.2	85.1	12.8

		Mo	Sn	Cs	Hf	Ta	W	Tl	Y	Be	U	Th	La	Ce	Pr	Nd	Sm	Eu	Gd	Tb	Dy	Ho	Er	Tm	Yb	Lu
Boot mud	KP 1	**2.50**	35.1	**5.16**	4.52	9.71	40.3	0.13	**30.8**	2.35	**2.66**	**10.3**	**29.6**	**54.2**	**7.07**	**27.6**	**5.61**	**1.22**	**4.70**	**0.73**	**4.24**	**0.90**	**2.45**	**0.36**	**2.28**	**0.34**
	KP 2	**2.27**	15.1	**4.62**	**4.49**	6.25	28.1	0	**27.2**	**2.01**	**2.47**	8.67	**27.2**	**50.3**	**6.59**	**26.0**	**5.41**	**1.20**	**4.69**	**0.73**	**4.24**	**0.92**	**2.51**	**0.36**	**2.30**	**0.34**
	KP 3	**2.25**	12.7	**5.53**	**5.06**	5.49	17.0	0	**29.9**	**2.14**	**2.63**	**9.90**	**30.6**	**56.6**	**7.43**	**29.0**	**5.92**	**1.33**	**5.05**	**0.80**	**4.70**	**1.01**	**2.72**	**0.40**	**2.54**	**0.37**
	KP 4	**2.57**	12.1	4.35	4.13	2.57	12.6	0	**27.5**	**1.74**	**2.06**	7.10	22.8	41.9	5.52	21.7	4.64	1.04	3.85	0.63	3.63	0.77	2.12	0.30	1.86	**0.29**
	KP 5	**2.17**	13.0	**4.66**	4.07	3.34	15.6	0	**30.3**	**2.08**	**2.32**	**8.34**	**26.8**	**49.5**	**6.48**	**25.3**	**5.26**	**1.12**	**4.37**	**0.71**	**4.10**	**0.88**	**2.50**	**0.35**	**2.31**	**0.34**
Control soils	KP 17	1.87	3.34	5.66	6.89	1.07	2.31	0	34.4	2.07	2.57	11.1	34.0	66.9	8.50	33.7	6.86	1.59	6.09	0.96	5.70	1.22	3.33	0.48	3.05	0.46
	KP 18	2.26	4.62	5.08	5.57	1.01	2.32	0	31.8	1.93	2.40	10.4	32.1	63.7	7.96	32.1	6.48	1.46	5.68	0.90	5.31	1.14	3.06	0.44	2.75	0.41
	KP 19	1.81	2.40	4.33	5.65	0.93	1.89	0	29.0	1.74	2.38	9.68	29.9	58.7	7.42	29.4	6.17	1.35	5.22	0.83	4.76	1.01	2.79	0.39	2.56	0.37
	KP 20	1.71	3.39	4.41	5.51	0.92	1.74	0	29.7	1.77	2.37	9.70	29.6	57.7	7.34	29.1	6.03	1.34	5.09	0.81	4.83	1.03	2.84	0.42	2.57	0.38
	KP 21	2.00	5.09	5.32	5.60	1.00	4.01	0	30.7	1.84	2.45	9.67	30.4	57.7	7.51	29.5	6.07	1.41	5.39	0.84	4.93	1.05	2.76	0.40	2.53	0.38
	KP 22	1.67	3.28	4.87	5.68	1.01	2.60	0	30.7	1.84	2.42	10.4	32.0	63.7	7.89	31.1	6.54	1.45	5.74	0.89	5.11	1.10	2.94	0.42	2.68	0.40
	KP 23	1.76	3.67	5.00	6.25	1.01	2.27	0	30.7	1.95	2.43	10.6	32.5	64.5	8.06	31.8	6.63	1.49	5.75	0.90	5.19	1.12	3.06	0.44	2.74	0.42
	KP 24	1.96	6.42	3.93	4.83	0.93	4.57	0	27.0	1.68	2.19	8.34	27.0	54.2	7.04	28.3	6.11	1.34	5.00	0.76	4.26	0.91	2.43	0.35	2.20	0.32
	KP 25	2.13	3.18	5.20	5.82	1.02	2.24	0	31.2	1.88	2.46	10.7	32.7	64.9	8.10	32.4	6.78	1.54	5.75	0.90	5.27	1.11	3.03	0.44	2.77	0.42

Table 2. *Selected major oxide and trace element ratios for mud taken from a suspect's boots compared with control soil samples from the crime scene (<150 μm fraction, analysed by ICP-OES and ICP-MS). Boot mud values which fall within ±10% of the range of values for the control soils are highlighted*

		$SiO_2/$ Al_2O_3	$SiO_2/$ Fe_2O_3	$Al_2O_3/$ Fe_2O_3	$Al_2O_3/$ K_2O	$SiO_2/$ CaO	$CaO/$ MgO	$K_2O/$ Na_2O	Cu/Pb	Zn/Pb	Rb/Sr	Ta/W	Ce/La	Nd/Sm	U/Th
Boot mud	KP 1	3.72	**8.20**	2.21	**5.45**	**5.20**	**5.01**	**1.79**	1.11	2.68	**0.34**	**0.24**	1.83	4.92	0.26
	KP 2	**4.21**	**8.70**	2.07	**5.71**	4.24	**5.38**	**2.24**	1.97	8.43	**0.37**	**0.22**	1.85	4.80	0.28
	KP 3	**3.95**	**9.03**	2.29	6.14	**5.59**	**4.74**	**2.19**	2.59	7.86	**0.40**	**0.32**	1.85	4.89	0.27
	KP 4	**4.24**	**9.33**	2.20	3.41	3.97	**5.40**	**1.75**	2.19	4.89	**0.30**	**0.20**	1.84	4.69	0.29
	KP 5	**3.96**	**8.85**	2.23	**5.20**	**5.49**	**4.47**	**1.65**	2.01	5.78	**0.37**	**0.21**	1.85	4.82	0.28
Control soils	KP 17	4.40	7.76	1.77	5.41	13.5	2.93	2.19	21.6	16.4	0.37	0.46	1.97	4.91	0.23
	KP 18	4.34	7.70	1.77	5.29	8.53	4.34	1.53	14.4	30.6	0.27	0.44	1.99	4.95	0.23
	KP 19	4.36	8.10	1.86	4.95	5.94	5.52	1.07	23.4	16.8	0.19	0.49	1.96	4.77	0.25
	KP 20	4.74	8.65	1.82	4.81	7.01	5.49	1.42	14.5	9.67	0.24	0.53	1.95	4.82	0.24
	KP 21	4.35	7.14	1.64	5.00	6.04	5.68	1.52	15.3	9.72	0.25	0.25	1.97	4.86	0.25
	KP 22	4.38	7.80	1.78	5.29	9.54	4.13	1.72	14.0	9.65	0.31	0.39	1.99	4.76	0.23
	KP 23	4.29	7.80	1.82	5.41	9.63	4.05	1.89	23.4	16.8	0.30	0.44	1.98	4.80	0.23
	KP 24	4.49	7.46	1.66	4.82	4.46	6.95	1.18	49.2	33.8	0.17	0.20	2.01	4.63	0.26
	KP 25	4.18	7.49	1.79	5.39	9.38	3.89	1.58	11.4	8.57	0.29	0.45	1.99	4.78	0.23

necessarily exclude a common source. It is important to consider whether any chemical contamination of the samples could have occurred to cause such differences, or whether differences are caused by truly different sources of material. Levels of heavy metals (such as Pb, Cu and Zn), for example, are often found to be higher in questioned forensic samples taken from vehicles or garden tools which contain these elements or which carry surface contaminants which do. Contaminant metals are also often concentrated in a relatively small number of particles, which may have an irregular distribution. Experience and adequate background information are necessary to judge whether small observed differences of this type are likely to be an artefact of transfer onto forensic exhibits or due to genuine differences in source material characteristics.

As well as comparing absolute concentrations of elements and normalized values, it is useful to compare samples in terms of the ratios of certain elements. A number of ratios have been found to be useful in our routine forensic case work, including: SiO_2/Al_2O_3, SiO_2/Fe_2O_3, Al_2O_3/Fe_2O_3, Al_2O_3/K_2O, SiO_2/CaO, CaO/MgO, K_2O/Na_2O, Cu/Pb, Rb/Sr, Ta/W, Ce/La, Nd/Sm and U/Th. Table 2 shows these ratios for our example dataset. In this case Al_2O_3/Fe_2O_3 and Cu/Pb are particularly diagnostic ratios, which clearly differentiate the two groups of samples. Additionally, rare earth elements can be normalized to a common reference standard, which mostly frequently comprises the concentrations in chondritic meteorites (e.g. Nakamura 1974; Boynton 1984), or average crustal composition (Taylor & McLennan 1985). This normalisation reduces the effects of natural abundance variations between the various rare earth elements.

In terms of presenting the data in graphical form, a simple but clear method is to present a bivariate plot of the concentrations of two elements (either absolute concentrations or normalised concentrations), or two elemental ratios. Any combination of these could potentially be useful, but in our experience good discrimination among soils and sediments can often be achieved by plots of Ce v. La, Nd v. Sm, U v. Th, Rb v. Sr, Zn v. Cu, Ta v. W, SiO_2/Al_2O_3 v. SiO_2/Fe_2O_3, Al_2O_3/Fe_2O_3 v. Al_2O_3/K_2O, Ce/La v. Nd/Sm, Ce/La v. U/Th and Nd/Sm v. U/Th. Four example plots are displayed in Figure 1 and demonstrate a clear separation between the mud on the suspect's footwear and the control soils. Results can also be plotted on a spider diagram (Fig. 2), although differences between samples can be difficult to identify when comparing more than a few broadly similar samples at once. Plots of chondrite-normalized rare earth element concentrations (Fig. 3) often provide a better visual means of identifying similarities or differences between samples.

Where forensic and control samples fall into known well-defined groups, it may be helpful to compute the mean, standard deviation and range of values of the groups of samples (Table 3). This exercise highlights the extent to which a group of forensic samples falls within, outside, or overlaps the range of control samples. This is particularly useful where a number of different groups have been identified, and can allow several to be quickly excluded. Simple statistical tests (such as the parametric *t*-test, or the non-parametric Wilcoxon or Mann-Whitney *U*-tests) can be used to determine whether there is a significant difference between groups of samples. However, due to the large numerical range of values in the full major and trace element dataset, these

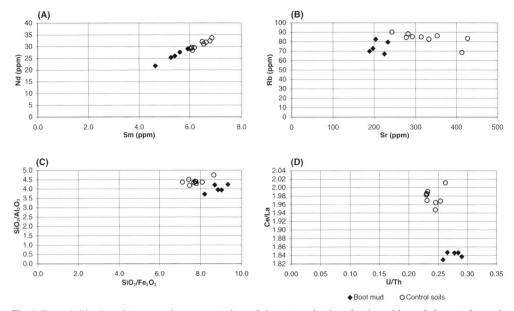

Fig. 1. Example bivariate plots comparing concentrations of elements and ratios of major oxides and elements for mud from the suspect's boots and control soil samples taken at the scene of the crime. (**a**) Nd v. Sm; (**b**) Rb v. Sr; (**c**) SiO_2/Al_2O_3 v. SiO_2/Fe_2O_3; and (**d**) Ce/La v. U/Th.

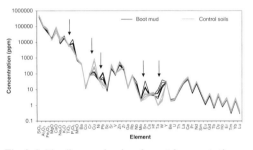

Fig. 2. Spider diagram showing elemental concentrations (log scale) in mud from the suspect's boots and control soil samples for all elements determined (the main differences are arrowed).

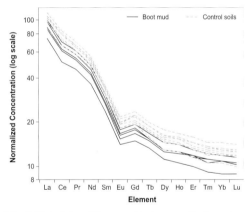

Fig. 3. Plot of chondrite-normalized rare earth element concentrations for mud from the suspect's boots and control soil samples.

tests should only be used to compare groups in terms of each element separately. Also, grouping of samples and comparison of the average elemental values can mask important differences between individual samples, and one very different sample (outlier) can skew the average values considerably. Consequently, simply because the mean values for a group of forensic samples are shown to be significantly different from control samples does not necessarily prove that they originate from a different source: the number and distribution of control samples may be insufficient to draw this conclusion validly, and the distribution of values for one or other dataset may be heavily skewed by one or more

'outlier' values, as noted above. For this reason, the use of average compositions of groups of samples is not a substitute for examining differences between individual samples.

Ideally, a statistical test is required to compare each sample against every other sample in the dataset in terms of all the elements and oxides determined. A widely used measure of similarity in this respect is Pearson's product-moment correlation coefficient, a dimensionless parameter which

Table 3. *Average chemical composition of mud taken from a suspect's boots compared with control soil samples from the crime scene (<150mm fraction, analysed by ICP-OES and ICP-MS)*

		SiO₂	Al₂O₃	Fe₂O₃	MgO	CaO	Na₂O	K₂O	TiO₂	P₂O₅	MnO	Ba	Co	Cr	Cu	Ni	Pb	Sc	Sr	V	Zn	Zr	Ga	Rb	Nb
Boot mud	Mean	41.4	10.4	4.71	1.71	8.57	1.09	2.05	0.62	0.84	0.08	385	12.5	93.8	137	40.2	74.8	10.1	210	91.7	379	179	13.7	74.4	20.5
	SD	2.80	1.06	0.42	0.08	1.02	0.29	0.34	0.06	0.43	0.01	30.0	1.24	15.4	53.6	3.34	36.5	1.64	18.7	10.2	72.8	19.2	1.77	6.48	6.49
	n	5	5	5	5	5	5	5	5	5	5	5	5	5	5	5	5	5	5	5	5	5	5	5	5
	Max.	45.6	11.5	5.08	1.81	9.76	1.49	2.61	0.70	1.54	0.09	424	14.5	116	227	44.2	123	11.9	233	102	507	208	15.8	82.5	31.0
	Min.	37.7	8.90	4.04	1.60	7.41	0.77	1.72	0.55	0.48	0.07	356	11.4	76.4	86.8	36.6	43.9	7.59	189	76.2	331	159	11.4	67.0	13.5
Control soils	Mean	50.5	11.5	6.52	1.39	6.64	1.48	2.23	0.69	0.36	0.09	375	15.0	92.7	423	45.9	20.4	12.9	326	97.4	347	236	13.9	83.7	12.2
	SD	3.47	0.97	0.51	0.06	1.76	0.30	0.09	0.06	0.05	0.01	45.1	1.55	6.67	241	4.79	4.72	1.02	62.3	8.20	205	22.6	1.32	6.20	1.15
	n	9	9	9	9	9	9	9	9	9	9	9	9	9	9	9	9	9	9	9	9	9	9	9	9
	Max.	55.8	12.7	7.19	1.48	9.76	2.11	2.35	0.76	0.46	0.10	467	16.7	102	927	52.6	27.7	14.5	428	109	658	282	15.7	90.1	13.7
	Min.	43.5	9.68	5.83	1.31	4.14	1.07	2.01	0.56	0.28	0.08	335	12.2	81.0	209	38.8	12.1	11.1	243	83.0	157	202	11.6	68.4	9.97

		Mo	Sn	Cs	Hf	Ta	W	Tl	Y	Be	U	Th	La	Ce	Pr	Nd	Sm	Eu	Gd	Tb	Dy	Ho	Er	Tm	Yb	Lu
Boot mud	Mean	2.35	17.6	4.86	4.46	5.47	22.7	0.13	29.1	2.06	2.43	8.87	27.4	50.5	6.62	25.9	5.37	1.18	4.53	0.72	4.18	0.90	2.46	0.36	2.26	0.34
	SD	0.17	9.86	0.47	0.39	2.81	11.4		1.66	0.22	0.25	1.29	3.04	5.62	0.72	2.74	0.48	0.11	0.45	0.06	0.38	0.09	0.22	0.04	0.24	0.03
	n	5	5	5	5	5	5	1	5	5	5	5	5	5	5	5	5	5	5	5	5	5	5	5	5	5
	Max.	2.57	35.1	5.53	5.06	9.71	40.3	0.13	30.8	2.35	2.66	10.3	30.6	56.6	7.43	29.0	5.92	1.33	5.05	0.80	4.70	1.01	2.72	0.40	2.54	0.37
	Min.	2.17	12.1	4.35	4.07	2.57	12.6	0.13	27.2	1.74	2.06	7.10	22.8	41.9	5.52	21.7	4.64	1.04	3.85	0.63	3.63	0.77	2.12	0.30	1.86	0.29
Control soils	Mean	1.91	3.93	4.87	5.76	0.99	2.66		30.6	1.85	2.41	10.1	31.1	61.6	7.76	30.8	6.41	1.44	5.52	0.87	5.04	1.08	2.92	0.42	2.65	0.40
	SD	0.20	1.23	0.55	0.56	0.05	0.96		2.04	0.12	0.10	0.82	2.11	4.12	0.46	1.81	0.32	0.09	0.36	0.06	0.41	0.09	0.25	0.04	0.23	0.04
	n	9	9	9	9	9	9	1	9	9	9	9	9	9	9	9	9	9	9	9	9	9	9	9	9	9
	Max.	2.26	6.42	5.66	6.89	1.07	4.57		34.4	2.07	2.57	11.1	34.0	66.9	8.50	33.7	6.86	1.59	6.09	0.96	5.70	1.22	3.33	0.48	3.05	0.46
	Min.	1.67	2.40	3.93	4.83	0.92	1.74	0.13	27.0	1.68	2.19	8.34	27.0	54.2	7.04	28.3	6.03	1.34	5.00	0.76	4.26	0.91	2.43	0.35	2.20	0.32

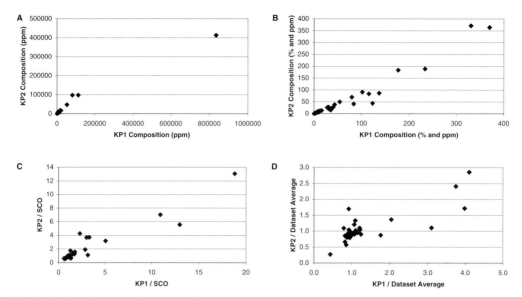

Fig. 4. Bivariate plots comparing two samples (KP1 and KP2) in terms of 49 oxides and elements: (**a**) with raw values expressed in p.p.m.; (**b**) with oxides expressed as wt % and all other elements expressed as p.p.m.; (**c**) with values normalized to composition of the SCO standard (marine shale); (**d**) with values normalized to the mean composition of the dataset.

ranges in value from -1 (perfect negative correlation) through 0 (no correlation) to $+1$ (perfect positive correlation). However, there is a problem with using this technique for comparing samples using geochemical datasets, resulting from the fact that the correlation coefficient is not a general measure of relationship between variables, but of the degree of a straight line tendency. Figure 4a shows a bivariate plot comparing two samples (KP1 and KP2) in terms of 49 raw elemental concentration values. The very wide range of numerical values encountered in ICP analysis ($>500\,000$ p.p.m. SiO_2 to 0.29 p.p.m. Lu) means that the degree of straight-line tendency is mainly determined by the most abundant elements (highest numerical values). As a result, the corresponding Pearson's correlation matrix (Table 4a) is dominated by the relative proportions of the single oxide SiO_2, which provides little discrimination between samples of the type considered here (note the shading in Table 4 which indicates higher correlation). In this situation, the concentrations of trace elements have little effect on the correlation values.

Differences between elements can be reduced by presenting the concentrations of major oxides as weight percents (wt %) (Fig. 4b). Numerical differences between the concentration values of elements are lower, and the correlation matrix (Table 4b) shows correspondingly greater discrimination between the mud samples from the suspect's boots and control soil samples. However, in this case the

correlation is disproportionately affected by the minor elements present in concentrations of 100s of p.p.m., such as Ba, Cu and Zn. Because of the scale of these effects, it is also desirable to compare samples in terms of separate groups of elements, i.e. majors (10), minors plus other traces (25), and rare earths (14). However, in every case the degree of correlation is dominated by the concentration of the most abundant element in the dataset, and in some cases the presence of outlier values for a single element (e.g. Cu, Zn or Pb) can result in 'anomalously' low correlation values, even where all other element concentrations show a high degree of similarity. Pearson correlation values determined using raw concentration data therefore need to be interpreted with care.

In order to provide equal emphasis on small differences in the rare earth elements and larger differences in the major oxides, the elemental data can be normalized to a reference material. To be suitable for normalization, the reference material used should be similar in composition to the samples under analysis. Spurious correlation values can be generated where one or more elements are appreciably more or less abundant in the reference material. Datasets such as the average composition of the upper continental crust or chondritic meteorites are therefore not recommended. Figure 4c shows a bivariate plot with values normalized to the composition of SCo1, the US Geological Survey reference material Cody Shale, whose use was first proposed by Jarvis and

Table 4. *Pearson's correlation values obtained by comparing mud taken from a suspect's boots with control soil samples in terms of 49 elements analysed by ICP. (**a**) with all elements expressed in p.p.m., and (**b**) with major elements expressed in weight percent and all other elements expressed in p.p.m. The highest correlation values are shown in bold*

		Boot mud					Control soils								
(a)		KP 1	KP 2	KP 3	KP 4	KP 5	KP 17	KP 18	KP 19	KP 20	KP 21	KP 22	KP 23	KP 24	KP 25
Boot mud	KP 1	**1.000**													
	KP 2	**0.998**	**1.000**												
	KP 3	**1.000**	**0.998**	**1.000**											
	KP 4	**0.997**	**0.999**	**0.997**	**1.000**										
	KP 5	**1.000**	**0.998**	**1.000**	**0.997**	**1.000**									
Control soils	KP 17	**0.993**	0.988	**0.994**	0.984	**0.994**	**1.000**								
	KP 18	**0.997**	**0.993**	**0.998**	**0.991**	**0.998**	**0.999**	**1.000**							
	KP 19	**0.999**	**0.997**	**0.999**	**0.996**	**0.999**	**0.996**	**0.999**	**1.000**						
	KP 20	**0.997**	**0.996**	**0.998**	**0.994**	**0.998**	**0.997**	**0.999**	**0.999**	**1.000**					
	KP 21	**0.999**	**0.997**	**0.999**	**0.995**	**0.999**	**0.996**	**0.999**	**1.000**	**0.999**	**1.000**				
	KP 22	**0.996**	**0.992**	**0.997**	0.989	**0.997**	**1.000**	**1.000**	**0.998**	**0.999**	**0.998**	**1.000**			
	KP 23	**0.996**	**0.992**	**0.997**	0.989	**0.997**	**1.000**	**1.000**	**0.998**	**0.999**	**0.998**	**1.000**	**1.000**		
	KP 24	**0.998**	**0.999**	**0.998**	**0.998**	**0.998**	0.989	**0.995**	**0.998**	**0.997**	**0.998**	**0.993**	**0.993**	**1.000**	
	KP 25	**0.996**	**0.992**	**0.997**	0.989	**0.997**	**0.999**	**1.000**	**0.998**	**0.999**	**0.998**	**1.000**	**1.000**	**0.993**	**1.000**

		Boot mud					Control soils								
(b)		KP 1	KP 2	KP 3	KP 4	KP 5	KP 17	KP 18	KP 19	KP 20	KP 21	KP 22	KP 23	KP 24	KP 25
Boot mud	KP 1	**1.000**													
	KP 2	0.979	**1.000**												
	KP 3	0.984	**0.995**	**1.000**											
	KP 4	0.968	0.972	0.966	**1.000**										
	KP 5	0.986	**0.992**	**0.991**	0.974	**1.000**									
Control soils	KP 17	0.838	0.814	0.842	0.886	0.830	**1.000**								
	KP 18	0.943	0.962	0.961	0.988	0.955	0.907	**1.000**							
	KP 19	0.876	0.831	0.869	0.830	0.854	0.866	0.850	**1.000**						
	KP 20	0.904	0.864	0.897	0.863	0.897	0.888	0.867	0.984	**1.000**					
	KP 21	0.890	0.845	0.884	0.855	0.872	0.917	0.870	0.990	**0.992**	**1.000**				
	KP 22	0.908	0.871	0.908	0.859	0.899	0.892	0.865	0.978	**0.995**	**0.992**	**1.000**			
	KP 23	0.834	0.805	0.833	0.882	0.825	**0.999**	0.903	0.876	0.895	0.923	0.894	**1.000**		
	KP 24	0.738	0.703	0.728	0.818	0.721	0.967	0.851	0.813	0.812	0.854	0.800	0.975	**1.000**	
	KP 25	0.908	0.873	0.910	0.844	0.899	0.855	0.848	0.975	0.990	0.982	**0.997**	0.856	0.751	**1.000**

(a) & (b) *Correlation values*, 0.950–0.990: *Correlation values*, **bold** 0.990–1.000

Jarvis (1985). Differences between elements now range between 0 and 20 units, and the resulting correlation matrix (Table 5a) shows a clear distinction between mud from the boots and control soil samples.

Another method of normalization is to use the average composition of the dataset under analysis, which constrains differences between elements to the minimum (Figure 4d). In the case example considered here, the correlation matrix (Table 5b) clearly discriminates between the boot mud samples and the control soil samples, with positive correlations among the boot mud samples and negative correlations between the boot muds and control soils. This normalization is most suitable when the dataset contains controls of a generally similar character

(e.g. all soils within a limited geographical area on the same parent material). The main drawback is that, if further samples are added to the dataset, such as an additional group of control samples, the average composition of the dataset and hence the degree of correlation between samples will also change. However, the technique can be useful as a screening tool which helps focus further analytical strategy.

An alternative to calculating Pearson's product moment correlation coefficient (a parametric test which makes an assumption that the data are normally distributed) is to use the non-parametric Spearman's Rank correlation coefficient, where elements are firstly ranked in order of abundance (Table 6). This method is effective in removing the influence

Table 5. *Pearson's correlation values obtained by comparing mud taken from a suspect's boots with control soil samples in terms of 49 elements analysed by ICP, with values normalised to (**a**) composition of the SCO standard (marine shale), and (**b**) average composition of the dataset. The highest correlation values are shown in bold*

(a)		KP 1	KP 2	KP 3	KP 4	KP 5	KP 17	KP 18	KP 19	KP 20	KP 21	KP 22	KP 23	KP 24	KP 25
		Boot mud					Control soils								
Boot mud	KP 1	**1.000**													
	KP 2	**0.949**	**1.000**												
	KP 3	**0.944**	**0.963**	**1.000**											
	KP 4	0.622	0.703	0.747	**1.000**										
	KP 5	**0.910**	**0.944**	**0.956**	0.876	**1.000**									
Control soils	KP 17	0.135	0.133	0.306	0.612	0.369	**1.000**								
	KP 18	0.164	0.203	0.375	0.693	0.451	**0.936**	**1.000**							
	KP 19	0.113	0.145	0.307	0.634	0.381	**0.955**	**0.912**	**1.000**						
	KP 20	0.132	0.155	0.318	0.659	0.399	**0.972**	**0.918**	**0.994**	**1.000**					
	KP 21	0.231	0.241	0.401	0.682	0.472	**0.970**	**0.913**	**0.984**	**0.990**	**1.000**				
	KP 22	0.156	0.178	0.340	0.649	0.411	**0.977**	**0.913**	**0.985**	**0.994**	**0.990**	**1.000**			
	KP 23	0.141	0.140	0.312	0.622	0.378	**0.999**	**0.937**	**0.962**	**0.978**	**0.976**	**0.980**	**1.000**		
	KP 24	0.184	0.184	0.350	0.657	0.421	**0.994**	**0.938**	**0.966**	**0.980**	**0.981**	**0.977**	**0.997**	**1.000**	
	KP 25	0.129	0.150	0.319	0.615	0.382	**0.963**	**0.905**	**0.986**	**0.988**	**0.986**	**0.995**	**0.966**	**0.961**	**1.000**

(b)		KP 1	KP 2	KP 3	KP 4	KP 5	KP 17	KP 18	KP 19	KP 20	KP 21	KP 22	KP 23	KP 24	KP 25
		Boot mud					Control soils								
Boot mud	KP 1	**1.000**													
	KP 2	**0.829**	**1.000**												
	KP 3	**0.811**	**0.873**	**1.000**											
	KP 4	0.370	0.403	0.126	**1.000**										
	KP 5	**0.798**	**0.820**	**0.704**	**0.702**	**1.000**									
Control soils	KP 17	−0.769	−0.823	−0.680	−0.582	−0.866	**1.000**								
	KP 18	−0.802	−0.736	−0.633	−0.404	−0.683	**0.719**	**1.000**							
	KP 19	−0.820	−0.742	−0.682	−0.488	−0.743	0.493	**0.576**	**1.000**						
	KP 20	−0.881	−0.812	−0.729	−0.454	−0.736	**0.610**	**0.614**	**0.929**	**1.000**					
	KP 21	−0.820	−0.788	−0.663	−0.611	−0.794	**0.636**	**0.572**	**0.896**	**0.901**	**1.000**				
	KP 22	−0.813	−0.739	−0.560	−0.613	−0.726	**0.692**	**0.598**	**0.809**	**0.893**	**0.893**	**1.000**			
	KP 23	−0.794	−0.853	−0.768	−0.533	−0.902	**0.973**	**0.722**	**0.537**	**0.627**	**0.637**	**0.635**	**1.000**		
	KP 24	−0.472	−0.525	−0.675	−0.185	−0.631	**0.535**	0.428	0.314	0.262	0.275	0.045	**0.687**	**1.000**	
	KP 25	−0.774	−0.727	−0.511	−0.624	−0.692	**0.640**	**0.594**	**0.816**	**0.867**	**0.907**	**0.975**	**0.574**	−0.008	**1.000**

(a) *Correlation values*, 0.800–0.900: *Correlation values*, **bold** 0.900–1.000

(b) *Correlation values*, 0.000–0.500: *Correlation values*, **bold** 0.500–1.000

of absolute abundance between elements and can provide effective discrimination for some datasets. However, the ranking procedure removes much of the detail in the dataset and therefore generates new problems. The concentration differences between many elements are so large that some will obtain the same rank in almost all samples considered (for example the major oxides will almost always obtain the highest ranks, often in the same order for broadly soil sample types), while small differences in trace elements can dramatically alter their ranks. In view of its low resolving power, use of Spearman's rank correlations is often of limited value in comparative soil studies.

Conclusions

The preceding discussion has outlined a number of simple methods which can be used to compare soils and sediments using geochemical data. More complex methods, involving multivariate statistical analysis, can also be used. Due to the variety of situations encountered in forensic studies, it is not possible to define a rigid protocol for the analysis of elemental concentration data. Instead, the process should involve a number of stages and be adapted to individual case circumstances. Care should be taken not to place overemphasis on the significance of statistics or apparent graphical similarities. Inspection of the tabulated data is an important

Table 6. *Spearman's rank correlation coefficients obtained by comparing mud from a suspect's boots with control soil samples in terms of 49 elements analysed by ICP (values expressed in p.p.m.)*

		Boot mud					Control soils								
		KP 1	KP 2	KP 3	KP 4	KP 5	KP 17	KP 18	KP 19	KP 20	KP 21	KP 22	KP 23	KP 24	KP 25
Boot mud	KP 1	**1.000**													
	KP 2	**0.995**	**1.000**												
	KP 3	**0.993**	**0.997**	**1.000**											
	KP 4	**0.992**	**0.993**	**0.995**	**1.000**										
	KP 5	**0.993**	**0.996**	**0.998**	**0.997**	**1.000**									
Control	KP 17	0.950	0.962	0.974	0.975	0.975	**1.000**								
soils	KP 18	0.951	0.963	0.975	0.975	0.976	**0.998**	**1.000**							
	KP 19	0.943	0.957	0.970	0.968	0.969	**0.997**	**0.998**	**1.000**						
	KP 20	0.949	0.961	0.974	0.973	0.974	**0.998**	**0.998**	**0.998**	**1.000**					
	KP 21	0.959	0.969	0.981	0.981	0.981	**0.997**	**0.997**	**0.997**	**0.998**	**1.000**				
	KP 22	0.952	0.965	0.976	0.976	0.977	**0.998**	**0.999**	**0.998**	**0.999**	**0.999**	**1.000**			
	KP 23	0.950	0.962	0.975	0.975	0.975	**0.999**	**0.999**	**0.998**	**0.998**	**0.997**	**0.999**	**1.000**		
	KP 24	0.966	0.976	0.984	0.985	0.985	**0.995**	**0.995**	**0.993**	**0.994**	**0.996**	**0.995**	**0.996**	**1.000**	
	KP 25	0.951	0.963	0.975	0.974	0.975	**0.998**	**0.999**	**0.999**	**0.999**	**0.999**	**0.999**	**0.998**	**0.994**	**1.000**

(a) *Correlation values, 0.950–0.990: Correlation values,* **bold** *0.990–1.000*

early stage in the evaluation procedure which may help to identify spurious correlations and/or or meaningful differences.

Great care must be taken with laboratory operating procedures and instrument maintenance and calibration to ensure that standards of precision and accuracy are maintained. This is particularly true for small bulk samples and separated size fractions, where errors due to sample weighing, environmental contamination, poor instrument calibration and poor signal to noise ratio become relatively more serious. Under ideal conditions, all samples should be analysed in replicate using several different size fractions. However, this is often limited by available sample size limitations and the frequently large number of control samples collected, all of which should be processed and analysed in the same way. Where the amount of critical 'forensic sample' material is strictly limited, it may often be better to obtain one analysis from a single sample rather than two or three analyses of poorer quality from separate but smaller subsamples. Where it is only possible to analyse a single prepared 'forensic' sample, several repeat determinations should be made to obtain a clear indication of the instrumental measurement error.

Analysis of major and trace element data has an important role to play in forensic comparisons of soils and sediments. It should, however, only be considered as one of a number of different methods by which samples can be compared. Conclusions should be based also on the results from several independent techniques, such as mineral composition, colour, grain size distributions, grain shape, surface textural properties and biological assemblages. An important use of major and trace element data analysis is as a screening tool that helps to focus the attention of the investigator on the samples which require more detailed study using other methods (e.g. optical microscopy, scanning electron microscopy, and X-ray microanalysis). Definitive association of a questioned sample with a particular known location, such as crime scene, normally requires the identification of one or more highly unusual particle types ('exotics') in common, and it is only in relatively rare circumstances that the existence of very unusual chemical signatures in the samples may serve the same purpose.

We gratefully acknowledge the assistance of D. Wray and L. Dyer at the Department of Earth and Environmental Sciences, University of Greenwich, UK. Useful comments on an earlier version of the manuscript were made by D. Wray and C. Jeans. Discussions with N. Walsh and D. Croft have also been of great value.

References

ACKERMANN, F., BERGMANN, H. & SCHIEICHERT, U. 1983. Monitoring of heavy metals in coastal and estuarine sediments – question of grain size: <20um versus <60mm. *Science and Technology Letters*, **4**, 317–328.

BOYNTON, W. V. 1984. Geochemistry of the rare earth elements: meteorite studies. *In*: HENDERSON, P. (ed.) *Rare Earth Element Geochemistry*. Elsevier, Amsterdam, 63–114.

BUTLER, L. R. P. (ed.) 1986. *Analytical Chemistry in the Exploration, Mining and Processing of Materials*. Blackwell Scientific Publications, Oxford.

DAVIS, J. C. 2002. *Statistical and Data Analysis in Geology*. 3rd edn. John Wiley & Sons. New York.

GILL, R. (ed.) 1997. *Modern Analytical Geochemistry*. Longman, Harlow.

JARVIS, I. & JARVIS, K. E. 1985. Rare earth element geochemistry of standard sediments: a study using inductively coupled plasma spectrometry. *Chemical Geology*, **53**, 335–344.

JARVIS, K. E., GRAY, A. L. & HOUK, R. S. 2003. *Handbook of Inductively Coupled Plasma Spectrometry*. Blackie, Glasgow. Reprinted by Viridian Publishing, Woking.

KETO, R. O. 1999. Analysis and comparison of bullet leads by inductively coupled plasma mass spectrometry. *Journal of Forensic Science*, **44**, 1020–1026.

MEREFIELD, J. R., STONE, I. M., ROBERTS, J., JONES, J., BARRON J. & DEAN, A. 2000. Fingerprinting airborne particles for identifying provenance. *In*: FOSTER, I. D. L. (ed.) *Tracers in Geomorphology*. Wiley, Chichester, 85–100.

MILLER, R. B. 1991. Identification of wood fragments in trace evidence. *In*: *Proceedings of the International Symposium on the Forensic Aspects of Trace Evidence*. FBI Academy, US Department of Justice, Quantico, 91–105.

NAKAMURA, N. 1974. Determination of REE, Ba, Fe, Mg, Na and K in carbonaceous and ordinary chondrites. *Geochimica et Cosmochimica Acta,* **38**, 757–775.

PARRY, S. J. 1997. Neutron activation analysis. *In*: GILL, R. (ed.) *Modern Analytical Geochemistry*. Longman, Harlow, 116–134.

PARRY, S. J. 2003. *Handbook of Neutron Activation Analysis*. [Reprinted by Viridian Publishing, Woking]

PIRRIE, D., POWER, M., ROLLINSON, G., CAMM, G. S., HUGHES, S. H., BUTCHER, A. R. & HUGHES, P. 2003. The spatial distribution and source of arsenic, copper, tin and zinc within the surface sediments of the Fal Estuary, Cornwall, UK. *Sedimentology*, **50**, 579–595.

RAMSEY, M. H. 1997. Sampling and sample preparation. *In*: GILL, R. (ed.) *Modern Analytical Geochemistry*. Longman, Harlow, 12–27.

RAWLINS, B. G., WEBSTER, R. & LISTER, T. R. 2003. The influence of parent material on topsoil geochemistry in eastern England. *Earth Surface Processes and Landforms*, **28**, 1389–1409.

ROLLINSON, H. 1993. *Using Geochemical Data: Evaluation, Presentation, Interpretation*. Longman, Harlow.

ROWAN, J. S., GOODWILL, P. & FRANKS, S. W. 2000.

Uncertainty estimation in fingerprinting suspended sediment sources. *In*: FOSTER, I. D. L. (ed.) *Tracers in Geomorphology*. Wiley, Chichester, 279–290.

SAPSFORD, R. & JUPP, V. (eds) 1996. *Data Collection and Analysis*. Sage Publications, London.

SHINOMIYA, T., SHINOMIYA, K., ORIMOTO, C., MINAMI, T., TOHNO, Y. & YAMADA, M. 1998. In- and out- flows of elements in bones embedded in reference soils. *Forensic Science International*, **98**, 109–118.

SKIRDA, M. A. 1991. Forensic glass comparisons. *In*: *Proceedings of the International Symposium on the Forensic Aspects of Trace Evidence*. FBI Academy, US Department of Justice, Quantico, 79–89.

SLATTERY, M. C., WALDEN, J. & BURT, T. P. 2000. Use of mineral magnetic measurements to fingerprint suspended sediment sources: results from a linear mixing model. *In*: FOSTER, I .D. L. (ed.) *Tracers in Geomorphology*. Wiley, Chichester, 309–322.

SWAN, A. R. H. & SANDILANDS, M. 1995. *Introduction to Geological Data Analysis*. Blackwell Scientific, Oxford.

TAYLOR, S. R. & MCLENNAN, S. M. 1985. *The Continental Crust: Its Composition and Evolution*. Blackwell, Oxford.

THOMPSON, M. & WALSH, J. N. 2003. *A Handbook of Inductively Coupled Plasma Atomic Emission Spectrometry*. Chapman & Hall, London. [Reprinted by Viridian Publishing, Woking.

WALSH, J. N. 1979. The simultaneous determination of the major, minor and trace constituents of silicate rocks using inductively coupled plasma spectrometry. *Spectrochimica Acta (B)*, **35**, 107–111.

WALSH, J. N. 1997. Inductively coupled plasma – atomic emission spectrometry (ICP-AES). *In*: GILL, R. (ed.) *Modern Analytical Geochemistry*. Longman, Harlow, 41–66.

WALSH, J. N., BUCKLEY, F. & BARKER, J. 1981. The simultaneous determination of rare earth elements in rocks using inductively coupled plasma source spectrometry. *Chemical Geology*, **33**, 141–153.

WEBSTER, R. & OLIVER, M. A. 2001. *Geostatistics for Environmental Scientists*. John Wiley & Sons, Chichester.

WILLARD, H. H., MERRITT, L. L., JR, DEAN J. A. & SETTLE, F. A., JR. 1988. *Instrumental Methods of Analysis*. Wadsworth Publishing Company, Belmont.

Investigating multi-element soil geochemical signatures and their potential for use in forensic studies

B. G. RAWLINS & M. CAVE

British Geological Survey, Kingsley Dunham Centre, Nicker Hill, Keyworth, Nottingham NG12 5GG, UK (e-mail: bgr@bgs.ac.uk)

Abstract: Data from a regional soil survey in eastern England have been used to determine whether samples over the same parent material can be discriminated on the basis of both individual and multi-element geochemistry. Discrimination was based on estimates of measurement uncertainty, which were calculated from the analysis of a series of duplicates and subsamples. In the multivariate analysis we estimated a covariance matrix for the two sources of uncertainty and compared this to Mahalanobis distances calculated for pairs of samples within each parent material group. For 12 of the 19 individual elements, it was possible on average to discriminate between more than 80% of the samples within parent material groups and typically between 15 and 17 of the 19 elements discriminated individual samples. In the multi-element analysis, typically more than 99.8% of samples within the same parent material group were discriminated from one another. Hence, the geochemistry of a natural soil sample, when collected and analysed according to a strict protocol, and compared to a database that adopted the same methods, could be used to help establish provenance within bedrock-derived soil types. However, there are significant differences between the nature of soil samples and the way they are collected or derived in soil surveys and forensic investigations. These questions need to be addressed thoroughly before any practical application to forensic cases in which an investigator is attempting to link a suspect to a location based on soil geochemical signatures.

When a sufficient quantity of soil can be collected from an evidentiary item in a forensic investigation, a comprehensive analysis of its inorganic geochemistry (or multi-element signature), and comparison with a series of analyses of topsoil samples in the vicinity of the crime scene, may help to constrain its provenance. In certain cases the presence of unusual or unique particles in a forensic soil sample may help to determine its original location more precisely. However, in the majority of cases, it is likely that the investigator will have to rely on the natural composition of the soil.

A forensic sample must, if it is to be of evidentiary value, be representative of some of the properties of the soil at the forensic site. The contribution that evidence based on natural soil composition can make to a particular case will vary greatly depending on the specific circumstances. For example, crime suspects may claim that the soil adhering to their shoes, or other items, was actually derived from a site some distance from an alleged crime scene. Subsequent analysis of particle size (Sugita & Marumo 2001), colour (Dudley 1975) or geochemical composition (Hiraoka 1994) may be undertaken on: (1) the forensic sample, (2) samples of soil collected from where the suspect claims to have been at the time of the crime, and (3) samples from the crime scene. If the former two groups of samples are wholly different, while the forensic sample and the crime scene samples appear to have comparable properties, doubt would be cast on the suspect's testimony and

this may be considered as significant circumstantial evidence. Murray & Tedrow (1992) cite a murder enquiry in Washington DC in which soil evidence was used to cast doubt on the validity of the suspect's alibi. By contrast, where the properties of the soil over a large part of the landscape, which comprises both the crime scene and the accepted movements of the suspect, are relatively consistent, forensic soil evidence might be of little value. Given the above, the prosecution would need to demonstrate that the forensic sample was either unique to the crime scene (or particularly unusual), based, for example, on its multi-element geochemical signature and other properties, for the soil evidence to be of significance. It is therefore important to know whether multi-element soil geochemical signatures are likely to be unique for a given geographic area.

The primary control on the geochemical composition of the relatively immature, natural soils of the UK is their parent material, the bedrock lithology or overlying Quaternary deposit from which they formed. Statistical analysis of data from a soil geochemical survey in eastern England (comprising around 4300 individual survey sites), showed that up to 48% of the variation in the total concentration of major and trace elements was explained by their parent material classification (Rawlins et al. 2003). However, the incompleteness of our understanding of the myriad factors which determine, for example, the concentration of an element in the soil at a specific location in the natural landscape has meant that

From: PYE, K. & CROFT, D. J. (eds) 2004. *Forensic Geoscience: Principles, Techniques and Applications*. Geological Society, London, Special Publications, **232**, 197–206. © The Geological Society of London, 2004.

its properties generally appear to be random (Webster 2000).

Based on semi-quantitative methods of analysis, for example scanning electron microscope (SEM) or X-ray diffraction (XRD), it would be relatively easy to distinguish between two soil samples derived from adjacent parent materials in the UK as they often have differing bulk properties (carbonate, silicate and clay mineral composition (Loveland 1984)). For example, there are clear differences between the mineralogy and geochemistry of soils derived from the Mercia Mudstone and the adjacent Sherwood Sandstone (Fig. 1). Soils derived from the same parent material have considerably less variation in these bulk properties, and attempts to discriminate between samples require more quantitative analytical methods, such as the determination of the total concentration of major and trace elements.

The total concentration of all elements above atomic number 3 (lithium) in the periodic table can theoretically be determined by X-ray fluorescence spectrometry (XRFS) (Bertin 1979). Comprehensive analysis of the inorganic geochemistry of a soil sample is one method which could be used to attempt to discriminate individual samples from different locations. Numerous studies have demonstrated that many natural soil properties are autocorrelated; samples taken close together are more similar than those further apart (many examples are given in Webster & Oliver 2001). If only one element is considered, approximately the same concentration might be reported at several locations in a specified area. The potential for discriminating individual samples from different locations is likely to improve with an increase in the number of elements for which the total concentration is determined, due to the greater total variance of the dataset. Several factors will affect the likelihood of discriminating between samples for individual chemical elements. Those which will increase discrimination include a broad concentration range, fine analytical precision and concentrations which are above the analytical limit of detection. The latter factors are largely dependent on the analytical method/instrument and the nature of the parent material from which the soil is derived, respectively. Two further conditions must be met for a comprehensive and accurate comparison of concentrations in different samples. The first concerns the support of the sample; this is a geostatistical term defined by Olea (1990) as 'an n-dimensional volume within which linear average values of a regionalized variable can be computed'. A specification of the support of a soil sample includes the geometrical shape, size and orientation of the volume. If a common property of two soil samples is to be compared statistically, the support upon which each soil sample is collected should be the same, as this is a primary factor in determining the variance of soil

properties in a series of samples. Larger sample supports have smaller *a priori* variances (Olea 1990). Secondly, in addition to providing a concentration for each element in each sample, it is imperative to quantify the measurement uncertainty of each element to assess the significance of the differences between concentrations determined in each sample.

In this paper we investigate the extent to which it is possible to discriminate between soil samples *within* a range of parent materials based on their individual and multi-element geochemical signatures, using data from a high-resolution (1 sample per $2 km^2$) soil survey. The soil geochemical data have been analysed statistically to address the following questions: (1) how effective are 19 elements in discriminating individual soil samples within parent materials, (2) is it easier to discriminate samples derived from particular parent material types, and (3) is it possible to discriminate individual samples from the same parent material based on their multi-element geochemistry? Finally, we discuss the potential problems concerning the use of soil geochemical signatures in forensic studies.

Methods

Soil geochemical survey and analysis

Between 1994 and 1996 a high-resolution soil geochemical survey comprising around 6500 sample locations was undertaken over eastern England covering a total area of around $13000 km^2$ (shown in Fig. 1). In this study we have chosen to consider only soils derived from six of the major bedrock types in the study region, for two reasons. Firstly, soils derived from bedrock are generally considered more chemically homogeneous than those over Quaternary deposits, the latter having been derived from a mixture of geological sources. This is certainly the case for the superficial glacial deposits along the northeastern coast of the study region (Catt & Penny 1966). The geochemistry of soils derived from a single Quaternary deposit may actually reflect several subpopulations with different geochemical signatures. Hence, initially, we considered it logical to test whether individual and multi-element signatures could be discriminated in soils derived from bedrock. Secondly, these six bedrock types cover larger, contiguous areas than the soils derived from Quaternary deposits in the region, which provided a larger sample size from the non-aligned sampling grid adopted in the survey.

To select soil samples over the six bedrock types the locations of the former were displayed in a geographical information system (GIS) along with digital maps of bedrock (British Geological Survey 1983) and Quaternary deposits (Institute of Geo-

Sample sites over the six lithologies

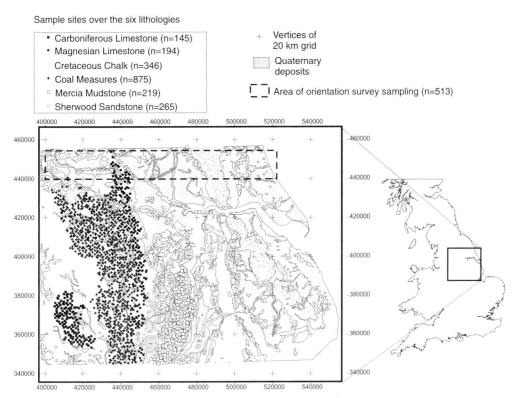

- Carboniferous Limestone (n=145)
- Magnesian Limestone (n=194)
- Cretaceous Chalk (n=346)
- Coal Measures (n=875)
- Mercia Mudstone (n=219)
- Sherwood Sandstone (n=265)

+ Vertices of 20 km grid

Quaternary deposits

Area of orientation survey sampling (n=513)

Fig. 1. The study region showing the location of soil samples collected over each of six parent materials and the region of the orientation study.

logical Sciences 1977). The codes for the bedrock types were joined to the soil samples identifiers. Each of the six groups had between 147 and 875 samples (Fig. 1). Sample sites were chosen from every second kilometre square of the British National Grid by simple random selection within each square, subject to the avoidance of roads, tracks, railways, domestic and public gardens, and other seriously disturbed ground. The samples were collected in the summer months of 1994, 1995 and 1996 in rural and peri-urban areas.

At each site a topsoil sample (0–15 cm depth) was taken from five holes augered at the corners and centre of a square with a side of length 20 m with a hand auger and combined to form a bulked sample (Fig. 2a) weighing c. 1 kg. Further deeper samples were collected in each auger hole at a depth of between 30 cm and 45 cm, and combined to form a bulked sample. Duplicate samples (Fig. 2b) were collected at one in every hundred of the original survey sites for the purposes of estimating measurement uncertainty. All samples of soil were dried and disaggregated; the topsoil samples were sieved to pass through a 2 mm sieve, the deeper samples were sieved to 150 μm. From each sample a 50 g subsam-

ple was ground in an agate planetary ball mill and pressed into pellets. The total concentrations of 23 major and trace elements (Ag, As, Ba, Bi, CaO, Cd, Co, Cr, Cu, Fe_2O_3, MgO, MnO, Mo, Ni, Pb, Rb, Se, Sr, TiO_2, U, V, Zn, Zr) were determined in each pellet by energy and wavelength dispersive XRFS. Numerous in-house reference materials were analysed with each batch of samples and proficiency testing was undertaken within the Wageningen Evaluating Programmes for Analytical Laboratories (WEPAL) International Soil Exchange scheme. Numerous sample concentrations for Ag, Cd, Bi, and Se had values below the lower limit of detection and these elements were removed from the database and are not considered further in this paper.

Comparison of coarse and fine soil size fraction geochemistry

Geochemical analysis of samples collected in the soil survey described above were undertaken on topsoil sieved to <2 mm, a size fraction commonly analysed in soil science in the UK, for example the National Soil Inventory (McGrath & Loveland

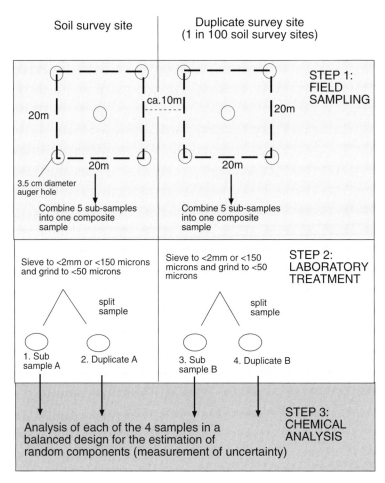

Soil survey site | Duplicate survey site (1 in 100 soil survey sites)

STEP 1: FIELD SAMPLING

ca.10m

20m · 20m

3.5 cm diameter auger hole

20m · 20m

Combine 5 sub-samples into one composite sample

Combine 5 sub-samples into one composite sample

STEP 2: LABORATORY TREATMENT

Sieve to <2mm or <150 microns and grind to <50 microns

Sieve to <2mm or <150 microns and grind to <50 microns

split sample

split sample

1. Sub sample A   2. Duplicate A

3. Sub sample B   4. Duplicate B

STEP 3: CHEMICAL ANALYSIS

Analysis of each of the 4 samples in a balanced design for the estimation of random components (measurement of uncertainty)

Fig. 2. Scheme for the collection of duplicate samples (and separation into subsamples) for the determination of analytical, sampling and measurement uncertainty for the topsoil samples collected in the regional soil survey.

1992). It could be argued that this fraction is on average coarser than that which might normally be analysed in a forensic study, and that an assessment of multi-element signatures based on the coarser fraction may give a different result to a finer (e.g. $<150\,\mu m$) fraction. There are likely to be significant differences between the quantities and representative nature of soil samples collected as part of a soil survey and forensic investigations (forensic and control samples). Forensic samples typically comprise small quantities of material (a few grains to a few grams), while there is no limitation on the quantity of material collected in a soil survey. Samples from the latter tend to be more representative of the bulk soil at a site as the material which is eventually analysed is based on homogenizing a larger quantity of soil (typically up to 1 kg). In addition, if a large soil particle is included in the geochemical analysis of a forensic sample, it may significantly bias the

result. To avoid this potential pitfall, the forensic investigator may prefer to compare the geochemistry of the fine fraction of control samples from the scene with the same size fraction of the forensic sample.

To address this question a comparison was undertaken between the total concentrations of elements in different size fractions from a subset of the soil survey samples collected as part of an orientation study. The start of the soil geochemical survey of the region shown in Figure 1 (beginning in 1994) marked a change in the approach to geochemical sampling in the British Geological Survey (BGS) G-BASE (Geochemical Baseline Survey of the Environment) project. Prior to 1994, soil sampling in areas of western England had been confined to deeper soil horizons at a consistent depth of 30–45 cm. Geochemical analysis of these samples was undertaken on the $<150\,\mu m$ size fraction; the same size fraction that has always been used in the

Table 1. *Statistics of the ratios of total element concentrations in soil samples (n = 513) from the orientation study (cf. Fig. 1)*

Element	Deeper soil (150 μm/2 mm)*			Deeper soil 2 mm/topsoil 2 mm[†]		
	Mean	SD	Median	Mean	SD	Median
As	1.23	0.49	1.14	0.73	0.25	0.71
Ba	1.17	0.23	1.15	1.01	0.93	0.83
CaO	1.26	1.69	1.12	1.33	1.98	0.91
Co	1.24	0.45	1.15	0.87	0.40	0.78
Cr	1.24	0.37	1.17	1.07	0.35	1.04
Cu	1.28	0.41	1.20	0.97	0.50	0.93
$Fe_2O_3 t$[††]	1.17	0.33	1.11	1.26	0.72	1.11
MgO	1.10	0.41	1.00	1.31	0.64	1.20
MnO	1.15	0.41	1.09	1.27	1.06	0.98
Mo	1.61	2.02	1.06	0.59	1.49	0.38
Pb	1.25	0.37	1.17	0.78	1.00	0.65
Rb	1.22	0.28	1.18	1.00	0.95	0.84
Sr	1.19	0.27	1.16	1.19	0.91	0.90
TiO_2	1.29	0.35	1.21	1.02	0.24	1.03
U	1.28	1.07	1.23	1.48	1.41	1.19
V	1.23	0.35	1.15	1.02	0.32	1.00
Zn	1.23	0.43	1.15	1.27	0.57	1.14
Zr	1.25	0.35	1.18	0.93	0.39	0.90
Zr	1.68	0.63	1.53	1.17	0.73	1.04

* Ratio of total concentration in deeper soil (35–50 cm) sieved to 150 μm and 2 μm
† Ratio of total concentration in deeper soil (35–50 cm) sieved to 2 mm and topsoil (0–15 cm) sieved to 2 mm
†† t refers to total

project for the analysis of stream sediments (British Geological Survey 2000). The aim of collecting samples from this depth was an attempt to avoid anthropogenic contamination which, it could be argued, is less likely to be detected in deeper soil samples. By 1994 it was clear that stakeholders also wanted access to topsoil geochemistry data. Hence, in addition to sampling soil at 30–45 cm depth, topsoil samples were also collected (0–15 cm depth). To be consistent with other UK soil surveys, most notably the National Soil Inventory (McGrath & Loveland 1992), it was decided that the topsoil samples would be analysed after sieving to less than 2 mm. To assess the significance of analysing two different soil size fractions (namely <150 μm and <2 mm), an orientation study encompassing 513 soil samples sites was undertaken in the north of the study region (shown in Fig. 1). Deeper soil samples were homogenized and subsamples sieved separately to both size fractions, and the concentrations of each element were determined in order to assess the magnitude of the differences. In addition, the <2 mm sieved topsoil samples were analysed to assess the magnitude of the differences between these and the deeper (2 mm sieved) soil samples.

The ratio of concentrations in the two size fractions from the deeper soils (30–45 cm), and the same size-fraction (<2 mm) in the topsoil (0–15 cm depth) and deeper soil were calculated for each of the 513 sample sites. Summary statistics of the results are presented in Table 1. In the deeper soils, the finer fraction has, on average, only slightly higher total element concentrations; mean ratios are generally between 1.1 and 1.3, most probably due to the higher quartz content of the coarser fraction. Comparison of the 2 mm sieved deeper soil and topsoil samples shows that they also, on average, have similar total element concentrations; mean ratios typically in the range 0.9–1.3. These data support the assertion that results of the subsequent statistical analysis based on the coarser (2 mm sieved), topsoil size fraction, would not be significantly different if the same approach had been applied to a finer (<150 μm) fraction.

Statistics

Calculation of measurement uncertainty: univariate analysis

We used analyses of 84 duplicate and 168 subsamples (see Fig. 2) from sites across the entire region depicted in Figure 1 to determine the measurement uncertainty (Ramsey 1998) associated with the concentrations of 19 major and trace elements. Duplicate samples (Fig. 2) were collected at a short distance (in this case 10 m) from the original survey location, representing the separation which might have occurred from an independent interpretation of the sampling protocol. Results from the chemical analysis of the duplicate and subsamples represents a balanced design which partitions the total variance into three components:

$$s_{total}^2 = s_{geochem}^2 + s_{samp}^2 + s_{anal}^2 \qquad (1)$$

in which $s_{geochem}^2$ is the variance attributable to the spatial variation in concentration of the analyte throughout the region, s_{samp}^2 is the variation at a site (which includes a component of spatial variance) and s_{anal}^2 is the analytical variance. The measurement uncertainty (u_a) with a 95% confidence interval (two standard deviations) was calculated using the following equation:

$$u_a = 2\sqrt{(s_{samp}^2 + s_{anal}^2)} \qquad (2)$$

The robust analysis of variance was undertaken using the method described by Ramsey (1998) with the computer program ROBCOOP4, cited in the original paper. Estimates of measurement uncertainty (calculated using equation 2) are based on single measurements and assume that analytical error and within-site variability are independent of one another. The measurement uncertainty for each element is shown in Table 2.

Element and sample discrimination within parent material groups

To determine whether the difference between two sites exceeds the combined measurement uncertainty with 95% confidence we need to calculate the sum of their individual variances (u_b in Equation 3),

$$u_b = 2\sqrt{2(s_{samp}^2 + s_{anal}^2)} \qquad (3)$$

For each of the 19 elements, for samples within each of the six lithological groups, we then calculated the number of times the envelope of its measurement uncertainty did not overlap with that of the other samples. This calculation is represented by equations (4) and (5):

$$x_i = \begin{cases} 1 & |t_i - t_k| > u_b \\ 0 & |t_i - t_k| \le u_b \end{cases} \qquad (4)$$

$$e_i = \sum_{k=1}^{N} x_i \qquad (5)$$

where t_i and t_k represent the concentration of the element at each of the N sites (Fig. 1), and e_i records the number of times the measurement uncertainty (u_b, defined in equation 3) is smaller than the difference between the concentrations at each site, and every other site over the same parent material. We then expressed the mean of e_i as a percentage by dividing it by the number of samples in each parent material type (N). Finally, we calculated the mean percentage discrimination over all the six parent material types for each element and the corresponding standard deviations (shown in Fig. 3).

We also determined, for each sample, how many of the 19 elements did not have overlapping measurement uncertainties when compared with every other sample from the same parent material group. This enabled us to compare the potential to discriminate samples within individual parent material types based on equations 6 and 7 (the notation is similar to that of equations 4 and 5):

$$x_i = \begin{cases} 1 & |t_i - t_k| > u_b \\ 0 & |t_i - t_k| \le u_b \end{cases} \qquad (6)$$

$$h_i = \sum_{j=1}^{19} x_i \qquad (7)$$

where the term h_i represents the number of elements (j) in which the difference between the concentrations at the ith site (t_i) and every other site over the same parent material (t_k) is greater than the measurement uncertainty (u_b). We summarized the data (h_i) for each of the 19 elements as relative density distributions for each of the six parent material types, each with N samples.

Investigating multi-element signatures

The investigation of multi-element geochemical signatures is more complex than the foregoing univariate analysis because there is likely to be some degree of correlation between analytical errors and within-site variations for different elements. Such correlations will lead to a smaller total uncertainty than a simple linear sum of the individual components when comparing the concentration of 19 elements in each pair of samples. We undertook a multivariate analysis to assess the extent to which pairs of soil samples from the same parent material could be discriminated. Firstly, we estimated a covariance matrix for the two sources of uncertainty from the available duplicate and subsample data. We then calculated Mahalanobis distances for every pairwise combination of samples within each parent material group. The Mahalanobis distance takes into account the correlation between variables when computing statistical distances.

A nested analysis of covariance based upon the analytical (subsamples) and sampling (duplicate samples) variances was conducted using the residual maximum likelihood (REML) algorithm of Calvin and Dykstra (1992) resulting in two separate $j \times j$ covariance matrices. The analytical and sampling variances extracted from these matrices are shown in Table 2. They are somewhat larger than the robust analysis of variance (ANOVA) estimates, in part because they are components of matrices constrained to be non-negative definite. These two components were added together to give $\mathbf{V}$, the $j \times j$ covariance matrix of a single soil specimen drawn from one site representing the variation around the mean site value and the analytical fluctuation around the true value at the site.

The calculation of Mahalanobis distances for sample pairs is represented by equations 8 and 9. Let $\mathbf{z}_i$ be a column vector with j entries for the trace element concentrations at site i, and $\mathbf{z}_k$ the same for site k. To compare these sites we computed a difference vector:

$$d_{i,k} = \mathbf{z}_i - \mathbf{z}_k. \qquad (8)$$

We then calculated the squared Mahalanobis distance as:

Table 2. *Detection limits, lower reporting limits and measurement uncertainties calculated using robust ANOVA (Ramsey 1998) and residual maximum likelihood estimation (REML; Calvin & Dykstra 1992)*

Element	Detection limit*	Lower reporting limit*	ANOVA sampling variance (s^2_{samp})	ANOVA analytical variance (s^2_{anal})	ANOVA Measurement uncertainty (u_a)	REML Sampling variance	REML Analytical variance
As	0.9	1	0.61	0.63	2.22	1.10	1.31
Ba	5.1	6	7.01	6.48	7.34	35.1	76.9
CaO (%)	–	0.1	0.08	0.031	0.66	0.25	0.22
Co	1.2	2	0.43	0.72	2.14	0.71	1.85
Cr	1.3	2	2.14	2.93	4.5	5.12	2.34
Cu	0.8	1	0.72	0.70	2.38	1.38	2.32
Fe$_2$O$_3$ (%)[†]	–	0.01	0.115	0.06	0.84	0.19	0.40
MgO (%)	–	0.1	0	0.049	0.44	0.09	0.04
MnO (%)	–	0.01	0.004	0	0.13	0	0.01
Mo	0.2	1	0.10	0.24	1.17	0.26	0.21
Ni	0.6	1	0.60	0.56	2.15	0.97	1.39
Pb	0.5	1	2.85	1.51	4.17	3.8	17.0
Rb	0.5	1	1.33	0.69	2.84	1.90	2.62
Sr	0.6	1	1.13	0.87	2.82	2.20	3.44
TiO$_2$ (%)	–	0.02	0.012	0.014	0.32	0.02	0.03
U	0.6	1	0.09	0.27	1.19	0.29	0.19
V	1.3	2	2.15	1.74	3.95	3.04	4.01
Zn	0.5	1	2.91	1.12	4.01	2.42	8.57
Zr	0.8	1	5.12	5.25	6.44	8.75	21.4

* Units mg/kg, unless element expressed as an oxide, in which case values are reported as a percentage.
† Refers to total.

$$D^2 = d^T_{i,k} V^{-1} d_{i,k.} \tag{9}$$

Under a null hypothesis that the two observations are drawn from the same population, $[D^2/2]$ is distributed as (χ^2) with j degrees of freedom. We would therefore regard the difference between two randomly drawn sites within the same parent material to be significant if D^2 is larger than twice the 5% point for the χ^2 distribution with 19 degrees of freedom, a numerical value of 60.3.

Results

Individual element and sample discrimination

For each of 13 elements, it was possible on average to discriminate between 80% of the samples within parent material groups (see Fig. 3). The most discriminatory elements differed for each parent material type and in some cases are clearly related to the nature of the bedrock. For example, 96% discrimination based on CaO in samples over the Cretaceous Chalk and 97% for MgO in samples over the Magnesian Limestone, these elements being the predominant cation in each of the respective carbonate lithologies. Magnesium was also the most discriminatory element for soils derived from the

Mercia Mudstone (97%) which may be related to the presence of dolomitic limestone and smectite clay minerals derived from this deposit (Leslie *et al.* 1993). For all parent materials the most powerful discriminatory elements, with the largest range of values and small measurement uncertainties, in decreasing order of average discrimination (%) were: Mn (93), Mg (90), Rb (89), Zr (89), Sr (87), Ni (87), Zn (86). Those elements which were least useful for discriminating samples (U and Mo, 50% and 55% respectively) were generally of lower abundance and poorer analytical resolution.

Results of the element discrimination (within parent material groups) showed that typically between 15 and 17 (median and mode statistics) of the 19 elements discriminated individual samples based on the measurement uncertainty (see Fig. 4). Five of the six bedrock types had similar sample discrimination frequency distributions, whereas it was more difficult to discriminate between element concentrations in samples over the Sherwood Sandstone Group. This can be explained by the lower total concentration and smaller variations in major and trace element contents in soils derived from the Sherwood Sandstone in comparison to soils over the other lithologies. These findings appear to suggest there is significant potential for using multi-element geochemical signatures to discriminate between soil

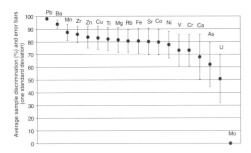

Fig. 3. Average sample discrimination (%) for six parent material groups for each of 19 elements and error bars showing ± one standard deviation.

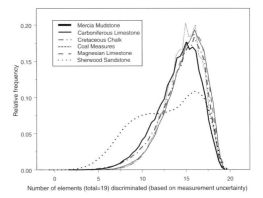

Fig. 4. Density plot showing number of elements ($n = 19$) discriminated in each sample comparison (h_i) for six parent material types.

survey samples from the majority of parent material types. However, it does not indicate whether individual soil survey samples are sufficiently different from those over the same parent material to be considered unique. This question is addressed in the next section.

Multi-element signatures

There is a need to be somewhat cautious in the interpretation of the results of the discriminant analysis. Firstly, the 5% significance level of 60.3 ignores the errors in the estimates of **V** (the covariance matrix); it is not clear how to account for these errors in the analysis. Hence the critical value is likely to be somewhat larger. Secondly, the analysis is not based upon independent random observations, and this could undermine assessment of overall rates of discrimination. To investigate this further, we therefore undertook an exercise in subsampling without replacement for each of the six series of Mahalanobis distances (within each parent material class) to assess its proportion of discrimination compared to the full dataset. In each of the six groups the effect was negligible and we have therefore chosen to present and interpret the latter.

The results of the multivariate discriminant analysis based on Mahalanobis distance estimates are

summarized in Table 3 and Figure 5. The number of paired comparisons shown for each parent material group is equal to $(N^2 - N)/2$, where N is the number of samples in each of the six groups (Table 3). With the exception of soils over the Sherwood Sandstone Group, more than 99.8% of samples could be discriminated from the other samples within the same parent material group (Table 3). Despite the caveats expressed above on being unable to incorporate errors from the covariance matrix and the result not being based on random samples, this is a very high degree of sample discrimination.

Frequency distributions for the Mahalanobis distances between samples over each parent material are shown in Figure 5. Each of the distributions is positively skewed; the largest distances are generally associated with the Coal Measures. These results are consistent with the findings from the univariate analysis in which fewer elements discriminated between soil samples over the Sherwood Sandstone (Fig. 5a), and which generally has the smallest Mahalanobis distances. Hence, we conclude that in soil samples collected from our regional survey, based on a consistent sampling protocol and the analysis of a 1 kg sample (reduced

Table 3. *Statistics for discriminant analysis based on Mahalanobis distance for soils from six parent material groups*

	Sherwood Sandstone	Mercia Mudstone	Magnesian Limestone	Cretaceous Chalk	Coal Measures	Carboniferous Limestone
Number of soil samples	265	219	194	346	875	145
Number of paired comparisons	34890	23871	18721	59685	382375	10440
Proportion of paired samples discriminated (%)*	95.65	99.84	99.95	99.85	99.99	100

* Paired samples with a Mahalanobis distance >60.3.

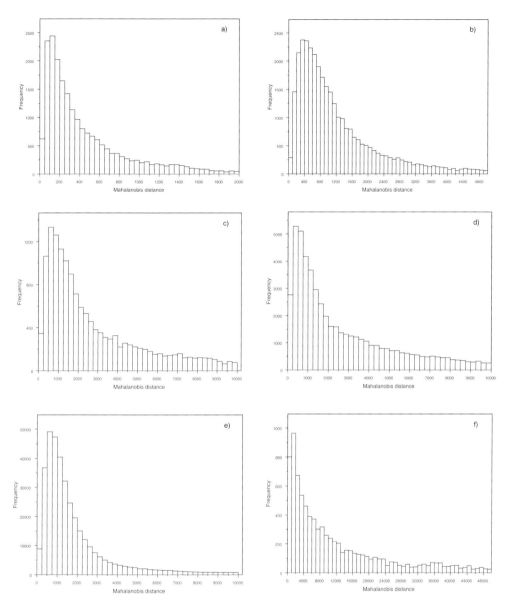

Fig. 5. Frequency distributions for Mahalanobis distances for paired soil samples within the six parent materials: (**a**) Sherwood Sandstone; (**b**) Mercia Mudstone; (**c**) Magnesian Limestone; (**d**) Cretaceous Chalk; (**e**) Coal Measures; (**f**) Carboniferous Limestone.

to a 50 g subsample), multi-element geochemical signatures can be established within most parent material groups.

The multivariate analysis has demonstrated that, in theory, the geochemistry of a natural soil sample, when compared with a comprehensive survey database, could be used to establish its provenance within several bedrock derived soil types throughout eastern England. The variation in soil geochemical properties throughout this region appears to be broadly representative of other soils throughout much of the UK. However, further analysis is needed to establish whether soil geochemical signatures can be established within a series of soil samples derived from a range of Quaternary deposits.

Implications for forensic studies

While the findings of our analysis suggest that multi-element soil geochemical signatures may be useful in establishing the provenance of 'unknown' (i.e. forensic) soil samples, there are significant differences between the nature of samples and the sampling methods in soil survey and forensic work. Firstly, the size (number of mineral grains) of the former tends to be much larger than the latter. Sampling error increases as the size of a sample selected to represent a bulk specimen decreases. Hence, the measurement uncertainty, which includes the sampling error associated with the smaller forensic samples, is liable to be significantly larger than for the soil survey data. Previous research has shown that, in the analysis of rock powders (consisting of the same minerals that weather to form soils), a sample mass of 1 g is necessary to provide an accurate representation of major element concentrations, while considerably more material is required to represent trace element contents adequately (Wilson 1964). Sufficient soil material may not be available in many forensic cases for a representative analysis to be undertaken, which reduces our confidence in comparing the geochemistry of forensic samples, other local (control) samples, and samples from a regional soil survey.

Secondly, in contrast to the soil survey, the support of the typical forensic sample (i.e. taken from the shoes of a suspect) is unknown. In general its provenance must be assumed to be an area of limited extent if the forensic sample is to have any evidential value. Comparison of its geochemistry with that of control samples is problematic as we do not know the support upon which the latter should be collected for comparison. This further undermines our confidence in comparing the forensic sample with a control sample from the same site. These questions need to be addressed and scrutinized before practical application to forensic cases in which an investigator is attempting to link a suspect to a specific location based on soil geochemical signatures.

The authors wish to thank all staff at BGS involved in each part of the G-BASE survey: sample collection, preparation, analysis and quality control. They would also wish to acknowledge M. Lark (Silsoe Research Institute) for providing helpful comments on an early version of the manuscript, and both M. Lark and R. White for their assistance with the Mahalanobis distance analysis. This paper is published with the permission of the Director of the British Geological Survey (National Environment Research Council).

References

BERTIN, E. P. 1979. *Principles and Practise of X-ray Spectrometric Analysis*. Plenum Press, New York.

BRITISH GEOLOGICAL SURVEY. 1983. *Humber-Trent (Solid Geology) Sheet 53N 02W*. British Geological Survey, Keyworth.

BRITISH GEOLOGICAL SURVEY. 2000. *Regional Geochemistry of Wales and Part of West-Central England: Stream Sediment and Soil*. British Geological Survey, Keyworth.

CALVIN, J. A. & DYKSTRA, R. L. 1992. An algorithm for restricted maximum likelihood estimation in balanced multivariate variance components models. *Journal of Statistical Computation and Simulation*, **40**, 233–246.

CATT, J. A. & PENNY, L. F. 1966. The Pleistocene deposits of Holderness, East Yorkshire. *Proceedings of the Yorkshire Geological Society*, **35**, 375–420.

DUDLEY, R. J. 1975. The use of colour in the discrimination of soil. *Journal of the Forensic Science Society*, **15**, 209–218.

HIRAOKA, Y. 1994. A possible approach to soil discrimination using X-ray fluorescence analysis. *Journal of Forensic Sciences*, **39**, 1381–1392.

INSTITUTE OF GEOLOGICAL SCIENCES. 1977. *Quaternary Map of the United Kingdom*. Ordnance Survey, Southampton.

LESLIE, A. B., SPIRO, B. & TUCKER, M. E. 1993. Geochemical and mineralogical variations in the upper Mercia Mudstone Group (Late Triassic) southwest Britain: correlation of outcrop sequences with borehole geophysical logs. *Journal of the Geological Society*, **150**, 67–75.

LOVELAND, P. J. 1984. The soil and clays of Great Britain. 1. England and Wales. *Clay Minerals*, **19**, 681–707.

MCGRATH, S. P. & LOVELAND, P. J. 1992. *The Soil Geochemical Atlas of England and Wales*, Blackie Academic and Professional, Glasgow.

MURRAY, R. C. & TEDROW, J. C. F. 1992. *Forensic Geology*. Prentice-Hall, New Jersey.

OLEA, R. A. 1990. *Geostatistical Glossary and Multilingual Dictionary*. Oxford University Press, New York.

RAMSEY, M. H. 1998. Sampling as a source of measurement uncertainty: techniques for quantification and comparison with analytical sources. *Journal of Atomic Spectrometry*, **13**, 97–104.

RAWLINS, B. G., WEBSTER, R. & LISTER, T. R. 2003. The influence of parent material on topsoil geochemistry in eastern England. *Earth Surface Processes and Landforms*, **28**, 1389–1409.

SUGITA, R. & MARUMO, Y. 2001. Screening of soil evidence by a combination of simple techniques: validity of particle size distribution. *Forensic Science International*, **122**, 155–158.

WEBSTER, R. 2000. Is soil variation random? *Geoderma*, **97**, 149–163.

WEBSTER, R. & OLIVER, M. A. 2001. *Geostatistics for Environmental Scientists: Statistics in Practice*. John Wiley & Sons, Chichester.

WILSON, A. D. 1964. The sampling of silicate rock powders for chemical analysis. *Analyst*, **89**, 18–30.

Bayesian sediment fingerprinting provides a robust tool for environmental forensic geoscience applications

INGRID F. SMALL[1], JOHN S. ROWAN[1], STEWART W. FRANKS[2], ADAM WYATT[2] &
ROBERT W. DUCK[1]

[1]*Environmental Systems Research Group, Department of Geography, University of Dundee,
Dundee DD1 4HN, UK (e-mail: i.f.small@dundee.ac.uk)*
[2]*School of Engineering, University of Newcastle, Newcastle, Callaghan 2308, NSW, Australia*

Abstract: Sediment fingerprinting is an approach for the quantitative determination of sediment provenance (both spatial sources and types of sediment supply) over a range of temporal and spatial scales. Though widely adopted, studies often vary in their attention to the underlying assumptions and in their treatment of modelling uncertainty. A Bayesian approach to the multivariate problem of 'unmixing' sediment sources is reported, showing the significance of source group variability and source group sampling density to the accuracy of model output. The model produces results as median source group contributory coefficients (and associated 95% quantiles). The model was applied to environmental data obtained from selected soil erosion studies reported within the peer-reviewed literature. Good correspondence ($r^2 = 0.89$) between reported mean source group contributory coefficients and median values were found when recalculated using the Bayesian analysis. However, confidence levels are highly variable, ranging from 2% to 97%. The robustness of any unmixing solution depends on factors such as the number of samples, the number of source groups and the variance of source group properties. It is concluded that 'forensic-style' investigations must recognize these uncertainties and be appropriately resourced to achieve tolerable accuracy and precision. The discussion considers additional confounding factors such as non-conservative tracer behaviour and enrichment/depletion during the sediment delivery process.

Sediment fingerprinting is now an established approach, with demonstrated value in elucidating the linkages between catchment erosion processes and the delivery of sediment downstream. Such work has applications in the study of transmission and delivery dynamics, reconstruction of sediment source variability over time, validation of deterministic erosion models and better focusing of the management of natural resources. The basis of the approach is to 'match' the properties of the 'target' sediment (suspended, bedload or sediment stored in sinks such as floodplain and lake sediment sequences) to equivalent values in potential sources.

Different properties have been successfully employed as sediment tracers, including clay mineralogy, sediment colour, sediment chemistry (Peart & Walling 1986), mineral magnetics (Yu & Oldfield 1989) and radionuclide concentrations (Wallbrink & Murray 1993). Composite tracer signatures involving different properties (e.g. chemical, radiometric and magnetic) are recognized to offer the best chance of unequivocal discrimination between source groups, but necessitate the use of multivariate unmixing models to quantify the relative contribution from each source (Collins *et al.* 1997, 1998; Krause *et al.* 2003).

While the range of tracer properties has grown dramatically, less attention has been paid to the performance of the unmixing models, all of which ultimately depend on optimization procedures to estimate source contributions (typically some form of constrained linear programming). However, these models are known to be subject to equi-finality problems (ie. non-uniqueness) whereby near equivalent levels of model performance, as measured by some likelihood function, can be achieved with widely differing sets of contributory coefficients (cf. Rowan *et al.* 2000). Questions thus remain about the robustness of the current generation of models, and there is a clear need for robust quantitative schemes fully inclusive of data and modelling uncertainties. These problems are compounded by the incorporation of additional uncertainties associated with analytical errors, non-linear additivity (Lees 1997) and grain size and organic matter enrichment effects (see Table 1). In environmental forensic applications where evidence is being provided to support a legal argument, and may therefore be subjected to cross-examination in a court of law, all forms of uncertainty will be scrutinized and potentially challenged. Thus results must be provided with reproducible confidence levels, that is defendable levels of accuracy and precision. Of course most environmental litigation investigations (cf. Danon-Schaffer 2002; Suggs *et al.* 2002) use a range of evidence, and quantitative analysis is often coupled with other lines of evidence, including characterization, description and qualitative appraisal.

From: PYE, K. & CROFT, D. J. (eds) 2004. *Forensic Geoscience: Principles, Techniques and Applications.* Geological Society, London, Special Publications, **232**, 207–213. © The Geological Society of London, 2004.

Table 1. *Principal sources of uncertainty within sediment fingerprinting schemes*

Nature of uncertainty	Key issues (and assumptions involved)
Problem formulation, i.e. how many source groups can be distinguished	Too few source groups compromises utility of approach; too many groups leads to problems of source group discrimination and spurious numerical solutions
Discriminating power of tracers to distinguish between source groups (dimensionality)	Depends partially on number, location and types of source groups to be distinguished, and laboratory resources available to the research team
Tracer bias	Order of magnitude variations in tracers
Characterization of source group variability	Discriminating between source groups depends on 'within group' variance relative to 'between group' variance
Measurement uncertainty of tracer properties	Radiometric measurements, e.g. ^{137}Cs associated with intrinsic uncertainties $\pm 5\%$, clay mineralogy typically semi−quantitative only
Tracer transformation	During transport and particularly sediment deposition diagenetic transformations may occur, e.g. synthesis of biogenic greigite in lake sediments (Dearing 2000)
Linear additivity	Some properties, such as mineral magnetic measurements present non−linear additivity problems (Lees 1997)
Enrichment	Preferential enrichment/selective deposition of fine/coarse fractions of the mineral sediment fraction and organic matter
Mixing models	Constrained linear programming (optimization based) – problems of equi-finality in prediction of estimated source contributions

Franks and Rowan (2000; henceforth FR2000) reported a Bayesian Monte Carlo modelling framework that explicitly addressed two key modelling issues, namely source group variability and source group sampling density. The FR2000 unmixing model yields median contributory coefficients with 95% confidence intervals and was validated using controlled laboratory mixtures and synthetic data (Small *et al.* 2002). This study utilizes an updated version of the code (Version 2) and further validates results, but the key contribution is to collate selected environmental datasets obtained from the geomorphological literature and to evaluate the levels of uncertainty contained therein. Studies were drawn from a range of environments and feature a diversity of tracer properties (geochemical, radiometric and magnetic). Problem complexity, as represented by the number of sediment sources sought to be 'unmixed' also varied from three to eight. It was thus an important requirement of the selection process that summary data were reported, for example sample number as well as tracer property mean and dispersion data (where available).

Unmixing model structure and validation

The Bayesian mixing model technique of FR2000 can be summarized briefly as follows:

$$\sum_{i=1}^{m} \hat{\bar{x}}_{i,j} \times A_j = X_i + \varepsilon_i \quad j = 1, m \tag{1}$$

with the added constraints of:

$$\sum_{j=1}^{n} A_j = 1 \quad 0 < A_j < 1 \tag{2}$$

where $\hat{\bar{x}}_{i,j}$ is the estimated population mean of trace property i, within the source group j, n is the number of source groups, m is the number of tracers, and A_j is the proportional contribution coefficient of source group j to the target sediment and ε_i is the error associated with the prediction of the sink trace characteristic X_i. Equation 2 provides additional constraints on the unknown source groups in that they must sum to unity.

To characterize the properties of a source group, it is assumed that the mean and associated uncertainty may be represented by a Student's t-distribution (t = student test with $_v$ degrees of freedom (d − 1)). Hence:

$$\hat{\bar{x}}_{i,j} \sim t_v \left(\mu, \sigma^2 \right) \tag{3}$$

where i is the specific source group, j is the tracer property, μ is the true value of the population mean (mean of $\hat{\bar{x}}_{i,j}$) and σ^2 is the variance of the probability

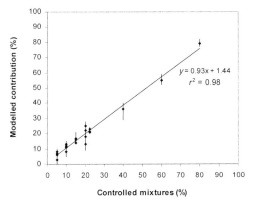

Fig. 1. Comparison between modelled and controlled contributory coefficients for an homogenized rock mixture (tracer type: geochemical).

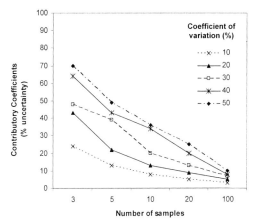

Fig. 2. Effect of source group sampling numbers and source group variability on model performance.

distribution of the population mean. The variance of the distribution is given by:

$$\hat{\sigma}^2 = \left(\frac{S}{\sqrt{d}} \right)^2 \qquad (4)$$

where S is the sample standard deviation, and d is the number of independent samples.

Using the above equations, the probability distribution of the population means of each of the tracer characteristics, from each of the source groups, can be calculated. Within the Bayesian framework the variability associated with each component tracer must be propagated into the final model output, and FR2000 achieves this through Monte Carlo sampling of the derived probability distributions for each source group and target sediment (e.g. suspended sediment obtained during a storm event). Equation 1 is solved through a robust global optimization routine (Shuffeld Complex Evolution; Duan *et al.* 1992). This produces the optimal estimate of the source groups $(A_j, j = 1, n)$ given the randomly sampled estimates of the true population means. In the Bayesian framework improved predictive performance is associated with more iterations and in the present study a minimum of 1000 realizations was completed for each solution.

The performance of the model is illustrated in Figure 1, which shows the results of a validation exercise using laboratory data. Five different rock samples (nominally sandstone, basalt, limestone, granite and slate) were milled to produce homogenized rock powders. X-Ray fluorescence (XRF) analysis (Phillips PW1400) was used to determine 18 different trace elements in both bulk samples (representing source groups) and a series of 'control mixtures' of known composition (representing target sediment). The median values lie on or near the equiline ($r^2 =$

0.98) indicating a 'good fit' between modelled and expected values. However, 95% uncertainty envelopes of 2–11% result from the relatively small number of samples ($n = 3$) used to determine source group and target tracer properties. Furthermore, even within nominally homogenized samples, a high degree of variability can remain, for example the La concentration in sandstone exhibited a CV% of 41% and variations between 5% and 20% are not uncommon (inclusive of measurement errors).

Model performance clearly depends on how well the source group tracer properties capture the true population mean and the magnitude of the sample standard deviation – itself a function of sample size. This is shown graphically in Figure 2 using synthetic data from a four source group and seven tracer numerical experiment using data presented in Franks and Rowan (2000). The suite of curves shows the interplay of sampling density and source group variability. By way of example, when tracers have a CV% of 20 determined from 20 independent samples then the resultant uncertainty envelope is 12%. Higher CV%s, or lower sample numbers, lead to considerably higher uncertainty in unmixing model performance. Therefore careful consideration should be given to the design of an appropriate sampling programme in order to collect sample numbers adequate to capture the source group variability.

Exploring the uncertainty within published data sets

The main contribution of this paper is to explore the levels of uncertainty inherent within the current generation of sediment fingerprinting models. The data obtained from the literature are summarized in Table

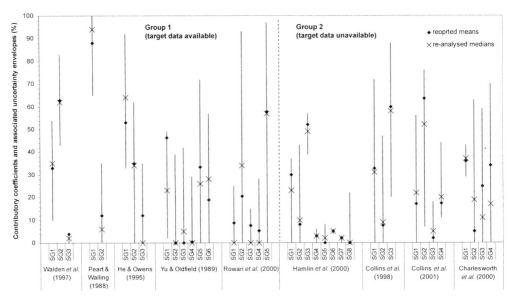

Fig. 3. Derived uncertainty bounds from peer-reviewed literature data sources, recalculated using number of source groups, number of tracers along with number of samples and source group variability.

2, and include the number of source groups (ranging from 2–8), the number and types of tracers (3–12), along with the number of samples (1–100) and source group variability (CV% 3–72). Studies were drawn from a range of environments and featured a diversity of tracer properties (geochemical, radiometric and magnetic). The first group of papers (Group 1) feature both summary source and target data, whereas Group 2 papers reported only source group variability and the results of the unmixing models (contributory coefficients).

FR2000 was reconfigured so that mean source group data and target values could be used to rerun the available data and generate median and 95% confidence intervals. In the case of Group 2 studies the model again produced 95% confidence intervals, but the uncertainties reported in these cases are minimum values because they are derived from the published contributory coefficients and the assumption that these values represented a unique solution to the unmixing problem.

The published contributory coefficients and the output obtained re-analysing these studies using FR2000 are illustrated in Figure 3. A strong relationship ($r^2 = 0.89$) was evident between the reported mean and modelled median values and, in most cases, both values were within 10% of each other. In all cases the reported values lay within the recalculated 95% confidence envelopes. The few larger deviations found can be attributed to the stochastic sampling approach adopted herein v. the original optimization procedures used to obtain a single

'best-fit' solution from a range of near equivalent statistical solutions drawn from across the parameter space (cf. Beven & Binley 1992).

What is most striking about Figure 3 is the spread of the confidence intervals calculated using the FR2000 model, ranging from 2–97% with a mean of 41%. Much of this spread is attributable to the interplay between relatively small number of samples used to describe source group variability (typically less than 20) and the characteristically high CV%s (upper ranges reported 50–72 %). A wider review of fingerprinting literature reveals the mean number of samples collected per source group to be $n = 34$, with a mean source group CV% of 38%. This combination entered into Figure 2 would indicate uncertainty envelopes of c. 20%.

In many environmental studies the relatively high costs of fieldwork, laboratory processing and analytical costs tend to militate against large numbers of samples being collected. This problem increases with the number of source groups involved. Moreover as the number of source groups increases, so the problem of unmixing becomes more complicated. In practical forensic applications this is a particularly important issue, where the objective is to obtain 'acceptable' confidence levels to the legal outcome. In this respect sampling uncertainties must be accounted for from the outset of the investigation. A sufficient number of samples should be collected to ensure that the quality of the data generated enables meaningful scientific statements to be made about the sampling target. Figure 3 is also significant in revealing the high degree of overlap that exists between the uncertainty envelopes in most of the studies includes. The study of Peart and Walling (1988) differs because, despite substantial uncertainty envelopes (e.g. 35%), there is sufficient discrimination between the two sources to unequivocally confirm the dominance of topsoil sources (SG1) over channel bank sources (SG2) in the catchment under investigation.

Lees (1994) suggests that, when using magnetic parameters for source modelling, only a small number (three or four) of source groups can realistically be identified using magnetic parameters. This may partially explain the wider uncertainty envelopes (29–72%) associated with Yu and Oldfield (1989), which involved six source groups, compared with Walden et al. (1997) who define only three. However, the former study also employed smaller sample numbers and was different because of its use of estuarine sediments requiring magnetic signatures to remain conservative during both transport and in storage.

Studies relying on one type of tracer (e.g. radionuclides) to distinguish between source groups have uncertainty envelopes of 35–62%. This would appear to suggest the importance of tracer selection (i.e. multi-parameter approach) in order to increase dimensionality of the data. Lack of dimensionality according to Lees (1994) can lead to groups of source samples that are 'numerical multiples' of one another. Bayesian sampling also implies that maximizing the number of tracers used in the analysis may potentially reduce uncertainty, but Rowan et al. (2000) argued that a minimum number of statistically robust tracers should be used to reduce problems of solution equifinality. Figure 3 additionally indicates that this is not necessarily the case, and that studies with many more tracers can continue to propagate wide uncertainty envelopes. Small et al. (2002) therefore concluded that the value of the individual tracers should be assessed according to their influence on inferred source group contributions, rather than discriminating power alone.

Discussion: further uncertainties in sediment fingerprinting

Sediment properties represent a means to apportion soil loss within a catchment system, but it has been shown that issues such as the number of source groups, the number of samples and the discriminatory power of the tracers involved all effect the quality of the model output. Further issues relate to the selective nature of the erosion and transportation process, which causes the enrichment of eroded sediment relative to the original in situ source (Novotny 1980). The consequence of such enrichment is that the sediment parameters of target sediments may differ in important ways from the source material from which they were derived without any actual chemical or physical change (Slattery et al. 1995).

Attempts to model sediment source linkages on the basis of their sediment signatures usually simplify the problem by: analysing the bulk source materials (Yu & Oldfield 1989); statistically correcting for particle size effects, for example for % clay and surface area (Collins et al. 1997); by functionally separating source materials into particle size fractions which approximately replicate those of the deposited materials (Yu & Oldfield 1993; Slattery et al. 1995); or analysing fractions which are thought to be representative of the different hydraulic transport components (Kelley & Nater 2000).

Such corrections ensure that differences in the tracer signatures between sources and sediments are due to 'true' differences in composition, rather than representing apparent differences due simply to differences in particle size distributions, which ultimately lead to coincidental or invalid sediment source ascription. Incorporation of such effects within the fingerprinting procedures is clearly very important. The contributory coefficients shown in Figure 3 were recalculated using the enrichment functions proposed by the authors (Table 2). Analysing this issue from an uncertainty perspective will form the basis of a subsequent paper.

Another potential area of uncertainty is tracer transformation, which, although widely perceived, has received little attention. Processes associated with this are mechanical and chemical alteration along with organic matter decomposition resulting in sediments no longer reflecting the original source material. Such effects must be recognized and accounted for in any robust modelling scheme (Walling et al. 1993).

Conclusion

A conceptually simple but robust Bayesian methodology has been developed for the assessment of uncertainty in sediment fingerprinting models. It is apparent that the accuracy and precision of the unmixing clearly depends on the number of samples obtained, combined with the inherent variability of tracer values within source groups. In academic research and applications of contracted sediment fingerprinting studies, the output of the model will form the basis of policy decision and associated resource expenditure, namely targeting the most significant sediment sources to focus limited mitigation budgets. Providing robust analysis with known confidence levels is thus very important. However, where sediment fingerprinting has real forensic applications, for example involving evidence

Table 2. *Data sets obtained from peer-reviewed literature*

Author	Number of source groups (*n*)	Number of tracers (m)	Type of tracers	Number of source group samples (d)	Source group variability (CV%)	Approach to grain size enrichment and approach to correction
Group 1 (target data available)						
He & Owens (1995)	3	3	radionuclides	9–17	20*	screened <2 mm + enrichment ratio
Peart & Walling (1988)	2	7	geochemical	50	20*	screened <63 μm + property ratio
Rowan *et al.* (2000)	5	7	magnetics, geochemical	10–20	3–63	screened <63 μm
Walden *et al.* (1997)	3	5	magnetics	3–12	20*	not significant
Yu & Oldfield (1989)	6	10	magnetics	1–6	20*	screened <63 μm
Group 2 (target data not available)						
Charlesworth *et al.* (2000)	4	6	geochemical, magnetics	12–30	20–70	screened <2 mm + <63 μm
Collins *et al.* (1998)	3	5	geochemical	80–96	5–25	screened <63 μm + ssa[†] ratio
Collins *et al.* (2001)	4	12	geochemical, radionuclides	20	18–65	screened <63 μm + ssa[†] ratio
Hamlin *et al.* (2000)	8	9	geochemical	3–18	3–72	screened <2 mm

* In absence of reported values tracer CV%s approximated at 20%
[†] Specific surface area ratio between source and target sediments

supporting a particular legal argument, the provision of results with known confidence levels is essential. Therefore tight confidence limits required for forensic geosciences investigations will require a sufficiently large sampling programme to be carried out to ensure that the environmental data is of 'acceptable' quality and usability for the purpose intended. The number of samples needed is itself a function of the natural variability of each source group and the resources available (i.e. analytical and financial). The demands by the legal profession for sediment fingerprinting are likely to intensify in the future, where criminal cases have an environmental dimension and pollution-related litigation involving individuals, corporations and public agencies become more common.

Results also re-emphasize the importance of the careful identification of all key sediment sources relevant to the research question, along with the rigorous identification of statistically significant tracers capable of distinguishing unequivocally between the different source groups. Evidently, reliable quantitative estimates of source–sediment linkages also require fingerprinting procedures to be conducted on a particle size related basis. Incorporation of such enrichment/depletion effects within the fingerprinting technique will thus help to reduce associated uncertainty and reduce the influ-

ence of spatial variability in tracer concentrations. This area of uncertainty is the main focus of present research and an updated modelling routine is currently being developed and will be reported in a future study.

I. F. Small acknowledges the award of a Carnegie Scholarship in funding a PhD at the University of Dundee and a Carnegie Trust Research Grant. S. W. Franks received an Academic Visitor award from the Royal Society of Edinburgh to conduct research at the University of Dundee.

References

BEVEN, K. J. & BINLEY, A. M. 1992. The future of distributed models: calibration and predictive uncertainty. *Hydrological Processes*, **6**, 279–298.
CHARLESWORTH, S. M., ORMEROD, L. M. & LEES, J. A. (2000) Tracing sediments within urban catchments using heavy metal, mineral magnetic and radionuclide signatures. *In:* FOSTER, I.D.L. (ed.) *Tracers in Geomorphology.* Wiley, Chichester, 345–368.
COLLINS, A. L., WALLING, D. E. & LEEKS, G. J. L. 1997. Source type ascription for fluvial suspended sediment based on a quantitative composite fingerprinting technique. *Catena*, 29, 1–27.
COLLINS, A. L., WALLING, D. E. & LEEKS, G. J. L. 1998. Use of composite fingerprints to determine the provenance

of the contemporary suspended sediment load transported by rivers. *Earth Surface Processes and Landforms*, **23**, 31–52.

COLLINS, A. L., WALLING, D. E., SICHINGABULA, H. M. & LEEKS, G. J. L. 2001. Suspended sediment source fingerprinting in a small tropical catchment and some management implications. *Applied Geography*, **21**, 387–412.

DANON-SCHAFFER, M. N. 2002. Investigation, remediation and cost allocation of contaminants from the Britannia mine in British Columbia: a case study. *Environmental Forensics*, **3**, 15–25.

DEARING, J. A. 2000. Natural magnetic tracers in fluvial geomorphology. *In*: FOSTER, I. D. L. (ed.) *Tracers in Geomorphology*. Wiley, Chichester, 279–291.

DUAN, Q., SOROOSHIAN, S. & GUPTA, V. K. 1992. Effective and efficient global optimisation for conceptual rainfall-runoff models. *Water Resources Research*, **28**, 1015–1031.

FRANKS, S. W. & ROWAN, J. S. 2000. Multi-parameter fingerprinting of sediment sources: uncertainty estimation and tracer selection. *In*: BENTLEY, L. R., BREBBIA, C. A., GRAY, W. G., PINDER, G. F. & SYKES, J. F. (eds) *Computational Methods in Water Resources*. Balkema, Rotterdam, 1067–1074.

HAMLIN, R. H. B., WOODWARD, J. C., BLACK, S. & MACKLIN, M. G. 2000. Sediment fingerprinting as a tool for interpreting long-term river activity: the Voidomatis Basin, north-west Greece. *In*: FOSTER, I. D. L. (ed.) *Tracers in Geomorphology*. Wiley, Chichester, 473–503.

HE, Q. & OWENS, P. 1995. Determination of suspended sediment provenance using caesium-137, unsupported lead-210 and radium-226: a numerical mixing model approach. *In*: FOSTER, I. D. L., GURNELL, A. M. & WEBB, B. W. (eds) *Sediment and Water Quality in River Catchments*. Wiley, Chichester, 207–227.

KELLEY, D. W. & NATER, E. A. 2000. Source apportionment of lake sediments to watersheds in an Upper Mississippi basin using chemical mass balance method. *Catena*, **41**, 277–292.

KRAUSE, A. K., FRANKS, S. W., KALMA, J. D., LOUGHRAN, R. J. & ROWAN, J. S. 2003. Multi-parameter fingerprinting of sediment deposition in a small gullied catchment in SE Australia. *Catena*, **53**, 327–348.

LEES, J.A. 1994. *Modelling the magnetic properties of natural and environmental materials*. PhD thesis, University of Coventry.

LEES, J.A. 1997. Mineral magnetic properties of mixtures of environmental and synthetic materials: linear additivity and interaction effects. *Geophysical Journal International*, **131**, 335–346.

NOVOTNY, V. 1980. Delivery of suspended sediment and pollutants from nonpoint sources during overland flow. *Water Resources Bulletin*, **16**, 1057–1065.

ROWAN, J. S., GOODWILL, P. & FRANKS, S. W. 2000. Uncertainty estimation in fingerprinting suspended sediment sources. *In*: FOSTER, I. D. L. (ed.) *Tracers in Geomorphology*. Wiley, Chichester, 279–291.

PEART, M. R. & WALLING, D. E. 1986. Fingerprinting sediment sources: the example of a small drainage basin in Devon, UK. *In*: HADLEY, R. F. (ed.) *Drainage Basin Sediment Delivery*. International Association for Hydrological Sciences Publications, **159**, 41–55.

PEART, M. R. & WALLING, D. E. 1988. Techniques for establishing suspended sediment sources in two drainage basins in the UK: a comparative assessment. *In*: BARDAS, M. P. & WALLING, D. E. (eds) *Sediment Budgets*. International Association for Hydrological Sciences Publications, **174**, 269–279.

SLATTERY, M. C., BURT, T. P. & WALDEN, J. 1995. The application of mineral magnetic measurements to quantify within-storm variations in suspended sediment sources. *In*: LEIBUNDGUT, C. (ed.) *Tracer Technologies for Hydrological Systems*. International Association for Hydrological Sciences Publications, **229**, IAHS Press, Wallingford, 143–151.

SMALL, I. F., ROWAN, J. S. & FRANKS, S. W. 2002. Quantitative sediment fingerprinting using a Bayesian uncertainty estimation framework. *In*: DYER, F. J., THOMS, M. C. & OLLEY, J. M. (eds) *The Structure, Function and Management Implications of Fluvial Sedimentary Systems*. International Association for Hydrological Sciences Publications, **276**, 433–442.

SUGGS, J. A., BEAM, E. W. *ET AL*. 2002. Guidelines and resources for conducting and environmental crime investigation in the United States. *Environmental Forensics*, **3**, 91–113.

WALDEN, J., SLATTERY, M. C. & BURT, T. P. 1997. Use of mineral magnetic measurements to fingerprint suspended sediment sources: approaches and techniques for data analysis. *Journal of Hydrology*, **202**, 353–372.

WALLBRINK, P. J. & MURRAY, A. S. 1993. Use of fallout radionuclides as indicators of erosion processes. *Hydrological Processes*, **7**, 297–304.

WALLING, D. E., WOODWARD, J. C. & NICHOLAS, A. P. 1993. A multi-parameter approach to fingerprinting suspended sediment sources. *In*: PERTERS, N. E., HOEHN, E., LEIBUNDGUT, C., TASE, N. & WALLING, D. E. (eds) *Tracers in Hydrology*. International Association for Hydrological Sciences Publications, **215**, 329–337.

YU, L. & OLDFIELD, F. 1989. A multi-variate mixing model for identifying sediment source from magnetic measurements. *Quaternary Research*, **32**, 168–181.

YU, L. & OLDFIELD, F. 1993. Quantitative sediment source ascription using magnetic measurements in a reservoir catchment system near Nijar, S.E Spain. *Earth Surface Processes and Landforms*, **18**, 441–451.

Isotope and trace element analysis of human teeth and bones for forensic purposes

KENNETH PYE[1,2]

[1]*Kenneth Pye Associates Ltd, Crowthorne Enterprise Centre, Crowthorne Business Estate, Old Wokingham Road, Crowthorne RG45 6AW, UK (e-mail: k.pye@kpal.co.uk)*
[2]*Department of Geology, Royal Holloway, University of London, Egham Hill, Egham TW20 0EX UK*

Abstract: Isotopic and elemental concentration data can be extremely useful in the identification of human remains. Archaeological, ecological and forensic investigations to date have primarily made use of $^{87}Sr/^{86}Sr$, $^{143}Nd/^{144}Nd$, $^{18}O/^{16}O$ and trace element data obtained from analysis of carbonate-hydroxyapatite in bones and teeth, and/or $^{12}C/^{13}C$, $^{14}N/^{15}N$, $^{18}O/^{16}O$ and $^{35}S/^{37}S$ ratios in bone collagen. However, a wide range of other chemical parameters are potentially useful for intersample comparison and environmental characterization, and increasing attention is being given to hair, nail and skin tissues ,which provide dietary and environmental information over shorter time periods than bones and teeth. This paper reviews some of the principles which underlie such work and the current position with regard to modern forensic applications.

Recent years have seen increasing interest in the use of isotopic and trace element analysis methods to determine the geographical origin of human remains and to assist in human identification, especially where DNA analysis and other standard procedures such as dental records and fingerprints have failed or cannot be applied. In the State of New South Wales, Australia, alone, over 200 unidentified bodies per annum are discovered (Gulson *et al.* 1997), and worldwide the figure runs into thousands. Teeth and bones have been of primary interest, although increasing attention is being given to other human tissues such as hair, nail and skin, which provide shorter time scale information. Interest is also growing in the 'fingerprinting' of a range of animal tissues, including ivory and rhino horn, to combat illegal trade and poaching (e.g. van der Merwe *et al.* 1990; Vogel *et al.* 1990), to identify migration patterns in animal and bird populations (e.g. Chamberlain *et al.* 1997), and to establish chemical 'traceability' criteria for a range of foodstuffs and drugs (Ehleringer *et al.* 1999, 2000; Heaton *et al.* 2003; Palhol *et al.* 2003).

Two broad groups of techniques are being used in such studies, the first focusing on isotopic ratios and the second on elemental concentrations and ratios. These techniques have previously been used in a variety of contexts by geoscientists and others, including archaeologists and ecologists concerned with the origins, migration and dietary characteristics of humans and animals. Consequently there is a wealth of background environmental and biological proxy information which is of considerable potential use to the forensic investigator. However, to date only a relatively small number of studies have been specifically concerned with modern populations and forensic applications.

The purpose of this paper is to provide an overview of the principles behind such work and to illustrate some of the opportunities and current limitations of the approach.

Potential information provided by different human tissues

Teeth

Humans have two dentitions (sets of teeth) during their lifetimes, one during childhood, called the *primary* or *deciduous* dentition, and one during adulthood, referred to as the *permanent* dentition. The primary dentition starts to form in the womb and the complete primary dentition of 20 teeth is normally present in a child between the ages of 2 and 6 years. The complete dentition consists of four incisors, two canines and four molars in both of the upper (*maxillary*) and lower (*mandibular*) jaws. Formation of the crown (including the enamel) is normally completed within the first year, while formation of the roots is normally completed by the age of 3 years (Hillson 1997; Woelfel & Scheid 2002; Table 1).

Between the ages of about 6 and 14 years the deciduous teeth are progressively lost, being replaced by permanent teeth. The complete permanent dentition consists of 32 teeth, consisting of four incisors, two canines, four premolars and six molars in each jaw. However, partial *anodontia* (teeth missing) is not uncommon, and some adults may have *supernumerary* teeth. With the exception of the

From: PYE, K. & CROFT, D. J. (eds) 2004. *Forensic Geoscience: Principles, Techniques and Applications.* Geological Society, London, Special Publications, **232**, 215–236. © The Geological Society of London, 2004.

Table 1. *Deciduous and secondary tooth formation and emergence. After Woelfel and Scheid (2002)*

	Tooth	Hard tissue formation begins	Crown completed	Emergence	Root completed
Deciduous maxillary teeth	Central incisor	4 mos. *in utero*	4 mon.	7½ mon.	1½ yr.
	Lateral incisor	4½ mon. *in utero*	5 mon.	9 mon.	2 yr.
	Canine	5 mon. *in utero*	9 mon.	18 mon.	3¼ yr.
	First molar	5 mon. *in utero*	6 mon.	14 mon.	2½ yr.
	Second molar	6 mon. *in utero*	11 mon.	24 mon.	3 yr.
Deciduous mandibular teeth	Central incisor	4½ mon. *in utero*	3½ mon.	6 mon.	1½ yr.
	Lateral incisor	4½ mon. *in utero*	4 mon.	7 mon.	1½ yr.
	Canine	5 mon. *in utero*	9 mon.	16 mon.	3 yr.
	First molar	5 mon. *in utero*	5½ mon.	12 mon.	2¼ yr.
	Second molar	6 mon. *in utero*	10 mon.	20 mon.	3 yr.
Permanent maxillary teeth	Central incisor	3–4 mon.	4–5 yr.	7–8 yr.	10 yr.
	Lateral incisor	10–12 mon.	4–5 yr.	8–9 yr.	11 yr.
	Canine	4–5 mon.	6–7 yr.	11–12 yr.	13–15 yr.
	First premolar	1½–1¾ yr.	5–6 yr.	10–11 yr.	12–13 yr.
	Second premolar	2–2¼ yr.	6–7 yr.	10–12 yr.	12–14 yr.
	First molar	birth	2½–3 yr.	6–7 yr.	9–10 yr.
	Second molar	2½–3 yr.	7–8 yr.	12–15 yr.	14–16 yr.
	Third molar	7–9 yr.	12–16 yr.	17–21 yr.	18–25 yr.
Permanent mandibular teeth	Central incisor	3–4 mon.	4–5 yr.	6–7 yr.	9 yr.
	Lateral incisor	3–4 mon.	4–5 yr.	7–8 yr.	10 yr.
	Canine	4–5 mon.	6–7 yr.	9–10 yr.	12–14 yr.
	First premolar	1¾–2 yr.	5–6 yr.	10–12 yr.	12–13 yr.
	Second premolar	2¼–2½ yr.	6–7 yr.	11–12 yr.	13–14 yr.
	First molar	birth	2½–3 yr.	6–7 yr.	9–10 yr.
	Second molar	2½–3 yr.	7–8 yr.	11–13 yr.	14–15 yr.
	Third molar	8–10 yr.	12–16 yr.	17–21 yr.	18–25 yr.

third molars ('wisdom teeth'), all of the permanent teeth are normally emergent by the age of 15 years (Table 1). Formation of the crowns, including enamel, of the permanent teeth except the third molars is normally complete by about the age of 8 years.

Teeth crowns are composed of a relatively thin outer layer of hard, white *enamel* which overlies softer yellowish tissue (*dentine*). Dentine also forms the bulk of the inner part of each root. The centre of the root is occupied by a pulp cavity containing blood vessels and nerves, while the outer part of the root is composed of a layer of *cementum*.

Enamel, dentine and cementum contain varying proportions of carbonate-hydroxy apatite. Carbonate-hydroxyapatite is a form of the mineral apatite $(Ca_{10}F_2(PO_4)_6)$, otherwise known as *dahllite* (McConnell 1973, 1981; Mann 2001), in which carbonate groups replace some of the phosphate and hydroxyl groups (LeGeros, 1969). The overall carbonate content is about 4–5%. Enamel is the hardest substance found in the body and is composed of 95% inorganic carbonate hydroxyapatite, *c*. 4% water and *c*. 1% organic matter. Cementum is composed of *c*. 65% inorganic carbonate-hydroxyapatite, 12% water

and *c*. 23% organic matter, mainly collagen (a protein composed of amino acids), while dentine consists of *c*.70% inorganic carbonate hydroxyapatite, 12% water and 18% organic matter, again mainly collagen (Woelfel & Scheid 2002).

Owing to its greater bond strength, enamel is less prone to environmental exchange and post-depositional diagenesis than dentine or cementum. For this reason, enamel is the material of choice for isotopic and trace element analysis in archaeological provenance and environmental reconstruction studies (e.g. Koch *et al.* 1997, 1999; Budd *et al.* 2000*b*, 2001; Montgomery *et al.* 2000, 2003). However, in the context of modern forensic investigations, where diagenesis is much less likely to be a serious problem in most cases, analyses performed on combined enamel and cleaned, primary dentine may often be perfectly satisfactory. However, where older forensic material or reactive soil/water environments are involved, it is preferable to analyse only enamel.

Because virtually all enamel is formed by the age of 15 years (Table 1), it potentially provides environmental information relating to the early years of life (from a few months before birth until the age of *c*. 15

years). Through analysis of different types of teeth, or by micro-sampling different parts of individual teeth, possible changes in diet and/or living environment on an annual or seasonal basis within this time period may be resolved (Fricke *et al.* 1998; Jones *et al.* 1999).

Bones

The adult human skeleton contains more than 200 separate bones which can be divided into three broad groups in terms of their basic shape: (1) long bones, which are found in the limbs and take the form of hollow tubes closed at both ends; (2) flat bones, which take the form of plates, for example those making up the skull vault; and (3) irregular bones, which include the vertebrae and bones at the base of the skull (Hancox 1972; White 1991; Mays 1998). Bone is composed of *c.* 70% mineral matter and 30% organic matter by dry weight, although there are considerable variations between different bones. Most of the organic matter consists of collagen. The mineral matter, which is mostly carbonate-hydroxyapatite, occurs as crystallites with a matrix of collagen fibres, and helps to give bones their rigidity. After death the organic material degrades, and the bones tend to become brittle with age. However, collagen can sometimes be well preserved, even in old archaeological bone.

Two different types of bone can be distinguished on the basis of gross structure: cortical bone and trabecular bone. Cortical bone is dense, and forms the outer layers of the bones, while trabecular bone is less dense, with a honeycomb-type structure, and forms much of the interior of the bones (White 1991). For chemical analytical work, dense cortical bone material from the shafts of the long bones is often preferred since it is least susceptible (although not immune) to environmental exchange and post-mortem diagenetic processes (Sillen 1989). However, a variety of other bones, including rib, ulna and clavicle, have also yielded useful results (e.g. Hoogewerff *et al.* 2001).

As noted above, many archaeological bones have been found to have experienced significant isotopic and other chemical changes as a result of weathering, biological degradation, environmental exchange and diagenesis which can affect both the inorganic and organic components (Sillen 1989; Price *et al.* 1992; Hedges 2002). The possible effects of these processes should also be considered in forensic investigations of more modern skeletal material whenever bones have been exposed to surface weathering, aquatic immersion or burial for any significant period (Pate *et al.* 1989). Detailed microscopic examination of the bone material will often aid in the assessment of

likely post-mortem chemical changes (Price 1989; Schultz 1997).

Initial development of the bones starts in the womb and continues throughout childhood and adolescence into adulthood. Different bones grow at different rates and in different ways, but most physical development is complete by the late teens. However, bone is remodelled by dissolution and reprecipitation of hydroxyapatite throughout life, the rate of turn-over depending on the type of bone and specific part of the skeleton involved (Jowsey 1971). In general, dense cortical bone has a slower turn-over than trabecular bone (Simmons & Grynaps 1989; Teitelbaum 2000). Long bones such as the femur and tibia, which have a high proportion of cortical bone, re-form over periods of decades, while bones with a high proportion of trabecular tissue, such as the ribs, have a turn-over rate of only a few years (Parfitt 1983; Simmons *et al.* 1991; Hill 1998). In the case of a normal 30-year-old male, dense femoral shaft bone would be expected to retain about 80% of the original bone material after 10 years, whereas trabecular bone-rich tissue, such as that forming the iliac crest, would be expected to retain only about 30% of the original bone material after the same time period (Price *et al.* 2000). Isotopic analysis of bulk femoral shaft tissue (or selected cortical bone) therefore provides an average measure over a longer period than does analysis of rib or similar tissue. Analysis of material taken from different bones, or different parts of bone, therefore provides the potential to provide greater time resolution of diet and possible geographical migration during later life (Sealy *et al.* 1995).

With increasing age, production of new bone material slows and may even cease, leading to possible loss of bone mass and size, especially in individuals affected by disease (Simmons *et al.* 1991). Data from older persons therefore needs to be interpreted carefully. In the case of children and adolescents, bone growth occurs rapidly, often episodically, and it may be possible to obtain time series data by profile sampling using a dentist's drill, by comparing samples from the unfused heads of the long bones with cortical shaft material, or by attempting density separation of older (more dense) and younger (less dense) bone after physical disaggregation of the bone material.

Hair, skin and nails

Since human nail and hair tissue grows relatively quickly, it offers the potential for providing information about diet and environmental exposure on time scales ranging from a few days to a few years (Nakamura *et al.* 1982; Katzenberg & Krouse 1989; Macko *et al.* 1999; O'Connell & Hedges 1999; O'Connell *et al.* 2001). Given that human hair typically grows at an average rate of *c*. 0.35 mm per day (although with considerable variation), generally 2–3 cm of a single hair is required for a single stable isotope analysis (equating to an average time period of *c*. 10 weeks). In order to obtain weekly time series data, a minimum of 10–12 human hairs is normally required (Ehleringer & Cerling 2003). Beard shavings collected every day may, in some cases, yield sufficient material for analysis.

Hair is composed of keratin, a protein composed mainly of the elements hydrogen, carbon, nitrogen, oxygen and sulphur which are derived predominantly from food and water in the diet (and to a lesser extent from inhaled air or moisture, or from surface contact). Trace elements are also present in varying concentrations, reflecting partly those in the diet and partly those from other forms of environmental exposure, including additives which may be deliberately or inadvertently applied to the skin and hair (e.g. shampoo, soap and other cleansers). The presence of dirt, oils, gels and sprays can significantly alter both the isotopic ratios and the trace element profiles obtained from human hair, and appropriate preparatory procedures need to be conducted before analysis is undertaken.

In principle, a wide range of isotopic ratios and trace element concentrations can be determined in hair and nail, but work to date has focused mainly on the stable isotopic ratios of oxygen, hydrogen, carbon, nitrogen and sulphur, and on concentrations of potentially toxic metals.

Hairs and similar materials obtained from cadavers are potentially prone to significant biodegradation (DeGaetano *et al.* 1992; Rowe 1997), and the possible effects of such processes on suitability for isotopic and trace element characterization in a forensic context should be carefully evaluated prior to analysis.

Isotope analysis

Many chemical elements occur as different types of atoms, or *isotopes*, which have the same number of protons but different numbers of neutrons in the nucleus. Some isotopes, known as *radioactive* isotopes, are unstable and undergo radioactive decay. Others, which do not undergo radioactive decay, are termed *stable* isotopes. The daughter products formed by radioactive decay of a parent radioactive isotope may be either stable or unstable and are termed *radiogenic* isotopes (Faure 1986; Dickin 1995).

Large variations exist in relative isotopic abundances between different reservoirs on the Earth. These variations are the result of two main factors: (1) initial abundance variations in the reservoir at the time of formation, and (2) the passage of time, which leads progressively to the increased abundance of radiogenic isotopes in the reservoir. Further changes in isotopic ratios occur as a result of weathering, transport and redeposition of primary rock material as sediments, which are usually mixtures of particles from different sources, and sometimes also of chemical precipitates. Once formed, sediments act as a distinct type of reservoir in which further radioactive decay takes place. Additional isotopic changes in near-surface sediments may arise due to bombardment by cosmic rays, post-depositional diagenesis, evaporation, and biological processes including photosynthesis, animal metabolism and microbial degradation.

In terms of geographical provenance studies of plants, animals and humans, the isotopes of elements that do not undergo changes in relative abundance (*fractionation*) as a result of biological processes are of particular interest because they offer the prospect of linking bone, teeth or other tissues to an area in which an organism lived, or at least derived the bulk of its diet. The radiogenic isotopes of strontium, neodymium and lead behave in this way and consequently have been extensively used as tracers. Isotopic ratios of elements which display ready isotope fractionation due to low temperature biotic and abiotic processes, such as carbon, oxygen, hydrogen, nitrogen and sulphur, may also be useful in provenance studies since the fractionation process is related to environmental variables, including temperature and evaporation, and the type of photosynthetic pathway used by different plant types. These stable light elements can provide useful indicators of diet, including drinking water, and of climate (Ambrose 1991; van der Merwe 1992). Radioactive isotopes, which can be either naturally occurring or fission products arising from human-induced nuclear reactions, may also provide useful information about geographical location and diet. For example, concentrations and isotopic ratios of uranium, thorium, lead, caesium, strontium, polonium, plutonium, and americium, can provide useful information about proximity to nuclear facilities, military installations, waste dumps and other areas of contamination (Popplewell *et al.* 1988; Noshkin *et al.* 1994; O'Donnell *et al.* 1997).

A further use of isotopes in relation to provenance studies involves radiometric dating. The rate of decay of a radioactive isotope to daughter products is a predictable statistical process, and measurement of the relative abundances of parent to daughter isotopes provides a means of dating. Several different radioactive decay series can be used to date bone, notably ^{14}C dating for older archaeological bone and ^{90}Sr, [210]Po and ^{210}Pb for more recent human remains (MacLaughlin-Black *et al.* 1992; Swift 1998; Neis *et al.* 1999; Swift *et al.* 2001). Other methods of radiometric dating, such as K/Ar, [40]Ar/[39]Ar, Rb/Sr, Nd/Sm and Th/Pb can also be useful in dating mineral particles such as micas, feldspars and zircon grains contained within the intestines, stomach and lungs of human beings. If the age of the grains is known, it may be possible to discriminate between alternative source areas of similar lithology but different geological age. For example, Muller *et al.* (2003) used $^{40}Ar/^{39}Ar$ dating to obtain ages for $100-400\,\mu$m-sized white micas in the intestinal contents of the Alpine Iceman. These micas are believed to have been ingested as a result of grinding of cereal or from drinking water containing suspended sediment. The age distribution of the micas suggested derivation from an area of gneiss lithology and excluded possible alternative source areas composed of geologically younger volcanics and phyllites (Muller *et al.* 2003).

Neodymium isotopes

Neodymium (atomic number 60) has seven naturally occurring stable isotopes (^{142}Nd, ^{143}Nd, ^{144}Nd, ^{145}Nd, ^{146}Nd, ^{148}Nd and ^{150}Nd) and several radioactive isotopes. ^{142}Nd and ^{143}Nd are radiogenic, and ^{143}Nd, which is formed by radioactive decay of ^{147}Sm, is of particular importance in Nd dating and tracer studies (DePaolo 1988).

The isotopic compositions of Nd and Sr are to some extent correlated and the two have been extensively used together in evolution and provenance studies of igneous and sedimentary rocks, sediments and dusts (Faure 1986; Grousset *et al.* 1988, 1992, 1998). Most studies have used the $^{143}Nd/^{144}Nd$ ratio plotted against $^{87}Sr/^{86}Sr$. Values of the $^{143}Nd/^{144}Nd$ ratio range from *c.* 0.5102 for some older Archaean continental crust rocks to *c.* 0.5133 for some of the youngest mid-ocean ridge basalts. The relationship between $^{143}Nd/^{144}Nd$ and rock age is not simple, however, owing to the fact that the isotopic composition of rocks at the time of their formation reflects the evolutionary history of the parent magma, and isotopic ratios can be significantly influenced by later tectonic and metamorphic events.

In terms of $^{143}Nd/^{144}Nd$ ratio values, Precambrian granitic-type rocks and metasediments of the continental crust generally have ratio values of 0.51285 or lower, older Phanerozoic rocks have ratios ranging from 0.51265 to 0.51285, and younger Phanerozoic rocks have values ranging from 0.51265 to 0.51325 (O'Nions *et al.* 1983; Faure 1986).

In the case of sedimentary rocks, the Nd isotope ratio depends not only on the decay of ^{147}Sm since deposition, but also on the ages of the detrital mineral particles and any chemical precipitates (e.g. formed from seawater). Those with little or no marine component generally all have $^{143}Nd/^{144}Nd$ ratio values of 0.51285 or higher. The $^{143}Nd/^{144}Nd$ ratios of seawater vary at the present day between different ocean basins

and have also varied over geological time (Piepgras & Wasserburg 1980; Miller & O'Nions 1985; Shaw & Wasserburg 1985). Modern seawater $^{143}Nd/^{144}Nd$ values range from 0.51190 to 0.51260 (McCulloch & Wasserburg 1978; O'Nions *et al.* 1978; Faure 1986). In general, the values in individual ocean basins reflect the composition and age of the rocks on the surrounding land masses. Consequently, ocean water $^{143}Nd/^{144}Nd$ values in the Atlantic are lower than those in the Indian Ocean and Pacific.

Chemical sediments, such as manganese nodules, phosphates and pure limestones, reflect the isotopic composition of the water from which they were precipitated (O'Nions *et al.* 1978; Shaw & Wasserburg 1985). The isotopic ratios of sediments and sedimentary rocks which are mixtures of detrital mineral matter and chemical or biochemical precipitates have intermediate bulk values, although the individual components may have quite markedly different values.

Many soils are composed mainly of weathered material inherited from the underlying and adjoining bedrock, and their main source of rare earth elements is weathering of accessory minerals derived from the parent material (Harlavan & Erel 2002). However, some soils are largely or wholly developed on allochthonous sediments which bear little or no compositional relationship to the local bedrock. Good examples include soils developed on glacial tills or outwash, and loess soils which may be formed of dust carried thousands of kilometres from the dust source (Pye 1987). Many soils (and also marine sediments) contain a significant airborne component derived either from natural or anthropogenic sources (e.g. Grousset *et al.* 1988, 1998; Borg & Banner, 1996).

Although there is a significant amount of data relating to surface sediments and soils, there are far fewer published $^{143}Nd/^{144}Nd$ values for surface terrestrial waters, vegetation, animal and human tissues. Studies of streams and rivers have mostly focused on the suspended sediment phase rather than the dissolved phase (e.g. Martin & McCulloch 1999), although there are exceptions (e.g. Aubert *et al.* 2002).

Isotope ratios are strictly only directly comparable if the samples are of the same age. The epsilon notation (DePaolo & Wasserburg 1976) provides a normalizing parameter for rocks of different age. It is a measure of the difference between the $^{143}Nd/^{144}Nd$ of a sample and a reference value, taken to be the $^{143}Nd/^{144}Nd$ ratio of CHUR (*ch*ondritic *u*niform *r*eservoir). For individual rocks at the present day, the epsilon neodymium value $\varepsilon_{Nd}(0)$ can be calculated from:

$$\varepsilon_{Nd}(0) = \left[\frac{(^{143}Nd/^{144}Nd)_{sample\ today}}{(^{143}Nd/^{144}Nd)_{CHUR\ today}} - 1 \right] \times 10^4 \quad (1)$$

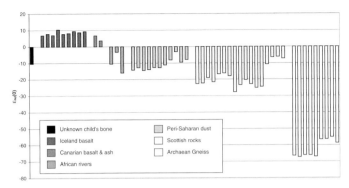

Fig. 1. $\varepsilon_{Nd}(0)$ values for some example rock and sediment types from different areas of the world, compared with bone (humerus) from an unidentified child's body found in the UK.

where $(^{143}Nd/^{144}Nd)_{CHUR\,today}$ is taken to be 0.512638 (Goldstein *et al.* 1984).

Values of $\varepsilon_{Nd}(0)$, which are easily compared, generally range from *c.* +11 for some very recent volcanic rocks to *c.* −68 for some of the oldest Archaean gneissic rocks (e.g. Figure 1).

While the $^{143}Nd/^{144}Nd$ ratios (and $\varepsilon_{Nd}(0)$ values) of rocks, sediments, and soils are not location specific, comparison with values obtained from bone or similar material can assist in eliminating potential source areas that have values which are either much too high or too low. An exact match between 'bedrock geology' and bone isotope values should not be expected, since, as discussed above, many soils contain material which is not derived from the underlying bedrock, the 'bioavailable' sources of various isotopes may not reflect the bulk soil or bedrock in terms of isotopic ratio, the dietary intake of most invertebrates is derived from a wider geographical area rather than a single 'spot' location, and the relative roles of diet and other environmental sources in determining neodynium uptake into the body remain a matter of some uncertainty.

A significant number of Nd isotope analyses have been performed in relation to illegal trading in ivory, rhino horn and similar materials, but most of the data are not published or widely available. High-quality data for human and animal teeth are few and far between, owing to the very low concentrations (usually much less than 1 p.p.m.) of neodymium in teeth and the small amount of material usually available for analysis. Consequently, most forensic attention has focused mainly on bone material.

Strontium isotopes

Strontium (atomic number 38) has four naturally occurring stable isotopes (^{84}Sr, ^{86}Sr, ^{87}Sr and ^{88}Sr). All except ^{87}Sr are non-radiogenic. ^{87}Sr is produced by β-decay of ^{87}Rb, which has a half-life of 48.8×10^9 years. There are also several radioactive isotopes of strontium, of which ^{90}Sr, which is formed as a nuclear fission product, is the most significant from a human point of view.

Spatial variations in the initial distribution of ^{87}Rb, combined with its decay to radiogenic ^{87}Sr, has produced significant variations in the abundance of ^{87}Sr at the present day. ^{87}Sr abundances are normally expressed relative to those of non-radiogenic ^{86}Sr. Owing to the long half-life of ^{87}Sr, changes in the $^{87}Sr/^{86}Sr$ are slight over relatively short geological time periods. However, measurements of the abundances of ^{87}Sr and ^{86}Sr can be made with great precision, and by comparison with quoted values of analytical error (which are of the order of ±0.00001–00003 or better) the differences in Sr isotope compositions from different parts of the Earth are large (Faure & Powell 1972; Faure 1986). As with neodymium, Sr isotopes do not undergo fractionation as a result of low temperature biological processes, making them useful as a biological and environmental tracer (Aberg 1996; Capo *et al.* 1998).

Some authors have preferred to use a simplified method of comparing $^{87}Sr/^{86}Sr$ ratios using a notation, referred to as the epsilon notation, which is analogous to the $\varepsilon_{Nd}(0)$ notation. The ε_{Sr} notation can be calculated using the equation (Beard & Johnson 2000):

$$\varepsilon_{Sr} = \left[\frac{(^{87}Sr/^{86}Sr)_{measured}}{(^{87}Sr/^{86}Sr)_{bulk\,Earth}} - 1\right] \times 10^4 \qquad (2)$$

where $(^{87}Sr/^{86}Sr)_{bulk\,Earth}$ is taken to be 0.7045.

However, opinion amongst geochemists is divided as to whether it is appropriate to use the epsilon notation in the context of strontium, since the $^{87}Sr/^{86}Sr$ ratio of the bulk Earth is not known pre-

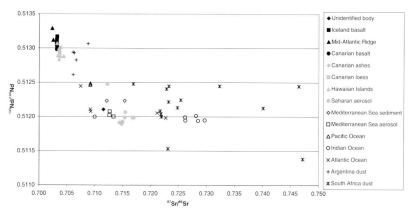

Fig. 2. Plot of $^{143}Nd/^{144}Nd$ v. $^{87}Sr/^{86}Sr$ for some example rock and sediment types, compared with a sample of bone from an unidentified body found in the UK.

cisely, and a majority of workers prefer to quote only the measured $^{87}Sr/^{86}Sr$ isotopic ratio values, often normalized to a reference standard.

There are a large number of published $^{87}Sr/^{86}Sr$ isotope ratio values relating to rocks, sediments and soils, although the data are scattered in the literature. Most of the data relate to bulk rock or sediment samples, although some relate to separated mineral fractions, particle size fractions, or leachate samples.

Young oceanic basalts and young volcanic arc-type rocks typically have $^{87}Sr/^{86}Sr$ ratios of between 0.7022 and 0.7060. Values for upper continental crust rocks and terrigenous sediments typically range from c. 0.710 to 0.725, while values for older continental crust rocks range up to 0.768.

The present-day isotopic composition of seawater varies only between 0.709241 ±32 and 0.709211 ±37, reflecting the fact that the average residence time of strontium in the oceans is about 4 Ma (Elderfield 1986). Seawater Sr isotope ratios have varied significantly over geological time, and since the Cretaceous have increased gradually from about 0.7077 to the present value. These changes are known in great detail, and Sr isotopes provide a valuable stratigraphic correlation tool for rocks containing marine carbonate (McArthur et al. 2001).

Sr isotope ratios in modern sediments and dusts vary significantly, reflecting the composition of the source material from which they are derived (e.g. Fig. 2). Similarly, Sr isotope ratios in modern soils vary considerably, reflecting both the underlying geology and atmospheric additions. By way of example, the range of bulk soil $^{87}Sr/^{86}Sr$ ratio values found in some modern Nigerian soils is shown in Table 2. These soils are developed on a range of bedrock types, ranging from Archaean Basement Complex igneous and metamorphics to Quaternary

sediments, and incorporate varying proportions of allochthonous components (Smith & Whalley 1981; McTainsh 1984; Vine 1987).

Owing to the effects of differential mineral weathering, leaching, and surface additions of dust and rainwater, the Sr isotopic composition of soil does not usually correspond exactly with that of the underlying bedrock, even in the absence of allochthonous components, although it is usually influenced significantly by it (Miller et al. 1993; Chadwick et al. 1999; Blum et al. 2000; Price et al. 2002). Similarly, soil and stream water in an area usually differ to a significant degree in terms of Sr isotope ratio and concentration compared with bulk soils and parent materials.

For these reasons, bulk soil and sediment values sometimes do not give a good indication of bioavailable strontium, and it may be more useful to compare the Sr isotope ratios of biological materials and those of different soil leachate samples. Table 3 shows Sr isotope values for acetic acid, nitric acid and hydrochloric acid leachate samples compared with solid residue bomb dissolution and bulk soil values for some of the Nigerian soils listed in Table 2. In many cases, acetic acid leachate samples provide a better indication of bioavailable strontium than bulk soil samples, although much depends on the soil mineralogy (see also Evans & Tatham 2004).

Difficulties may also arise in comparing data for individual rock, sediment, soil and even plant samples with human bone or tooth data owing to the fact that such 'control' samples generally relate only to a very small geographical area, whereas the dietary foodstuffs, including drinking water, from which the Sr isotopes and other constituents of human tissues are largely derived, is drawn from a much wider area (Sillen & Sealy 1995; Sillen et al. 1998). Sediment samples, including river alluvium

Table 2. *Strontium isotope ratios for the bulk <2 mm fractions of some soil samples from Nigeria, developed on different geological parent materials*

Location	Sample	Bedrock age	Bedrock type*	Lab†	$^{87}Sr/^{86}Sr$	2 se	$^{\varepsilon}Sr$	Sr (p.p.m.)
Kaduna	RF1	Precambrian	UB	1	0.733400	0.000010	410.22	23
Kaduna	RF4	Precambrian	UB	1	0.728140	0.000010	335.56	13
Kano	RF6	Precambrian	OG	1	0.727910	0.000010	332.29	10
Kano	RF12	Precambrian	OG	1	0.736950	0.000010	460.61	33
Zaria	RF16	Precambrian	OG	2	0.743706	0.000005	556.51	ND
Jos	RF18	Precambrian	UB	2	0.749771	0.000007	642.60	ND
Bauchi	RF20	Precambrian	OG	1	0.718670	0.000010	201.14	178
Yankari	RF23	Tertiary	TS	1	0.717650	0.000010	186.66	16
Jos	RF24	Precambrian	UB	1	0.728970	0.000020	347.34	50
Jos	RF26	Precambrian	OG	1	0.720650	0.000020	229.24	10
Abuja	RF48	Precambrian	UB	1	0.727320	0.000010	323.92	15
Abuja	RF49	Precambrian	UB	1	0.717650	0.000010	186.66	71
Lokoja	RF53	Precambrian	UB	1	0.718070	0.000010	192.62	37
Lokoja	RF54	Precambrian	UB	1	0.718070	0.000020	192.62	105
Ife	RF58	Precambrian	UM	1	0.721440	0.000020	240.45	23
Oshogbo	RF64	Precambrian	UM	1	0.735650	0.000010	442.16	13
Oshogbo	RF66	Precambrian	UM	1	0.733470	0.000010	411.21	16
Oyo	RF67	Precambrian	UM	1	0.736040	0.000010	447.69	51
Oyo	RF68	Precambrian	UM	1	0.715670	0.000010	158.55	28
Abeokuta	RF69	Precambrian	UB	1	0.727980	0.000010	333.29	34
Abeokuta	RF73	Precambrian	UB	1	0.763010	0.000030	830.52	40
Ilorin	RF76	Precambrian	UB	1	0.736910	0.000010	460.04	28
Ilorin	RF80	Precambrian	UB	1	0.723470	0.000010	269.27	33
Ilorin	RF81	Precambrian	UB	1	0.724540	0.000010	284.46	102
Ogbomosha	RF82	Precambrian	UB	1	0.726210	0.000010	308.16	17
Jebba	RF83	Precambrian	UB	1	0.721950	0.000010	247.69	98
Jebba	RF84	Precambrian	UB	1	0.724660	0.000010	286.16	52
Jebba	RF85	Precambrian	UB	1	0.766990	0.000010	887.01	31
Ilorin	RF86	Precambrian	UB	1	0.735490	0.000010	439.89	9
Ibadan	RF90	Precambrian	UM	2	0.714609	0.000005	143.49	ND
Ibadan	RF92	Precambrian	UM	2	0.705873	0.000007	19.49	ND
Akure	RF97	Precambrian	UB	1	0.746640	0.000010	598.15	59
Akure	RF98	Precambrian	UB	1	0.733060	0.000010	405.39	91
Ifon	RF99	Cretaceous	CS	1	0.721900	0.000010	246.98	13
Benin City	RF101	Quaternary	CP	2	0.713666	0.000006	130.11	ND
Benin City	RF107	Quaternary	CP	2	0.713834	0.000050	132.49	ND
Ore	RF108	Precambrian	UB	2	0.740896	0.000003	516.62	ND
Ijebu-ode	RF109	Cretaceous	CS	2	0.716278	0.000003	167.18	ND
Lagos	RF116	Quaternary	CP	2	0.724060	0.000011	277.64	ND
Enugu	RF117	Cretaceous	CS	2	0.715644	0.000006	158.18	ND
Jander	RF118	Cretaceous	CS	2	0.725302	0.000003	295.27	ND
Makurdi	RF119	Cretaceous	CS	2	0.733333	0.000003	409.27	ND

* Parent material codes: CD, Coastal plain sands and Chad Formation; CS, Cretaceous sediments; CG, older granite; TS, Tertiary sediments; UB, undifferentiated basement complex; UM, undifferentiated metasediments.

† Laboratory codes: 1, Carleton University, Canada; 2, NIGL, Keyworth, UK.

and airborne dust deposits, which represent 'average' values for a wider catchment or regional area, are often a more useful comparator than individual surface soil samples in this respect.

Although data relating to rocks, sediments and bulk soils can provide a useful preliminary basis for evaluation, detailed comparison between a questioned sample and human tissue or other suitable biological material, such as animal bone or teeth, is required, and for modern forensic purposes there is no real substitute for comparison with data relating to modern humans whose life histories and dietary/cultural behaviour are well known.

There is a considerable amount of published and unpublished data relating to $^{87}Sr/^{86}Sr$ ratios in modern and fossil vegetation and animal tissues, and in archaeological bone, teeth and similar materials (e.g. Sealy *et al.* 1991; Price *et al.* 1994a, b, 2000,

Table 3. *Strontium isotope values for (**a**) bulk samples and (**b**) sequential leachate extractions from some Nigerian soils (see Table 2 for locations)*

Sample	$^{87}Sr/^{86}Sr$	2 SE±	$^{87}Sr/^{86}Sr$ normalized to 0.710240
Bulk Samples			
RF 90s	0.714582	5	0.714567
RF 90 (bomb dissolution)	0.714582	5	0.714567
RF 92s	0.705873	7	0.705858
RF 92 (bomb dissolution)	0.705829	3	0.705814
RF 101s	0.713666	6	0.713651
RF 107s	0.713834	5	0.713819

Sample	$^{87}Sr/^{86}Sr$	2 SE±	$^{87}Sr/^{86}Sr$ normalized to 0.710240
Sequential leachate extractions			
Acetic acid leach			
RF 90s	0.716516	3	0.716505
RF 92s	0.707379	4	0.707368
RF 101s	0.711402	4	0.711391
RF 107s	0.711464	4	0.711453
1M nitric acid leach			
RF 90s	0.720875	4	0.720890
RF 92s	0.707367	5	0.707382
RF 101s	0.713437	5	0.713452
RF 107s	0.716126	8	0.716141
1M hydrochloric acid leach			
RF 90s	0.714330	4	0.714345
RF 92s	0.706912	5	0.706927
RF 101s	0.711641	5	0.711656
RF 107s	0.715363	4	0.715378
Residue bomb dissolution HF1HNO3			
RF 90s	0.714363	4	0.714352
RF 92s	0.705767	4	0.705756
RF 101s	0.714144	5	0.714133
RF 107s	0.714616	3	0.714605

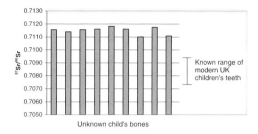

Fig. 3. $^{87}Sr/^{86}Sr$ ratio values obtained from several different bulk bone samples from the same unidentified child's body, compared with the known range from modern UK childrens' teeth. The lower values relate to the latest-formed bone material.

2002; Aberg *et al.* 1998; Budd *et al.* 2000*b*, 2003; Hoogewerff *et al.* 2001; Montgomery *et al.* 2003; Evans & Tatham 2004). However, far fewer data are readily available relating to modern bones and teeth. Most recent studies have generally aimed to identify and reassociate separated human body parts in mass graves or battlefield situations (e.g. Fulton *et al.* 1986; Beard & Johnson 2000), or relate to investigations of immigrant populations (e.g. Gulson *et al.* 1997), and very few systematic studies have been conducted to assess the variation which exists in radiogenic and stable isotope ratios in modern populations.

However, in the context of specific investigations, useful comparative data can be obtained from the bones, teeth or other tissues of small animals, for example mice, rats, rabbits or snails, which live within reasonably well-defined geographical areas but whose feeding areas are wider than a single 'spot' sampling location (Price *et al.* 2002). Farmed livestock, such as sheep, goats and cattle, can also be useful in this respect although their geographical range and diet are often more strictly regulated, and care needs to be taken to avoid domestic animals that are fed wholly or largely on imported feedstuffs.

Sr isotope values, like Nd isotope ratios, are not unique to any given area or individual but can nonetheless be useful as a basis on which to assess the likelihood of a particular individual originating from, or having spent significant time in, a given geographical area or areas. For example, Figure 3 shows $^{87}Sr/^{86}Sr$ values obtained from several different bones relating to the body of an unidentified male child found in the UK, compared with the known range of isotopic ratios for modern UK children's teeth. The values for the unidentified individual lie well outside this range, suggesting a low probability that he had spent much of his life in the UK. However, it is rarely possible to draw conclusions on the basis of a single isotopic ratio, and several different isotopes should be considered in combination. Where possible, data for teeth and bones should be considered in conjunction in order to address the issue of possible migration (e.g. Fig. 4).

Lead isotopes

Lead (atomic number 82) has four stable isotopes (^{204}Pb, ^{206}Pb, ^{207}Pb and ^{208}Pb) of which all but ^{204}Pb are radiogenic. ^{208}Pb has the greatest natural abundance (52.4%) and ^{204}Pb the lowest (1.4%). There

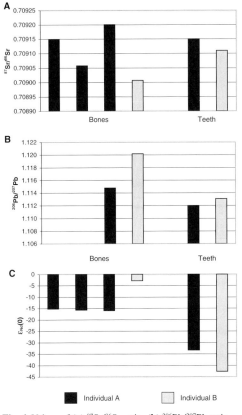

Fig. 4. Values of (**a**) $^{87}Sr/^{86}Sr$ ratio, (**b**) $^{206}Pb/^{207}Pb$ ratio, and (**c**) $\varepsilon_{Nd}(0)$ for bones and teeth taken from two unidentified individuals found at different locations in England.

are also several radioactive isotopes of lead, of which ^{210}Pb is of greatest significance in the environment. The abundance of ^{204}Pb has essentially been constant since the Earth was formed, and this isotope is commonly used as a reference against which the abundances of the radiogenic isotopes are compared. Since the abundances of the latter have increased over geological time, $^{206}Pb/^{204}Pb$, $^{207}Pb/^{204}Pb$ and $^{208}Pb/^{204}Pb$ ratios are generally higher in younger rocks than in older rocks. For example, the $^{206}Pb/^{204}Pb$ ratio in the c. 1700 Ma-old lead-zinc-silver deposit at Broken Hill, New South Wales, Australia, which was a major source of lead added to British and other European petrol until the 1990s has a $^{206}Pb/^{204}Pb$ ratio of c. 16.0, whereas 400–500 Ma-old rocks on the same continent have $^{206}Pb/^{204}Pb$ ratios of 18.1–18.3 (Gulson et al. 1997).

Lead may enter the body from a variety of sources, including food, drinking water and air. Food and drinking water will contain lead from the original source environment but may also be contami-

nated with lead from cooking ware, lead pipes or 'industrial' lead generally dispersed in the environment. The lead in teeth and bones is therefore usually a mixture from different sources, acquired over differing time periods from different sources, beginning in utero (Gulson & Wilson 1994; Gulson 1996).

In Britain, lead has been mined for use in alloys and in association with silver since ancient times, and was extensively traded throughout much of Europe, North Africa and Asia by the Romans. Lead levels in many Romano-British bones and teeth were very high by modern standards, reflecting the widespread use of lead-based cooking ware (Waldron et al. 1976; Whittaker & Stack 1984). In more recent times, however, additional sources of environmental lead have included lead water distribution pipes, airborne contamination resulting from metal smelting, waste incineration and internal combustion engines.

The major source of environmental lead in the mid to late twentieth century has been airborne lead added as an anti-knock agent in petrol. As noted above, in the UK most of this 'petrol' lead originated from the Australian Broken Hill mines. However, since widespread introduction of lead-free petrol in the early 1990s, the importance of this source of lead has declined. Studies of lead concentrations in children's tooth enamel have shown that overall levels in the UK have been declining since the early 1980s (Delves et al. 1982; Alexander et al. 1993; Farmer et al. 1994; Budd et al. 2000b). Correspondingly there have been changes in the lead isotope ratios of teeth and bones over time. For example, Budd et al. (2000b) reported $^{207}Pb/^{206}Pb$ ratios of 0.8279–0.8498 for four Neolithic individuals from the Monkton Up Wimborne area of Dorset. $^{208}Pb/^{206}Pb$ ratios in these (pre-metallurgical expansion) individuals ranged from 2.0391 to 2.0786 and lead concentrations from 0.15 to 0.68 p.p.m. The values obtained show relatively close isotopic agreement with Chalk and Chalk soil leachate values from the areas in which these individuals lived. By comparison, $^{207}Pb/^{206}Pb$ ratios in modern children's teeth from the UK, sampled in the period 1998–2003, range from 0.8740 to 0.8935 and $^{208}Pb/^{206}Pb$ values from 2.108 to 2.141, while $^{206}Pb/^{204}Pb$ ratios range from 17.35 to 17.95 and lead concentrations from 0.16 to 2.10 p.p.m. (Figs 5 & 6).

Gulson et al. (1997) first demonstrated the potential value of lead isotopes for forensic human provenance purposes in a study of the lead isotope composition of immigrants to the Sydney region of Australia. This study showed that the immigrants from eastern and southern Europe had completely different $^{207}Pb/^{206}Pb$ and $^{206}Pb/^{204}Pb$ characteristics compared with native Australian subjects. Lead concentrations in some of the immigrants' teeth from countries such as Bulgaria, Poland and the former Yugoslavia were also much higher. However, it was

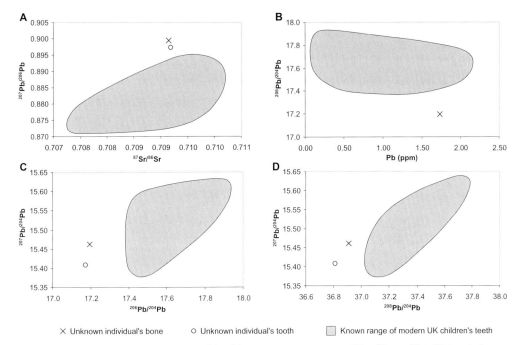

Fig. 5. Plots of (**a**) $^{207}Pb/^{206}Pb$ v. $^{87}Sr/^{86}Sr$, (**b**) $^{206}Pb/^{204}Pb$ v. Pb concentration, (**c**) $^{207}Pb/^{204}Pb$ v. $^{206}Pb/^{204}Pb$ and (**d**) $^{207}Pb/^{204}Pb$ v. $^{208}Pb/^{204}Pb$ showing known range of values for modern UK childrens' teeth enamel and tooth and bone samples from an unidentified murder victim found in southern England.

not possible to discriminate between all of the immigrants' home countries individually on this basis.

Although Pb isotope ratios are not unique to any specific region or country, they can also assist in assessing the likelihood that an unidentified deceased individual is native to the county in which the body was found. For example the unidentified child's body, whose bone Sr isotope ratios are shown in Figure 3, also yielded a $^{206}Pb/^{204}Pb$ ratio of 18.048, which is well outside the known range for modern UK children's teeth. Values of $^{208}Pb/^{204}Pb$, $^{207}Pb/^{204}Pb$, $^{206}Pb/^{207}Pb$, $^{208}Pb/^{207}Pb$ and lead concentration for this unidentified individual also lay well outside the known UK range (Fig. 5 & 6). In the case of two other unidentified individuals for whom selected isotopic data are shown in Figure 4, differences in $^{206}Pb/^{207}Pb$ and other isotopic ratios between teeth and bones suggest that both had migrated some significant distance during their lifetimes.

Oxygen and hydrogen isotopes

Oxygen (atomic number 8) has three stable isotopes (^{16}O, ^{17}O and ^{18}O) whose relative abundances are 99.63%, 0.0375% and 0.1995%, respectively (Faure 1986). Oxygen is the most abundant chemical element in the Earth's crust and, together with

hydrogen, which has two isotopes ^{1}H and ^{2}H (or D, deuterium), comprises the water molecule H_2O. The isotopic ratios of oxygen and hydrogen are highly sensitive to temperature and evaporation, with the result that they have been widely used as climatic and palaeo-climatic indicators. The isotopic compositions of both oxygen and hydrogen are usually reported in terms of $^{0}/_{00}$ (per mil) differences of $^{18}O/^{16}O$ and D/H ratios relative to SMOW (standard mean ocean water; Craig 1961a, b) and sometimes (especially when relating to carbonate rocks) relative to the isotopic composition of carbon dioxide produced from belemnites in the Cretaceous Peedee Formation in South Carolina (values expressed as $\delta^{18}O$ % PDB).

When water evaporates from the surface of the ocean, the water vapour is enriched in ^{16}O and H relative to the source; consequently the water vapour in the atmosphere has $\delta^{18}O$ and δD values which are negative. When raindrops form by condensation of the water vapour, the liquid phase is enriched in ^{18}O and D, so that the composition of the first-formed raindrops is most similar to that of ocean water (Craig 1961b; Dansgaard 1964). As removal of ^{18}O and D continues as a result of further precipitation, the remaining water vapour becomes more isotopically negative. Consequently, as a generalization, the $\delta^{18}O$ and δD values of precipitation become

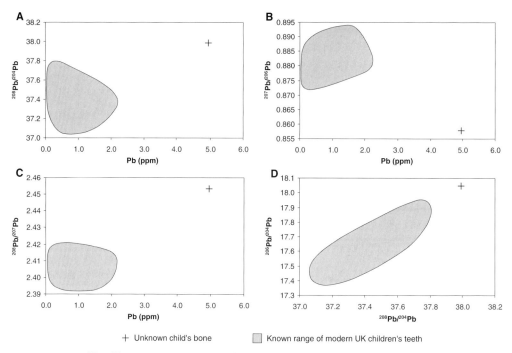

Fig. 6. Plots of (**a**) $^{208}Pb/^{204}Pb$ v. Pb concentration, (**b**) $^{207}Pb/^{206}Pb$ v. Pb concentration, (**c**) $^{208}Pb/^{207}Pb$ v. Pb concentration, and (**d**) $^{206}Pb/^{204}Pb$ v. $^{208}Pb/^{204}Pb$ showing known range for modern childrens' teeth enamel and a bone sample from an unidentified child murder victim found in southern England.

more negative with increasing latitude, altitude and distance from the sea.

Surface waters which have not been significantly influenced by evaporation usually have a strong isotopic similarity with the source precipitation. Shallow groundwaters often also show a fairly close association with precipitation, although deep groundwaters and surface waters fed from artesian sources sometimes show significant differences compared with modern precipitation due to water–rock interaction and the effects of climatic changes.

$\delta^{18}O$ values of meteoric waters range from $c. -40$ to $c. +5.7\%_{00}$ relative to SMOW. δD values of meteoric waters show a larger range, from $c. -200$ to $c. +20\%_{00}$ relative to SMOW (Hoefs, 1987; Rollinson 1993). In low latitude near-oceanic areas, most meteoric waters have $\delta^{18}O$ and δD values fairly close to zero, while values become increasingly negative at higher latitudes, at higher altitudes and towards continental interiors. An exception to this pattern occurs where there are high levels of evaporation, as in arid regions, or in some rainforest areas where high rates of interception loss and evapo-transpiration in the vegetation canopy significantly influence the isotopic composition of waters on the ground. Formation waters (groundwaters) in sedimentary basins show a considerable range in both $\delta^{18}O$ and δD values,

reflecting diverse histories of water–rock interaction, but typically are intermediate between low latitude and high latitude meteoric water extremes.

$\delta^{18}O$ and δD ratios have been extensively used as temperature, precipitation and evaporation indicators in palaeoclimate studies. A wide range of biological materials has been investigated to provide data, including marine foraminifera (Shackleton & Opdyke 1973), landsnails (Lecolle 1985; Goodfriend 1999), rodent teeth (Lindars *et al.* 2001), herbivore teeth (Fricke *et al.* 1998), deer, elephant and other animal bones (Longinelli 1984; Luz *et al.* 1984, 1990; Ayliffe *et al.* 1994).

The O isotope compositions of phosphate and structural carbonate in mammalian tooth enamel and bone apatite have been shown to be linked to that of body water at constant body temperature near 37°C by way of a known isotopic fractionation factor (Luz & Kolodny 1985; Bryant *et al.* 1996). In most circumstances the main source of body water is drinking water, although some water is also absorbed from food and inhaled. Longinelli (1984) found experimentally that the $\delta^{18}O$ of body water varies almost linearly with the mean $\delta^{18}O$ of local meteoric water, and that the relationship is very similar to that between bone phosphate and local meteoric water. However, more recent work has refined these relationships, which have been shown to vary somewhat between vertebrate species.

In the case of humans, a similar strong association between local meteoric water $\delta^{18}O$ composition of bones and teeth might be expected in most archaeological populations, where consumption of imported drink and foodstuffs was relatively limited. Consequently oxygen isotopes have been extensively used in studies of migrant archaeological populations and individuals (e.g. White *et al.* 1998; Budd *et al.* 2001; Hoogewerff *et al.* 2001; Muller *et al.* 2003). However, the results of such studies on fossil tooth and bone potentially may be subject to error due to the effects of diagenesis involving isotopic exchange, and great care needs to be taken in identification and pretreatment of suitable material for analysis (Nelson *et al.* 1986; Koch *et al.* 1997; Sharp *et al.* 2000; Trueman *et al.* 2003).

The potential use of oxygen, hydrogen and other light stable isotopes for the identification and geographical tracing of modern humans was noted by Katzenberg & Krouse (1989). They reported that the $\delta^{18}O$ and δD values of natives of Calgary citizens (measured in urine samples) were not as negative as their drinking water, which has an average δD of $-145\,^{0}/_{00}$ and $\delta^{18}O$ of $-19\,^{0}/_{00}$, respectively. They suggested that one reason for this was that a significant portion of the body water intake originates from vegetables and fruit grown in lower British Columbia or the southwestern USA and Mexico, where the surface water δD and $\delta^{18}O$ values are less negative. This illustrates a widespread complexity in studies of modern human populations, particularly those in the more developed world, where a large proportion of food and drink in the diet is drawn from beyond the immediate geographical area. The 'globalization' of food supply, and the increasing frequency and geographical range of individual travel, which has become much more marked in the second half of the twentieth century, acts to reduce the degree of isotopic 'identity' of regional populations.

For some parts of the world, including the British Isles, relatively good data exist relating to spatial and temporal variations in the $\delta^{18}O$ and δD values of modern precipitation, surface waters and shallow groundwaters (e.g. Darling & Talbot 2003; Darling *et al.* 2003). Surface waters originate mainly from rainfall, from groundwater discharge, or a combination of the two. Recorded $\delta^{18}O$ values for surface flowing waters in the UK range from -8 to $-4\,^{0}/_{00}$, in line with recorded values for precipitation. Values for surface standing waters show a range of -7.7 to $-1.25\,^{0}/_{00}$ and reflect enhanced evaporitic fractionation (Darling *et al.* 2003). $\delta^{18}O$ values for modern groundwaters range from less then $5\,^{0}/_{00}$ in the extreme west of the country to *c.* $8.5\,^{0}/_{00}$ in the Grampian mountains of northeast Scotland.

The relative contribution of surface waters and

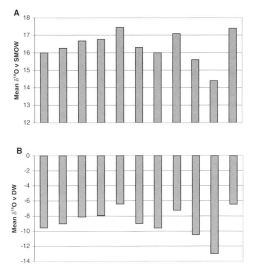

Fig. 7. Values of (**a**) mean $\delta^{18}O$ v. SMOW and (**b**) mean $\delta^{18}O$ DW (calculated 'drinking water') values for a number of modern UK children's teeth enamel samples. 'Drinking water' values were calculated according to the method of Levinson *et al.* (1987). Analysis was performed according to the method of O'Neil *et al.* (1994).

groundwaters to drinking water supplies in the UK varies from region to region. Moreover, in some parts of the country much of the tapwater is brought from some distance away. However, the results of a preliminary study have indicated reasonably close agreement between shallow groundwaters and tapwaters, except for some northern and southwestern localities (Darling *et al.* 2003). Larger differences occur mainly where mountain lakes or reservoirs, which are subject to enhanced evaporation, provide a major source.

However, at present the relationships between oxygen isotope ratios in tapwaters and the teeth and bone of modern populations are poorly known. Preliminary studies in the UK have indicated that there is no simple relationship between the two: values of $\delta^{18}O$ in children's teeth, corrected for drinking water to body water fractionation using the method of Levinson *et al.* (1987), show a wider range of values than that observed in modern tapwaters (Figure 7) and no simple pattern of geographical correlation. Better agreement may exist between surface-water values and archaeological tooth enamel values, although data are limited (e.g. Budd *et al.* 2003).

Carbon isotopes

Carbon (atomic number 6) has three isotopes, two of which are stable (^{12}C and ^{13}C) and one (^{14}C)

radioactive. The ratio of ^{13}C to ^{12}C, expressed by the notation $\delta^{13}C$ in parts $^0/_{00}$ (per mil) relative to the PDB international standard, in human bone and teeth has long been known to be a sensitive indicator of diet. Plants which use the C_3 (Calvin-Benson) photosynthetic pathway discriminate more markedly against ^{13}C during CO_2 fixation than do plants which use the C_4 (Hatch-Slack) photosynthetic pathway. C_3 plants (e.g. wheat, rice, potato, barley, cassava) have isotopically 'light' carbon with $\delta^{13}C$ values of -23 to $-32^0/_{00}$, while C_4 plants (e.g. sugar cane, corn, sorghum and millet) have relatively 'heavier' carbon, with $\delta^{13}C$ values of between -10 and -16 $^0/_{00}$ (Deines 1980; Nakamura et al. 1982). The $^{13}C/^{12}C$ ratios in herbivore (and vegetarian human) tissues may be expected to reflect the proportions of C_3 and C_4 vegetation eaten, allowing for some further fractionation, and in turn the $^{13}C/^{12}C$ ratios of predators should reflect the diets of their main prey. In general, the proportion of C_4 to C_3 grasses decreases with increasing latitude, and consequently there are latitudinal differences in the isotopic composition of hair and milk in grazing animals, a trend also reflected in human populations (Webb et al. 1980; Katzenberg & Krouse 1989).

Information about diet in the short term can be provided by analysis of hair, nails and skin, in the medium term by analysis of bone and tooth collagen, and in the longer term by analysis of bone carbonate-hydroxapatite. Published data relate to both extracted collagen, bulk protein, lipid and carbonate fractions (e.g. Nakamura et al. 1982; Ambrose, 1990; Vogel et al. 1990), and also to 'bulk' bone, tooth, hair and nail samples (e.g. Hoogewerff et al. 2001). Care therefore needs to be taken to ensure comparison of like with like.

Nakamura et al. (1982) compared $\delta^{13}C$ values in human hair from individuals in Chicago, Tokyo and Munich and found values of -16.4, -18.0 and -20.4 respectively, which they suggested were related to variations in diet. Minson et al. (1975) reported latitudinal variations in the isotopic composition of hair and milk in cattle grazing tropical and temperate pastures in Queensland, while Webb et al. (1980) reported similar latitudinal variation in the $\delta^{13}C$ ratios of hair in both vegetarians and omnivores in eastern Australia. Latitudinal $\delta^{13}C$ trends were also reported in the hair of Canadian citizens by Katzenberg & Krouse (1989).

Nitrogen isotopes

Nitrogen (atomic number 7) has two stable isotopes, ^{14}N and ^{15}N, which have relative abundances of 99.64% and 0.36%, respectively. The isotopic composition of nitrogen in a sample is expressed by the $^{15}N/^{14}N$ ratio in the sample relative to the $^{15}N/^{14}N$ ratio in standard N_2 which comprises the atmosphere (Mariotti 1984). The isotopes of nitrogen are fractionated extensively by biological reactions and by some abiotic isotopic exchange mechanisms. $\delta^{15}N$ values of 'total soil nitrogen' range widely, from c. -4.4 to $+17.0^0/_{00}$ (Letolle 1980).

Nitrogen isotopes in human and animal tissues reflect diet, but $\delta^{15}N$ values at any trophic level are about $2.7^0/_{00}$ more positive than those in the diet (Schoeninger & DeNiro 1983, 1984). In terrestrial ecosystems nitrogen-fixing plants such as peas and beans have the lowest $\delta^{15}N$ values, while non-nitrogen-fixing plants, herbivores and carnivores have successively higher values. An association between $\delta^{15}N$ values and climate has also been observed, with higher $\delta^{15}N$ ratios in animals living in drier environments (Vogel et al. 1990).

Studies of elephants and other animals have demonstrated that populations in different parts of southern Africa can be distinguished on the basis of $\delta^{15}N$ values, although with some overlap (van der Merwe et al. 1988; Vogel et al. 1990). Many marine organisms can also be readily distinguished from their terrestrial counterparts on the basis of $\delta^{15}N$ ratios (Schoeninger & De Niro 1984). Numerous studies have also shown that stable isotope ratios of carbon and nitrogen, in combination with C/N ratios, can provide useful information about prehistoric human diet (van der Merwe 1992; Richards & Hedges 2000). However, in the case of modern diets a close association with local food production only applies in some relatively underdeveloped countries, and the diverse nature of most diets in developed societies makes interpretation much more difficult. $\delta^{15}N$ values, may, however, still provide a useful indicator in communities that have a high proportion of dietary marine fish, or vegetables grown on soils heavily influenced by marine spray of seabird guano (natural or used as artificial fertilizer).

Sulphur isotopes

Sulphur (atomic number 16) has four stable isotopes, ^{32}S, ^{33}S, ^{34}S and ^{35}S, whose relative abundances are 95.02%, 0.75%, 4.21% and 0.02%, respectively (Faure 1986). The most widely used isotopic ratio is that of ^{34}S to ^{32}S, expressed as values of $\delta^{34}S$ relative to the $^{34}S/^{32}S$ ratio of troilite (FeS) in the Canyon Diablo iron meteorite standard ($^0/_{00}$ per mil, CDT). S isotopes are very sensitive to microbiological as well as some forms of abiotic fractionation, and have been found to provide useful indicators in a wide range of provenance and environmental process contexts. With regard to human origins and dietary studies, a widespread use of S isotopes has been as an indicator of marine influence. The $\delta^{34}S$ value of modern seawater ranges from approximately $+18.5$ to $21^0/_{00}$, although there has been considerable variation over geological time (Claypool *et al.* 1980). By comparison, freshwater and modern atmospheric sulphur is relatively light, although values of the latter vary considerably, depending on the degree of atmospheric pollution contributed by the burning of fossil fuels and smelting of sulphide ores (Nielsen 1979; Krouse, 1980). Sulphur isotope ratios in foods are related to origin, reflecting relative contributions of sulphur from marine, lithospheric and atmospheric sources. However, S isotope fractionation moving along a food chain is relatively small (Katzenberg & Krouse 1989).

S isotope ratios have so far been used in relatively few archaeological investigations of palaeodiet, human migration and trading patterns, but their potential in modern forensic investigations was highlighted by Katzenberg and Krouse (1989). In a pilot study these authors found that residents of Canberra, Australia had $\delta^{34}S$ values in hair and nails of $+12$–$+15$, while residents of Calgary, Canada, had $\delta^{34}S$ values closer to zero.

Radioactive isotopes

Concentrations of radionuclides, including ^{239}Pu, ^{240}Pu, ^{90}Sr, ^{210}Po, ^{210}Pb, ^{226}Ra and ^{137}Cs, in modern children's teeth have been intensively investigated as part of assessment of radiological hazard (Popplewell *et al.* 1988; Henshaw *et al.* 1994; Yamamoto *et al.* 1994; O'Donnell *et al.* 1997). As noted above, measurement of the concentrations and ratios of such isotopes may provide information about geographical proximity and environmental exposure to nuclear installations, weapons testing, or areas with naturally high geological radioactivity, and may also give indications of post-mortem interval (Swift 1998; Swift *et al.* 2001). However, selection and processing of material for analysis is critical for reliable interpretation, since different bone tissues absorb and retain radionuclides in differing proportions (Hamilton 1971). Plutonium and strontium show a tendency to lodge in bone because their chemistry mimics that of calcium, while cesium, which mimics potassium, spreads through all the body tissues. ^{239}Pu is primarily associated with nuclear fuels, while ^{90}Sr, ^{131}I and ^{137}Cs are fission products associated mainly with nuclear weapons testing. In most parts of the world, radiation exposure overwhelmingly originates from natural sources.

Igneous intrusive rocks such as granites often contain high levels of uranium which can give rise to relatively high levels of radioactive and radiogenic isotopes in the overlying soils, air and surface waters. These are then taken up in local food and drink, or otherwise ingested by the indigenous human population. Of particular importance in this context are products of the ^{238}U to ^{206}Pb decay chain, which include ^{222}Rn and ^{210}Pb.

Major and trace elements

A *major element* can be defined as one which occurs in $>1.0\%$ abundance, a *minor element* one which occurs in abundance of 1.0–0.1% and a *trace element* as one which has abundance of $<0.1\%$ (or <1000 p.p.m.) in a sample (Rollinson 1993). In practice, an element which is a major constituent in one sample may be a minor or trace constituent in another. However, it is conventional when determining geochemical concentrations to identify ten major elements whose abundances are normally reported as wt % oxides, namely the elements Si, Al, Fe, Ti, Na, K, Ca, Ti, Mn and P, and to refer to the remainder as 'trace' elements (Pye & Blott 2004).

Several techniques are available for simultaneous multi-element chemical analysis, including X-ray fluorescence (Fitton 1997), neutron activation (Parry 1997) and various forms of inductively coupled plasma spectrometry (Walsh 1997; Jarvis 1997; Sylvester 2001). Each of these techniques has the capacity to provide a chemical profile, or chemical fingerprint, which can be used to characterize the material in question and to compare it with other samples or reference materials. Comparison may be based on the full suite of elements analysed, or on specific subsets of major elements and/or trace elements, such as the transition metals or rare earth elements.

A variety of trace elements are present in teeth and bones, reflecting diet and other forms of environmental exposure (Priest & van de Vyver 1999). It has long been recognized that concentrations of metals in bone, especially trace elements, can provide useful indicators of diet, nutrition and disease (Mahanti & Barnes 1983; Jaworowski *et al.* 1985; Samuels *et al.* 1989; Baranowska *et al.* 1995).

Considerable attention has focused on the concentrations of lead, combined with lead isotope ratios, in both archaeological and modern teeth, partly in terms of assessing recent levels of lead pollution in the modern environment compared with historical 'background' values (e.g. Samuels *et al.* 1989; Gulson 1996; Aberg *et al.* 1998; Budd *et al.* 1998, 2000*a, b*; Stack 1999). High concentrations of other elements, such as copper, zinc or aluminium, in teeth, bones, hair, nails and excreta may also provide useful information about environmental exposure, diet and geographical origins (e.g. Price 1989; Steenkamp *et al.* 2000; Hoogewerff *et al.* 2001; Onianwa *et al.* 2001). Trace element spider diagrams (e.g. Fig. 8) provide a rapid means of comparing samples and identifying similarities or otherwise

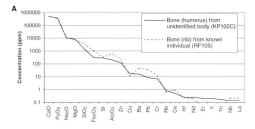

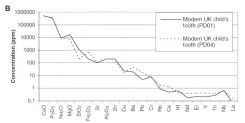

Fig. 8. Spider plots (log scale) comparing (**a**) concentrations of selected elements in bone samples from an unidentified body and an individual from a known location, and (**b**) two modern UK children's teeth enamel samples. Elemental concentrations were determined by ICP-AES and ICP-MS.

between two or more individuals, one of whom may be of unknown origin, and of identifying particular chemical anomalies which may be of diagnostic value and warrant further investigation. More precise quantitative comparisons can also be made using a variety of statistical methods (e.g. Rollinson 1993).

Ba/Ca and Sr/Ca ratios have been used in archaeological studies as indicators of dietary, and hence environmental, exposure, although the results have often been equivocal (Burton *et al.* 1999). However, such ratios, and others such as Ba/As, Sr/Zn and As/Zn (Hoogewerff *et al.* 2001) can provide useful comparative indicators for assessing the homogeneity of populations or the likely geographical association of a questioned individual. Trace elements have also been used to reassemble scattered and mixed human bones in multiple gravesites. Fulton *et al.* (1986) found the Mb/Zn ratio, supplemented by Zn/Na, Mg/Na and Cr/Na ratios, to be most useful in this respect.

Fluoride concentrations in teeth

Fluoride levels in teeth (and to a lesser extent bones) are also of potential interest from a forensic provenance point of view. Natural levels of fluoride in stream and groundwaters vary significantly, depending on local geology. For example, in the UK, measured concentrations in streamwaters range from

<0.05 p.p.m. in much of Wales, Cumbria and upland Scotland to >0.4 p.p.m. in the sedimentary basins of eastern and southeastern England (British Geological Survey, unpublished data). However, the public water supplies to approximately 5.5 million people in the UK are now naturally or artificially fluoridated at a level (0.7–1.0 p.p.m.), considered sufficient to provide significant protection against tooth decay, with the West Midlands and the northeast being the most extensively fluoridated regions. Local fluoridation schemes also exist in parts of Yorkshire, the East Midlands, Lincolnshire, west Cumbria and elsewhere. However, large parts of the country, including Scotland, Wales, Northern Ireland, southwestern and northern England, have both low natural fluoride levels and only localized artificial fluoridation schemes.

Two compounds are permitted for artificial fluoridation in the UK: hexafluorosilicic acid (H_2SiF_6), and disodium hexafluorosilicate (Na_2SiF_6). They are added to achieve a target concentration of fluoride of 1 p.p.m. (Jackson *et al.* 2002). Fluoridation on a significant scale in the UK began in the 1960s, following trials. The Water (Fluoridation) Act 1985, now consolidated in the Water Industry Act 1991, gives water companies the power to add fluoride at the request of a health authority. There are proposals to extend the practice, but opinion regarding its dental and other health benefits remains divided (McDonagh *et al.* 2000). Experimental studies have shown that presence of the fluoride ion in drinking water encourages the formation of fluorohydroxyapatite in developing tooth enamel, which is then more resistant to acid attack than hydroxyapatite. The fluoride ion substitutes irreversibly for hydroxyl groups in the hydroxyapatite lattice. However, excessive fluoride intake has long been known to be toxic and can cause dental fluorosis (brownish or black discolouration) in children.

Worldwide, an estimated 210 million people drink artificially fluoridated water, mostly in North America, and a further 103 million drink water whose natural fluoride level is sufficiently high (>0.3 p.p.m.) to provide some level of protection against tooth decay. In Europe the permitted upper limit for fluoride in drinking water is 1.5 p.p.m., but 99% of the water in western continental Europe is not fluoridated. Relatively high fluoride levels in teeth may therefore be expected to be the exception rather than the norm, and may assist in highlighting certain geographical areas as possible source areas for an unidentified individual.

Conclusions

Studies of isotopic ratios and elemental concentrations can play an important role in forensic investigations of unidentified human remains. Analysis of different bones, teeth, and tissues such as hair, nail and skin has the potential to provide information about both long-term and short-term diet and environmental exposure, and may provide useful indicators of human origins and migration. At the present time, however, the utility of such methods is limited by the restricted availability of appropriate database information that can be used for comparison purposes. Existing data are scattered across a diverse scientific literature, and problems of interpretation often arise as a result of different procedures being used for sample pre-treatment and analysis.

Problems associated with post-mortem isotopic and chemical change (diagenesis) are less severe when dealing with modern human skeletal remains than older archaeological bone and tooth material. However, additional complications of interpretation arise with modern teeth and bones due to the increasing tendencies for 'globalization' in food supplies and more frequent human travel, especially in developed societies. However, recent studies have provided encouraging indications that regional, and even local, dietary and environmental signals can still be identified in human tissues, providing that geographical mobility during life has not been excessive.

A great deal of further basic research is required to quantify fully the range of isotopic and trace element variation which exists within and between different human populations, and the effects of dietary behaviour, cultural practices and migration on these chemical characteristics. Ideally, reference database information used to assist interpretation of data obtained from unidentified modern skeletal remains should relate to well-documented human bone or dental tissue, but such material is often difficult to obtain. In such circumstances, newly dead small mammals can act as a useful biological proxy, and human hair and nail clippings can provide suitable comparative material for many, although not all, isotopic and trace element attributes.

I thank M. Thirlwall, J. Evans, F. Darbyshire, C. Chenery, B. Cousens, N. Walsh and D. Wray for generating data presented in this paper. However, all interpretations and opinions remain those of the author. Financial support for the work was provided by several UK police forces. Assistance with global database compilation was provided by S. Blott and D. Croft.

References

ABERG, G. 1996. The use of natural strontium isotopes as tracers in environmental studies. *Water, Air and Soil Pollution*, **79**, 309–322.

ABERG, G., FOSSE, G. & STRAY, H. 1998. Man, nutrition and mobility: a comparison of teeth and bone from the medieval era and the present day using Pb and Sr isotopes. *Science of the Total Environment*, **224**, 109–119.

ALEXANDER, L. M., HEAVEN, A., DELVES, H. T., MORETON, J. & TREENOUTH, M. J. 1983. Relative exposure of children to lead from dust and drinking water. *Archives of Environmental Heath*, **48**, 392–400.

AMBROSE, S. H. 1990. Preparation and characterization of bone and tooth collagen for isotopic analysis. *Journal of Archaeological Science*, **18**, 293–317.

AMBROSE, S. H. 1991. Effects of diet, climate and physiology on nitrogen isotope abundances in terrestrial foodwebs. *Journal of Archaeological Science*, **18**, 293–317.

AUBERT, D., STILLE, P., PROBST, A., GAUTHIER-LAFAYE, F., POUCELOT, L., & DEL NERO, M. 2002. Characterization and migration of atmospheric RRR in soils and surface waters. *Geochimica et Cosmochimica Acta*, **19**, 3339–3350.

AYLIFFE, L. K., CHIVAS, A. R. & LEAKEY, M. G. 1994. The retention of primary oxygen isotope compositions of fossil elephant skeletal phosphate. *Geochimica et Cosmochimica Acta*, **58**, 5291–5298.

BARANOWSKA, I., CZERNICKI, K. & ALEKSANDROWICZ, R. 1995. The analysis of lead, cadmium, zinc, copper and nickel content in human bones from the Upper Silesian industrial district. *Science of the Total Environment*, **159**, 155–162.

BEARD, B. L. & JOHNSON, C. M. 2000. Strontium isotope composition of skeletal material can determine the birth place and geographic mobility of humans and animals. *Journal of Forensic Sciences*, **45**, 1049–1061.

BLUM, J. D., TALIAFERRO, E. H., WEISSE, M. T. & HOLMES, R. T. 2000. Changes in Sr/Ca, Ba/Ca and $^{87}Sr/^{86}Sr$ ratios between two forest ecosystems in the northeastern USA. *Biogeochemistry*, **49**, 87–101.

BORG, L. E. & BANNER, J. L. 1996. Neodymium and strontium isotopic constrains on soil sources in Barbados, West Indies. *Geochimica et Cosmochimica Acta*, **60**, 4193–4206.

BRYANT, J. D., KOCH, P. L., FROELICH, P. N., SHOWERS, W. J. & GENNA, B. J. 1996. Oxygen isotope partitioning between phosphate and carbonate in mammalian apatite. *Geochimica et Cosmochimica Acta*, **24**, 5245–5148.

BUDD, P., MONTGOMERY, J., COX, A., KRAUSE, P., BARREIRO, B. & THOMAS, R. G. 1998. The distribution of lead within ancient and modern human teeth: implications for long-term and historical exposure monitoring. *Science of the Total Environment*, **200**, 121–136.

BUDD, P., MONTGOMERY, J., BARREIRO, B. & THOMAS, R. G. 2000*a*. Differential diagenesis of strontium in archaeological human dental tissues. *Applied Geochemistry*, **15**, 687–694.

BUDD, P., MONTGOMERY, J., EVANS, J. & BARREIRO, B. 2000*b*. Human tooth enamel as a record of the comparative lead exposure of prehistoric and modern people. *Science of the Total Environment*, **263**, 1–10.

BUDD, P., MONTGOMERY, J., EVANS, J. & CHENERY, C. 2001. Combined Pb-, Sr- and O-isotope analysis of human dental tissue for the reconstruction of archaeological residential mobility. *In*: HOLLAND, G. & TYANNER, S. D. (eds) *Plasma Source Mass Spectrometry. The New Millenium*. Royal Society of Chemistry, Cambridge, 311–323.

BUDD, P., CHENERY, C., MONTGOMERY, J., EVANS, J. & POWESLAND, D. 2004. Anglo-Saxon residential mobility at west Heslerton, North Yorkshire, UK from combined O- and Sr-isotope analysis. *In*: TANNER, S. & HOLLAND, G. (eds) *Proceedings of the 8th International Conference on Plasma Source Mass Spectrometry*. Royal Society of Chemistry, Cambridge. (in press)

BURTON, J. H., PRICE, T. D. & MIDDLETON, W. D. 1999. Correlation of bone Ba/Ca and Sr/Ca due to biological purification of calcium. *Journal of Archaeological Science*, **23**, 557–568.

CAPO, R. C. STEWART, B. W. & CHADWICK, O. A. 1998. Strontium isotopes as tracers of ecosystem processes: theory and methods. *Geoderma*, **82**, 197–225.

CHADWICK, O. A., DERRY, L. A., VITOUSEK, P. M., HUEBERT, B. J. & HEDIN, L. O. 1999. Changing sources of nutrients during four million years of ecosystem development. *Nature*, **397**, 491–497.

CHAMBERLAIN, C. P., BLUM, J. D., HOLMES, R. T., XIAHONG FENF, SHERRY T. W. & GRAVES, G. R. 1997. The use of isotope tracers for identifying populations of migratory birds. *Oecologia*, **109**, 132–141.

CLAYPOOL, G. E., HOLSERT, W. T., KAPLAN, I. R, SAKAI, H. & ZAK, I. 1980. The age curves of sulfur and oxygen isotopes in marine sulphate and their mutual interpretation. *Chemical Geology*, **28**, 199–260.

CRAIG, H. 1961*a*. Standard for reporting concentrations of deuterium and oxygen-18 in natural waters. *Science*, **13**, 1833–1934.

CRAIG, H. 1961*b*. Isotopic variations in meteoric waters. *Science*, **133**, 1702–1703.

DANSGAARD, W. 1964. Stable isotopes in precipitation. *Tellus*, **16**, 436–468.

DARLING, W. G. & TALBOT, J. C. 2003. The O and H stable isotopic composition of fresh waters in the British isles. 1. Rainfall. *Hydrology and Earth System Sciences*, **7**, 163–181.

DARLING, W. G., BATH, A. H. & TALBOT, J. C. 2003. The O and H stable isotopic composition of freshwaters in the British Isles. 2. Surface waters and groundwater. *Hydrology and Earth System Sciences*, **7**, 183–195.

DEGAETANO. D. H., KEMPTON, J. B. & ROWE, W. E. 1992. Fungal tunnelling of hair from a buried body. *Journal of Forensic Sciences*, **37**, 1048–1054.

DEINES, P. 1980. The isotopic composition of reduced organic carbon. *In*: FRITZ, P. & FONTES, I. CH. (eds) *Handbook of Environmental Isotope Geochemistry*. Vol. 1A. *Terrestrial Environments*. Elsevier, Amsterdam, 329–406.

DELVES, H. T., CLAYTON, B. E., CARMICHAEL, B. E., BUBEAR, M. & SMITH, M. 1982. An appraisal of the analytical significance of tooth-lead measurements as possible indices of environmental exposure of children to lead. *Annals of Clinical Biochemistry*, **19**, 329–337.

DEPAOLO, D. J. 1988. *Neodymium Isotope Geochemistry: An Introduction*. Springer Verlag, New York.

DEPAOLO, D. J. & WASSERBURG, G. J. 1976. Nd isotopic variations and petrogenesis models. *Geophysical Research Letters*, **3**, 249–252.

DICKIN, A. P. 1995. *Radiogenic Isotope Geology*. Cambridge University Press, Cambridge.

EHLERINGER, J. & CERLING, T. 2003. Stable isotope analy-
sis: forensic applications with hair. In *Background Notes, NITECRIME Workshop on Natural Isotopes in Criminalistics and Environmental Forensics, 3rd European Academy of Forensic Sciences Conference, Istanbul, Turkey, 22 September 2003*, 1–12.

EHLERINGER, J. R., CASALE, J. F., LOTT, M. J. & FORD, V. L. 2000. Tracing the geographical origin of cocaine. *Nature*, **408**, 311–312.

EHLERINGER, J. R., COOPER, D. A., LOTT, M. J. & COOK, C. S. 1999. Geolocation of heroin and cocaine by stable isotope ratios. *Forensic Science International*, **106**, 27–35.

ELDERFIELD, H. 1986. Strontium isotope stratigraphy. *Palaeogeography, Palaeoclimatology and Palaeoecology*, **57**, 71–90.

EVANS, J. A. & TATHAM, S. 2004. Defining 'local signature' in terms of Sr isotope composition using a tenth- to twelfth-century Anglo-Saxon population living on a Jurassic clay-carbonate terrain, Rutland, UK. *In*: PYE, K & CROFT, D. J. (eds) *Forensic Geoscience: Principles, Techniques and Applications*. Geological Society, London, Special Publications, **232**, 237–248.

FARMER, J. G., SUGDEN, C. L., MACKENZIE, A. B., MOODY, G. H. & FULTON, M. 1994. Isotopic ratios of lead in human teeth and sources of exposure in Edinburgh. *Environmental Technology*, **15**, 593–599.

FAURE, G. 1986. *Principles of Isotope Geology*. Wiley, New York.

FAURE, G. & POWELL, T. 1972. *Strontium Isotope Geology*. Springer Verlag, New York.

FITTON, G. 1997. X-ray fluorescence spectrometry. *In*: GILL, R. (ed.) *Modern Analytical Geochemistry*. Longman, Harlow, 87–115.

FRICKE, H. C., CLYDE, W. & O'NEIL, J. R. 1998. Intra-tooth variations in $d^{18}O$ (PO_4) of mammalian tooth enamel as a record of seasonal variations in continental climate variables. *Geochimica et Cosmochimica Acta*, **62**, 1839–1850.

FULTON, B. A., MELOAN, C. E. & FINNEGAN, M. 1986. Reassembling scattered and mixed human bones by trace element ratios. *Journal of Forensic Sciences*, **31**, 1455–1462.

GOLDSTEIN, S. L., O'NIONS, R. K. & HAMILTON, P. J. 1984. A Sm-Nd isotopic study of atmospheric dusts and particulates from major river systems. *Earth and Planetary Science Letters*, **70**, 221–236,

GOODFRIEND. G. A. 1999. Terrestrial stable isotope records of Late Quaternary paleoclimates in the eastern Mediterranean region. *Quaternary Science Reviews*, **18**, 501–513.

GROUSSET, F. E., BISCAYE, P. E., REVEL, M., PETIT, J-R., PYE, K. JOUSSAUME, S. & JOUZEL, J. 1992. Antarctic (Dome C) ice-core dust at 18 kyr B.P.: isotopic constraints on origins. *Earth and Planetary Science Letters*, **111**, 175–182.

GROUSSET, F. E., BISCAYE, P. E., ZINDLER, A., PROSPERO, J. & CHESTER, R. 1988. Neodymium isotopes as tracers in marine sediments and aerosols: North Atlantic. *Earth and Planetary Science Letters*, **87**, 367–378.

GROUSSET, F. E., PARRA, M., BORY, A., MARTINEZ, P., BERTRAND, P., SHIMMIELD, G. & ELLAM, M. 1998. Saharan wind regimes traced by the Sr-Nd isotopic composition of subtropical Atlantic sediments: Last

Glacial maximum versus today. *Quaternary Science Reviews*, **17**, 395–409.

GULSON, B. L. 1996. Tooth analyses of sources and intensity of lead exposure in children. *Environmental Health Perspective*, **104**, 306–312.

GULSON, B. L. & WILSON, D. 1994. History of lead exposure in children revealed from isotopic analyses of teeth. *Archives of Environmental Health*, **49**, 279–283.

GULSON, B. L., JAMESON, W. & GILLINGS, B. R. 1997. Stable lead isotopes in teeth as indicators of past domicile – a potential new tool in forensic science? *Journal of Forensic Sciences*, **42**, 787–791.

HAMILTON, E. I. 1971. The concentration and distribution of uranium in human skeletal tissues. *Calcified Tissue Research*, **7**, 150–162.

HANCOX, N. M. 1972. *Biology of Bone*. Cambridge University Press, Cambridge, Maryland.

HARLAVAN, Y. & EREL, Y. 2002 The release of Pb and REE from granitoids by the dissolution of accessory phases. *Geochimica et Cosmochimica Acta*, **66**, 837–848.

HEATON, K., KELLY, S. & HOOGEWERFF, J. 2003. The geographical origin of beef: possibilities for traceability schemes based on light and heavy stable isotope analysis. *Abstracts of the Stable Isotope Mass Spectrometry Users Group, 14–16th April 2003, School of Chemistry, University of Bristol*, 59.

HEDGES, R. E. M. 2002. Bone diagenesis: an overview of processes. *Archeometry*, **44**, 319–328.

HENSHAW, D. L., ALLEN, J. E., KEITCH, P. A. & RANDLE, P. H. 1994. Spatial distribution of naturally occurring [210]Po and [226]Ra in children's teeth. *International Journal of Radiation Biology*, **66**, 815–826.

HILL, P. A. 1998. Bone remodelling. *British Journal of Orthopaedics*, **25**, 101–107.

HILLSON, S. 1997. *Dental Anthropolology*. Cambridge University Press, Cambridge.

HOEFS, J. 1987. *Stable Isotope Geochemistry*, 3rd edition. Springer Verlag, Berlin.

HOOGEWERFF, J., PAPESCH, W. *ET AL.* 2001. The last domicile of the Iceman from Hauslabjoch: a geochemical approach using Sr, C and O isotopes and trace element signatures. *Journal of Archaeological Research*, **28**, 983–989.

JACKSON, P. J., HARVEY, P. W. & YOUNG, W. F. 2002. *Chemistry and Bioavailability Aspects of Fluoride in Drinking Water*. WRc-NSF Ltd, Marlow, Report No. CO 5037.

JARVIS, K. E. 1997. Inductively coupled plasma-mass spectrometry. *In*: GILL, R. (ed.) *Modern Analytical Geochemistry*. Longman, Harlow, 171–187.

JAWOROWSKI, Z., BARBALAT, F. & BLAIN, C. 1985. Heavy metals in human and animal bones from ancient and contemporary France. *Science of the Total Environment*, **43**, 103–126.

JONES, A. M., IACUMIN, P. & YOUNG, E. D. 1999. High-resolution d18O analysis of tooth enamel phosphate by isotope ratio monitoring gas chromatography mass spectrometry and laser fluorination. *Chemical Geology*, **153**, 241–248.

JOWSEY, J. 1971. The internal remodeling of bones. *In*: BOURNE, G. H. (ed.) *The Biochemistry and Physiology of Bone. III. Development and Growth*. Academic Press, New York, 201–238.

KATZENBERG, M. A. & KROUSE, H. R. 1989. Application of

stable isotope variation in human tissues to problems of identification. *Canadian Society of Forensic Science Journal*, **22**, 7–19.

KOCH, P. L., SCHOENINGER, M. J. & BARKER, W. W. 1999. Altered states: effects of diagenesis on fossil tooth chemistry. *Geochimica et Cosmochimica Acta*, **63**, 2737–2747.

KOCH, P. L., TUROSS, N. & FOGEL, M. L. 1997. The effects of sample treatment and diagenesis on the isotopic integrity of carbonate in biogenic hydroxylapatite. *Journal of Archaeological Science*, **24**, 417–429.

KROUSE, H. R. 1980. Sulfur isotopes in our environment. *In*: FRITZ, P. & FONTES, J. CH. (eds) *Handbook of Environmental Isotope Geochemistry*. Vol. 1A. *The Terrestrial Environment*. Elsevier, Amsterdam, 435–471.

LECOLLE, P. 1985. The oxygen isotope composition of land snail shells as a palaeoclimatic indicator: applications to hydrogeology and palaeoclimatology. *Chemical Geology (Isoptope Geoscience Section)*, **58**, 157–181.

LEGEROS, R. Z. 1969. Two types of carbonate substitution in apatite structure. *Experimena*, **24**, 5–7.

LETOLLE, R. 1980. Nitrogen-15 in the natural environment. *In*: FRITZ, P. & FONTES, J. CH. (eds) *Handbook of Environmental Isotope Geochemistry*. Vol. 1A. *The Terrestrial Environment*. Elsevier, Amsterdam, 407–434.

LEVINSON, A. A., LUZ, B. & KOLODNY, Y. 1987. Variations in oxygen isotope compositions of human teeth and urinary stones. *Applied Geochemistry*, **2**, 367–371.

LINDARS, E. S., GRIMES, S. T., MATTEY D. P., COLLINSON, M. E., HOOKER, J. J. & JONES, T. P. 2001. Phosphate d[18]O determination of modern rodent teeth by direct laser fluorination: an appraisal of methodology and potential application to palaeoclimate reconstruction. *Geochimica et Cosmochimica Acta*, **65**, 2535–2548.

LONGINELLI, A. 1984. Oxygen isotopes in mammal bone phosphate: a new tool for paleohydrological and paleoclimatological research? *Geochimica et Cosmochimica Acta*, **48**, 385–390.

LUZ, B. & KOLODNY, Y. 1985. Oxygen isotope variations in phosphate of biogenic apatites. IV. Mammal teeth and bones. *Earth and Planetary Science Letters*, **75**, 29–36.

LUZ, B., CORMIE, A. B. & SCHWARCZ, H. P. 1990. Oxygen isotope variations in phosphate of deer bones. *Geochimica et Cosmochimica Acta*, **54**, 1723–1728.

LUZ, B., KOLODNY, Y. & HOROWITZ, M. 1984. Fractionation of oxygen isotopes between mammalian bone-phosphate and environmental drinking water. *Geochimica et Cosmochimica Acta*, **48**, 1689–1693.

MACKO, S. A., ENGEL, G., ANDRUSEVICH, G., LUBEC, G. & O'CONNELL, T. C. 1999. Documenting the diet in ancient human populations through stable isotope analysis of hair. *Philosophical Transactions of the Royal Society of London (B)*, **354**, 65–76.

MAHANTI, H. S. & BARNES, R. M. 1983. Determination of major, minor and trace elements in bone by inductively coupled plasma emission spectrometry. *Analytica Chimica Acta*, **151**, 409–417.

MANN, S. 2001. *Biomineralization. Principles and Concepts in Bioinorganic Materials Chemistry*. Oxford University Press, Oxford.

MARIOTTI, A. 1984. Natural 15N abundance measurements

and atmospheric nitrogen standard calibration. *Nature*, **311**, 251–252.

MARTIN, C. E. & McCULLOCH, M. T. 1999. Nd–Sr isotopic and trace element geochemistry of river sediments and soils in a fertilized catchment, New South Wales, Australia. *Geochimica et Cosmochimica Acta*, **63**, 287–305.

MAYS, S. 1998. *The Archaeology of Human Bones*. Routledge, London and New York.

McARTHUR, J. M., HOWARTH, R. J. & BAILEY, T. R. 2001. Strontium isotope stratigraphy: LOWESS Version 3. Best-fit line to the marine Sr-isotope curve for 0 to 509 Ma and accompanying look-up table for deriving numerical age. *Journal of Geology*, **109**, 155–169.

McCONNELL, D. 1973. Biomineralogy of phosphates and physiological mineralization. *In*: GRIFFITHS, E. J. (ed.) *Environmental Phosphorus Handbook*. Wiley, New York, 425–442.

McCONNELL, D. 1981. Human and vertebrate minerals. *In*: FRYE, K. (ed.) *The Encyclopedia of Mineralogy*. Hutchinson Ross, New York, 200–203.

McCULLOCH, M. T. & WASSERBURG, G. J. 1978. Sm–Nd and Rb–Sr chronology of continental crust formation. *Science*, **200**, 1003–1011.

McDONAGH, M., WHITING, P. *ET AL*. 2000. *A Systematic Review of Public Water Fluoridation*. NHS Centre for Reviews and Dissemination, University of York.

McLAUGHLIN-BLACK, S. M., HERD, R. J. M., WILLSON, K., MYERS, M. & WEST I. E. 1992. Strontium-90 as an indicator of time since death: a pilot investigation. *Forensic Science International*, **57**, 51–56.

McTAINSH, G. 1984. The nature and origin of aeolian mantles in central northern Nigeria. *Geoderma*, **33**, 13–37.

MILLER, R. G. & O'NIONS, R. K. 1985. Source of Precambrian chemical and clastic sediments. *Nature*, **314**, 325–330.

MILLER, E. K., BLUM, J. A. & FRIEDLAND, A. J. 1993. Determination of soil-exchangeable-cation loss and weathering rates using Sr isotopes. *Nature*, **362**, 438–441.

MINSON, D. J., LUDLOW, M. M. & TROUGHTON, J. H. 1975. Differences in natural carbon isotope ratios of milk and hair from cattle grazing tropical and temperate pastures. *Nature*, **256**, 602.

MONTGOMERY, J., BUDD, P. & EVANS, J. 2000. Reconstructing the lifetime movements of ancient people: a Neolithic case study from southern England. *European Journal of Archaeology*, **3**, 370–385.

MONTGOMERY, J., EVANS, J.A. & NEIGHBOUR, T. 2003. Sr isotope evidence for population movement within the Hebridean Norse community of NW Scotland. *Journal of the Geological Society, London*, **160**, 649–653.

MULLER, W., FRICKE, H., HALLIDAY, A. N., McCULLOCH, M. T. & WARTHO, J.-A. 2003. Origin and migration of the Alpine Iceman. *Science*, **302**, 862–866.

NAKAMURA, K., SCHOELLER, D. A., WINKLER, F. J. & SCHMIDT, H.-L. 1982. Geographical variations in the carbon isotope composition of the diet and hair in contemporary man. *Biomedical Mass Spectrometry*, **9**, 390–394.

NEIS, P., HILLE, R., PASCHKE, M., PILWAT, G., SCHNABEL, A., NIESS, C. & BRATZSKE, H. 1999. Strontium-90 for determination of time since death. *Forensic Science International*, **99**, 47–51.

NELSON, B. K., DeNIRO, M. J., SCHOENINGER, M. J. & DePAULO, D. J. 1986. Effects of diagenesis on strontium, carbon, nitrogen and oxygen concentration and isotopic composition of bone. *Geochimica et Cosmochimica Acta*, **50**, 1941–1949.

NIELSEN, H. 1979. Sulfur isotopes. *In*: JAGER, E. & HUNZIKER, J. C. (eds) *Lectures in Isotope Geology*. Springer Verlag, Berlin, 283–312.

NOSHKIN, V. E., ROBISON, W. L. & WONG, K. M. 1994. Concentration of ^{210}Po and ^{210}Pb in the diet of the Marshall Islands. *Science of the Total Environment*, **155**, 87–104.

O'CONNELL, T. C. & HEDGES, R. E. M. 1999. Investigations into the effect of diet on modern human hair isotopic values. *American Journal of Physical Anthropology*, **108**, 409–425.

O'CONNELL, T. C., HEDGES, R. E. M., HEALEY, M. A. & SIMPSON, A. H. R. W. 2001. Isotopic comparison of hair, nail and bone: modern analyses. *Journal of Archaeological Science*, **28**, 1247–1255.

O'DONNELL, R. G., MITCHELL, P. I., PRIEST, N. D., STRANGE, L., FOX, A., HENSHAW, D. L. & LONG, S. C. 1997. Variations in the concentration of plutonium, strontium-90, and total alpha-emitters in human teeth collected within the British Isles. *Science of the Total Environment*, **201**, 235–243.

O'NEIL, J. R., ROE, L. J., REINHARD, E. & BLAKE, R. E. 1994. A rapid and precise method of oxygen isotope analysis of biogenic phosphate. *Israel Journal of Earth Sciences*, **43**, 203–212.

ONIANWA, P. C., ADEYEMO, A. O., IDOWU, O. E. & OGABIELA, E. E. 2001. Copper and zinc contents of Nigerian foods and estimates of the adult dietary intakes. *Food Chemistry*, **72**, 89–95.

O'NIONS, R. K., CARTER, S. R., COHEN, R. .., EVENSEN, N. M. & HAMILTON, P. J. 1978. Pb, Nd and Sr isotopes in oceanic ferromanganese deposits and ocean floor basalts. *Nature*, **273**, 435–438.

O'NIONS, R. K., HAMILTON, P. J. & HOOKER, P. J. 1983. A Nd isotope investigation of sediments related to crustal development in the British isles. *Earth Planetary Science Letters*, **63**, 229–240.

PALHOL, F., LAMOUREUX, C. & NAULET, N. 2003. N-15 isotopic analyses: a powerful tool to establish links between seized 3,4-methylenedioxymethamphetamine (MDMA) tablets. *Analytical and Bioanalytical Chemistry*, **376**, 486–490.

PARFITT, A. M. 1983. The physiologic and clinical significance of bone data. *In*: RECKER, R .R. (ed.) *Bone Histomorphometry: Techniques and Interpretation*. CRC Press, Boca Raton, 143–223.

PARRY, S. J. 1997. Neutron activation analysis. *In*: GILL, R. (ed.) *Modern Analytical Geochemistry*. Longman, Harlow, 116–134.

PATE, F. D., HUTTON, J. T. & NORRISH, K. 1989. Ionic exchange between soil solution and bone: toward a predictive model. *Applied Geochemistry*, **4**, 303–316.

PIEPGRAS, D. J. & WASSERBURG, G. J. 1980. Neodymium isotopic variation in seawater. *Earth Planetary Science Letters*, **50**, 128–138.

POPPLEWELL, D. S., HAM, G. J., DODD, N. J. & SHUTLER, S. D. 1988. Plutonium and Cs-137 in autopsy tissues in

Great Britain. *Science of the Total Environment*, **70**, 321–334.

PRICE, T. D. (ed.) 1989. *The Chemistry of Prehistoric Human Bone*. Cambridge University Press, Cambridge.

PRICE, T. D., BLITZ, T. J., BURTON, J. H. & EZZO, J. A. 1992. Diagenesis in prehistoric bone: problems and solutions. *Journal of Archaeological Science*, **19**, 513–529.

PRICE, T. D., GRUPE, G. & SCHRORTER, P. 1994a. Reconstruction of migration patterns in the Bell Beaker period by stable strontium isotope analysis. *Applied Geochemistry*, **9**, 413–417.

PRICE, T. D., JOHNSON, C. M., EZZO, J. A., ERICSON, J. & BURTON, J. H. 1994b. Residential mobility in the prehistoric southwest United States: a preliminary study using strontium isotope analysis. *Journal of Archaeological Science*, **21**, 315–330.

PRICE, T. D., MANZANILLA, L. & MIDDLETON, W. D. 2000. Immigration and the ancient city of Teotihuacan in Mexico: a study using strontium isotope ratios in human bone and teeth. *Journal of Archaeological Science*, **27**, 903–913.

PRICE, T. D., BURTON, J. H. & BENTLEY, R. A. 2002. The characterization of biologically available strontium isotope ratios for the study of prehistoric migration. *Archaeometry*, **44**, 117–135.

PRIEST, N.D. & VAN DE VYVER, F. (eds) 1999. *Trace Metals and Fluoride in Bones and Teeth*. CRC Press, Boca Raton.

PYE, K. 1987. *Aeolian Dust and Dust Deposits*. Academic Press, London.

PYE, K. & BLOTT, S. J. 2004. Comparison of soils and sediments using major trace element data. *In*: PYE, K. & CROFT, D. J. (eds) *Forensic Geoscience: Principles, Techniques and Applications*. Geological Society, London, Special Publications, **232**, 183–196.

RICHARDS, M. P. & HEDGES, R. E. M. 2000. FOCUS: Gough's Cave and Sun Hole Cave human stable isotope values indicate a high animal protein diet in the British Upper Paleolithic. *Journal of Archaeological Science*, **27**, 1–3.

ROLLINSON, H. 1993. *Using Geochemical Data: Evaluation, Presentation, Interpretation*. Longman, Harlow.

ROWE, W. F. 1997. Biodegradation of hairs and fibres. *In*: HAGLUND, W.D. & SORG, M.H. (eds) *Forensic Taphonomy: The Postmortem Fate of Human Remains*. CRC Press, Boca Raton, 337–351.

SAMUELS, E.R., MERANGER, J.C., TRACY, B.L. & SUBRAMANIAN, K. S. 1989. Lead concentrations in human bones from the Canadian population. *Science of the Total Environment*, **89**, 261–269.

SHACKLETON, N. J. & OPDYKE, N. D. 1973. Oxygen isotope and palaeomagnetic stratigraphy of equatorial Pacific core V28-238: oxygen isotope temperatures and ice volumes on a 10^5 and 10^6 time scale. *Quaternary Research*, **3**, 39–55.

SCHOENINGER, M. J. & DENIRO, M. J. 1983. Stable nitrogen isotope ratios of bone collagen reflect marine and terrestrial components of prehistoric human diet. *Science*, **220**, 1381–1383.

SCHOENINGER, M. J. & DENIRO, M. J. 1984. Nitrogen and carbon isotope composition of bone collagen from marine and terrestrial animals. *Geochimica et Cosmochimica Acta*, **48**, 625–639.

SCHULTZ, M. 1997. Microscopic investigation of excavated skeletal remains: a contribution to palaeopathology and forensic medicine. *In*: HAGLUND, W. D. & SORG, M. H. (eds) *Forensic Taphonomy: The Postmortem Fate of Human Remains*. CRC Press, Boca Raton, 201–222.

SEALY, J. C., ARMSTRONG, R. & SCHRIRE, C. 1995. Beyond lifetime averages: tracing life histories through isotopic analysis of different calcified tissues from archaeological human skeletons. *Antiquity*, **69**, 290–300.

SEALY, J. C., VAN DER MERWE, N. J., SILLEN, S., KRUGER, F. J. & KRUEGER, H. W. 1991. $^{87}Sr/^{86}Sr$ as a dietary indicator in modern and archaeological bone. *Journal of Archaeological Science*, **18**, 399–416.

SHARP, Z. D., ATUDOREI, V. & FURRER, H. 2000. The effects of diagenesis on oxygen isotope ratios of biogenic phosphates. *Science*, **300**, 222–237.

SHAW, H. F. & WASSERBURG, G. J. 1985. Sm–Nd in marine carbonates and phosphates: implications for Nd isotopes in seawater and crustal ages. *Geochimica et Cosmochimica Acta*, **49**, 503–518.

SILLEN, A. 1989. Diagenesis of the inorganic phase of cortical bone. *In*: T. D. PRICE (ed.) *The Chemistry of Prehistoric Human Bone*. Cambridge University Press, Cambridge, 211–229.

SILLEN, A. & SEALY, J. C. 1995. Diagenesis of strontium in fossil bone: a reconsideration of Nelson *et al.* (1986). *Journal of Archaeological Science*, **22**, 313–320.

SILLEN, A., HALL, G., RICHARDSON, S. & ARMSTRONG, 1998. $^{87}Sr/^{86}Sr$ ratios in modern and fossil food webs of the Sterkfontein Valley: implications for early hominid habitat preference. *Geochimica et Cosmochimica Acta*, **62**, 2463–2478.

SIMMONS, D. J. & GRYNAPS, M. D. 1989. Mechanism of bone formation *in vivo*. *In*: HALL, B. K. (ed.) *Bone*. Vol. 1. *The Osteoblast and Osteocyte*. Telford Press, Caldwell, 193–302.

SIMMONS, E. D., PRITZEKER, K. P. H. & GRYNAPS, M. D. 1991. Age-related changes in the human femoral cortex. *Journal of Orthopaedic Research*, **9**, 155–167.

SMITH, B. J. & WHALLEY, W. B. 1981. Late Quaternary drift deposits of northern Nigeria examined by scanning electron microscopy. *Catena*, **8**, 345–368.

STACK, M. V. 1999. Lead in human bones and teeth. *In*: PRIEST, N.D. & VAN DE VYVER, F. (eds) *Trace Metals and Fluoride in Bones and Teeth*. CRC Press, Boca Raton, 191–218.

STEENKAMP, V., VON ARB, V. & STEWART, M. J. 2000. Metal concentrations in plants and urine from patients treated with traditional remedies. *Forensic Science International*, **114**, 89–95.

SWIFT, B. 1998. Dating human skeletal remains: investigating the viability of measuring the equilibrium between 210Po and 210Pb as a means of estimating the post-mortem interval. *Forensic Science International*, **98**, 119–126.

SWIFT, B., LAUDER, I., BLACK, S. & NORRIS, J. 2001. An estimation of the post-mortem interval in human skeletal remains: a radionuclide and trace element approach. *Forensic Science International*, **117**, 73–87.

SYLVESTER, P. (ed.) 2001. *Laser-Ablation-ICPMS in*

the Earth Sciences: Principles and Applications. Mineralogical Association of Canada, St John's, Short Course, **29**.

TEITELBAUM, S. L. 2000. Bone resporption by osteoblasts. *Science,* **289**, 1504.

TRUEMAN, C., CHENERY, C., EBERTH, D. A. & SPIRO, B. 2003. Diagenetic effects on the oxygen isotope composition of bones of dinosaurs and other vertebrates recovered from terrestrial and marine sediments. *Journal of the Geological Society, London,* **160**, 895–901.

VAN DER MERWE, N. J. 1992. Light stable isotopes and the reconstruction of prehistoric diets. *Proceedings of the British Academy,* **77**, 247–264.

VAN DER MERWE, N. J., LEE-THORP, J. & BELL, R. H. V. 1988. Carbon isotopes as indicators of elephant diets and African environments. *African Journal of Ecology,* **26**, 163–172.

VAN DER MERWE, LEE-THORP, J. A., *ET AL*. 1990. Source area determination of elephant ivory by isotopic analysis. *Nature,* **346**, 744–746.

VINE, H. 1987. Wind-blown materials and W. African soils: an explanation of the 'ferallitic soil over loose sandy sediment' profile. *In*: FROSTICK, L. E. & REID, I. (eds.) *Desert Sediments Ancient and Modern.* Geological Society, London, Special Publications, **35**, 171–183.

VOGEL, J. C., EGLINGTON, B. & AURET, J. M. 1990. Isotope fingerprints in elephant bone and ivory. *Nature,* **346**, 747–749.

WALDRON, H. A., MACKIE, A. & TOWNSHEND, A. 1976. The lead content of some Romano-British bones. *Archaeometry,* **18**, 221–227.

WALSH, J. N. 1997. Inductively coupled plasma-atomic emission spectrometry (ICP-AES). *In*: GILL, R. (ed.) *Modern Analytical Geochemistry.* Longman, Harlow, 41–66.

WEBB, Y., MINSON, D. J. & DYE, E. A. 1980. A dietary factor influencing ^{13}C content of human hair. *Search,* **11**, 200–201.

WHITE, C. D., SPENCE, M. W., STUART-WILLIAMS, H. LEQ. & SCHWARCZ, H. P. 1998. Oxygen isotopes and the identification of geographical origins: the Valley of Oaxaca versus the Valley of Mexico. *Journal of Archaeological Science,* **25**, 643–655.

WHITE, T. D. 1991. *Human Osteology.* Academic Press, San Diego.

WHITTAKER, D. K. & STACK, M. V. 1976. The lead, cadmium and zinc content of some Romano-British teeth. *Archaeometry,* **26**, 37–42.

WOELFEL, J. B. & SCHEID, R. C. 2002. *Dental Anatomy: Its Relevance to Dentistry.* 6th edition. Lippincott Williams & Wilkins, Philadelphia.

YAMAMOTO, M., HINOIDE, M., OHKUBO, Y. & UENO, K. 1994. Concentration of ^{226}Ra in teeth. *Heath Physics,* **67**, 535–540.

Defining 'local signature' in terms of Sr isotope composition using a tenth- to twelfth-century Anglo-Saxon population living on a Jurassic clay–carbonate terrain, Rutland, UK

J. A. EVANS[1] & S. TATHAM[2]

[1]NERC Isotope Geosciences Laboratory, British Geological Survey, Kingsley Dunham Centre, Nicker Hill, Keyworth, Nottingham NG12 5GG, UK
[2]School of Archaeology & Ancient History, University of Leicester, Leicester LE1 7RH, UK

Abstract: The Sr isotope ratios and Sr concentration in tooth enamel from a rural tenth–twelfth century Anglo-Saxon population living on a Jurassic clay-carbonate terrain in eastern England gives the following mean values: $^{87}Sr/^{86}Sr = 0.7098 \pm 0.0018$ (2σ, $n = 22$) and Sr concentrations = 74 ± 62 p.p.m. (2σ). The isotope data are taken to be representative of Anglo-Saxon biosphere values in the area of study. The Sr isotope composition of soil leachates, plant material, riverwaters and animal tooth enamel associated with the burial site were all analysed to see which gave the best approximation to these local Anglo-Saxon values, the aim being to define the best method of predicting the local Sr signature of areas for archaeological purposes. The Sr isotope composition of acetic acid soil leachates were dominated by the carbonate soil component and gave 0.7085 ± 0.0020 water leachates gave 0.7090 ± 0.0014 and plant material gave 0.7092 ± 0.0018 (all at 2σ, $n = 12$). All of these materials were less radiogenic that those of the Anglo-Saxon population. Riverwater gave the same result as the plants at 0.7092 ± 0.0012 (2σ, $n = 3$). The Anglo-Saxon animal tooth enamel gave the best match with a value of 0.7099 ± 0.0017 (2σ, $n = 13$). Analysis of variance (ANOVA) tests show that there is a high probability (>70% probability, 2SD) that the animals and the humans sampled were from the same population with respect to Sr isotope composition. Thus animal tooth enamel proved to be the best proxy, in this study, for the local human population.

Many archaeological studies use skeletal Sr isotope characteristics in an attempt to determine whether a group of people comprise a single population, whether there is an exotic component in the group, or whether or not the entire group come from outside the area in question (Price *et al.*, 1994; Budd *et al.* 2000; Ezzo *et al.* 1997; Sillen *et al.* 1998; Hoogewerff *et al.* 2001; Montgomery *et al.* 2003, 2004; Schweissing & Grupe 2003). When data are tightly clustered, with only a few obvious outliers, it is relatively straightforward to identify the outliers as migrants. However, as described in a recent paper (Price *et al.* 2002), the population distribution is often not tightly defined and determining which individuals are local and which are not is more difficult. It soon becomes apparent, when undertaking such studies, that the geographical definition of 'local' is important, coupled with the ability to predict the expected value and natural variation of Sr isotope composition in the biosphere of the area defined as 'local'. With this in mind, we undertook this study with the aim of assessing the natural variation of Sr isotope composition in a human population by testing various methods of sampling the biosphere of a 'local area' in order to see what provided the best 'match' to the human population. In other words, we wished to establish the best predictive approach for defining the Sr isotope composition and variation in

an area so that, when a burial site is excavated, the data for human remains can be compared with a reliable 'local environment' benchmark.

We approached the problem of how to define 'local Sr signature' for a particular area by taking a human population that, based on all available archaeological evidence, was a local agricultural community. We started from the assumption that we had a local community and then compared this dataset with a variety of local environmental data to see which would provide the best match with the human population. The sample set with closest match provided the best method of defining the expected variation in a human population for comparison with an excavated population of unknown origin.

A number of different materials exist that can be used to characterize the Sr isotope characteristics of the local environment (Table 1). In this study we used present-day water and dilute acetic acid soil leachates, present-day plant material, riverwater samples and archaeological skeletal animal remains found at the sites for comparison with the human skeletal material.

Stratigraphically, the study area is a simple sequence of interbedded carbonate-rich and clay-rich horizons but the heterogeneity of the lithologies provides a range of Sr isotope signatures. This

From: PYE, K. & CROFT, D. J. (eds) 2004. *Forensic Geoscience: Principles, Techniques and Applications*. Geological Society, London, Special Publications, **232**, 237–248. © The Geological Society of London, 2004.

Table 1. *A summary of the advantages and disadvantages of using the composition of $^{87}Sr/^{86}Sr$ isotopes ratios in certain materials for determining the 'local' Sr isotope signature of a region*

Material	Advantages	Disadvantages
Soil leachates	Guaranteed provenance, easily handled.	An estimate of strontium contribution to biosphere, not a direct measurement
Plant material	Guaranteed provenance, biosphere sample.	Very localized sample, difficult to dissolve.
Riverwater samples	Average of a large catchments area.	There may be seasonal change and the composition may be biased by most soluble components. Could contain a considerable component of modern fertilizers
Roadkill and invertebrates	Average of larger strontium pool.	Less certain provenance and possibly affected by modern pollutants and foodstuffs.
Modern domestic animals	Well provenanced and averaging a comparable amount of material to a human from the same area.	May have been given feed from unknown, non-local sources.
Modern humans	Directly comparable positions in the food chain and with similar omnivorous diet.	Material difficult to obtain/ ethical and health and safety issues, the modern western diet appears to be to a large extent divorced from the local area of upbringing. May be possible in modern rural communities.
Preserved plants	Plant compositions from era under study. No modern chemical effects.	Subject to secondary alteration. They provide a possible option in desiccating environments but are rarely/never preserved, unaltered, in the UK.
Preserved animal remains	From the same era/ food chain	May be of uncertain origin and relationship to community
Other archaeological humans	Would be the best comparison to use.	It is usually difficult to find a suitable community of known locals in the same area as the population under study.

means that, in defining the values and standard deviation (SD) of this population, we would be providing a generous estimate of variance in a population and would expect populations living on monolithological terrains, such as the Chalk, to have a more restricted range in values. Our aims were: (1) to provide an example of the values and variance of a rural population living in a clay-carbonate environment; (2) to find the best approach to defining the 'local' Sr isotope signature in an area, and (3) to assess the soil-leaching methods as a means of defining the biosphere values.

Tooth enamel as an archive for $^{87}Sr/^{86}Sr$ life signatures

Tooth enamel provides a unique archive for isotope signatures that have been preserved from the time when the tooth originally mineralized during childhood because the enamel is composed predominantly (>96%) of inorganic biogenic apatite, which is resistant to secondary diagenetic alteration (Trickett *et al.* 2003). Permanent human teeth form and are mineralized during early childhood, and the

crown of the tooth is completely formed between 3.5 and 8 years of age, depending on the tooth. Strontium, which is derived from dietary sources, is incorporated into the tooth enamel in the calcium sites. For a fuller discussion of human tooth formation and mineralization see Montgomery (2002). The isotope composition of the strontium ($^{87}Sr/^{86}Sr$) reflects the average isotope composition of dietary intake during the formation of the enamel and, as the $^{87}Sr/^{86}Sr$ composition is dependent upon the age and rubidium content of the soils, and ultimately upon the underlying geology of an area, the Sr isotope ratio provides information about the nature of the land that supplied the main dietary components to an individual during early childhood ((Price *et al.* 2002)

Geology of the study area

The village of Ketton [SK 980043] is located in Rutland, UK, southeast of the market town of Oakham. Geologically the area comprises gently eastward-dipping sedimentary rocks of the Jurassic series with recent deposits of boulder clay and alluvium (British Geological Survey 1978) (Fig. 1). The

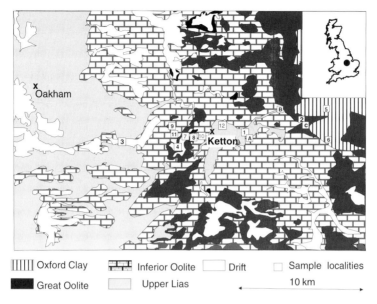

Fig. 1. A sketch map of the geology of the Stamford 1:50000 map showing the site of the Ketton excavation and sample localities of this study.

sedimentary rocks comprise the Lias, the Great and Inferior Oolite series and Oxford Clay. This results in an interbedded sequence of sandstones, limestones and mudrocks. Easterly flowing rivers, which cut through the geology, produce a spatially varied pattern of surface solid geology. The area is dominated by the clay deposits of Lower and Upper Lias; however, the relatively thin layers of Lower and Upper Lincolnshire Limestone cover about a third of the surface area and therefore contribute significantly to the soil composition. Deposits of glacial boulder clay lie mostly to the north and west of Ketton, and recent alluvium deposits are found in the river valleys. In summary, the area of Ketton is geologically a structurally simple sedimentary sequence. However, the mixture of clay, sand and carbonate deposits suggests that the soils may vary considerably in terms of Sr release mechanisms into the soil because of mixed silicate-carbonate nature of the rocks (Graustein 1989).

Archaeology of Ketton

The archaeological site of Ketton is 6.5 km to the southeast of Rutland Water and is situated in fields c. 2 km away from the village of Ketton. Located on land belonging to Castle Cement Quarry, its existence was discovered during quarry excavations. Human burials and evidence of buildings suggested a settlement, which was excavated in its entirety in 1998 (Cooper 2000). The cemetery contains 72 indi-

viduals: 11 males, 19 females, 5 adults of indeterminate sex and 37 juveniles. The settlement comprises a single-cell church, a cemetery surrounding it, and four timber halls, one of which is larger and aisled. The settlement is small and was probably composed of only a few farmsteads. Although there is a church and cemetery, it is not believed to have been an independent parish. The settlement was, in all probability, linked to the village of Ketton. Stamford-ware pottery suggests a tenth–twelfth-century occupation, and archaeologists interpreted the site as a manorial complex, inhabited by an extended family with its labourers and servants (Tatham unpublished data).

The cemetery layout suggested two separate burial groups: one to the north and the other to the south. The northern group is clearly separated from the southern one and is clustered around a former tree hole. Radiocarbon dating of four of the individuals produced late ninth-century to mid-eleventh-century dates from both burial zones. This not only demonstrated that the separate burial zones were contemporary with each other, but also that the manorial complex predated, and ceased to exist, or at least to be used as a burial site, by the time of the Norman Conquest.

In medieval times, the northern side of a cemetery was usually the least popular, people preferring to be interred on the southern and eastern sides, and unbaptized children typically being found closest to the church walls. There was often male/female segregation, but this was not observed at Ketton. Apart from the distinct north–south zonation, the cemetery

Table 2. $^{87}Sr/^{86}Sr$ *isotope ratios of soil leachates and associated plant samples from sample sites within the area of Ketton*

Locality	Plant sample	Soil sample	$^{87}Sr/^{86}Sr$ of water leachate	$^{87}Sr/^{86}Sr$ of acetic acid leachate	$^{87}Sr/^{86}Sr$ of plant leaves	Common plant name	National Grid reference	Underlying geological formation
1	KET 13	KET 14	0.709782	0.709607	0.710412	cow parsley	TF 009062	River alluvium
2	KET 20	KET 19	0.708659	0.707898	0.708817	elder leaves	TF 065069	River alluvium
3	KET 26	KET 27	0.709669	0.709654	0.710200	elder leaves	SK 915051	Boulder clay
4	KET 8	KET 7	0.708671	0.707580	0.708633	cow parsley	SK 966047	Boulder clay
5	KET 24	KET 23	0.709889	0.709728	0.710218	wild geranium	TF 076076	Oxford Clay
6	KET 21	KET 22	0.708547	0.708400	0.708759	cow parsley	TF 075055	Kellaway Sands
7	KET 1	KET 2	0.708367	0.707621	0.708516	cow parsley	SK 972047	Blisworth Limestone
8	KET 3	KET 4	0.708115	0.707458	0.708379	cow parsley	SK 973046	Upper 'Estuarine' series
9	KET 12	KET 11	0.710037	0.709997	0.709928	unknown	SK 956062	Upper 'Estuarine' series
10	KET 5	KET 6	0.709021	0.708084	0.709053	cow parsley	SK 976044	Upper Lincolnshire Limestone
11	KET 9	KET 10	0.708301	0.707700	0.708290	cow parsley	SK 958057	Upper Lincolnshire Limestone
12	KET 17	KET 16	0.708400	0.707971	0.708897	grass	SK 997069	Northampton Sand Ironstone

layout was typical of a small Christian settlement. As the entire cemetery had been excavated, the entire buried population was available for osteological analysis. Unfortunately, soil composition and the fact that the burials were close to the surface meant that most of the bones were damaged, eroded and fragmented; however, there was sufficient skeletal material available to subdivide the buried population into possible familial groups using non-metric traits. These traits are minor anomalies of skeletal anatomy, used to predict possible family groups, usually in conjunction with metric analyses, such as skull measurements.

Non-metric traits were observed in the Ketton buried population and,given the small size of the population, it would be unlikely for two individuals with no genetic link to share these traits. Four possible family groups can be identified according to these non-metric variations, adding to the evidence that the Ketton population comprised family groups consistent with the model of locally raised individuals (Tatham unpublished data). Neither artefactual nor physiological evidence supports the presence of exotic or immigrant individuals.

Methods

The soil samples comprised *c.* 500 g of material collected from the top 10 cm section of soil. An associated plant leaf sample was then collected a few centimetres away from the soil site. Plant sampling was generally restricted to non-woody, wild flowers, such as wild geranium or cow parsley, but at two places (Localities 2 and 3) the leaves from elder trees

were sampled because low-growing wildflowers were not available (see Table 2). Much of the sampled area was tilled fields but samples were taken from hedgerows where the soil had not been obviously ploughed. Samples were collected primarily across an east–west *c.* 18 km traverse, centred around Ketton and cross-cutting the main geological units in the area, Data are thus discussed within the context of being local to a *c.* 9 km radius around Ketton.

A sample of c. 1 g of soil was placed in a centrifuge tube and approximately 10 ml of either deionized water or 10% acetic acid was added. Each sample was agitated gently throughout the day and then left overnight. The fluid was poured off and centrifuged to remove particulate matter. This supernatant fluid was then dried down and converted to the chloride form using quartz distilled 6M HCl.

500 ml of riverwater was collected and a 100 ml sample was centrifuged to remove particulate matter. 5 ml of 6M Hcl was added to the supernatant water to convert the dissolved salts to the chloride form and this was evaporated to dryness.

The plant samples were washed in de-ionized water to remove possible surface contamination, dried for 20 min at 200 °C in an oven and then reduced to a powder. About 0.5 g of plant material was dissolved using microwave digestion in 10 ml of 16M Teflon-distilled HNO_3. The dissolution vessels were pre-cleaned using HCl and blank values for the final 10 ml HCl cleaning stage averaged 380 pg ($n = 3$).

Tooth enamel from 16 adults and seven children was analysed. All of the teeth were either permanent premolars or molars with the exception of three deciduous molars (KCC 98-57, 98-68a and 98-71).

Human tooth samples were prepared as follows. Teeth were halved and the exterior crown surface of the tooth samples was abraded from the surface to a depth of $>100\,\mu$m using a tungsten carbide dental burr. Dentine was drilled out and the reject material was discarded; then thin enamel slices were cut from the tooth using a flexible diamond-edged rotary dental saw. The teeth of grazing animals mineralize incrementally so that a zonation can develop within the teeth if the animal is exposed to different food or water compositions through the growth period (Balasse et al. 2002). To overcome the possibility of Sr isotope zonation within the animal teeth, the full length of each tooth was sampled. All saw surfaces were mechanically cleaned with a tungsten carbide burr and any adhering dentine was also removed. The resulting enamel samples were transferred to a clean (class 100, laminar flow) working area for further preparation.

In a clean laboratory, the enamel samples were first cleaned ultrasonically in high purity water to remove dust, rinsed twice in water then in high-purity acetone, and then weighed into precleaned Teflon beakers. An aliquot of ^{84}Sr tracer solution was added to each sample, which was then dissolved in Teflon-distilled 16M HNO_3.

Strontium was collected from all samples using conventional ion exchange methods. The average strontium blank value during analysis was 67pg, $n=7$.

The Sr isotope composition and concentrations were determined by thermal ionization mass spectrometry (TIMS), using a Finnigan Mat 262 multicollector mass spectrometer. The tabulated ^{87}Sr/^{86}Sr ratios were determined to a precision of ±0.000007 1 SE (standard error) or better. All strontium ratios have been corrected to a value of 0.710240. Typical external precision of this machine for National Bureau for Standards, NBS-987 ^{87}Sr/^{86}Sr ratio is 0.004% (2s). The results are presented in Tables 2 and 3.

Results

A comparison of soil-leachates and plant composition

A study of Sr isotope compositions in forested areas (Blum et al. 2000) showed no significant difference in the Sr isotope composition from leached soils up the trophic levels from plants, through insects, to birds. Leaching of soil to remove the labile components is commonly used to estimate the Sr isotope composition of the local biosphere (Budd et al. 2002). However, this study shows that such a practice can lead to an underestimation of the biosphere values. Figure 2 shows data for 12 localities covering

Table 3. Sr concentrations and ^{87}Sr/^{86}Sr isotope ratios of tooth enamel from animal and human molar teeth at the Ketton site and isotope ratios for riverwater samples sampled in the area

Sample number	Sr (p.p.m.)	^{87}Sr/^{86}Sr
Sheep/goat molars		
KCC 98–1	147	0.709685
KCC 399	105	0.710469
KCC 98–2	96	0.709685
KCC 348	115	0.709352
KCC 98 401	213	0.711188
466*	74	0.708921
416	104	0.710994
Cattle molars		
KCC 98 384	95	0.708814
KCC 98	146	0.711259
KCC 401	511	0.709703
KCC 98	88	0.710285
416*	80	0.708938
Pig molars		
KCC 401	46	0.709322
Human adult premolars		
KCC 98 58	67	0.709392
KCC 98 67	123	0.710261
KCC 98 43	67	0.708578
KCC 98 9	69	0.709349
KCC 98 7	66	0.708966
KCC 98 64	105	0.711407
KCC 98 65	60	0.710489
KCC 98 66	81	0.709375
KCC 98 17	58	0.709505
KCC 98 47	72	0.709372
KCC 98 14	153	0.712663
KCC 98 60	123	0.710273
KCC 98 52	60	0.709535
KCC 98 34	62	0.709213
KCC 98 55	75	0.710566
KCC 98 63	86	0.710267
KCC 98 54	126	0.709835
KCC 98 56	97	0.710997
Human juvenile molars		
KCC 98 6	104	0.709275
KCC 98 32	51	0.709244
KCC 98 40	79	0.709564
KCC 98 13	44	0.709460
KCC 98 68b	58	0.709878
Human juvenile deciduous molars		
KCC 98 68a	53	0.709588
KCC 98 71	86	0.709358
KCC 98 57	44	0.709553
Modern river waters		
Locality A TL 009 062		0.709618
Locality B TL 045 078		0.708451
Locality C TL 065 069		0.709395

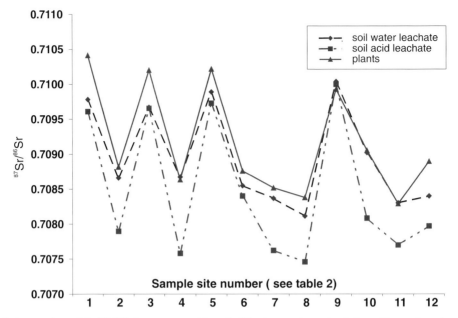

Fig. 2. A comparison of the $^{87}Sr/^{86}Sr$ isotope composition of soil leachates and associated plants. The numbers along the x-axis refer to sample localities given in Table 2.

most of the geological formations and rocks types within a *c.* 9 km radius of Ketton.

At each site, plant and soil samples were taken and the Sr isotope composition determined for the plant sample and for both the acetic acid and de-ionized water leachates of the soil. Figure 2 shows that the acetic acid leachate gave the lowest $^{87}Sr/^{86}Sr$ isotope ratio composition in all cases except Locality 9. The limestones of this area would preserve the Jurassic seawater strontium composition of *c.* 0.707 (McArthur *et al.* 2001). It is thus not surprising that, in the soils formed above carbonate-rich formations, such as the Blisworth Limestone Formation (Locality 7), the soluble carbonate dominates the acetic acid leachate solution. However, the results from the acetic acid soil leachates show that all the soils (with the exceptions of Localities 3 and 9) contain a carbonate component that is preferentially removed by weak acidic leaching. The de-ionized water leachate provides a closer approximation to the biosphere values, as represented by the plant data, but the results are not consistent insofar as, at some sites, the water leachates agree with the plant data (e.g. Localities 9–11) whereas at other Localities it is closer in composition to the acid leachate values (e.g. Localities 1, 3 & 12). This means that it is not easy to predict when the water leachate is likely to give a value that provides a good estimate of the plant values. Both water and acetic acid leachate data agree with the plant data at Locality 9. This site is in a well-established woodland with a thick undisturbed,

organic-rich soil. It is speculative, but worth suggesting, that the reason why the $^{87}Sr/^{86}Sr$ isotope compositions of the three samples from this locality agree is because this is a mature and possibly well-equilibrated soil. There are no obvious carbonate fragments in the soils and, if they were once present, the carbonate component in this soil may have been washed out over time by the slightly acidic conditions of the organic-rich soil, thus leaving a more homogenous soil. The conclusion of this experiment is that, in clay-carbonate terrains, direct sampling of plant material is the most reliable method of deriving the isotope composition of biosphere in a particular area.

Plants and soils have the advantage of being readily available and can be precisely sourced; however, their disadvantage is that they provide the average value for only the few centimetres of ground from which they derived their nutrients. A further disadvantage of plants is that they sample the present-day environment and do not take into account possible changes in land use that may effect the conditions of soil, fertilizer use, present-day contamination, erosion and hence strontium release. To overcome the problem of using 'point sources' to estimate the biosphere isotope composition of the 'local area', we also analysed riverwater, which gives an average value for the catchment upstream of the sample points. Finally we analysed the tooth enamel of domestic Anglo-Saxon animals that were contemporary with the period of use of the graveyard. This

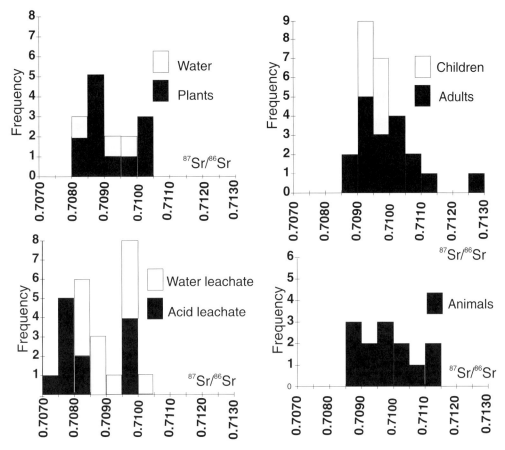

Fig. 3. Comparative histograms of ^{87}Sr/^{86}Sr isotope composition of samples taken as part of this study.

provides an average Sr isotope composition for the area over which the animals grazed during the formation of their tooth enamel, along with a method of looking at land composition in the past.

Data from the Anglo-Saxon human population

The ^{87}Sr/^{86}Sr isotope compositions of enamel from the human tooth samples are presented as histograms in Figure 3 and as averages and SDs in Figure 4 . These samples represent the composition of a population living on UK Jurassic lithologies during the Anglo-Saxon period, and they yield a mean value of 0.7098 ± 0.0018 (2σ, $n=22$). Within this group, data from the juveniles form a very tight cluster of 0.7094 ± 0.0002 (2σ, $n=7$). (Fig. 4) This may highlight the concept of 'local signature' as it is probable that the children were born and raised in, or very near, the manor, and they may typify the average Sr isotope composition of the land close to the settlement.

The dataset includes one outlier (KCC 98 14), that is more radiogenic than all the other samples. This sample is from a man for whom there is no archaeological evidence to suggest that he is 'exotic'. The diet of a rural community would be dominated by local crops and drinking water, and by locally raised animals used for either meat and dairy products. It is assumed that, within such a community, there would be a limited/negligible amount of imported food and drink, and that anything brought in would not constitute a major part of the diet. Permanent tooth enamel is mineralized during childhood, and it is unlikely that imported food and drink would constitute a significant part of a child's diet. Therefore, even where there is evidence of exotic foodstuffs, it seems reasonable to assume that the content of the childhood diet of an individual would be derived mainly from local sources. The ranges that are seen in these individuals are taken to represent the average value and variation within a community living within a radius of c. 9 km of the settlement site.

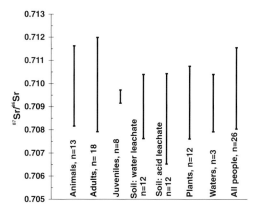

Fig. 4. A comparison of the average and 2σ range of $^{87}Sr/^{86}Sr$ isotope composition of populations from the area of study.

Comparison of human data with other datasets

The data on tooth enamel from animals (one pig, seven sheep and five cattle) may provide the best proxy for 'local signature' with respect to humans. The animal population is likely to have been dominated by animals that were born and bred in the area (O'Sullivan pers comm.), and because they grazed in several fields they would average the $^{87}Sr/^{86}Sr$ isotope composition of the plant matter they ingested, and the meat and dairy products they provided would constitute a considerable part of the human diet. The Sr isotope data support this hypothesis. The tooth enamel data for animals show the same range of values as those for humans (Table 3). Both have a comparable mean value and SD of 0.7099 ± 0.0017 $(2\sigma, n = 13)$.

By contrast to the two archaeological populations of human and animal tooth enamel, the present-day data from soil leachates and plants are all shifted to less radiogenic values. The plant leaves and riverwater samples have mean values of 0.7092 ± 0.0016 $(2\sigma, n = 12)$ and 0.7091 ± 0.0012 $(2\sigma, n = 3)$ respectively, and the soils give average values of 0.7085 ± 0.0020 $(2\sigma, n = 12)$ for acid leachates and 0.7090 ± 0.0014 $(2\sigma, n = 12)$ for water leachates (Fig. 4).

Population comparisons using statistical methods of analysis.

To test the hypothesis that there are significant differences between some or all of the sample groups in this study, we applied statistical tests using analysis of variance (ANOVA). The results are given in Table 4. The acid leachate from the soils shows a low prob-

ability of sampling common populations with all other samples. The riverwater shows a high probability (18–76%) of sampling the same source of $^{87}Sr/^{86}Sr$ as all the other sample sets, except acid leachate of soils. However, the riverwater data should be treated with some caution as only three samples were analysed. The data from present-day riverwaters, the water leachates of soils and plant leaves show a high probability of a common parent population and suggest that these sample sets provide a good estimate of the values of present-day biosphere $^{87}Sr/^{86}Sr$ values. The acetic acid leachates of the soils favour the carbonate component in the soil too strongly and produce data that is less radiogenic than the other present-day indicators.

The data from the Anglo-Saxon human adults and Anglo-Saxon animals show a high probability of sampling the same $^{87}Sr/^{86}Sr$ source in the past. There is a 75% probability that the sample populations come from the same background population when tested at the 2σ level. There appears to be a compositional gap between present-day data and Anglo-Saxon data because the null hypothesis is rejected when comparing the adults or animals with present-day plants or with soil leachates of either type. However, the data for the children bridge the time gap. Because the data for the children are so tightly grouped, they fall within all the other datasets and, in the ANOVA test, sample a common population with all except the soil leachates.

The data show that the tooth enamel from the Anglo-Saxon animal population is a good match for the tooth enamel of the Anglo-Saxon adult humans, probably because both populations sourced their diet from predominantly the same geographical area. The data for the children appears to reflect a diet sourced from a much smaller geographical area. The data for the children and the present-day riverwaters appear to give the best estimate for the median values of the area.

Environmental differences between Anglo-Saxon times and the present day

Several factors could account for the less radiogenic nature of the present-day soil and plant leaf data in comparison with Anglo-Saxon tooth enamel data. It is possible that the carbonate component in the soils is now contributing more to the biosphere than it did in Anglo-Saxon times. Modern ploughing methods are likely to dig deeper and bring more of the limestone to the ground surface. Acid rain is favouring the dissolution of the limestone, and the generally greater extent of ploughed land is removing the cover of mature soils, such as those still found in woodland sites (Locality 9), where the soils leachates and the plant leaf values all agree. It seems

Table 4. *Results of analysis of variance using ANOVA of the Sr isotope ratios for different sample populations from the study area. The results in bold highlight data pairs where the probability is greater than 10% that they sample the same population*

	animals					
children	**0.21**	children				
adults	**0.75**	**0.164**	adults			
acid soil	0.001	0.01	0.0003	acid soil		
water soil	0.007	0.05	0.004	**0.18**	water soil	
plants	0.04	**0.29**	0.022	0.07	**0.48**	plants
rivers	**0.19**	**0.18**	**0.17**	**0.28**	**0.66**	**0.97**

Table 5. *A comparison of the probability that the animal data from Ketton sample the same Sr biosphere as other UK sites. In all cases except Mangotsfield the hypothesis is rejected; the Ketton teeth data distribution seem unique to the Jurassic clay-carbonate assemblage*

	West Heslerton	Monkton	Winchester	Gloucester	Mangotsfield
ANOVA probability	0.06	0.019	0.010	0.02	0.86

important, therefore, to try to use contemporary estimates of biosphere strontium when looking at archaeological populations.

Can overlapping British populations be distinguished?

Many past human populations in the British Isles will have strongly overlapping $^{87}Sr/^{86}Sr$ values but it may be possible to characterize populations from different geological environments by looking at the overall distribution of data from a site. To assess this possibility the animal tooth enamel dataset, which defines the nature of strontium distribution in the Jurassic site at Ketton, has been compared with archaeological human populations from other sites in the British Isles. These include: (1) individuals from an early Anglo-Saxon cemetery in an area of chalk and clay at West Heslerton in Yorkshire (Budd *et al.* 2002; Montgomery 2002); (a) a late Roman sarcophagus from Mangotsfield near Bristol, located on Lower Lias rocks (Montgomery 2002); (3) a late medieval Dominican Friary in Gloucester (Budd *et al.* 2000; Montgomery 2002), also located on Lower Lias rocks; (4) a fourth-century Roman cemetery on the site of the Eagle Hotel in Winchester (Montgomery 2002), which is on Chalk; and (5) a Neolithic henge monument at Monkton Up Wimborne (Montgomery *et al.* 2000) in Dorset, which is also on Chalk. We have no conclusive evidence that the individuals excavated at these sites were raised in the local area, but this is not critical because the test demonstrates the robustness of the approach when comparing datasets from various settings. The data from these populations overlap with the Ketton data, sometimes considerably. The results for ANOVA assessment of these comparisons are given in Table 5.

The null hypothesis that the animals from Ketton sample the same population as the humans can be rejected for the Chalk-based sites of Monkton, Winchester and the Dominican Friary site. Although the datasets overlap, the less radiogenic nature of the chalk-based populations and the overall more radiogenic nature of the Dominican Friary population show that samples could not have come from the same geological environment as the Ketton animals.

The data from West Heslerton show a 10% probability of samples being from the same parent population as the Ketton Anglo-Saxon animals. This result is inconclusive. There is, however, a very strong probability that the data from Roman sarcophagus at Mangotsfield, near Bristol, do sample the same $^{87}Sr/^{86}Sr$ population as the animals at Ketton. Given the hundreds of kilometres between the two sites, this suggests a flaw in the approach; however, it may demonstrate the viability of the method. The individuals were buried a few kilometres west of the Upper Lias formations of the Jurassic, extending in a northeasterly direction through central England to the Ketton area. It is therefore possible that the $^{87}Sr/^{86}Sr$ distribution in the burials at Mangotsfield is similar to that in the animal tooth enamel from Ketton because their diets reflect the same geology; in fact the two groups do sample the same $^{87}Sr/^{86}Sr$ population, despite being many kilometres apart. Further data is needed to test this hypothesis fully but, if true, it would mean it is possible to create statistical fingerprints for the main geological subdivisions of the UK against which populations of unknown origin can be tested.

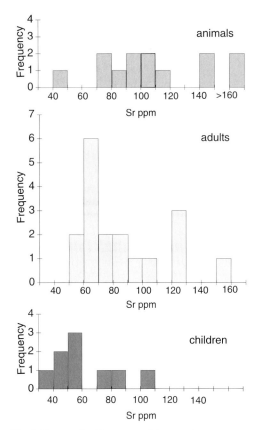

Fig. 5. Comparative histograms of the strontium concentrations in the animals and human adults and children of this study.

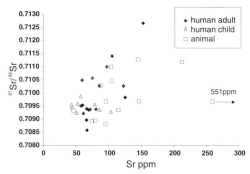

Fig. 6. $^{87}Sr/^{86}Sr$ plotted against the strontium concentration for tooth enamel samples analysed in this study.

Variation in strontium concentrations in tooth enamel

The variation of strontium concentration in teeth also has the potential to provide information about diet, dietary range and geographic/climatic regions. The strontium concentration in tooth enamel and bone is dependent upon the strontium content of the diet and the trophic level of the individual (Burton *et al.* 2003). It is generally observed that herbivores have higher strontium concentrations, than carnivores and this is attributed to the strontium-rich nature of plants, coupled with the body's tendency to take up calcium in preference to strontium, thus reducing the Sr/Ca ratio with increasing trophic level (Blum *et al.* 2000). However, there appear to be no absolute levels that are diagnostic of certain diets as there are large differences in populations, which appear to be geographically and climatically controlled. Archaeological tooth data from the UK suggest strontium concentration range for UK residents of 68 ± 23 p.p.m. (2σ, $n = 126$) (Montgomery

2002). A study of modern UK teeth (Brown *et al.* in press) from children in southeastern England yielded 118 ± 78 ppm (2σ, $n = 27$). In contrast, work in Arizona on archaeological tooth enamel of the Pueblo Native Americans found much higher overall strontium concentrations: 360 ± 578 p.p.m. (2σ, $n = 68$), with the highest value in the populations recorded at 1100 p.p.m. (Ezzo *et al.* 1997). It is not currently known whether this difference between the UK and the Arizona data is due to much higher strontium concentrations in the soil being passed into the biosphere, or whether, perhaps, the much fleshier, water-retentive nature of plants in the UK maritime climate dilutes the strontium uptake when compared to desert vegetation. Whatever the reason for this large difference in population strontium, it suggests that discussion of strontium concentrations, and possibly Sr/Ca ratios, should be done relative to similar communities and not be taken as absolute values and dietary indicators.

The strontium concentration data from this study of Anglo-Saxon Ketton provides a point of reference for UK studies (Fig. 5). data shows a mean of 86 ± 58 p.p.m. (2σ, $n = 18$) for the adults, 65 ± 44 p.p.m. (2σ, $n = 8$) for the children and 140 ± 236 p.p.m. (2σ, $n = 13$) for the animals. The large SD in the animals is caused by the single high concentration results of 515 p.p.m. If this is excluded, the animals show a range of 109 ± 86 (2σ, $n = 12$).

The individual with the highest strontium concentration (KCC 98 14) also has the most radiogenic isotope signature; therefore, although there is no archaeological reason to suggest that he is not part of the population, it appears that he could have come from a different geological area and/or had a different dietary regime.

It was beyond the scope of this study to subdivide the teeth in order to assess possible strontium variation along the length of a tooth, but the close isotope correlation between humans and animals argues against a significant bias caused by sampling within this study of the animal tooth population.

Comparison of Sr p.p.m. v. Sr isotope composition within the three groups

The data are displayed as $^{87}Sr/^{86}Sr$ v. Sr p.p.m. in Figure 6. The data from the children's tooth enamel form a well-defined cluster which we take to represent the very local conditions around the settlement. Seven of the adults fall within this field, as does one sample of tooth enamel from a pig. This grouping suggests that it may be possible to further subdivide the humans into those individuals that fall within the field of the children, and may have been raised exclusively in the settlement, and those individuals that come from slightly further a field. However, within the remit of this study, the human data match the animal data and are assumed to come from the same geographical area as the animals.

Conclusions

This study concludes that the best method of predicting the isotope composition of an historic/ prehistoric human population is to define the Sr isotope composition of the area of interest using animal teeth from the same period. An understanding of land usage and coverage is useful in this context, and a definition of 'local' within the confines of a study is needed. By analysing a significant number of human individuals (or animals) that can be assumed to have been raised in a particular area, it should be possible to create 'fingerprint populations' for certain geological environments that can be used for comparative purposes when assessing populations of unknown provenance. Strontium concentrations broadly follow a pattern of diminishing concentration with rising trophic level but such comparisons must be restricted to similar geographical and climatic regions. The vast difference in strontium concentration range between British populations and desert-dwelling Native Americans suggests that local conditions play an important role in determining the strontium concentrations in tooth enamel.

In this study, data from modern sources tended to give slightly less radiogenic datasets. The reason for this is uncertain but it may result from deeper and more extensive ploughing bringing more carbonate to the surface, with the possibility of more acidic present-day rain leaching out the carbonate preferentially.

We thank J. Wakley and D. O'Sullivan for discussions and J. Montgomery for constructive comments on an early version of the manuscript, which was undertaken as part of NIGL project IP/735/1001.

References

BALASSE, M., AMBROSE, S. H., SMITH, A. B. & PRICE, T. D., 2002. The seasonal mobility model for prehistoric herders in the south-western Cape of South Africa assessed by isotopic analysis of sheep tooth enamel. *Journal of Archaeological Science*, **29**, 917–932.

BLUM, J. D., TALIAFERRO, E. H., WEISSE, M. T. & HOLMES, R. T. 2000. Changes in Sr/Ca, Ba/Ca and $^{87}Sr/^{86}Sr$ ratios between trophic levels in two forest ecosystems in the northeastern USA. *Biogeochemistry*, **49**, 87–101.

BROWN, C. J., CHENERY, S. R. N. *ET AL.* (in press). Environmental influences on the trace element content of teeth: implications for disease and nutritional status. *Archives of Oral Biology*.

BUDD, P., MONTGOMERY, J., BARREIRO, B. & THOMAS, R. G. 2000. Differential diagenesis of strontium in archaeological human dental tissues. *Applied Geochemistry*, **15**, 687–694.

BUDD, P., CHENERY, C., MONTGOMERY, J., EVANS, J. & POWESLAND, D. 2002. Anglo-Saxon residential mobility at West Heslerton, North Yorkshire from combined O- and Sr-isotope analysis. *In*: TANNER, S. & HOLLAND G. (eds) *Proceedings of the 8th International Conference on Plasma Source Mass Spectrometry*. Royal Society of Chemistry, Durham, 195–208.

BURTON, J. H., T. D. PRICE, CAHUE, L. & WRIGHT, L. E. 2003. The use of barium and strontium abundances in human skeletal tissues to determine their geographic origins. *International Journal of Osteoarchaeology*, **13**, 88–95.

COOPER, N. J. 2000. *The Archaeology of Rutland Water*. Leicester Archaeology Monographs, 6.

EZZO, J. A., JOHNSON, C. M. & PRICE, T. D. 1997. Analytical perspectives on prehistoric migration: a case study from east-central Arizona. *Journal of Archaeological Science*, **24**, 447–466.

GRAUSTEIN, W.C., 1989. $^{87}Sr/^{86}Sr$ ratios measure the sources and flow of strontium in terrestrial ecosystems. *In*: RUNDEL, P. W., EHLERINGER, J. R. & NAGY, K. A. (eds) *Stable Isotopes in Ecological Research: Ecological Studies*. Springer Verlag, New York, 491–512.

HOOGEWERFF, J., PAPESCH, W. *ET AL.* 2001. The last domicile of the Iceman from Hauslabjoch: a geochemical approach using Sr, C and O isotopes and trace element signatures. *Journal of Archaeological Science*, **28**, 983–989.

MCARTHUR, J.M., HOWARTH, R.J. & BAILEY, T.R., 2001. Strontium isotope stratigraphy: LOWESS version 3: best fit to the marine Sr-isotope curve for 0–509 Ma and accompanying look-up table for deriving numerical age. *Journal of Geology*, **109**, 155–170.

MONTGOMERY, J. 2002. *Lead and strontium isotope compostions of human dental tissues as an indicator of ancient exposure and population dynamics*. Bradford University, PhD thesis.

MONTGOMERY, J., BUDD, P. & EVANS, J. A. 2000. Reconstruction of the lifetime movements of ancient people: a Neolithic case study from southern England. *European Journal of Archaeology*, **3**, 407–422.

MONTGOMERY, J., EVANS, J. A. & NEIGHBOUR, T. 2003. Sr isotope evidence for population movement within the Hebridean Norse community of NW Scotland. *Journal of the Geological Society*, **160**, 649–653.

MONTGOMERY, J., EVANS, J. & ROBERTS, C. A. 2004. Continuity or colonization in Anglo-Saxon England? Isotope evidence for mobility, subsistence practice and status at West Heslerton. *American Journal of Physical Anthropology*.

PRICE, T. D., JOHNSON, C. M., EZZO, J. A., ERICSON, J. & BURTON, J. H. 1994. Residential-mobility in the prehistoric southwest United States – a preliminary study using strontium isotope analysis. *Journal of Archaeological Science*, **21**, 315–330.

PRICE, T. D., BURTON, J. H. & BENTLEY, R. A. 2002. The characterization of biologically available strontium isotope ratios for the study of prehistoric migration. *Archaeometry*, **44**, 117–135.

SCHWEISSING, M. M. & GRUPE, G. 2003. Tracing migration events in man and cattle by stable isotope analysis of appositionally grown mineralized tissue. *International Journal of Osteoarchaeology*, **13**, 96–103.

SILLEN, A., HALL, G., RICHARDSON, S. & ARMSTRONG, R. 1998. $^{87}Sr/^{86}Sr$ ratios in modern and fossil food-webs of the Sterkfontein Valley: Implications for early hominid habitat preference. *Geochimica et Cosmochimica Acta*, **62**, 2463–2473.

TRICKETT, M.A., BUDD, P., MONTGOMERY, J. & EVANS, J. 2003. An assessment of solubility profiling as a decontamination procedure for the $^{87}Sr/^{86}Sr$ analysis of archaeological human skeletal tissue. *Applied Geochemistry*, **18**, 653–658.

Forensic geology of bone mineral: geochemical tracers for post-mortem movement of bone remains

CLIVE N. TRUEMAN

School of Earth and Environmental Sciences, University of Portsmouth, Burnaby Building, Burnaby Road, Portsmouth PO1 3QL, UK (e-mail: clive.trueman@port.ac.uk)

Abstract: The trace element chemistry of bone is rapidly altered post-mortem, and the post-mortem chemistry of bone mineral has been used successfully to discriminate between fossil bones removed from stratigraphically and spatially separate excavations. These techniques can be used to identify and protect scientifically or culturally sensitive artefacts. Measurable changes in the trace element composition of bone may occur within a few years of death. These changes show considerable potential for use in forensic investigations, particularly where post-mortem movement or scattering of bone remains has occurred. Early post-mortem alteration is complex, governed by a number of environmental variables, and measuring subtle changes in bone chemistry presents considerable analytical challenges. More work is needed in this area.

Previously interred bones may be moved either through natural processes, such as erosion and reworking, or by human activities, such as excavation, reburial and trading. It may be necessary to trace the original burial location of such disturbed bones, for instance to investigate suspected associations in scattered or disarticulated skeletal remains.

As bone is interred, its elemental composition is altered (e.g. Parker & Toots 1970; Price 1989; Radosevich 1993; Trueman & Tuross 2002). Many elements are added to bone in concentrations that far exceed *in vivo* levels. These diagenetic elements could potentially be used to provide a record of the original location and environment of burial.

This paper shows, with case studies, how the post-mortem alteration of bone chemistry may be exploited to protect ancient bone assemblages from commercial exploitation. Rates and mechanisms of early post-mortem chemical alteration of bone mineral are then discussed, together with the possibility that such changes could aid modern forensic investigations.

Trace elements in ancient bone

Why should we care about ancient bone?

Ancient bone remains represent a valuable and restricted cultural resource. Public interest in these rare and often striking artefacts has led to a thriving commercial market akin to the trade in fine art, with comparable prices attached to the finest specimens. It is seldom possible to dig up a grand master, but palaeontological and archaeological resources are vulnerable to unrestricted commercial exploitation, particularly in geographical areas with low incomes. Unrestricted exploitation of fossils is potentially very damaging to science, and to the cultural heritage of an area, but some public access to fossils is also necessary. Fossils are uniquely able to inspire public interest in science, and restricting ownership of fossils to a scientific élite can stifle this and foster resentment towards science. Furthermore, many valuable specimens would never be found without the dedicated and skilled work of amateur and commercial collectors. A balance is therefore needed, restricting collection from sensitive sites but allowing collection from other, less scientifically sensitive localities.

Unfortunately, the only way to impose restrictions on the collection of fossils from specific individual sites is to monitor the site physically. This is expensive and difficult to enforce, and there are many examples of illegal removal of fossils from restricted localities. It is therefore much easier and cheaper either to impose a blanket ban on private collection or to impose no restriction at all. Neither of these solutions is ideal. These problems could be lessened or overcome if a fossil bone could be traced to its original source locality at any time after excavation. Suspected wrongful excavation of bone artefacts could then be tested by enforcement agents at the point of recovery, or could be insisted upon as part of a legitimate transaction. The diagenetic chemistry of bone potentially provides a natural chemical 'tag' that could be used to link bones to their original source locality. As this tag is integral to the bone material, it cannot be subsequently erased or altered without destroying the fossil.

Natural post-mortem chemical tags in bone

The controls affecting the trace element composition of any bone at time *t* after death, may be expressed as:

From: PYE, K. & CROFT, D. J. (eds) 2004. *Forensic Geoscience: Principles, Techniques and Applications*. Geological Society, London, Special Publications, **232**, 249–256. © The Geological Society of London, 2004.

$$X_{i\,(bone\,final)} = f(X_{i\,(bone\,initial)}, X_{i\,(pore\,water)},$$
$$KD_i, D_i, C, H, M) \qquad (1)$$

where X_i is the concentration of trace element (i) in system X, KD is the apatite fluid adsorption coefficient, D is the diffusion coefficient, C is the chemistry of the microenvironment of burial, H is the hydrology of the microenvironment of burial, and M is the bone microstructure (after Trueman 1999).

This is clearly a complex system. Variations in the chemistry and hydrology of the local burial environment may alter the speciation of the trace element under study, and therefore alter both its diffusion and adsorption coefficients. Minor differences in soil properties may influence the rate and relative extent of uptake of different trace metals into bone from pore waters. The post-mortem trace element chemistry of bone mineral is therefore potentially a sensitive indicator of the early depositional environment and hence, the burial locality.

Many elements are incorporated into bone post-mortem. However to be useful as a tracer for the burial locality, any target element or elements must meet several criteria;

(1) They must vary significantly between environments.
(2) They must not be common (present) in living tissue.
(3) They must be incorporated rapidly and easily into bone post-mortem.
(4) They must not be susceptible to appreciable fractionation after initial incorporation into bone.

A limited number of elements fulfill all these criteria. Presently the rare earth elements (REEs) show the most promise. The REEs, or lanthanide elements, are typically present in living bone in concentrations of between 0.1 and 10 parts per billion (Trueman & Tuross 2002). They have no known physiological function and are not concentrated in any major foodstuff. Controlled feeding experiments show that REEs pass almost quantitatively through the gut with little biological uptake (e.g. Austreng et al. 2000). Consequently there is no known difference in in vivo concentrations of REEs between taxonomic or dietary groups. The REEs have an extremely strong affinity for calcium sites in the apatite lattice, and fossil bone is typically enriched in REEs >5 orders of magnitude over in vivo concentrations (Bernat 1975). The rate of incorporation of REEs into bone varies with depositional environment, but uptake appears to be essentially complete, and stabilized within 1–10 ka in many depositional settings (Trueman & Tuross 2002). Variation of REE composition in bones between and across depositional environments has been demon-

strated in several studies (Trueman & Benton 1996; Trueman 1999; Staron et al. 2001; Patrick et al. 2002). Consequently the REE composition of fossil bone has been identified as a possible tracer to identify original burial location in ancient fossil bones.

Using chemical tracers to establish the provenance of fossil bone

Trace elements are incorporated into bone from groundwaters and are not transferred directly from the soil minerals. Therefore, the trace element composition of a buried bone reflects the chemistry of the groundwater and may be very different from the bulk soil chemistry. This is a critical observation, as it means that the trace element composition of a fossil bone cannot be linked directly to the chemistry of the soil at a particular suspect site. All bones interred in a common burial environment, however, will inherit a common trace element signal. The trace element composition of fossil bones can therefore be used to investigate original burial associations in scattered bone remains by comparing the chemical composition of the bones.

The evidential value of chemical signals in buried bones must be assessed with reference to natural levels of variation in the case in question. Chemical signals will vary within individual bones (Williams & Potts 1988), and between bones within a single undisturbed skeleton (Samoilov & Benjamini 1996). It is therefore essential to assess the level of variation in a known sample against which the test sample can be compared (Fig. 1).

In practice, investigators would like to know whether a particular bone or set of bones has been removed from a specific locality. In order to test this, a sample of bone fragments should be removed from the suspected excavation site (reference population), and the concentration of a suite of trace elements (e.g. REEs, U and Th) determined in these bones. The suspect bone or bones can then be analysed geochemically and compared statistically to this population (e.g. using analysis of variations (ANOVA) or discriminant analysis). A quantitative level of confidence may then be ascribed to any interpretations regarding bone provenance. If the suspect bone falls within the range of bones from the reference population, however, then the null hypothesis that the two bones were removed from the same site cannot be rejected (Fig. 1A). If the suspect bone or population of bones falls outside of the range of the reference population, then it is unlikely that the suspect bone was removed from the same site (Fig. 1B). Chemical similarity between two bones or groups of bones does not prove that these bones were originally associated. The trace element signal inherited by bones is a function of the availability of trace elements in pore waters, and it is

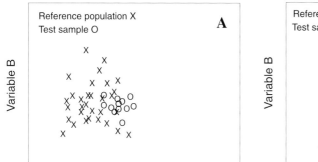
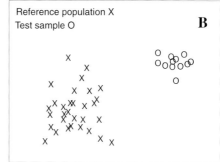

Fig. 1. Establishing the provenance of fossil bones using chemical signals. The provenance of a suspect test sample can be tested by comparing its chemical composition with a reference sample taken from a suspected or claimed site of origin. (**a**) Test and reference populations cannot be separated on the basis of chemical variables A and B. In this case the null hypothesis (that the test sample was removed from the same locality as the reference sample), cannot be rejected. This does not prove that the two populations were removed from the same locality, as other untested populations could yield similar signals. (**b**) The test and reference populations yield distinct values for variables A and B. In this case it is unlikely that the test sample was removed from the same locality as the reference sample.

quite possible that unrelated depositional environments will impose an analytically indistinguishable chemical signal on their respective interred bones. The likelihood of this occurring is a function of the chemical homogeneity of the depositional setting. Deep marine environments, for instance, are relatively buffered and well mixed, and unrelated depositional settings are likely to yield indistinguishable trace element signals in interred bones. Soil environments tend to be more spatially and temporally heterogenous, and the evidential value of chemical similarity between bones in these settings will be higher. Unfortunately it is currently impossible to quantify these environmental differences because very few detailed studies of spatial variation in bone chemistry within single depositional environments have so far been published (Trueman *et al.* 2003).

Geochemical comparisons provide a powerful method to test specific claims of provenance, but the strength of the method relies on adequate sampling of a defined reference population. Samples removed from the likely source for the suspect bone (or the site claimed as the exact source) must contain the full range of chemical variation across that site. In some cases, variation in REEs, U and Th compositions may be extremely high across the site, and in these cases it may be possible to choose different elemental or isotopic tags with lower spatial variation.

The trace element pattern developed in a fossil bone will reflect the aqueous transport chemistry of trace elements in local groundwaters. The REE are fractionated during weathering and transport, so that, broadly speaking, the REE chemistry of a fossil bone reflects the depositional setting (Fig. 2). Within

each depositional environment, however, local environmental conditions produce variation that can be used to distinguish bones from different stratigraphic horizons. In some situations, bones buried less than 20 m apart laterally and over less than 1 ka in time may be distinguished using REE chemistry (Trueman *et al.* unpublished data).

Case studies

The power of trace element chemistry to discriminate between bones from different burial horizons or localities has been demonstrated in several recent publications (Plummer *et al.* 1994; Trueman & Benton 1997; Trueman 1999; Staron *et al.* 2001; Trueman *et al.* 2001; Patrick *et al.* 2002; Trueman and Tuross 2002; Trueman *et al.* 2003). These are reviewed briefly below.

The provenance of a fossil human mandible was investigated by Plummer *et al.* (1994). The mandible had been recovered from surface weathering of exposed rock surfaces in Kanjera, Kenya. The skull could potentially have been derived from one of several original depositional horizons, ranging in age from the Pliocene to the Recent. If the mandible were derived from the older units in this succession it would represent the earliest occurrence of anatomically modern humans. Plummer *et al.* (1994) determined the concentration of a suite of trace elements in *c.* 300 fossil bone and tooth fragments nondestructively using X-ray fluorescence spectroscopy. A principal components analysis showed that bones and teeth recovered directly from six

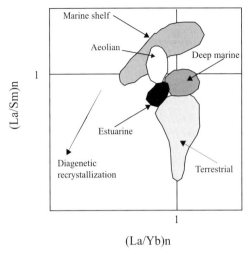

Fig. 2. REE ratios of bone samples broadly reflect the dominant transport mechanism of REEs in specific environments. Figure represents a composite of *c.* 1000 individual analyses of bones and teeth from a wide range of geological ages and environmental contexts. Adapted from Trueman & Tuross (2003) and Lécuyer *et al.* (2003).

lithologies which could have been the source for the disputed mandible could be distinguished on the basis of their trace element composition. Subsequent analysis of the disputed mandible showed that it grouped closely with bones and teeth derived from overlying Holocene sediments, and therefore previous assignments of early age to the bone were doubtful. Evidently this methodology could be applied equally to test whether a bone, or bones, had been removed illegally from one of the six sampled horizons, assuming that these six horizons reflected all possible local sources.

Trueman & Benton (1997) demonstrated that the REE signal inherited in bones during early diagenesis was resistant to later reworking into a different environmental setting. In so doing, they established that the REE signal could be used as a natural chemical tag, identifying the early depositional environment. This method is an improvement over that of Plummer *et al.* (1994), as the REEs are unequivocally associated with the calcium sites in the bone mineral lattice. The REE signal in bone is therefore related only to movement of the bone itself and is not affected by contamination from detrital or authigenic minerals. The application of REE geochemistry to problems of bone provenance was extended further by Staron *et al.* (2001), who used correspondence analysis to quantify the strength of geochemical associations between groups of fossil bones. Trueman *et al.* (2003) demonstrated that the variation in REE composition of bones within single assemblages varied significantly between assemblages and could in fact be used as a taphonomic character indicating the relative degree of mixing or time averaging in bone assemblages.

These studies demonstrate convincingly that the trace element chemistry and particularly the REE chemistry of fossil bone can discriminate between bones from differing early depositional settings. However, as suggested by Trueman (1999) and demonstrated by Trueman *et al.* (2003), the degree of spatial and temporal resolution available to these techniques is a function of the degree of chemical variation in groundwater at the time of early diagenesis and the amount of post-depositional movement of bone before permanent burial. Both of these variables will vary between assemblages. In a recent study, Trueman *et al.* (2001) determined REE compositions of *c.* 450 bone fragments from four stratigraphic horizons in the Pleistocene Olorgesailie Formation. This sample suite included 150 bones recovered from 11 quarries within a 500 m transect across a single palaeosol representing 1 ka in time. While the REE composition of bones from excavations within the palaeosol overlapped, in total >75% of the bones could be assigned to their original quarry based on discriminant analysis. Despite this relatively high level of variation within a single horizon, bones from each of the four stratigraphic horizons were separated with a total classification success of 94% (Fig. 3; Table 1). This study showed that, in ideal circumstances, REE chemistry can pinpoint the original excavation location of a fossil bone with high accuracy to within a few metres laterally and to the precise stratigraphic horizon. These methods could provide a powerful test of the provenance of suspected contraband artefacts.

Bone chemistry in modern settings

The incorporation of trace elements into bone post-mortem is rapid on palaeontological and archaeological time scales, but little is known about the rates of trace element change in bone on decadal time scales. Consequently, the post-mortem trace element composition of modern bone has not often been used to aid forensic investigations. If the trace element composition of bone is altered significantly within a few years of death, this very early post-mortem alteration may contain information that could be used in a forensic context. If not, the bone chemistry might reliably be used as an indicator of the provenance and life history of the living individual (Amin *et al.* 2003; Pye 2004). Bone crystals are among the smallest known biologically synthesised crystals, averaging around $20 \times 30 \times 5$ nm, with a correspondingly high surface area (Weiner & Price 1986). Bone crystals therefore have a high surface energy and efficiently remove many trace elements from pore

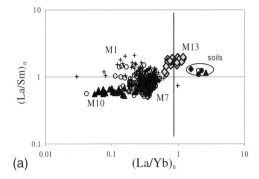

(a)

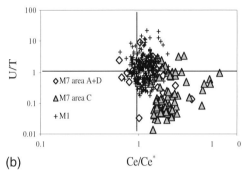

(b)

Fig. 3. Discriminating bones from separate stratigraphic horizons using their trace element composition. Plots of shale-normalized ratios of rare earth element compositions of c.400 bones from the Pleistocene Olorgesailie Formation, southern Kenya. Four main stratigraphic horizons were sampled within the Olorgesailie Formation. Bones from these horizons possess distinct trace element compositions. (**a**) Bones from two horizons (M7 & M10) show distinct rare earth element ratios, whereas bones from M1 & M7 cannot be separated on the basis of simple REE ratios alone. The REE composition of soil samples from all four horizons are circled, and do not correspond to the interred bones. (**b**) Redox sensitive trace element ratios successfully distinguish between most bones from M1 and M7. Geochemical separation was tested and quantified by discriminant analysis (Table 1). 10% error bars are smaller than plotted symbols.

Table 1. *Results of jackknifed discriminant analyses of bones shown in Figure 3. Individual bones could be assigned to their known original source with c. 90% accuracy*

Member	1	7	10	13	% correct
1	**242**	2	2	1	**98**
7	12	**95**	5	1	**84**
10	1	0	**22**	0	**96**
13	1	0	0	**19**	**95**
Total	**256**	**97**	**29**	**21**	**94**

waters by adsorption. Bonemeal is currently under investigation as a natural remediation agent that could be used to remove dissolved or colloidal metals from contaminated land (Valsami-Jones *et al.* 1996). The high cation exchange capacity shown by bone crystals suggests that their trace element chemistry will be altered readily whenever bone crystals come into contact with pore waters containing dissolved metals.

In a study of bones left to weather on the surface of the Amboseli Plain, southern Kenya, Tuross *et al.* (1989) noted that concentrations of strontium in bone mineral increased markedly within 10 years. These increases were associated with growth of bone mineral crystals rather than growth of new mineral phases within the pore spaces of bone, and were present throughout the sampled thickness of the bone. Shinomiya *et al.* (1998) also reported an influx of elements such as Fe, Al, and Ba in bones buried experimentally for 2 years. These two studies show that measurable changes in the trace element composition of bone can occur within a few years post-mortem.

Recently, Trueman *et al.* (2004) re-examined bones from the Amboseli Plain, now exposed for up to 40 years post-mortem. As demonstrated by Tuross *et al.* (1989), all bones showed significant changes in trace element (REE, Ba) composition during exposure and weathering. The concentrations of REEs in these bones increased by 400–1000%, and Ba concentrations increased by 100–700% over 15–25 years (Fig. 4). Unlike the previous studies mentioned, Trueman *et al.* (2004) sampled different portions of each bone, and the degree of mineralogical and chemical alteration differed significantly between different portions of the same bone. The REEs showed maximum post-mortem increases nearest to the bone/sediment interface, whereas increases in Ba concentrations showed no systematic distribution throughout the bone. The alteration of these bones was driven by high rates of evaporative transport of groundwater from the soil through the bone, and consequently bones acquired trace metals from the soil pore waters despite remaining unburied on the soil surface. The extreme rates of uptake of trace elements seen in bones from Amboseli reflect the high rates of evaporation and also relatively rapid degradation of collagen in this environment, freeing bone crystal surfaces (Trueman *et al.* 2004). Where pore water flow and /or rates of collagen decomposition are slower, uptake of trace metals will be reduced. Rates of collagen decomposition vary with temperature and microbial activity (e.g. Collins *et al.* 2002) and pore water flow is dependent on the local hydrological conditions. Clearly, it is difficult to predict the extent of trace element alteration likely to be suffered by a given bone without detailed knowledge of the

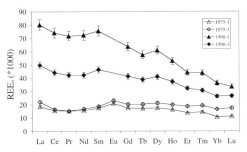

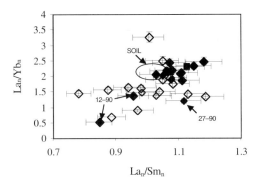

Fig. 4. Increase in REE concentrations in exposed bones (ribs) from the skeleton of an African elephant (*Loxodonta africana*) after 15 years of weathering in Amboseli National Park, Kenya. Open symbols, REE contents in bone recovered in 1975; closed symbols, REE contents in bone recovered in 1990. Note 100–400% increase in REE contents in 15 years of exposure. 10% error bars shown based on repeated analyses of standard reference materials.

Fig. 5. Shale normalized REE ratios of bones from Amboseli National Park, Kenya. Open symbols, bones collected from skeletons in 1975; closed symbols, bones collected from the same skeletons in 1990. Note variation in REE ratios is reduced during weathering, and all bones trend towards composition of REE adsorbed onto soil particles (arrowed ellipse). 10% error bars shown on repeated analyses of standard reference materials. Outlier 12–90 contained a mineral crust, and outlier 27–90 showed no weathering effects. Data taken from Trueman *et al.* (2004).

environment of deposition. Where alteration is rapid, however, post-mortem changes in the trace element composition of bones have the potential to provide information about the early post-mortem location of bone remains.

Several taxa were incorporated in the Amboseli study, with distinct (but very low) *in vivo* REE compositions, presumably related to minor differences in the REE composition of diet and/or drinking water. As the REEs were adsorbed onto bone crystal surfaces post-mortem, however, the total REE compositions of all bones trended towards a common diagenetic signal (Fig. 5). This observation supports the argument that the early post-mortem trace element chemistry of bone may provide a signal that is related to its location. This is a very preliminary result, however, and a number of caveats should be attached to any interpretation.

(1) *Analytical errors* The concentrations of many trace elements, such as the REEs, are relatively low in fresh and recently buried bone, and differences between individual bones are extremely low. The analytical errors associated with measuring trace elements in modern bone are therefore relatively high, leading to a low 'signal to noise' ratio. Recent advances in analytical technology, particularly development of cell technology; and multicollector inductively coupled plasma mass spectrometry (ICP-MS), will doubtless improve this situation, but interpreting chemical evidence can be difficult when operating close to the limits of reliable analytical performance.

(2) *Spatial variation.* The bones in the Amboseli study were recovered from a relatively wide

area ($>1 \text{km}^2$) covering a range of microenvironments. All these bones trend towards a similar post-mortem REE signal. This signal would therefore appear to be of limited practical use in forensic applications. Other trace elements (e.g. transition metals) have more complex aqueous chemistry, however, and these elements may possibly show greater spatial variation on the decadal time scales that are relevant to forensic studies.

(3) *Hydrology.* Post-mortem increases in trace elements are not distributed evenly throughout a single bone. It is therefore critical to select the appropriate portion of bone for sampling, and this will vary according to the hydrological setting and the element(s) of interest. Generally, external surfaces of bones are most rapidly altered, and are therefore more likely to record a post-mortem chemical signal after relatively short (decadal) post-mortem intervals.

Despite these challenges, the study described above does indicate the potential for using trace elements added to bone soon after death to indicate the early burial location, and/or to test whether disarticulated or scattered remains shared a common early post-mortem history.

It is important to recall that the trace element composition of a bone in contact with soil and pore waters will be determined by a large number of variables (equation 1), so very careful recording of the burial conditions may be needed to interpret post-mortem chemical signals in fossil bones (c.f. Hanson

2004; Pye 2004). More experimental work needs to be done before post-mortem alterations in the trace element chemistry of recent bone can be used reliably as an evidential tools in forensic cases.

Conclusions

(1) The trace element chemistry of bone is readily altered after death, the speed of this alteration being dependent on the trace element in question, local hydrology and chemistry. Many elements only sparingly present in bone *in vivo* are concentrated in bone post-mortem via adsorption onto crystal surfaces.

(2) The trace element signal acquired in bone after death can be related to the early post-mortem environment. This provides a potential method to link a bone to its early depositional source.

(3) Fossil or archaeological bones derived from the same depositional environment may be grouped according to their diagenetic trace element composition, particularly their REE composition. This provides a powerful method to test the provenance of excavated bones, and potentially to regulate exploitation of fossil and archaeological resources. As the signal acquired in bone is not directly related to the local soil or sediment, however, bones can only be compared to other bones, and a carefully constructed reference collection is necessary to test the provenance of a fossil bone.

(4) The trace element composition of bone may be altered within a year of death, and may therefore provide sensitive data linking scattered remains to single depositional settings. Presently, however, very little is known about such early alteration in trace element chemistry in bone, and more work must be done to investigate its potential use in forensic investigations.

This paper builds upon work conducted through the Smithsonian Institution Postgraduate Fellowship Programme and the Burch Fellowship Programme of the Weizmann Institute of Science, Israel. I am indebted to A.K. Behrensmeyer, N. Tuross, S. Weiner and R. Potts for support, access to materials and equipment, and expertise.

References

AMIN, J., BRAMER, M. & EMSLIE, R. 2003. Intelligent data analysis for conservation: experiments with rhino horn fingerprint identification. *Knowledge-Based Systems*, **16**, 329–336.

AUSTRENG, E., STOREBAKKEN, T., THOMASSEN, M. S., REFSTIE, S. & THOMASSEN, Y. 2000. Evaluation of selected trivalent metal oxides as inert markers used to estimate apparent digestibility in salmonids. *Aquaculture*, **188**, 65–78.

BERNAT, M. 1975. Les isotopes de l'uranium et du thorium et les terres rares dans l'environnement marin. *Cahiers ORSTOM, Series Geologie*, **7**, 65–83.

COLLINS, M. J., NIELSEN-MARSH, C. M. ET AL. 2002. The survival of organic matter in bone: a review. *Archeometry*, **44**, 383–394.

HANSON, I. D. 2004. The importance of stratigraphy in forensic investigation. In: PYE, K. & CROFT, D. J. (eds) *Forensic Geoscience: Principles, Techniques and Applications*. Geological Society, London, Special Publications, **232**, 39–47.

LÉCUYER, C., BOGEY, C. ET AL. 2003. Stable isotope composition and rare earth element content of vertebrate remains from the Late Cretaceous of northern Spain (Laño): did the environmental record survive? *Palaeogeography, Palaeoclimatology, Palaeoecology*, **193**, 457–471.

PARKER, R. B. & TOOTS, H. 1970. Minor elements in fossil bones. *Bulletin of the Geological Society of America*, **81**, 925–931.

PATRICK, D., MARTIN, J. E., PARRIS, D. C. & GRANDSTAFF, D. E. 2002. Rare earth element signatures of fossil vertebrates compared with lithostratigraphic subdivisions of the Upper Cretaceous Pierre Shale, central South Dakota. *Proceedings of the South Dakota Academy of Science*, **81**, 161–179.

PLUMMER, T. W., KINUYUA, A. M. & POTTS, R. 1994. Provenancing of hominid and mammalian fossils from Kanjera, Kenya, using EDXRF. *Journal of Archaeological Science*, **21**, 553–563.

PRICE, T. D. 1989. Multi-element studies of diagenesis in prehistoric bone. In: PRICE, T. D. (ed.) *The Chemistry of Prehistoric Human Bone*. Cambridge University Press, Cambridge, 126–154.

PYE, K. 2004. Isotope and trace element anlysis of human teeth and bones for forensic purposes. In: PYE, K. & CROFT, D. J. (eds) *Forensic Geoscience: Principles, Techniques and Applications*. Geological Society, London, Special Publications, **232**, 215–236.

RADOSEVICH, S. C. 1993. The six deadly sins of trace element analysis: a case of wishful thinking in science. In: SANFORD, M. K. (eds) *Investigations of Ancient Human Tissue: Chemical Analyses in Anthropology. Food and Nutrition in History and Anthropology*, Vol. 10. Gordon and Breach, Langhorne, 269–332.

SAMOILOV, V. S. & BENJAMINI, C. 1996. Geochemical features of dinosaur remains from the Gobi Desert, south Mongolia. *Palaios*, **11**, 519–531.

SHINOMIYA, T., SHINOMIYA, K., ORIMOTO, C., MINAMI, T., TOHNO, Y. & YAMADA, M. 1998. In- and out-flows of elements in bones embedded in reference soils. *Forensic Science International*, **98**, 109–118.

STARON, R., GRANSTAFF, B., GALLAGHER, W. & GRANDSTAFF, D. E. 2001. REE signals in vertebrate fossils from Sewel, NJ: implications for location of the K-T boundary. *Palaios*, **16**, 255–265.

TRUEMAN, C. N. 1999. Rare earth element geochemistry and taphonomy of terrestrial vertebrate assemblages, *Palaios*, **14**, 555–568.

TRUEMAN, C. N. & BENTON, M. 1997. A geochemical method to trace reworking in vertebrate assemblages. *Geology*, **25**, 263–266.

TRUEMAN, C. N. & TUROSS, N. 2002. Trace elements in modern and ancient bone. *In*: KOHN, M. J., RAKOVAN, J. & HUGHES, J. M. (eds) *Phosphates – Geochemical, Geobiological and Materials Importance.* Mineralogical Society of America, Washington DC, Reviews in Mineralogy and Geochemistry, **48**, 489–522.

TRUEMAN, C. N., BEHRENSMEYER, A. K., POTTS, R. & TUROSS, N. 2001. Trace element and micro-damage features in bones as records of meter-scale landscape variability. *Journal of Vertebrate Paleontology,* **21**, 108A.

TRUEMAN, C. N., BEHRENSMEYER, A. K., TUROSS, N. & WEINER, S. (2004). Mineralogical and compositional changes in bones exposed on soil surfaces in Amboseli National Park, Kenya: diagenetic mechanisms and the role of sediment pore fluids. *Journal of Archaeological Science*, **31**, 721–739.

TRUEMAN, C. N., BENTON, M. J. & PALMER, M. R. 2003. Geochemical taphonomy of shallow marine vertebrate assemblages. *Palaeogeography, Palaeoclimatology, Palaeoecology*, **197**, 151–169.

TUROSS, N., BEHRENSMEYER, A. K. & EANES, E. D. 1989. Strontium increases and crystallinity changes in taphonomic and archaeological bone. *Journal of Archaeological Science*, **16**, 661–672.

VALSAMI-JONES, E., RAGNARSDOTTIR, K. V, CREWE-READ, N. O., MANN, T., KEMP, A. J. & ALLEN, G. C. 1996. An experimental investigation of the potential of apatite as radioactive and industrial waste scavenger. *In*: BOTTRELLS, S. H. (ed.) *Fourth International Symposium on the Geochemistry of the Earth's Surface*, Yorkshire, UK. University of Leeds, 686–689.

WEINER, S. & PRICE, P. A. 1986. Disaggregation of bone into crystals. *Calcified Tissue International*, **39**, 365–375.

WILLIAMS, C. T. & POTTS, P. J. 1988. Element distribution maps in fossil bones. *Archeometry*, **30**, 237–247.

Stable carbon and nitrogen isotope variations in soils: forensic applications

DEBRA J. CROFT & KENNETH PYE

Kenneth Pye Associates Ltd, Crowthorne Enterprise Centre, Crowthorne Business Estate, Crowthorne, RG45 6AW, UK (e-mail: d.croft@kpal.co.uk)

Abstract: A study has been undertaken to assess the significance of carbon and nitrogen content and stable isotopic variation in soils in relation to their practical use as a tool in forensic soil investigations. It forms part of a wider study to assess a range of techniques in a forensic context. Carbon and nitrogen abundance and $\delta^{13}C$ and $\delta^{15}N$ values have been determined in soil samples from six locations, using continuous flow–isotopic ratio mass spectrometry, to quantify: (1) stability over short time periods up to 2 years, (2) variation over short-scale distances, and (3) variability during primary transfer and mixing. Over a 2-year time period, variation was found to be largest in the elemental abundance, with the isotopic ratios being more stable. Used in combination, stable isotope analysis can be diagnostic and useful for discriminating between sites. No statistically significant differences at the 95% confidence level (using analysis of variance, ANOVA) were found for one-stage primary transfer in three of the four soils tested; the fourth sediment from an estuarine environment did show statistically significant difference at the 95% confidence level.

Context and background

Forensic geoscience is an emerging subdiscipline within forensic science. There is only limited published work in the field, although technical studies have been undertaken for more than a century and, in the past 10 years, have become widely used in criminal investigations (Murray & Tedrow 1972; Pye & Croft 2004). Geological trace evidence can be highly valuable, both for 'intelligence' and evidentiary purposes, especially where DNA and other types of direct evidence are unavailable. The range of techniques available to the geoscientist is very wide but not all have yet been applied or tested in a forensic context.

Stable isotope analysis has been applied in a number of different forensic areas, including studies of bone and teeth, man-made fibres, explosives, various arson-related chemicals, and drug quality and sourcing. Stable isotopes (mainly of oxygen, carbon, nitrogen and sulphur, and, less usually, of hydrogen and chlorine) have also been used in sedimentological and soil science investigations of provenance, diagenesis and palaeoclimate, but their use in forensic soil investigations has been largely neglected.

A key question is whether carbon and nitrogen isotopes can provide an additional means of discrimination between similar soils from different locations, or from the same location taken at different times. A further question concerns whether or not any significant changes occur during or after the transfer of soil from its source to an object (e.g. footwear, digging implements etc.). This paper presents the results of three pilot experiments carried out to address these issues. The work formed part of a wider study to evaluate a number of different techniques which are, or could be, used in current forensic soil investigations (Croft 2003).

Theory

In the wider fields of geology, and environmental science, stable isotope analysis has been used extensively for many years (Faure 1986; Mattey 1997; Neilson *et al.* 1998; Amiotte-Suchet *et al.* 1999; Lobe *et al.* 2000). Applications include sedimentology (Arthur *et al.* 1983), environmental and ecological processes (Lajtha & Michener 1994), palaeontology, and the study of the early life in the Archaean (Grassineau *et al.* 2001*b*), climate change studies and the tracking of methane gas sources (Lowry *et al.* 2001).

The use of isotopic ratio mass spectrometry provides data for both percentage element content and isotopic ratio for up to five elements (carbon, hydrogen, nitrogen, oxygen and sulphur). In this study, attention has been focused on the analysis of carbon and nitrogen.

Carbon

Carbon is one of the most abundant elements on the planet and can be found in a variety of forms: as the native element in graphite and diamond; in reduced form in organic compounds and coal; and in oxidized form as carbon dioxide, aqueous carbonate ions and carbonate minerals (Faure 1986).

The two stable isotopes are ^{12}C (comprising 98.89%) and ^{13}C (comprising 1.11%). The isotopic

From: PYE, K. & CROFT, D. J. (eds) 2004. *Forensic Geoscience: Principles, Techniques and Applications.* Geological Society, London, Special Publications, **232**, 257–267. © The Geological Society of London, 2004.

Table 1. *Examples of δ¹³C ratios found in nature (after Faure 1986; Lathja & Michener 1994; Rundel* et al. *1998; Century Research Group 2000)*

Source	$\delta^{13}C$ (‰)
Normal terrestrial range	
Biogenic methanes	−60 to
Carbonates (organic-rich systems)	+20
Current atmospheric CO_2	−7.7
Plants	−8 to −30
Organic soils/sediments (recent)	−10 to −30
Marine organisms	−5 to −30
Current seawater	+2.5
Coal	−20 to −30

Table 2. *Examples of δ¹⁵N ratios found in nature (after Faure 1986; Lathja & Michener 1994; Century Research Group 2000)*

Source	$\delta^{15}N$(‰)
Normal terrestrial range	−20 to +30
Bacterial denitrification	
Some igneous rocks	
Atmospheric nitrogen	0.00
Plants	−8 to +10
Organic soils:	−4 to +20
surface	−4 to +2
20–40 cm depth	+6 to +10
Soil with nitrates	+2 to +14
Fresh forest litter	−5 to +2

composition of carbon is expressed in terms of 'delta-13-C' ($\delta^{13}C$) measured in parts per thousand (‰, or 'per mil'). This parameter is defined as follows:

$$\delta^{13}C = \left[\frac{(^{13}C/^{12}C)_{sample} - (^{13}C/^{12}C)_{standard}}{(^{13}C/^{12}C)_{standard}} \right] \times 10^3 \quad (1)$$

where the standard is carbon dioxide gas obtained from the reaction of a Peedee Formation belemnite (*Belemnitella Americana*, Peedee Formation, Cretaceous, S. Carolina) with acid. More usually this is a working standard that has been compared to the Peedee Belemnite. The PDB ratio is zero and measured sample values are compared to this, being enriched in (more positive, or heavier) or depleted in (more negative, or lighter) ^{13}C (e.g. Table 1).

Plants often produce a significant input to surface soils and basically form a bimodal distribution, with the C_3 (Calvin cycle) plants with $\delta^{13}C$ values generally in the range −23 to −27‰ and the C_4 (Slack-Hatch cycle) and CAM (crassulacean acid metabolism) plants with a $\delta^{13}C$ range from −11 to −15 ‰. It is possible to further distinguish between CAM and C_4 by using hydrogen isotopes (Craig 1957; Smith & Epstein 1971; Lajtha *et al*. 1994). C_3 plants include wheat, rye, oaks, and beech, C_4 plants include maize, sugar cane and millet, and the CAM plants include succulents and epiphytes. These observed differences are a result of a number of factors, the major influences being the different photosynthetic pathways used by different plant groups and different transpiration rates and mechanisms.

Carbon dioxide is converted into carbohydrates by plants. The heavier molecules, ^{13}C, pass into the leaf structure more slowly than ^{12}C, producing discriminatory fractionation. The methods of incorporation, and also leaf transpiration, vary between plant types, producing different isotopic fractionation rates. Air temperature, humidity, soil moisture

and the presence of bacteria can all affect these processes. Variation in $\delta^{13}C$ can be broadly predicted for different plant systems, climatic regimes; that is, different ecosystems might be expected to provide characteristic $\delta^{13}C$ fingerprints.

Nitrogen

Nitrogen, in particular, is stated by Faure (1986) to be useful in the study of geological environments, including soils, where biological activity combined with geological processes potentially produces a distinctive isotopic ratio or fingerprint.

The two stable isotopes of nitrogen are ^{14}N (comprising 99.64%) and ^{15}N (comprising 0.36%). The isotopic composition of nitrogen is expressed in terms of 'delta-15–N' ($\delta^{15}N$) measured in parts per thousand (‰, or 'per mil'). This parameter is calculated as follows:

$$\delta^{15}N = \left[\frac{(^{15}N/^{14}N)_{sample} - (^{15}N/^{14}N)_{standard}}{(^{15}N/^{14}N)_{standard}} \right] \times 10^3 \quad (2)$$

where the standard is atmospheric nitrogen, with value zero. Nitrogen values in soils and sediments are very variable, correlated primarily with cultivation and growing systems. These differences are accentuated by the growth of 'fixing plants' (peas, beans, lentils, clover etc.) as well as by the introduction of nitrate/nitrite fertilisers, and biological activity (Faure 1986; Neilson *et al*. 1998; Rundel *et al*. 1998). The 'fixing' plants have specific bacteria which live in the root nodules and fix atmospheric nitrogen by combining it with other elements, thereby making it available to the plant. Other plants are specialized to take up nitrogen from organic matter. Table 2 shows typical values.

In general, the $\delta^{15}N$ values are higher (more enriched, or more positive) with the increased presence of woody plants and decomposed organic matter, particularly at depth. Fresh organic rich material (including fresh litter and manure) is likely to show lower (more depleted, or more negative) $\delta^{15}N$ values. (Letolle 1980; Lajtha & Michener 1994).

Methods

The sample locations and collection method is described below in the relevant experimental sections. As part of a standard routine for forensic analysis in our laboratory, the $>150\,\mu m$ and $<150\,\mu m$ size fractions are used for a range of analytical techniques. It was decided to use the $<150\,\mu m$ size fraction for isotopic analysis, being the clay, silt and fine sand fraction which, based on earlier work, is generally the representative size fraction in soils. For each soil sample, a subsample was separated by wet sieving using distilled, de-ionized water through disposable Netlon™ mesh (Lockertex 150) on a polycarbonate frame. The meshes were retained with the coarser size fractions and stored for microscope analysis. The finer size fractions (in suspension) were allowed to settle and then evaporated at low temperature ($c.\ 30$–$40\,°C$) before grinding by hand in an agate pestle and mortar. Previous tests (Croft 2003) carried out for chemical element analysis using inductively coupled plasma atomic emission spectroscopy (ICP-AES) has proved this preparation method to have no statistically significant effects on the bulk chemical content of standard materials. The resulting powder was analysed for carbon and nitrogen content and isotopic ratios.

Analysis was carried out at the Geology Department of Royal Holloway, University of London. The equipment consisted of an elemental analyser (EA) attached to a mass spectrometer (MS). The EA is a Fisons Instruments NA1500, with a Perkin-Elmer Optima MS system attached. The system uses a continuous flow technique for both elemental and isotopic ratio analysis, where a stream of helium carrier gas transports the combusted sample gas directly into the ion source of the mass spectrometer. The instrument is stabilized using a standard gas and blanks. Calibration standards are then used before sample analysis commences. For carbon, two international standards are used, IAEA (International Atomic Emissions Agency) CO9, and NBS (National Bureau for Standards) NBS-21, as well as two laboratory standards (GF graphite and Royal Holloway and Bedford New College carbonate). For nitrogen, two international standards (IAEA N1 and N2) and a laboratory standard, sulphanilamide, are used. The calibration standards produce a range of values and calibration lines to check for linearity and against which sample results are corrected. Five blanks are run before analysis commences and a blank and a standard are run after

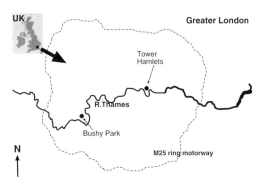

Fig. 1. Plan of London, UK, showing the two experimental sites: Tower Hamlets Cemetery (THC) to the east and Bushy Park (BP) to the west.

every eight samples to check for drift during the batch. Any samples containing very high levels of carbon or nitrogen are followed by a blank (or several blanks in some cases) to prevent memory effects and carry-over in the MS. In these cases, the sample has to be reweighed to an adjusted lower weight. The prepared samples were weighed into pure tin capsules, which were subsequently introduced into the combustion chamber of the analyser and ignited at high temperature (1500–$1800\,°C$) to achieve full combustion. The resultant gas was carried by the helium flow directly into the mass spectrometer. Carbon abundance and $\delta^{13}C$ values and nitrogen abundance and $\delta^{15}N$ values are measured in two separate batches using this method, but the analyses are relatively rapid: 60–70 a day after set-up and stabilization. Typical precision and accuracy figures for this instrument are routinely calculated at $\pm0.1\%$ or better (Grassineau *et al.* 2001*a, b*).

Spatial and temporal variation

Samples were collected quarterly from two experimental sites: Tower Hamlets Cemetery in east London for 2 years, and Bushy Park to the west of London for 1 year (see Fig. 1, showing sampling locations). Both sites have similar public usage, underlying geology (post-Anglian fluvial deposits on London Clay) and stable surface soil (with no recent major works or soil importation). They differ in vegetation input. Bushy Park is stable mixed grassland, with some grazing by deer. The three trees which define the sampling area are old common hawthorn, with no significant input. Tower Hamlets Cemetery has mixed grasses, combined with significant input from plane trees, brambles, nettles etc. There is die-back in autumn and winter, with organic matter lying on the ground and degrading, not collected by park authorities.

Fig. 2. Tower Hamlets Cemetery sampling site (same location taken from different directions): (**a**), March 2001; (**b**) September 2001.

Fig. 3. Bushy Park sampling site: (**a**), March 2001; (**b**) September 2001. The tree in (**b**) appears front right in (**a**).

At the Tower Hamlets Cemetery site, 15 samples were taken on a grassed transect (17 m overall), adjacent (*c.* 1 m) to a copse of plane trees and mixed undergrowth (see Fig. 2 for Tower Hamlets Cemetery sampling site). At the Bushy Park site, 15 samples were taken from a grassed triangle (side length approximately 12 m), with no other direct vegetation input (see Figure 3 for the Bushy Park sampling site). The same sample sites were sampled every 3 months: the beginning of March, June,

September and December. Splits of these samples were processed as detailed above and the ground powder weighed into tin capsules for analysis.

Primary transfer and mixing

Two experiments were carried out to investigate primary transfer, one involving the simple, one-stage, direct transfer of soil from soil surface to foot-

Table 3. *Details of soils sampled and footwear / implements used in the transfer experiments*

Samples and media reference ID	Details
Soil A	Cultivated allotment, Middlesex, UK
Soil B	Grassland/wasteland, Surrey, UK
Soil C	Tidal-dominated estuarine floodplain (R. Severn tributary), Somerset, UK
Soil D	Stable woodland, Berkshire/Wiltshire, UK
Footwear DM	'Dr Marten' style boots (worn)
Footwear LS	'Lacoste' shoes (new)
Footwear TR	Non-labelled trainers (very worn)
Footwear WB	'Dickies' steel toe cap work boots (worn)
Footwear WL	Wellington boots (lightly worn)
Implement 1	Trowel (stainless steel)
Implement 2	Border spade (stainless steel)
Implement 3	Spade (stainless steel)
Implement 4	Fork (stainless steel)
Implement 5	Rake (enamelled)

wear soles, and one involving transfer and mixing of soil from a partial soil profile (0–50 cm) using digging implements. In both cases four soils were used, chosen to give a range of underlying geology and soil types, although they were sampled at different times for the two experiments. Another study may have chosen a larger range of sites and soil types, but this research is part of a wider study (Croft 2003) whose aim was the application of a range of techniques to a small number of sites, thereby restricting the number of parameters evaluated. Table 3 presents the details of the soils and the media used for transfer.

Soil A is from a cultivated allotment (vegetable-growing) in Middlesex, UK. The vegetation cover was grass (a fallow area for the previous 2 years), the underlying geology is Thames River gravels/sands and the soil type (Soil Survey of England and Wales 1983 throughout) is 571W (river terrace drift). Soil B is from grassland/wasteland in Surrey, UK, which may have had building waste incorporated in the last 5 years. The underlying geology is London Clay and the soil type is 711g (drift over Tertiary clay/loam). Soil C is from an estuarine floodplain of a tributary of the River Severn, Somerset, UK, with sparse common cord grass (*Spartina Anglica*). The underlying geology is river alluvium over Upper Lias, and the soil type is 814c (marine alluvium). Lastly, Soil D is mixed deciduous woodland, but because of access problems, samples have slightly differing underlying geology and soil types from each other. For the simple, one-stage transfer using shoes, the soil (from the Berkshire/Wiltshire border, UK) has an underlying geology of Reading Beds over Chalk,

with soil type 582B (plateau and glacio-fluvial drift). For the transfer and mixing experiment, the soil from Wiltshire, UK, has underlying geology of greensand and gault over Chalk, with soil type 341 (calcareous, well-drained, humose).

For the footwear experiment, a large (*c.* 10 kg) surface soil sample (from 0–5 cm below surface depth) was removed to the laboratory, homogenized with a trowel and placed in a large shallow container. Each of the five pairs of footwear was then worn and 'walked' on the spot for 3 min. The soil in the container was homogenized between each 'walking' period. At the end of this time, the shoes and boots, with the adhering soil, were laid on paper sheets in a cool, unused, ventilated laboratory. The left foot soil sample was removed after 24 h and the right foot sample after 72 h. All of the soil present was removed in each case with a stainless steel spatula, tweezers and dental pick, and placed in plastic containers for subsequent testing. At the start of the experiment a subsample of each of the four bulk soils was taken for comparison with the material removed from the footwear.

Subsamples for analysis were taken from each sample by coning and quartering the sample, chosing a quarter at random and taking further subsamples from throughout that quarter using a spatula. Only the right foot samples were used for carbon and nitrogen analysis (after wet sieving, drying and grinding of the <150 μm fraction) in order to maximize any changes in elemental content or isotopic ratio. The ground powder was weighed into tin capsules for analysis as a complete batch. Further analyses were undertaken using both left and right footwear samples in order to characterize the soils chemically and physically and will be reported elsewhere.

For the experiment using digging implements, a 50+ cm deep pit was dug at each of the four locations, A–D. The pits were dug to represent a typical grave or burial pit of an average depth, as reported in forensic case histories. A sample section from a side wall was then taken using a 50 cm length of plastic, box section ducting which had one side removed to function as a lid, adapted from a system used for salt marsh sediment profile sampling (Stoodley 1998). The rigid, square section box was driven into the sidewall and then cut away from behind to remove a true length section, avoiding problems of vertical compaction.

The boxed sections were stored in a refrigerator before subsampling in the laboratory at 10 cm intervals. The implements, with soil adhering, were left in the open air (in a ventilated, unused laboratory) for 72 h before sampling. Splits of all the samples were processed as detailed above and the ground powder weighed into tin capsules for analysis as a complete batch.

Table 4. *Quarterly mean and standard deviation for percentage carbon and nitrogen content and δ¹³C and δ¹⁵N values; Tower Hamlets Cemetery (THC) (a) year 1 and (b) year 2; and (c) Bushy Park (BP)*

Quarter	Parameter	Carbon (%)	$\delta^{13}C$ (‰)	Nitrogen (%)	$\delta^{15}N$ (‰)
(a) THC year 1					
Jun. 00	Mean ($n = 15$)	8.25	−26.64	0.58	5.65
	SD	1.51	0.32	0.12	0.55
Sep. 00	Mean ($n = 15$)	8.55	−26.51	0.53	6.31
	SD	1.52	0.42	0.19	0.85
Dec. 00	Mean ($n = 15$)	5.82	−26.05	0.41	6.18
	SD	3.25	0.46	0.23	0.78
Mar. 01	Mean ($n = 15$)	4.34	−25.52	0.44	5.98
	SD	1.60	0.90	0.19	0.68
ANNUAL	**Mean ($n = 60$)**	**6.74**	**−26.18**	**0.38**	**4.54**
SUMMARY	**SD**	**2.01**	**0.51**	**0.09**	**0.35**

Quarter	Parameter	Carbon (%)	$\delta^{13}C$ (‰)	Nitrogen (%)	$\delta^{15}N$ (‰)
(b) THC year 2					
Jun. 01	Mean ($n = 15$)	9.58	−26.74	0.53	5.26
	SD	1.85	0.36	0.08	0.54
Sep. 01	Mean ($n = 15$)	8.54	−26.49	0.56	5.69
	SD	1.10	1.09	0.06	0.51
Dec. 01	Mean ($n = 15$)	8.96	−26.93	0.63	5.62
	SD	1.31	0.33	0.10	0.53
Mar. 02	Mean ($n = 15$)	10.83	−26.93	0.71	5.71
	SD	0.96	0.33	0.08	0.56
ANNUAL	**Mean ($n = 60$)**	**9.48**	**−26.77**	**0.61**	**5.57**
SUMMARY	**SD**	**1.57**	**0.63**	**0.11**	**0.55**

Quarter	Parameter	Carbon (%)	$\delta^{13}C$ (‰)	Nitrogen (%)	$\delta^{15}N$ (‰)
(c) BP					
Jun. 00	Mean ($n = 15$)	20.30	−28.60	1.40	4.07
	SD	7.11	0.27	0.47	0.57
Sep. 00	Mean ($n = 15$)	19.55	−28.61	1.60	3.30
	SD	8.96	0.39	0.55	0.95
Dec. 00	Mean ($n = 15$)	18.99	−28.45	1.38	4.22
	SD	10.20	0.27	0.75	0.48
Mar. 01	Mean ($n = 15$)	14.48	−28.66	0.96	4.28
	SD	9.68	0.27	0.50	0.81
ANNUAL	**Mean ($n = 60$)**	**18.33**	**−28.58**	**1.34**	**3.97**
SUMMARY	**SD**	**2.62**	**0.09**	**0.27**	**0.45**

Results and discussion

Spatial and temporal variation

The results are summarized in Table 4, quarter by quarter for the Tower Hamlets Cemetery (THC) site (over 2 years) and for the Bushy Park (BP) site (for 1 year).

The isotopic ratios for both carbon and nitrogen are well constrained at both locations, with mean values for $\delta^{13}C$ of −26.18‰ (SD=0.51) for THC year 1, −26.77‰ (SD=0.63) for THC year 2 and −28.58‰ (SD=0.09) for BP. Mean values for $\delta^{15}N$ are 4.54‰ (SD=0.35) for THC year 1, 5.57‰ (SD=0.55) for THC year 2 and 3.97 ‰ (SD=0.45) for BP. The values and variations seen are consistent with previously published data (Neilson *et al.* 1998). Spatial variation at the local

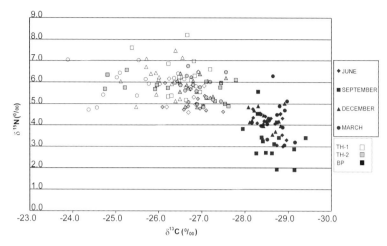

Fig. 4. $\delta^{13}C$ values against $\delta^{15}N$ values for samples from Tower Hamlets Cemetery (years 1and 2) and Bushy Park (year 1 only). All values are $\%_{00}$.

scale (<20m), indicated by the SD, is very similar between the two sites and for the quarter-by-quarter data. At BP, the carbon data exhibit no statistically significant difference between the seasons (using ANOVA, $p = 0.95$), but the nitrogen data are significantly different for the September quarter compared to the other three quarters. At THC, the situation is reversed, with no statistically significant difference across the seasons in the nitrogen data and a significant difference in the carbon data for the March quarter, compared with the other quarters. It is unclear, with the relatively short overall sampling period, whether these anomalous sampling periods are random fluctuations or a more significant seasonal trend. Possible reasons for this might be the result of deciduous leaf fall and a pulse of organic matter, or differential bacterial activity between the winter and summer, or a combination of these and other explanations.

Figure 4 presents a bivariate plot of the $\delta^{13}C$ values against $\delta^{15}N$ values for both experimental sites, showing clear discrimination between the two sites.

The results all fall within the expected values for vegetated soil in this part of the northern hemisphere (Faure 1986; Neilson *et al.* 1998; Rundel *et al.* 1998). The carbon values are characteristic of reduced organic carbon, with the BP values slightly more depleted (lighter, or more negative) than the THC values. The BP site exhibits less variation and the carbon content and isotopic ratio may have reached a stable plateau, with more variation at THC due to cyclical vegetation input and degradation from a wider assemblage of species, which included plane trees, mixed grasses, common bramble, nettles, ragwort and members of the dock family.

The higher $\delta^{15}N$ values at THC relative to BP are also likely to be a reflection of the mixed vegetation and the presence of woody plants and trees on the margins of the sampling strip, with these higher figures resulting in marginally less seasonal variation. At BP, the nitrogen values have a greater range relative to THC, but this greater range is still within reported values for mixed northern hemisphere grassland (Rundel *et al.* 1998).

Primary transfer and mixing

A summary of results from the simple, one-stage primary transfer of surface soil to footwear experiment is presented in Table 5.

For soils A, B and D there are no statistically significant differences at the 95% confidence level between the carbon content and $\delta^{13}C$ values for the surface soil and the soils removed from footwear (using ANOVA). Soil C does show a significant difference at the 95% confidence level. This reflects both complex variation in the estuarine soil analysed, with the observed differences due to rapid diagenetic changes occurring near the sediment surface in a spatially irregular manner (Pye *et al.* 1997; Crooks & Pye, 2000) and the high levels of carbonate present in the sample, making analysis more complex. The carbonate may be in patches or nodules which do not transfer, or, because of the structure, may affect bacterial activity and therefore isotopic variation. Further investigations would benefit from repeating the analysis of the soil, after the removal of carbonate from the sample, as a pre-treatment combined with a wider sampling plan to elucidate the causes of the differences. The

Table 5. *Carbon and nitrogen percentage content and δ¹³C and δ¹⁵N values (⁰/₀₀) for experimental soils A–D (surface samples) and samples transferred to five footwear types*

Samples		Carbon (%)	$\delta^{13}C$ (⁰/₀₀)	Nitrogen (%)	$\delta^{15}N$ (⁰/₀₀)
Soil A (surface)		5.23	−27.13	0.43	8.55
Footwear ($n=5$)	Mean	5.35	−27.06	0.41	8.75
	SD	0.39	−0.13	0.01	0.59
Soil B (surface)		1.66	−24.92	0.12	6.05
Footwear ($n=5$)	Mean	2.01	−24.42	0.09	5.82
	SD	0.31	−0.33	0.01	0.83
Soil C (surface)		4.62	−15.09	0.23	6.42
Footwear ($n=5$)	Mean	5.86	−17.10	0.23	5.58
	SD	0.35	−0.57	0.01	0.92
Soil D (surface)		30.04	−27.66	1.77	−2.27
Footwear ($n=5$)	Mean	31.83	−27.57	1.71	−1.57
	SD	1.20	−0.08	0.10	−0.46

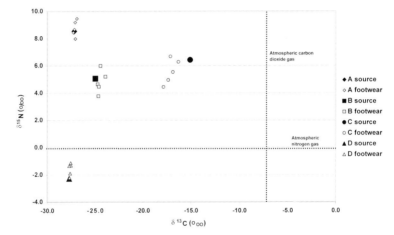

Fig. 5. $\delta^{13}C$ values against $\delta^{15}N$ values for the experimental soils A–D, (surface samples) and the samples removed from the footwear. All values are ⁰/₀₀.

nitrogen content and $\delta^{15}N$ values show no statistically significant difference at the 95% confidence level between the surface sample and the samples removed from footwear for any of the four soil types.

Figure 5 presents a bivariate plot of $\delta^{13}C$ and $\delta^{15}N$ values for the four surface soils and the soil samples removed from footwear soles after 72 h. Atmospheric carbon dioxide and atmospheric nitrogen are shown on the plot as dotted lines for comparison.

The plot shows good discrimination between the four soil types, with close correspondence between values for the surface soils and the samples removed from the footwear, with the exception of soil C (as previously stated).

The data for the partial soil profile and digging implements experiment are presented in Table 6.

For soils A, B and C, there are no statistically significant differences (using ANOVA, $p=0.95$) between the carbon content and $\delta^{13}C$ values for the mean of the profile soil samples and the mean of the soil samples ($n=6$) removed from the digging implements ($n=5$). Soil D (woodland soil) had a surface layer of organic-rich material, which was at a very early stage of decomposition, resulting in extremely high levels of carbon (21.2%). This resulted in a statistically significant difference ($p=0.95$) for carbon content and $\delta^{13}C$ value. However, if this was corrected for (by substituting the mean carbon value for soil D), then no statistically significant difference at the 95% confidence level was

Table 6. *Carbon and nitrogen percentage content and $\delta^{13}C$ and $\delta^{15}N$ values ($^{0}/_{00}$) for experimental soils A–D (surface samples) and samples transferred to five digging implements*

Samples		Carbon (%)	$\delta^{13}C$ ($^{0}/_{00}$)	Nitrogen (%)	$\delta^{15}N$ ($^{0}/_{00}$)
Soil A (0–50cm)	Mean	3.15	−25.74	0.21	9.10
(n=6)	SD	1.69	−1.43	0.13	0.85
Implements	Mean	4.10	−26.16	0.28	8.41
(n=5)	SD	0.65	−0.36	0.05	0.80
Soil B (0–50cm)	Mean	1.14	−22.46	*bd*	*bd*
(n=6)	SD	0.43	−0.55		
Implements	Mean	1.07	−22.36	*bd*	*bd*
(n=5)	SD	0.12	−0.67		
Soil C (0–50cm)	Mean	5.23	−16.50	0.17	6.49
(n=6)	SD	0.16	−0.55	0.03	0.52
Implements	Mean	5.03	−15.80	0.13	6.20
(n=5)	SD	0.35	−1.37	0.01	0.41
Soil D (0–50cm)	Mean*	4.46	−25.90	0.11	3.94
(n=6)	SD	8.24	−0.64	0.16	3.78
Implements	Mean	2.51	−26.37	0.10	2.08
(n=5)	SD	0.73	−0.31	0.03	1.61

* *Soil D mean value is skewed by a relatively high surface soil value for carbon content (21.22%)*
bd = below level of accurate detection

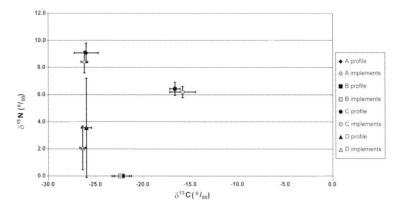

Fig. 6. Mean $\delta^{13}C$ against mean $\delta^{15}N$ for the soils A–D (profile samples) and samples removed from the digging implements. All values are $^{0}/_{00}$. (Note : the carbon isotopic ratio for soil B is plotted on the *x*-axis (*y*-axis = 0); however, this is for completeness, as the nitrogen content was below reliable detection and does not represent a value of $0^{0}/_{00}$.)

observed. Also, the nitrogen content and $\delta^{15}N$ values show no statistically significant differences between the sample sets for the soils A, C and D. Nitrogen levels in soil B were too low for reliable determination of the $\delta^{15}N$ value.

Figure 6 presents a bivariate plot of the mean $\delta^{13}C$ values and the mean $\delta^{15}N$ values for the three soils (A, C and D) and the soil samples removed from implements. The $\delta^{13}C$ values for soil B have been plotted on the *x*-axis (*y*-axis = 0), as the nitrogen content was below reliable detection, and the

$\delta^{15}N$ values obtained have not been used (this does not imply that the $\delta^{15}N$ is set at zero). The SDs for each mean value are shown as error bars. Once, again there is good discrimination between the soils, but the variation (as shown by the error bars) is relatively large for soil D, where the surface soil was markedly different from the subsurface soil samples, due to the organic-rich layer. However, the soil D samples taken from the implements still fall within the mean ±1 SD of the soil profile samples.

Conclusions

The following conclusions can be drawn from this preliminary study:

(1) The discrimination between the soil types used in the experiments is good, particularly when using both carbon and nitrogen data in combination.

(2) The variation found on the >20 m spatial scale is within the range of natural sampling variation. It is of the same order as the seasonal variation found at the two experimental sites. Comparison of soil samples taken from different points over this short distance scale and at different time periods up to several months apart can be undertaken in a meaningful way. It is recommended that several samples (minimum 5) should be taken on each occasion to obtain a representative average for each sample point.

(3) For simple, one-stage primary transfer of soils, as shown in the footwear experiment, a close agreement was found between 'source soil' and 'transferred soil' values, with the exception of soil C, where differences in carbon content and $\delta^{13}C$ values appear to reflect complex fractionation in the surface estuary mud and the relatively large amounts of carbonate present. The values obtained from the footwear do not appear to have been affected by the primary transfer mechanism per se.

(4) For complex mixing and transfer, as in the digging implements experiment, greater variation can be expected in the values obtained from transferred soil, but for the soils examined, reliable comparisons could still be made for carbon content and $\delta^{13}C$ values. Nitrogen is more variable than carbon, both in terms of content and $\delta^{15}N$ values, and provides a less precise method for comparison.

(5) The technique has the power to discriminate between soil types and, when used in combination with other analytical techniques, carbon and nitrogen isotopic analysis is a useful tool for soil sample comparisons, enabling the inclusion (or exclusion) of people, sites and implements (which have already been identified) in criminal investigations. The complex nature of isotopic variation and lack of a comprehensive database (even if this was possible with seasonal differences and soil importation/excavation problems) means that intelligence uses of the technique are precluded.

(6) More work is needed to provide further detailed information about the magnitude of variation which exists within a larger range different soil and sediment types.

References

AMIOTTE-SUCHET, P., AUBERT, D., PROBST, J. L., GAUTHIER-LAFAYE, F., PROBST, A., ANDREUX, F., & VIVILLE, D. 1999. $\delta^{13}C$ pattern of dissolved inorganic carbon in a small granitic catchment: the Strengbach case study. *Chemical Geology (including Isotope Science)*, **159**, 129–145.

ARTHUR, M. A., ANDERSON, T. F., KAPLAN, I. R., VEIZER, J., & LAND, L. S. 1983. *Stable isotopes in Sedimentary Geology*. SEPM, Dallas, Short Course Notes, **10**.

CENTURY RESEARCH GROUP 2000. Carbon 14C and 13C. World Wide Web Address: http://www.nrel.colostate.edu/projects/century5/reference/labeled.htm

CRAIG, H. 1957. Isotopic standards for carbon and oxygen and correction factors for mass spectrometric analysis of carbon dioxide. *Geochimica et Cosmochimica Acta*, **12**, 133–149.

CROFT, D. J. 2003. *Forensic geoscience: development of techniques for soil analysis*. PhD Thesis, Royal Holloway and Bedford New College, University of London.

CROOKS, S. & PYE, K. 2000. Sedimentological controls on the erosion and morphology of saltmarshes: implications for flood defence and habitat recreation. *In*: PYE, K. & ALLEN, J. R. L. *Coastal and Estuarine Environments: Sedimentology, Geo-morphology and Geo-archaeology*. Geological Society, London, Special Publications **175**, 207–222.

FAURE, G. 1986. *Principles of Isotope Geology*. John Wiley & Sons, New York.

GRASSINEAU, N. V., MATTEY, D. P. & LOWRY, D. 2001a. Rapid sulphur isotope analysis of sulphide and sulphate minerals by continuous flow-isotope ratio mass spectrometry (CF-IRMS). *Analytical Chemistry*, **73**, 220–225.

GRASSINEAU, N. V, NISBET, E. G. ET AL. 2001b. Antiquity of the biological sulphur cycle: evidence from sulphur and carbon isotopes in 2700 million-year-old rocks of the Belingwe Belt, Zimbabwe. *Proceedings of the Royal Society of London, Series B*, **268**, 113–119.

LAJTHA, K & MICHENER, R. H. 1994. *Stable isotopes in Ecology and Environmental Science*. Blackwell Scientific Publications, Oxford.

LETOLLE, R. 1980. Nitrogen-15 in the natural environment. *In*: FRITZ, P. & FONTES, J. (eds) *Handbook of Environmental Isotope Geochemistry*, **1A**, 407–434.

LOBE, I., AMELUN, W. & DU PREEZ, C. C. 2000. Losses of carbon and nitrogen with prolonged arable cropping from sandy soils of the South African Highveld. *European Journal of Soil Science*, **52**, 93–101.

LOWRY, D., HOLMES, C. W., RATA, N. D., NISBET, E. G. & O'BRIEN, P. 2001. London methane emissions: use of diurnal changes in concentration and $\delta^{13}C$ to identify sources and verify inventories. *Journal of Geophysical Research*, **106**, 7427–7448.

MATTEY, D. P. 1997. Gas source mass spectrometry: isotopic composition of lighter elements. *In*: GILL, R. (ed.) *Modern Analytical Geochemistry*. Addison Wesley Longman, Harlow, 154–170.

MURRAY, R. C. & TEDROW, J. C. F. 1992. *Forensic Geology*. Prentice Hall, New Jersey.

NEILSON, R., HAMILTON, D. ET AL. 1998. Stable isotope

natural abundances of soil, plants and soil inverte-brates in an upland pasture. *Soil Biology and Biochemistry*, **30**, 1773–1782.

PYE, K. & CROFT, D. J. (eds) 2004. *Forensic Geoscience: Principles, Techniques and Applications.* Geological Society, London, Special Publications, **232**.

PYE, K., COLEMAN, M. L. & DUAN, W. M. 1997. Microbial activity and diagenesis in saltmarsh sediments. *In*: JICKELLS, T. D. & RAE, J. E. (eds) *Biogeochemistry of Intertidal Sediments.* Cambridge University Press, Cambridge, 119–151.

RUNDEL, P. W., EHLERINGER, J. R. & NAGY, K. A. 1998.

Stable Isotopes in Ecological Research. Ecological Studies Series, Springer-Verlag, New York, **69**.

SMITH, B. N. & EPSTEIN, S. 1971. Two categories of 13C/12C ratios for higher plants. *Plant Physiology*, **47**, 380–384.

SOIL SURVEY OF ENGLAND AND WALES. 1983. *1:250000 Soil Maps and Legend.* Soil Survey of England and Wales (now Soil Survey and Land Research Centre), National Soil Research Institute, Cranfield University.

STOODLEY, J. A. 1998. A monolith sampler for saltmarsh sediments. *Journal of Sedimentary Research*, **68**, 1046–1047.

The use of plant hydrocarbon signatures in characterizing soil organic matter

LORNA A. DAWSON, WILLIE TOWERS, ROBERT W. MAYES, JULIE CRAIG, R. KATARIINA VÄISÄNEN & E. CLARE WATERHOUSE

The Macaulay Institute, Craigiebuckler, Aberdeen AB15 8QH, UK
(e-mail: l.dawson@macaulay.ac.uk)

Abstract: Resistant compounds associated with vegetation have potential for understanding and uniquely describing soil. Although much of the forensic identification of soils has focused on the mineral component, this study illustrates how the origin of the organic component can be a useful tool in soil identification. The epicuticular wax of most plants contains mixtures of hydrocarbons (mainly *n*-alkanes) and plant species differences are persistent. Evidence from three separate studies is compiled to show the validity of this approach. In the first example, on upland grassland vegetation, the *n*-alkane pattern of the soil at one site reflected that of the overlying grass, whereas at another site, it reflected that of the previous vegetation, heather. In the second study, *n*-alkane analysis data indicated the presence of heather in a buried horizon, matching independent evidence from pollen identification. The third study was one covering the whole of Scotland, using an unbiased grid-sampling strategy. Results show that the patterns in the soil *n*-alkane profiles reflected the overlying vegetation. Where this was not the case, the profiles matched previously grown vegetation. Such biomarker information, derived from plant wax signatures, coupled with soil spatial information, has potential in the unique identification of soils.

The use of earth materials, including soil, in criminal investigations has been recognized for over a century. More recently the scope has broadened into the area often referred to as environmental forensics, which Morrison (2000) defined as 'the systematic examination of environmental information used in litigation'. Forensic geology is accepted by courts of law as a valid source of scientific evidence in the US (Murray & Tedrow 1992), and is now being considered as a valid approach in the UK. With the exception of work relating to pollen and other botanical evidence, soil analysis for forensics has primarily relied on the assessment of soil colour, particle size and mineral examination or analysis. Occasionally, the forensic scientist can, in addition, look for an unusual fragment in a soil sample, that is a microfossil, a chemical contaminant or an uncommon mineral. Geochemical techniques using isotope ratios, and geochemical signatures have also been utilized in forensic work (e.g. Trueman 2004). These methods, however, do not always discriminate samples effectively (Sugita & Marumo 1996).

In addition to the above methods, the organic component in soil can be used for forensic comparison. Fourier transform infrared (FTIR) spectra can characterize soil organic matter according to soil type (Chapman *et al.* 2001). However, its potential use in forensics to discriminate between soils of similar colour has only been investigated on a limited number of soils (Cox *et al.* 2000). The use of *n*-alkanes can add another dimension to soil discrimination in that specific types of plant origin can be determined in addition to soil signature matching. As yet there have been no reports of such plant wax compounds being used in forensic work. The work presented here explores the use of *n*-alkanes in the identification of soil organic matter origin and history, and can give an indication of land use and vegetation cover. This consequently adds to the range of techniques available to enable the forensic scientist to discriminate between soils and identify sample origin.

Although hydrocarbons (*n*-alkanes) are only one of a range of long-chain aliphatic compounds present in plant waxes (such as esters, acids, alcohols and ketones), they were selected as the first marker to investigate due to their relative ease of analysis compared with the other compounds. The *n*-alkanes have from 21 to 37 carbon atoms. They are discrete, readily analysable and identifiable components, which appear to have a robust persistence in soils. Species differences in plant *n*-alkane patterns were used originally to determine the diet composition of grazing animals (Dove & Mayes 1996). This enabled the proportion of an animal's diet to be determined using the *n*-alkane signature in the animal's faeces. The method has subsequently been used to determine the species composition of root mixtures (Dawson *et al.* 2000). The *n*-alkane signature of the roots of five closely related grasses could be individually distinguished using canonical variate analysis. In addition, *n*-alkane patterns have been used in the quantitative and semi-quantitative apportioning of sources of hydrocarbons found in recent aquatic sediments polluted with fossil fuels (Volkman *et al.* 1992).

From: PYE, K. & CROFT, D. J. (eds) 2004. *Forensic Geoscience: Principles, Techniques and Applications*. Geological Society, London, Special Publications, **232**, 269–276. © The Geological Society of London, 2004.

Previous studies investigating *n*-alkanes in soil have shown that the distribution of *n*-alkanes at some sites can reflect the *n*-alkane pattern in the overlying vegetation (Almendros *et al.* 1996). However, one report suggested that, in some samples under juniper, the *n*-alkane signature was degraded and substituted by microbial breakdown products (Andreyev *et al.* 1980). A more recent study on soils under beech, spruce and grass (Marseille *et al.* 1999) found that the *n*-alkane signatures of the current vegetation was preserved in the top litter layer, but that in the fragmentation and humification layers below, the *n*-alkane signatures were progressively replaced by a signature composed of C_{27} and C_{25}. While the authors suggested that this could have been due to the effect of fungal activity, we suggest the change in signature may also be due to differences in the past vegetation cover. The lipid profile of a soil largely represents the product of the synthesis, polymeric and degradative processes on the vegetation, all of which are determined by the soil environment. Although plant hydrocarbons have a low concentration in soils, they can be useful due to their persistence. The present study was initiated to investigate whether wax markers in the soil could be used to identify past or present vegetation.

Methods

N-*alkane analysis*

Hydrocarbon analyses were carried out using the method of Mayes *et al.* (1986) modified by Salt *et al.* (1992) in which reductions were made to sample size and the use of two internal standards was introduced. Samples of freeze-dried and ground soil and vegetation (0.5 g and 0.2 g respectively for soil and shoot material) were, after the addition of *n*-alkane internal standards (*n*-docosane, C_{22} and *n*-tetratriacontane, C_{34}), saponified in 56.11 g dm^{-3} potassium hydroxide dissolved in pure ethanol. This was after having established that the levels of the standards present in soil were negligible. Any large visible roots were removed and analysed separately. The hydrocarbons were extracted into *n*-heptane, purified by passage through silica gel columns (1 cm^3 bed volume, added to empty column as a slurry in *n*-heptane and washed with *n*-heptane) and separated by gas chromatography on a non-polar glass capillary column, 30 m × 0.75 mm internal diameter and having a 1 μm film thickness. The chromatograph was fitted with a splitless injector and a flame ionization detector; the carrier gas was helium. The *n*-alkanes in the samples were identified and quantified by reference to a standard mixture containing all *n*-alkanes in the range C_{21} to C_{36}. All results are expressed on an organic matter (OM) basis; to deter-

mine OM, samples were ashed in a muffle furnace overnight at 700 °C for soil and 500 °C for vegetation.

Pollen analysis

The soil monoliths in case study 2 were analysed for pollen within 1 cm-thick layers at 2 cm intervals down the soil profile. For each sample, pollen was extracted from 1 cm^3 of soil using standard hydrogen chloride, sodium hydroxide, hydrogen fluoride and acetolysis treatments (Faegri & Iversen 1989), suspended in silicon oil and mounted on a slide. Counting and identification of pollen and spores were conducted using a light microscope at a magnification of ×600. For each depth, a minimum of 300 land pollen grains were counted and overall pollen composition (subdivided into trees, shrubs, heaths, grasses, sedges and spores) calculated as a percentage of the total to provide an indication of past vegetation.

Results and discussion

Background

Most grass species, for example *Lolium* spp., have predominantly C_{29} to C_{33} *n*-alkanes in their leaf material (Dove & Mayes 1991) (Fig. 1), while *Calluna vulgaris* shows relatively more C_{33} than C_{29}. *Trifolium repens*, on the other hand, has lower concentration values and C_{29} dominates the pattern. In the foliage of many trees, for example *Betula* sp., the shorter chain *n*-alkanes predominate (C_{23}, C_{25} and C_{27}) (Fig. 1).

Case study 1

As an initial preliminary survey, individual samples of soil (to a depth of 3 cm) and associated overlying vegetation were sampled from two sites located at the Macaulay Institute's Glensaugh Research Station in northeastern Scotland, where historic land use records have been kept for many decades. At the first site, which was permanent grassland (minimum of 100 years) on a mineral soil (grid reference NO 665793), the *n*-alkane pattern in the soil closely reflected that of the overlying vegetation (improved pasture) (Fig. 2a). However, at a hill-grazing site close by, on an organic soil (iron podzol) (grid reference NO 672782), the *n*-alkane signature in the soil did not reflect the current vegetation (a mixture of upland grasses), but closely resembled that of heather (*Calluna vulgaris*). Land use records indicated that, until about 30 years ago, this site was

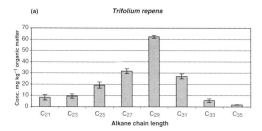

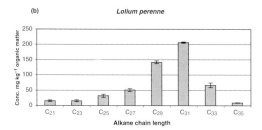

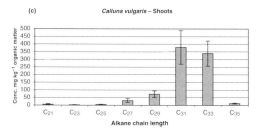

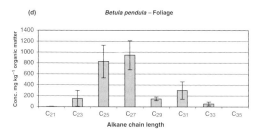

Fig. 1. Patterns of *n*-alkane signatures of: (**a**) *Trifolium repens* (white clover), *n* = 12; (**b**) *Lolium perenne* (perennial ryegrass), *n* = 7; (**c**) *Calluna vulgaris* (ling heather) n = 6; and (**d**) *Betula pendula* (silver birch), *n* = 6. Values are mean ±standard deviation.

covered by *Calluna vulgaris* heath (Fig. 2*b*). These preliminary findings highlight the importance of knowing the land use history when determining the role these signatures play in soil identification.

Case study 2

At a site on the Trotternish Ridge, Skye, northwestern Scotland (grid reference NG 472582) buried

organic-rich palaeosols were originally sampled to determine the relationship between vegetation and slope development. The slopes are currently used for rough grazing. The vegetation history of the slopes is thought to span around 6000 years, based on radiocarbon dates obtained by Hinchliffe (1999). The buried organic horizons indicate periods of slope stability and contain fossil pollen grains indicative of the plant species present at that time.

This study was initiated to compare results on vegetation distribution (as had contributed to the buried soil organic matter) obtained from pollen analysis with that determined using *n*-alkane analysis. Buried palaeosols were sampled at two locations on a slope at depths of 37–38 cm, 47–48 cm and 51–52 cm for the upper slope and 67–68 cm, 73–74 cm and 79–80 cm for the lower slope for *n*-alkane analysis. The locations selected matched those already sampled and analysed for fossil pollen analysis, providing the opportunity to compare two different sources of information relating to past vegetation. Current surface vegetation was characterized along the slope transect and had a uniform composition from the top to the bottom of the slope, consisting of *Agrostis-Festuca-Thymus/Alchemilla* herb-rich grassland.

Comparison of pollen content in the buried soil samples suggested that a heather dominated heath community was once present on the upper part of the slope and that a more mixed vegetation community was present on the lower part of the slope (Fig. 3*a, b*). Each graph relates to one sample location along the transect, and shows changes in pollen composition with depth (cm). There were no dramatic changes in the pollen composition down individual profiles but there were differences along the transect, from the top to the bottom of the slope. The *n*-alkane analysis of soil suggested a similar effect, with evidence of a more heather dominated community on the upper slopes in the past (Fig. 4*a, b*). The mean soil *n*-alkane concentration values for C_{31} reached 200 mg kg^{-1} in the upper slopes, compared to a mean value of 100 mg kg^{-1} in the lower slopes. Typical values for heather are around 400 mg/kg for this C_{31} *n*-alkane (Fig. 1), with mixed grass having values from 50–100 mg kg^{-1} (Dawson *et al.* 2000). Although these concentrations do overlap, when the ratios of $C_{29}:C_{33}$ are compared, the lower slope has a higher ratio (mean 0.922, SD 0.236) than the upper slope (mean 0.576, SD 0.170). This change in ratio again reflects the shift to a greater proportion of grass on the lower slope.

Case study 3

Another approach was to further test the technique on a range of soils with contrasting properties and vegetation. Table 1 provides some basic soil and

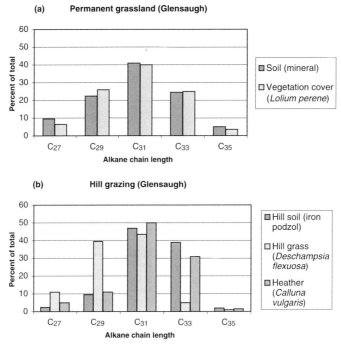

Fig. 2. Patterns of *n*-alkane signatures in soil at Glensaugh reflecting: (**a**) the *Lolium perenne* current vegetation signature, and (**b**) the previous land cover of *Calluna vulgaris* dominated heathland.

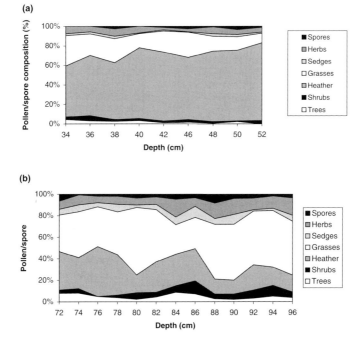

Fig. 3. Percentage of pollen grains and or spores by plant species in: (**a**) the upper part of the slope, and (**b**) the lower part of the slope.

(a)

Buried horizons – upper slope – mean

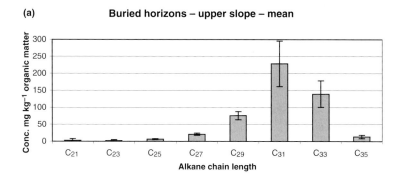

(b)

Buried horizons – lower slope – mean

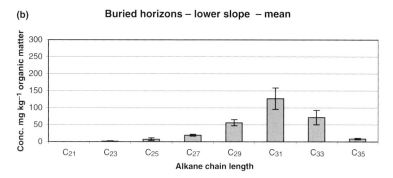

Fig. 4. The soil *n*-alkane concentration values for: (**a**) upper part of the slope, and (**b**) the lower part of the slope, $n = 3$. Values are ±standard deviation.

Table 1. *Soil and vegetation types*

Grid reference	Soil classification	Vegetation	Soil depth (cm)	Soil pH (CaCl$_2$)	Carbon content (%)
NJ 400100	Humus-iron podzol	Sitka spruce plantation	4–9	3.05	16.6
NO 500900	Peaty podzol	Boreal heather moor	1–10	3.00	46.5
NO 400400	Brown earth	Improved pasture	5–15	5.63	4.63

land use data on the three soils chosen. These soil/vegetation combinations were picked to reflect three of the main land uses in Scotland. Improved pasture is the predominant crop on the improved agricultural land of Scotland, heather moorlands are extensive in the eastern and central Highlands, and Sitka spruce is the most common tree species currently growing in Scotland. There is close correspondence between the distribution of the *n*-alkane signatures of the dominant aboveground vegetation species (Fig. 5*a*) and that of the soil (Fig.5*b*). Heather (*Calluna vulgaris*) is the dominant species within boreal heather moor, showing a peak at the C$_{31}$ *n*-alkane chain length and a close secondary peak at C$_{33}$. The corresponding gas chromatogram traces for heather shoots and underlying soil show these peaks before processing with SpectraWinner software (Fig. 6). This pattern differs from ryegrass (*Lolium perenne*), the dominant species within improved pasture, where the dominant *n*-alkane chain length is also C$_{31}$ but the secondary peak is at C$_{29}$. This pattern is not replicated precisely in the underlying soil, but the relatively high value for C$_{29}$, in relation to the other chain lengths, does provide evidence for ryegrass. In addition, both Sitka spruce foliage and the underlying soil show very low concentrations of *n*-alkanes compared to the other two sites. The *n*-alkane distribution pattern does not

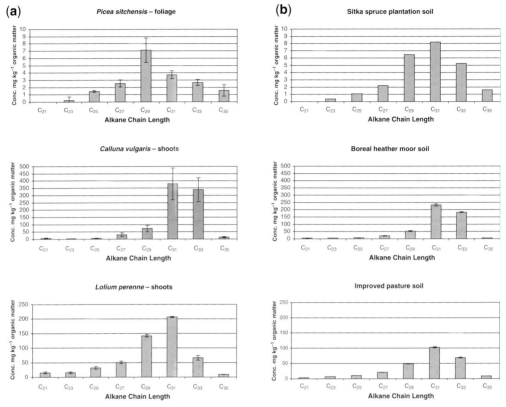

Fig. 5. (**a**) The *n*-alkane concentration distribution patterns of corresponding aboveground vegetation (bar represents mean and range of duplicate values); and (**b**) the soil *n*-alkane concentration distribution at three contrasting sites (bar represents the mean of six values obtained from the analysis of three separate samples in duplicate). The Sitka spruce and heather samples were collected from sites at grid references NJ 283542 (Teindland, Moray), NY 002915 (Ae, Dumfries) and NY 654923 (Kielder, Northumberland). The ryegrass was collected from the Macaulay Institute's Hartwood Research Station (grid reference NS 848600).

match that of the current vegetation but could reflect the previous heather moor vegetation which was present approximately 80 years ago, indicated by the higher C_{31} and C_{33} peaks.

Conclusions and future work

Results concur with findings of Almendros *et al.* (1996) in that patterns and concentrations of *n*-alkanes in the soil organic matter do reflect, in general, the signatures found in the overlying vegetation. However, our data also show a strong resistance of the signal *n*-alkane from previous vegetation. In order to build up a more complete picture of the relationships between soil, vegetation and *n*-alkane signatures, we intend to extend the approach outlined in this paper to include more replicates of the principal soil/vegetation combinations found in Scotland. Each sample will be taken from a

different site and will include arable, grassland, semi-natural vegetation and woodland ecosystems. This will ultimately provide a database which could be of use to the forensic scientist. We intend to run a compositional model to determine species contribution to the *n*-alkane signatures reflected in the soil organic matter at both case studies 2 and 3. In addition, work will consider the use of a wider range of markers, for example long-chain alcohols and fatty acids. These are especially important for conifers, where concentrations of *n*-alkanes are low and concentrations of long-chain alcohols have been shown to be relatively high. Soil biomarker information, derived from plant wax signatures, coupled with soil spatial information, could potentially assist in the identification of soil, providing another tool for use by the forensic scientist.

Funding by the Scottish Executive Environment and Rural Affairs Department is gratefully acknowledged.

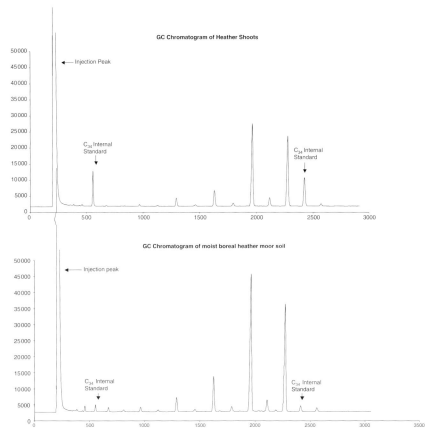

Fig. 6. Gas chromatogram traces from *Calluna vulgaris* shoots and from the underlying heather moor soil.

References

ALMENDROS, G., SANZ, J. & VELASCO, F. 1996. Signatures of lipid assemblages in soils under continental Mediterranean forest. *European Journal of Soil Science*, **47**,183–196.

ANDREYEV, L. V., NEMIROVSKAYA, I. B., NIKITIN, D. I., TOMASHCHUK, A. Y. & KHMELNITSKIY, R. A. 1980. Lipid composition of humus. *Soviet Soil Science*, **12**, 406–412.

CHAPMAN, S. J., CAMPBELL, C. D., FRASER, A. R. & PURI, G. 2001. FTIR spectroscopy of peat in and bordering Scots pine woodland: relationship with chemical and biological properties. *Soil Biology & Biochemistry*, **33**, 1193–1200.

COX, R. J., PETERSON, H. L., YOUNG, J., CUSIK, C. & ESPINOZA, E. O. 2000. The forensic analysis of soil organic by FTIR. *Forensic Science International*, **108**, 107–116.

DAWSON, L. A., MAYES, R. W., ELSTON, D. A. & SMART, T. S. 2000. Root hydrocarbons as potential markers for determining species composition. *Plant, Cell and Environment*, **23**, 743–750.

DOVE, H. & MAYES, R. W. 1991. The use of plant wax alkanes as marker substances in studies of the nutrition of herbivores – a review. *Australian Journal of Agricultural Research*, **42**, 913–952.

DOVE, H. & MAYES, R.W. 1996. Plant wax components: a new approach to estimating intake and diet composition in herbivores. *Journal of Nutrition*, **126**, 13–26.

FAEGRI, K. & IVERSEN, J. 1989. *Textbook of Pollen Analysis*. John Wiley, Chichester.

HINCHLIFFE, S. 1999. Timing and significance of talus slope reworking, Trotternish, Skye, northwest Scotland. *The Holocene*, **9**, 483–494.

MARSEILLE, F., DISNAR, J. R., GULIIET, B. & NOACK, Y. 1999. n-N-alkanes and free fatty acids in humus and Al horizons of soil under beech, spruce and grass in the Massif-Central (Mont-Lozere), France. *European Journal of Soil Science*. **50**, 433–441.

MAYES, R. W., LAMB, C. S. & COLGROVE, P. M. 1986. The use of dosed and herbage n-n-alkanes as markers for the determination of herbage intake. *Journal of Agricultural Science*, **107**, 161–170.

MORRISON, R. D. 2000. *Environmental Forensics*. CRC Press, Boca Raton.

MURRAY, R.C. & TEDROW, J. C. F. 1992. *Forensic Geology*. Prentice Hall, Upper Saddle River, New Jersey.

SALT, C. A., MAYES, R. W. & ELSTON, D. A. 1992. The effects of season, grazing intensity and diet composition on the radiocaesium intake by sheep on reseeded hill pasture. *Journal of Applied Ecology*, **29**, 378–387.

SUGITA, R. & MARUMO, Y. 1996. Validity of colour examination for forensic soil identification. *Forensic Science International*, **83**, 201–210.

TRUEMAN, C. 2004. Geochemical profiling of palaeonto-logical resources-and a cautionary study of spatial variation in trace element patterns. *In*: PYE, K. & CROFT, D. (eds) *Forensic Geoscience: principles, techniques and applications*. Geological Society, London, Special Publications, **232**, 249–256.

VOLKMAN, J. K., HOLDSWORTH, D. G. & BAVOR, H. J. 1992. Identification of natural, anthropogenic and petroleum hydrocarbons in aquatic sediments. *Science of the Total Environment*, **112**, 203–219.

The use of diatom analysis in forensic geoscience

N. G. CAMERON

Environmental Change Research Centre, Department of Geography, University College London, 26 Bedford Way, London WC1H OAP, UK (e-mail: ncameron@geog.ucl.ac.uk)

Abstract: Diatoms are unicellular, siliceous algae that are common in most aquatic environments. Their species composition is strongly related to water quality and aquatic habitats and, because of their silica skeletons, diatom valves can be well preserved and provide a record of past and present environmental conditions. Diatom remains are often diverse and can be identified with high taxonomic precision. These factors allow diatoms to be used in a range of applications in forensic geoscience. These include: the matching of environmental samples with items that have been in contact with water, the investigation of cases of drowning, and the identification of traces of diatomaceous materials used in the manufacture of materials or liquids. Recent developments include an assessment of the potential to use the succession of colonizing, attached diatom species in the determination of time of death. Advances in analytical quality control and the use of multivariate statistical techniques will improve the analysis, presentation and interpretation of diatom data in forensic investigations.

The diatoms are a group of unicellular algae most clearly characterized by their silica cell walls. Over 60% of the dry weight of the diatom cell may be silica. However, silica uptake and deposition involves less energy expenditure than formation of the equivalent organic walls found in other classes of algae and higher plants (Round *et al.* 1990). Each living cell is contained within two valves, one slightly larger than the other and both fitting together like a box and lid. The diatoms are yellow-brown pigmented and photosynthetic, although exceptionally a few species are heterotrophic. The classification of the huge number, perhaps as many as 10^5 (Mann & Droop 1996), of extant and fossil diatom species is predominantly based on the structure of the valve, its shape, intricate patterning and ornamentation (Round *et al.* 1990). Living diatoms are distributed in almost all aquatic and damp terrestrial habitats and in many of these the diatoms represent the most abundant and diverse algal class. They are abundant in the phytoplankton, benthos and attached algal communities of marine and freshwaters. Different diatom species are highly sensitive to water quality and many species are habitat-specific. Further, diatoms grow rapidly and under favourable conditions produce many cells within a short period. They colonize submerged surfaces quickly and many taxa show seasonality of growth. For these reasons they have been widely used in such applications as palaeoecological reconstruction (Battarbee *et al.* 2001), water quality monitoring (Stevenson & Pan 1999) and in forensic science (Peabody 1999). The diatom valve allows the majority of both living and fossil taxa to be identified to the species or subspecies level. For microscopic examination, fossil and living cells are usually cleaned to remove organic contents or sediment and to allow details of the silica cell wall structure to be revealed (Battarbee 1986; Battarbee *et al.* 2001). Diatom cells are mainly solitary but some colonial taxa produce special separation valves during vegetative growth. The formation of resting stages and auxospores can also produce different valve morphologies. These characteristics lead to a number of actual and potential applications of diatoms in forensic science.

The 'diatom test' for drowning is probably the most often applied and studied application of diatom analysis in forensic investigations (e.g. Pollanen 1998*a*) and has become an established forensic technique. The earliest use of diatoms in the diagnosis of drowning was by workers in Hungary and Germany during the 1940s (references cited in Peabody & Burgess 1984). Drowning occurs when water enters the lungs and then the bloodstream via ruptures in the alveoli. Any small particles suspended in the water may also be carried into the bloodstream and potentially throughout the body. Such particles are likely to be deposited in the capillaries of the major organs and muscles. In most natural fresh and saltwater bodies this particulate material will include diatoms. However, not all water sources contain diatoms in suspension. For example domestic water supplies in the UK contain few diatoms (Peabody 1999) although Pollanen (1998*b*) has reported matching diatom assemblages from domestic bathwater and femoral bone marrow from a drowning case in Canada.

There have been conflicting findings on the diatom content of organs from the bodies of non-drowned subjects. A number of factors may lead to variable recovery of diatoms from drowned and non-drowned subjects (e.g. Peabody & Burgess 1984;

From: PYE, K. & CROFT, D. J. (eds) 2004. *Forensic Geoscience: Principles, Techniques and Applications*. Geological Society, London, Special Publications, **232**, 277–280. © The Geological Society of London, 2004.

Krstic *et al.* 2002). These factors include the potential for contamination during sample preparation where a relatively large volume of organic matter, a low diatom concentration and low species diversity are involved. Diatoms may also be deposited in human organs before death. Peabody & Burgess (1984) compiled the findings of 11 researchers who had examined the organs of drowned and non-drowned subjects. In about one-third of non-drowning cases, diatoms were found in the lungs or liver. Lower values were reported for kidney and bone marrow. Conversely the absence of diatoms does not necessarily exclude the possibility of drowning. Pollanen *et al.* (1997) have analysed the outcome of the diatom test in over 700 cases of drowning. They found that diatom frustules were present in the bone marrow in only approximately one-third of freshwater drownings. Months of the year where there were high concentrations of diatoms (e.g. when spring and autumn diatom blooms occur) correlated with the presence of diatoms in the bone marrow. In appropriate cases, where the circumstances of death and post-mortem findings lead to the suspicion of drowning as a cause of death, diatom analysis can therefore provide useful supporting evidence. In a novel application, diatom analysis has been used experimentally in forensic pathology to determine the time of death (Casamatta & Verb 2000). Algal colonization of rat carcasses in a stream indicated that both diatom species diversity and the sequence of species colonizing the bodies have some potential for estimating the period of submersion of a body or material.

Perhaps of greater interest to the geoscientist is the use of diatoms in identifying the provenance of individuals, clothing or materials from sites of investigation. However, in this context, in addition to the taxonomy and ecology of diatoms, an important consideration in the forensic context is the taphonomy of diatom assemblages. This involves assessing the fidelity with which test samples reflect the environment from which they were derived, either where an attempt is being made to identify a source environment or where comparison is made between a test sample and an environmental sample to determine the degree of similarity. In many cases it may be adequate to be aware of the likelihood for modification of the diatom assemblage, but in others more critical analysis is required. Diatom taphonomy has been studied in the context of Holocene palaeoecology using experimental studies of dissolution in saline environments (Barker *et al.* 1994; Ryves *et al.* 2001), the representative quality of recent fossil assemblages compared with living diatom communities (Cameron 1995) and the transport of diatoms in estuaries (Juggins 1992). These factors are relevant to forensic diatom studies in circumstances where the alteration of diatom assemblages is of concern.

Where materials have been submerged or there is contact with littoral or riparian sediment or vegetation, diatom analysis of sediments or other diatomaceous traces present on clothing or footwear can be used to identify the type of habitat involved. In cases of drowning the range of locations at which the person may have drowned may be inferred from diatom analysis of suspended material that has entered the body at the time of drowning (Ludes *et al.* 1999). The comparison of external diatom assemblages is of particular interest in cases where a body has been transported a significant distance by currents, or where circumstances suggest that a body has been deliberately moved from the site of death. In rivers or estuaries where there are strong environmental gradients of factors, such as pH, nutrient concentrations or salinity, diatom analysis will reflect this variation in water quality. In lakes too the types of diatom communities and sediment assemblages are characteristic of the water quality. In a case study from New England, USA, Siver *et al.* (1994) were able to link suspects with a lake that was the scene of a serious assault. They compared the diatom content of mud taken from suspect's and victim's shoes and freshwater pond sediment from the site of the crime. Diatom abundance, presence or absence, along with population ratios for a group of *Eunotia* spp., showed marked similarities between the samples. It was therefore possible to state that the shoes were exposed to a common, if not necessarily the same, locality. In some circumstances the silica scales of scaled chrysophytes (and chrysophyte stomatocysts), a class of algae related to the diatoms and utilized in palaeoecology (Duff *et al.* 1995), may also prove useful in the forensic context, although these assemblages are usually less species-rich than diatom assemblages. In this investigation Siver *et al.* (1994) also recorded the presence of abundant chrysophyte scales from a single species, *Mallomonas caudata*, in all three classes of sample. A similar diatom analytical approach, using percentage diatom analysis has been used to associate a suspect's clothing and footwear with a stretch of the River Avon, Wiltshire, UK, in which the body of a murder victim was found. Here the multivariate statistical techniques of ordination and analogue matching (Flower *et al.* 1997) were employed as a means of presenting these data graphically and to show the best analogues from amongst a group of test samples (Fig. 1).

The use of diatoms in manufacturing processes and products leads to their potential use in forensic investigations. Raw materials containing diatoms include diatomaceous clays used in the manufacture of ceramics. Pottery, tile or brick for example may retain diatoms where firing temperatures are low. However, the most important diatomaceous raw material is diatomite. This is a porous, low-density,

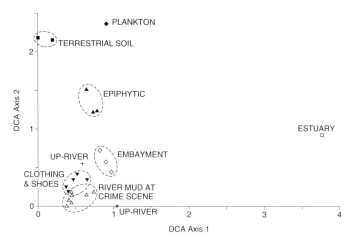

Fig. 1. Scatterplot of diatom assemblages on the first two axes of variation from a detrended correspondence analysis (DCA). The distance between points shows the degree of similarity of the diatom assemblages. This illustrates that diatom assemblages taken from the clothing and shoes of a suspect is most similar to river mud where a murder victim was found. Diatom assemblages from other locations along the same river, the riverbank and from vegetation and plankton samples provide poorer analogues. These diatom data suggest that the clothing and footwear was in direct contact with submerged mud in the river.

sedimentary rock formed by the accumulation and compaction of diatoms. The physical properties of diatomite have led to its use in many commercial applications, the most common being for the filtration of liquids or as a filler, for example in paints. Other uses of diatomite include insulation (fire bricks), fine abrasion (in some toothpaste and polishes) and as a pesticide. Freshwater and marine diatomites are mined in various parts of the world and range in age from the Late Cretaceous to the Holocene so there may be a great deal of variation in the diatom assemblages present. Estimated total world production was 1.5 million tonnes in 1995 (Harwood 1999).

Diatomite is used in a number of materials that have been employed by forensic scientists to provide evidence of association. At present commercial sources of diatomite are restricted to a few large sites. Consequently the same raw material, potentially having a similar fossil diatom flora, can be used in the manufacture of a large number of products. However, Peabody (1999) notes that forensic scientists have examined diatoms from paint and polishes, and these may show evidence of association. Peabody (1999) also reports that diatomite was formerly used as a fireproof packing material between the outer casing and inner compartment of safes. If this type of safe is broken open diatomite may contaminate clothing with a high concentration of diatom valves. The comparison of diatom assemblages in such cases will show strong evidence of association because diatomite safe ballast is likely to have been taken from extinct diatomite workings,

and it is unlikely that an individual would have this material on clothing by chance (Peabody 1999).

In a different context, the appearance in the environment of exotic or unusual species derived from diatomite can also be used as a tracer. Clarke (1991) deduced that *Cyclotella sevillana*, a species very rare in the UK, which appeared in an area of the Norfolk Broads, was discharged from diatomite filters at a nearby cider factory. Clarke comments that one of the few other records of this particular species in the UK, in a mine adit on the west coast of Scotland, might have its origin in the dynamite formerly used for blasting. In the past nitroglycerine was absorbed in diatomite as a means of safe transport and this form of the explosive is still used occasionally (Harwood 1999). These instances illustrate the potential to use contamination from diatomite in a forensic context. Other examples of natural diatom contamination in the environment that are of potential use for the forensic scientist as tracers are the transport of diatoms in the atmosphere (Harper 1999) and in the ballast water of ships (e.g. Hallegraeff & Bolch 1992).

A number of characteristics of diatoms, including their widespread presence in water, sensitivity to environmental water quality, high species diversity, habitat specificity, and good potential for preservation, suggest that this group of algae has further potential for use in the field of forensic geoscience. For criminal investigations in particular, the invisibility of valves to the naked eye and the fact that diatoms are not well know to the layman are useful attributes. Standardization of diatom taxonomy and

numerical techniques developed in the field of
Holocene palaeoecology (e.g. Battarbee *et al.*
2000) will improve the comparison and numerical
analysis of diatom assemblages in the forensic
context and the presentation of diatom data to a lay
audience.

Thanks to J. Quinn for redrawing Figure 1 and to D. Ryves,
V. Jones and N. Branch for comments on the manuscript.

References

BARKER, P., FONTES, J. C., GASSE, F. & DRUART, J. C.
1994. Experimental dissolution of diatom silica in
concentrated salt solutions and implications for
palaeoenvironmental reconstruction. *Limnology and
Oceanography*, **39**, 99–110.

BATTARBEE, R. W. 1986. Diatom analysis. *In*: BERGLUND,
B. E. (ed.) *Handbook of Holocene Palaeoecology and
Palaeohydrology*. John Wiley, Chichester, 527–570.

BATTARBEE, R. W., JONES, V .J., FLOWER, R. J., CAMERON,
N. G., BENNION, H, CARVALHO, L. & JUGGINS, S. 2001.
Diatoms. *In*: SMOL, J. P., BIRKS, H. J. B. & LAST, W.
M. (eds) *Tracking Environmental Change Using Lake
Sediments*. Vol. 3. *Terrestrial, Algal, and Siliceous
Indicators*. Kluwer Academic Publishers, Dordrecht,
155–202.

BATTARBEE, R. W., JUGGINS, S., GASSE, F., ANDERSON, N. J.,
BENNION, H. & CAMERON, N. G. 2000. European
Diatom Database (EDDI). An Information System
for Palaeoenvironmental Reconstruction. European
Climate Science Conference, Vienna City Hall,
Vienna, Austria, 19–23 October 1998, pp 1–10.

CAMERON, N. G. 1995. The representation of diatom communities by fossil assemblages in a small acid lake.
Journal of Palaeolimnology, **14**, 185–223.

CASAMATTA, D. A. & VERB, R. G. 2000. Algal colonization
of submerged carcasses in a mid-order woodland
stream. *Journal of Forensic Sciences*, **45**, 1280–1285.

CLARKE, K. B. 1991. The search for a Norfolk lake deposit
containing *Cyclotella*: a cautionary tale and some
observations on *C. sevillana* and *C. sexpuntata* Deby.
Diatom Research, **6**, 211–221.

DUFF, K. E., ZEEB B. A. & SMOL J. P. 1995. *Atlas of
Chrysophycean Cysts*. Kluwer Academic Publishers,
Dordrecht.

FLOWER, R. J., JUGGINS, S. & BATTARBEE, R. W. 1997.
Matching diatom assemblages in lake sediment cores
and modern surface sediment samples: the implications for lake conservation and restoration with
special reference to acidified systems. *Hydrobiologia*,
344, 27–40.

JUGGINS, S. 1992. *Diatoms in the Thames Estuary,
England: Ecology, Palaeoecology, and Salinity
Transfer Function*. Bibliotheca Diatomologica, **25**.

HALLEGRAEFF, G. M. & BOLCH, C. J. 1992. Transport of
diatom and dinoflagellate resting spores in ship's
ballast water: implications for plankton biogeography

and aquaculture. *Journal of Plankton Research*, **14**,
1067–1084.

HARPER, M. A. 1999. Diatoms as markers of atmospheric
transport. *In*: STOERMER, E. F. & SMOL, J. P. (eds) *The
Diatoms: Applications for the Environmental and
Earth Sciences*. Cambridge University Press,
Cambridge, 429–435.

HARWOOD, D. M. 1999. Diatomite. *In*: STOERMER, E. F. &
SMOL, J. P. (eds) *The Diatoms: Applications for the
Environmental and Earth Sciences*. Cambridge
University Press, Cambridge, 436–443.

KRSTIC, S., DUMA, A., JANEVSKA, B., LEVKOV, Z.,
NIKOLOVA, K. & NOVESKA, M. 2002. Diatoms in
forensic expertise of drowning: a Macedonian experience. *Forensic Science International*, **127**, 198–203.

LUDES, B., COSTE, M., NORTH, N., DORAY, S., TRACQUI, A.
& KINZ, P. 1999. Diatom analysis in victim's tissues as
an indicator of the site of drowning. *International
Journal of Legal Medicine*, **112**, 163–166.

MANN, D. G. & DROOP, S. J. M. 1996. Biodiversity, biogeography and conservation of diatoms. *Hydrobiologia*, **336**, 19–32.

PEABODY, A. J. 1999. Forensic science and diatoms. *In*:
STOERMER, E. F. & SMOL, J. P. (eds) *The Diatoms:
Applications for the Environmental and Earth
Sciences*. Cambridge University Press, Cambridge,
413–418.

PEABODY, A. J. & BURGESS, R. M. 1984. Diatoms in the
diagnosis of death by drowning. *In*: MANN, D. G. (ed.)
*Proceedings of the Seventh International Diatom
Symposium*. Otto Koeltz, Koenigstein, 537–541.

POLLANEN, M. S. (ed.) 1998*a*. *Forensic Diatomology and
Drowning*. Elsevier, Amsterdam.

POLLANEN, M. S. (ed.) 1998*b*. Diatoms and homicide.
Forensic Science International, **91**, 29–34.

POLLANEN, M. S., CHEUNG, L. & CHAISSON, D. A. 1997.
The diagnostic value of the diatom test for drowning.
I. Utility: a retrospective analysis of 771 cases of
drowning in Ontario, Canada. *Journal of Forensic
Sciences*, **42**, 281–285.

ROUND, F. E., CRAWFORD, R.M. & MANN, D.G. 1990. *The
Diatoms: Biology and Morphology of the Genera*.
Cambridge University Press, Cambridge.

RYVES, D. B., JUGGINS, S, FRITZ, S. C. & BATTARBEE, R. W.
2001. Experimental diatom dissolution and the quantification of microfossil preservation in sediments.
Palaeogeography, Palaeoclimatology, Palaeoecology,
172, 93–113.

SIVER, P. A., LORD, W. D. & MCCARTHY, D. J. 1994.
Forensic limnology: the use of freshwater algal community ecology to link suspects to an aquatic crime
scene in southern New England. *Journal of Forensic
Sciences*, **39**, 847–853.

STEVENSON, R. J. & PAN, Y. 1999 Assessing environmental
conditions in rivers and streams with diatoms. *In*:
STOERMER, E. F. & SMOL, J. P. (eds) *The Diatoms:
Applications for the Environmental and Earth
Sciences*. Cambridge University Press, Cambridge,
11–40.

The right way and the wrong way of presenting statistical and geological evidence in a court of law (a little knowledge is a dangerous thing!)

WAYNE C. ISPHORDING

Department of Earth Sciences, University of South Alabama, Mobile, AL 36688, USA
(e-mail: wisphord@jaguar1.usouthal.edu)

Abstract: On 21 March 1981 a young black male was abducted in Mobile, Alabama, and taken to a site across Mobile Bay where he was beaten and murdered. The act was in apparent revenge for the mis-trial of a black man accused of killing a white police officer from Birmingham, Alabama. Three members of the notorious Ku Klux Klan were apprehended and charged with murder. All were found guilty and the ringleader was executed in 1997; the others are now serving life sentences.

Ironically, the perpetrators were successful in having evidence tying them to the crime scene completely discredited. Statistical evidence purporting to show similarities in soil chemistry from samples taken from the victim, the defendants, and the crime scene was totally invalidated because improper statistical analyses were used. Following completion of the trial, the defence's expert witness was contacted by the district attorney and asked to describe the appropriate statistical tests that should have been offered and that would have supported his case. He was further able to take solace by learning that strong mineralogical evidence could also have been used. It simply had not been reviewed by anyone who possessed the proper expertise.

One of the most heinous racial crimes that ever took place in the USA occurred on 21 March 1981. A 19-year-old black male, Michael Donald, was abducted from a street in downtown Mobile, Alabama, and taken across Mobile Bay to a site in Baldwin County, Alabama, (Fig. 1) where he was beaten and murdered. Subsequently, he was brought back to Mobile and hung from a tree directly across the street from the house of one of the defendants where he was found the next morning. The young man had walked to the store to purchase cigarettes when he was randomly abducted and killed in apparent revenge for a mis-trial that had been granted a black man charged with murdering a white police officer from Birmingham, Alabama. Two members of the notorious Ku Klux Klan, Henry Francis Hays and James 'Tiger' Knowles, were apprehended and charged with the murder. A third individual, Benjamin Franklin Cox, was charged as an accessory. Several years later the father of Henry Hays was also indicted as an accomplice. Hays, Knowles and Cox were found guilty of capital murder in 1983. Hays was executed in the electric chair in Holman Prison (Alabama) in 1997; Knowles is serving a life sentence for violating Donald's civil rights. The third co-defendant (Cox) is serving a 99-year prison sentence. Hay's father, Bennie Jack Hays, a noted Ku Klux Klan figure, was extradited in 1993 from Ohio and prosecuted in the same year for aiding and abetting the crime and found guilty. He died from natural causes before beginning his life sentence.

In 1987 the mother of the slain black man filed a civil suit against the Invisible Empire, Knights of the Ku Klux Klan. A Mobile, Alabama, jury returned a verdict for $7 million in damages, the largest legal blow ever rendered against the Klan and the biggest award in the history of the local US District Court. The United Klans of America (UKA), whose former headquarters was in Alabama, was shut down by the huge judgement rendered by the all-white federal court jury. Beulah May Donald thereby bankrupted the UKA. The judgement against the UKA also included its national headquarters building in Tuscaloosa, Alabama, and other Klan property. Similar suits followed which devastated the internal operations of Klan activity throughout the USA.

In spite of the fact that justice was apparently served, it is ironic to note that the three perpetrators of the crime were successful in having the major evidence tying them to the crime scene wholly discredited by the defence's expert witness. Statistical evidence offered by the prosecution aimed at showing similarities in the chemistry of soil samples found on the victim's clothing, the defendant's shoes and car and the actual crime scene were totally invalidated because inappropriate (and incorrect) statistical methods were used and mistakes were made in the treatment of samples on which chemical analyses were carried out. The lack of all but a cursory knowledge of statistical principles by the prosecution's witness was made clearly apparent to the jury, and Mark Twain's famous statement that 'There are three kinds of lies: lies, damned lies, and statistics!' was hammered home. Furthermore, the prosecution

From: PYE, K. & CROFT, D. J. (eds) 2004. *Forensic Geoscience: Principles, Techniques and Applications*. Geological Society, London, Special Publications, **232**, 281–288. © The Geological Society of London, 2004.

Fig. 1. Location map showing major geological units present in Mobile and Baldwin Counties, Alabama, the location of the defendant's home and the crime scene.

witnesses were also shown to have little knowledge of soil mineralogy and the chemical variability associated with soils. Whereas excellent evidence could have been offered by the prosecution, had they chosen to use the differences between the heavy mineral and/or the clay mineral 'fingerprint' of soils at the murder scene and those at the defendant's place of residence, the prosecution instead attempted to show 'similarities' in the chemistry of various soil samples. Fortunately, the co-defendants, desiring to avoid a possible death sentence, admitted their involvement and testified that Hays was the chief conspirator. Following completion of the trial, the defence's expert witness was contacted by the district attorney and queried as to what statistical tests should have been used (if any!) and what other evidence might have been offered to support his case.

The 'wrong' way of presenting evidence

Chemical analyses of soil were carried out by the prosecution on samples collected from the victim, from the defendant's shoes and vehicle, from the crime site in Baldwin County and from the principal defendant's home. Additional soil samples were also collected by the Federal Bureau of Investigation (FBI). The results of these analyses are shown in Table 1.

While the defence accepted that the chemical analyses had been carried out by a reputable laboratory, the use of chemical analytical data in a court of law permits a number of questions to be raised by the opposing attorney. These include (but are not restricted to) the following.

(1) What quality assurance/quality control (QA/QC) protocols are operated in the laboratory where the samples were processed, and were these followed? Assuming the protocols were followed, the laboratory is then required to produce all the records for the day(s) on which the analyses were run (including data for the reporting limit and method detection limit for each element, the results for the calibration blank, mid-calibration standard and high calibration standard, and the calibration curves obtained for each element). It is also required to explain any aberrant results that were present when the machine calibration procedures were being carried out and to produce the results of all check standards tests that were run, including the results for the low calibration standard and the continuing calibration verification (CCV) standard. Note that failure of a laboratory to retain such records not only prevents the laboratory from being certified by the National Environmental

Table 1. *Comparison of heavy metal analyses from soil samples from defendant with those from the victim (mg kg^{-1})*

	Samples from Defendant			Samples from Victim				
Element	Sample 1	Sample 2	Sample 3	Sample 1	Sample 2	Sample 3	Sample 4	Sample 5
As	<0.001	<0.001	<0.001	NA	<0.001	<0.001	NA	NA
Pb	8.6	10.51	10.46	3.66	17.68	145.1	6.22	8.14
Cd	0.41	0.61	0.21	0.23	0.93	0.73	0.19	0.41
Cr	<0.05	<0.05	<0.05	<0.05	<0.05	<0.05	<0.05	<0.05
Al	1,394	2,418	504	<0.1	<0.01	<0.1	<0.1	<0.1
Ba	<0.1	<0.1	<0.1	<0.1	<0.1	<0.1	<0.1	<0.1
Ca	389	1,195	844	1,111	558	786	1,177	1,498
Fe	1,506	2,365	526	87	0.08	574	249	375
K	412	683	389	1,428	396	453	1,287	1,221
Na	1,497	2,404	1,789	6,031	1,917	1,305	11,100	7,003
Sr	<0.04	<0.04	<0.04	<0.04	<0.04	<0.04	<0.04	<0.04
Zn	279	38	31	40	639	154	30	44
Mg	66	118	75	133	195	224	57	81

Laboratory Accreditation Council (NELAC) and the US Environmental Protection Agency (EPA), it also immediately allows the opposing attorney to challenge the use of such results in court and to demand their exclusion from testimony and consideration of the results by a jury.

(2) Was US EPA procedure 6010B (EPA 1999) followed completely in the analysis of the samples? This procedure is applicable to the analysis of a wide variety of sample types (soils, sludges, sediments etc.) that are analysed by inductively coupled plasma atomic emission spectrometry (ICP-AES) and describes applicable elements that can be analysed, recommended wavelength settings, evaluation of interference effects, inter-element corrections, reagent and standards preparation etc. Failure to follow these procedures exactly again provides the opposing attorney with an opportunity to legitimately challenge any results and demand their exclusion.

(3) During the actual analysis of the samples, how many samples were analysed and what spikes were used? The US EPA requires that sample batches be prepared for no more than 20 samples and that five individual blanks be prepared with each batch (see EPA 1999).

(4) How often was the CCV run while samples were being analysed? How many duplicate analyses were carried out? What final procedures were run at the conclusion of the analyses to ensure that the instrument was performing properly (the low calibration standard and high calibration standards should be within specific limits when compared with their original values at the start of the analyses)?

Because of the strict procedures that are required in order for a laboratory to retain its national accreditation, it is nearly impossible for any results carried out by non-accredited labs to be accepted in court. Consequently, expert witnesses are well advised *not* to run the analyses themselves but, instead, to testify as to the meaning and significance of analyses that were carried out by an accredited laboratory.

While the results of the laboratory analyses of the soil samples were accepted as having been performed by a certified laboratory, the prosecution made a serious mistake in permitting the analyst to testify as to the significance of the results. The analyst was a chemist and was not skilled in soil science, statistics or mineralogy. As such, he was vulnerable and was attacked by the defence in each of these areas and his results totally discredited. Specific criticisms are described in the following paragraphs.

Soil analysis

The prosecution's analyst could not explain the large discrepancy between levels of aluminum observed on samples from the defendants and those from the victim. It was quickly shown that:

(1) The analyst did not first determine the size frequency distribution of each sample (which could have been done, even on very small samples, using any of a number of electronic particle size analytical methods).

(2) The analyst was not aware that differences in size analyses strongly control the levels of various metals that can be present. Sand size

Table 2. *Results of Wilcoxon-Mann-Whitney test for median values of various heavy metals for soil samples from defendant and victim*

	Test of Ho (Md$_1$ = Md$_2$)	Result
Pb$_1$ = Pb$_2$	Is significant at $p = 0.766$	Cannot reject at $\alpha = 0.05$
Cd$_1$ = Cd$_2$	Is significant at $p = 0.882$	Cannot reject at $\alpha = 0.05$
Ca$_1$ = Ca$_2$	Is significant at $p = 0.766$	Cannot reject at $\alpha = 0.05$
Fe$_1$ = Fe$_2$	Is significant at $p = 0.074$	Cannot reject at $\alpha = 0.05$
K$_1$ = K$_2$	Is significant at $p = 0.223$	Cannot reject at $\alpha = 0.05$
Zn$_1$ = Zn$_2$	Is significant at $p = 0.766$	Cannot reject at $\alpha = 0.05$
Mg$_1$ = Mg$_2$	Is significant at $p = 0.371$	Cannot reject at $\alpha = 0.05$

particles will always show low levels of metals; samples with high contents of clay size material will be substantially higher (the low levels of aluminum seen for defendant's samples in Table 1 simply reflected a far lower content of clay in these samples).

(3) The analyst, similarly, had no answer for the marked differences in iron observed in samples from the victim compared with samples from the defendant. Again, these are simply a function of differences in the particle size distributions of the individual samples and reflect differences in the quantity of oxy-hydroxide compounds (goethite-limonite and hematite) that are ubiquitous in most soils.

Mineralogical analysis

It was also determined that the prosecution's witness had no fundamental understanding of the variability that can occur in the chemistry of soil samples; these are associated with differences in major constituents (quartz, feldspar, clay minerals etc.) and heavy minerals that are also present in soils. Even for samples that showed similar metal levels for both the victim and defendants, the analyst was forced to admit that, if the similar quantities that were observed were, coincidentally, from different minerals, then the different minerals would certainly indicate a different source for the samples, for example iron from goethite ($HFeO_2$), as opposed to iron from vivianite ($Fe_3(PO4)_2 \cdot 8H_2O$)), pyrite (FeS_2) etc. The witness admitted that no attempt had been made to confirm that the mineralogy of soil samples from the victim and defendants was the same.

An additional error made in the analysis of the soil samples involved the basic digestion methodology that had been used. The analyst admitted on the stand that he had used a 'partial digestion procedure' ($HCl–HNO_3$) and that this was 'standard practice' in his laboratory for dealing with soil and sludge samples. While this method is widely used by many analytical labs, it should not be used for testimony in a capital murder trial. A complete digestion method, using either lithium metaborate fusion, perchloric acid digestion, or high temperature microwave digestion using $HF–HNO_3–HCl$, should be used to ensure complete digestion. The analyst was forced to agree that metals in sediments are adsorbed and partition in different forms (exchangeable sites, adsorbed by iron and manganese oxy-hydroxide compounds, as sulphide phases, as carbonate phases, as structural phases etc.). He could not adequately explain the degree to which his digestion procedure extracted metals from these phases and was forced to admit that metals held in defect lattice sites and octahedral and tetrahedral sites in the lattices of certain silicates, and strongly refractory minerals (e.g. zircon), might not be extracted by the method he used. Because the defence demonstrated that there were differences in the mineralogy of the samples, he was then forced to admit that comparing the chemical analysis of one sample to that of another might be analogous to 'comparing apples with oranges!'

Statistical analysis

It was further obvious that the prosecution's witness had only a cursory knowledge of statistical procedures, and that these were limited to a few parametric tests.

In an attempt to show that the chemical analyses of samples taken from the defendant and from the victim were 'similar', the witness ran a standard (Pearson) correlation analysis. This was done by first averaging the three individual values obtained for each metal from samples taken from the defendants and the five values for each element obtained from the victim. The pair of averaged values obtained for each metal were then treated as a 'sample' and the correlation was carried out on the eight samples (actually eight different metals). The resulting correlation coefficient of 0.77 was deemed 'significant' by the prosecution and interpreted by their witness as showing a 'strong likelihood' that the samples were derived from the same location (i.e. the crime scene).

The best that can be said for the above interpretation and methodology is that it is a clear violation of basic statistical theory. It encompasses so many misuses of statistics that it required little effort on the part of the defence to totally destroy the prosecution's entire statistical argument. The results were cited by the defence attorney as an example of Mark Twain's quotation that there are 'lies, damned lies, and statistics!'

Among the mistakes made by the witness in attempting to make a 'statistical' case for his data were the following.

(1) First and foremost, a correlation analysis (either parametric OR non-parametric) cannot be used to demonstrate association (or lack of association). This is not a two-sample case. Rather, it consists of two groups on which eight variables have been measured. The results of the chemical analyses for each metal were averaged, and then the pairs of averages 'correlated'. The distribution of averages is unknown, hence the Pearson correlation, which assumes that a normal distribution would be inappropriate from the beginning. Further, the Pearson correlation assumes homoscedastiscity (equal variances). Even a cursory examination of Table 1 reveals that the results for some of the metals are highly skewed and that the variances are certainly not equal for the two groups (see especially iron, potassium and sodium).

(2) Notwithstanding the fact that correlation is a wholly inappropriate test for the data, the witness admitted that no attempt had been made statistically to test the significance of the correlation coefficient and to assign a $\pm$ value to the coefficient. Further, because of the very small number of samples that were analysed , the power of any statistical test would be greatly reduced and any correlation coefficient (parametric or non-parametric) would be viewed with distrust (see below).

(3) The witness cited Sokal and Rohlf (1969) as his 'authority' for carrying out the correlation analyses. Had the witness read a few more pages past the section describing the procedure for carrying out a Pearson correlation, he would have encountered an important caveat applicable to small samples. Namely (p. 519): 'For small size samples, the calculation of exact probabilities (for statistical significance) is difficult . . . it is not known how small a sample size is adequate but, most likely, *it should not be used for $n < 10$*' (author's italics).

The 'right' way of presenting evidence

After completion of the trial, the defence's expert witness was contacted by the district attorney and asked in a friendly manner how he might have dealt with the physical evidence that had been compiled. He was told that the use of both proper statistical analyses and mineralogical data would have supported his case.

Statistical analysis

The defence's witness had been asked by the prosecution while under cross-examination, in light of his strong criticism of the defence's use of correlation analysis, whether there were statistical procedures that would have been appropriate. The witness answered in the affirmative that the non-parametric Wilcoxon-Mann-Whitney test could have been used to compare the median values of each metal from the two groups, but that even this method would suffer from the small sample size. The district attorney asked the witness if he had carried out the test. The witness, on answering 'yes' to the question, then expected the district attorney to ask him the results of his analysis. The results of the test, shown in Table 2, could have been used to show that there was a statistical likelihood that the samples obtained from the defendants and those from the victim were similar and were also similar to analyses of soils from the crime scene. The operable null hypothesis that was tested was that the median value of each metal from group 1 (the defendant) was equal to the median value from group 2 (the victim). Stated differently, the hypothesis, tested at the 0.05 level of significance, was that the median value of samples from group 1 (defendants) could have come from a population from which the median value obtained from group 2 (victim) was also obtained. Naturally, none of the metals whose test results were 'less than' a specific level(Table 1) were used for the comparison. For each of the metals that were used, there was no statistical basis for rejecting the null hypothesis. Unfortunately, however, one of the cardinal rules in the legal profession is 'Never ask a witness a question to which you do not already know the answer'. The district attorney stated that, although he wanted to ask the question, he could not take the chance of the answer from the defence witness being damaging to his case (it should be noted here that the defence's expert witness was testifying as a 'hostile witness'). When this witness had first been approached by the defence, he had reviewed the prosecution's data without knowing to which case the information was applicable. After pointing out the numerous mistakes, he was told the case to which the information pertained. He vehemently

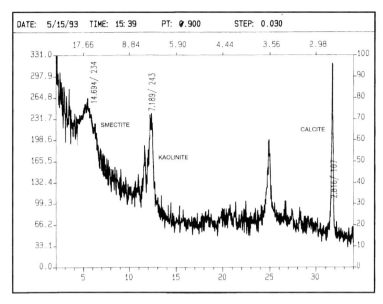

Fig. 2. X-ray diffractogram of clay mineral fraction (<4 μm) from soil sample from defendant's residence (Cu K-α radiation).

stated that he had acquired, from the newspaper accounts, a 'pre-conceived opinion' as to the guilt of the defendants and in no way wished to assist, or be associated with, the defence. He was then advised that a subpoena could, and would, be issued and that, under oath, he would have to answer truthfully and affirm each of the 'errors' that he had previously identified. This he did, with reluctance.

Mineralogical analysis

Overlooked by the prosecution was mineralogical evidence that would have clearly linked the defendants to the crime scene. When arrested the day after the murder, the three defendants stated that they had not been in Baldwin County, Alabama, for several weeks and had spent the evening when the murder took place at the principal defendant's residence and had not ventured elsewhere in the city of Mobile. A geological map showing the location of the defendant's residence is shown in Figure 1. Note that the residence lies on Quaternary floodplain deposits adjacent to Mobile Bay whereas the crime scene was across the bay at a site on the Plio-Pleistocene Citronelle Formation. Each of these units has a distinctive clay mineral signature and each is characterized by a different heavy mineral suite. An X-ray diffractogram of the <4 μm fraction from the soil at the defendant's home is shown in Figure 2. These soils consist largely of kaolinite but also contain subordinate smectite (montmorillonite) and calcite

(from shell material). Samples from the crime scene are distinctly different. The Citronelle Formation is a fluvial deposit, whose clay mineral suite consists wholly of kaolinite (with minor illite). Smectite and calcite are completely absent (Fig. 3). Soil samples from the defendant's car, shoes and clothing, and from the victim, were characterized by a clay mineral suite whose mineralogy, and clay mineral proportions, were typical of samples from the Citronelle Formation.

Heavy minerals extracted from the soil samples also confirmed the fact that the defendants lied about their whereabouts on the night of the murder. The heavy mineral suite of the terrace deposits (Table 3), consists largely of black opaques (ilmenite), staurolite, kyanite and leucoxene but also contains small quantities of garnet and hornblende. Feldspar is common as a minor constituent in the light mineral fraction. The Citronelle Formation, in contrast, contains significantly higher quantities of ilmenite, kyanite and zircon, and lesser amounts of leucoxene (Isphording 1971). Hornblende and garnet are totally absent in the Citronelle Formation, as is also feldspar in the light mineral fraction. The ratio of the percentage of zircon plus tourmaline to rutile (the ZTR index) is also characteristically higher in the Citronelle Formation and will consistently serve to differentiate the two units (see Isphording 1976). Heavy mineral analyses from soil samples collected from the defendants clearly displayed a Citronelle signature and could have been presented to the jury as further evidence that the two principal defendants

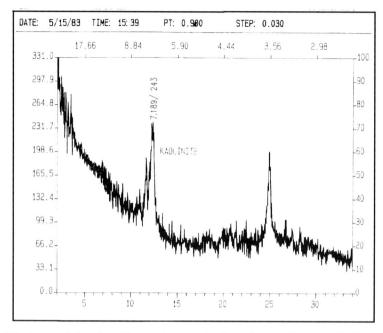

Fig. 3. X-ray diffractogram of clay mineral fraction ($<4\,\mu m$) of soil sample from crime scene (Cu K-α radiation).

Table 3. *Heavy mineral content of Citronelle sediments and Terrace deposits and ZTR index*

	Citronelle Formation				Terrace deposits			
	Sample 1	Sample 2	Sample 3	Sample 4	Sample 1	Sample 2	Sample 3	Sample 4
Black opaques (ilmenite)	34	37	32	36	27	27	28	23
Tan opaques	4	6	5	7	12	10	13	15
Leucoxene	8	10	8	6	13	14	13	13
Zircon	10	8	11	9	5	2	2	4
Tourmaline	12	10	10	7	8	10	11	11
Staurolite	9	10	13	9	16	18	15	15
Rutile	6	4	3	4	5	7	6	7
Kyanite	17	15	18	20	12	9	10	8
Hornblende	–	–	–	–	Tr	1	1	2
Garnet	–	–	–	–	1	3	Tr	1
ZTR index	3.6	4.5	7.0	4.0	2.6	1.7	2.2	2.1

had lied and, at the very least, had recently been to a site underlain by the Citronelle Formation.

Conclusions

Several lessons applicable to forensic geology, and especially trial testimony, from this investigation should be emphasized.

(1) If chemical analytical data are to be used, the witness should insist that the analyses be

carried out by an accredited laboratory. Any questions regarding the QA/QC procedures employed by the laboratory should be answered by the laboratory supervisor or the actual analyst. Under no circumstances should the geoscientist introduce analyses that he/she ran themselves unless their laboratory is fully certified by regulatory authorities. The geoscientist's task is to answer questions involving interpretation of analytical results and they should confine themselves to this area.

(2) The use of statistical tests should be carried out

by professionals who not only have a proper understanding of statistical analytical procedures, but also understand when the resulting 'numbers' have true statistical significance. Under most circumstances a jury will not understand the actual procedures. Therefore presentation of statistical results to a jury, while useful to the scientist, should generally be avoided.

(3) Mineralogical evidence is generally well received by juries. The writer's experience is that jurors accept the notion of mineral suites and X-ray diffraction data as reliable and legitimate 'fingerprints' of the provenance of samples.

References

EPA 1999. *United States Environmental Protection Agency Method 6010B.* (Modified for sequential P-E inductively coupled plasma-atomic emission spectrometry, version 1.3 (January 1999).

ISPHORDING, W. C. 1971. Provenance and petrography of Gulf Coast Miocene Sediments. *In: Geological Review of some North Florida Mineral Resources.* Fifteenth Field Conference Guidebook, Southeastern Geological Society, 43–55.

ISPHORDING, W. C. 1976. Multivariate mineral analysis of Miocene-Pliocene Coastal Plain sediments. *In:* BALDERAS, J. (ed.) *Transactions of the Gulf Coast Association of Geological Societies,* **26,** 326–331.

SOKAL, R. R. & ROHLF, F. J. 1969. *Biometry.* W.H. Freeman & Company, New York.

Using geological information to identify apparent 'misrepresentation' of facts in a litigious situation

WAYNE C. ISPHORDING

Department of Earth Sciences, University of South Alabama, Mobile, AL 36688, USA
(e-mail: wisphord@jaguar1.usouthal.edu)

Abstract: On 7 November 1980, a motorcyclist entered a curve in a fog-shrouded, marshy area. The cyclist subsequently stated that his helmet visor suddenly 'fogged up',and that he was unable to raise the visor. The motorcycle slid out of control, throwing the rider into the path of an oncoming car; the impact led to the surgical removal of his left leg. The victim sued the local motorcycle dealer and the motorcycle visor manufacturer, alleging that he had not been advised that the visor could not be raised.

A number of lines of evidence argued against the plaintiff's version of the events. Index of refraction, X-ray diffraction and fluorescence analyses clearly showed that the visor on the plaintiff's helmet was not of the same material as visors sold by the defendants. Furthermore, although the plaintiff claimed he had purchased the visor the night before the accident, the visor had a curvature that could have been acquired only from being attached to the helmet for a number of weeks. Scanning electron microscopy (SEM) disclosed scratches that were found to contain orthoclase feldspar. The accident site, however, was on the Plio-Pleistocene Citronelle Formation, which is completely devoid of feldspar. Each of these points was used to show that the visor had not been purchased the night before the accident. The outcome of the trial, however, showed that nothing is certain in a court of law.

'Do you swear that the testimony you are about to give is the truth, the whole truth, and nothing but the truth?' Ideally, an affirmative reply from a witness would allow one to assume that all subsequent questions would be answered in a wholly truthful manner. However, only the most naive would presume that human nature and human prejudices are not overprinted on most answers. Therefore, one of the frequent tasks of an expert witness is to refute information that he or she knows to be false.

The case reported in this paper involved an incident that took place on 7 November 1980 at 07:10 in Mobile County, Alabama (USA). A young man driving a motorcycle on a rural highway descended a hill that led into a fog-shrouded, marshy area. His testimony was that, as he reached the bottom of the hill, the visor on his motorcycle helmet suddenly 'fogged up', severely reducing his range of vision. He then lost control of the motorcycle, ran off the road onto the right hard shoulder, then came back across the road and was struck by an oncoming car (Fig. 1). The accident resulted in the loss of his left leg. Several days later, while he was still recovering in hospital, a suit was filed on his behalf alleging that he had not been told that the visor, which he stated had been purchased the night before the accident, could not be pivoted upward in the event of an emergency. The defendants in the suit were the local motorcycle dealer, from whom the plaintiff alleged that he had purchased the visor, and the national motorcycle manufacturer who had supplied the visor to the local dealer.

Fig. 1. Accident site on Roberts Road, between State Highway 45 and Celeste Road, Mobile County, Alabama.

Problems and inconsistencies with the plaintiff's allegations

Several problems were immediately apparent when the defence attorney began accumulating the case facts.

(1)　The plaintiff had stated that his wife had purchased the visor the night before the accident. None of the sales personnel at the motorcycle dealership, however, remembered serving the woman. The local dealer argued strenuously that the visor must have been purchased elsewhere.

From: PYE, K. & CROFT, D. J. (eds) 2004. *Forensic Geoscience: Principles, Techniques and Applications.* Geological Society, London, Special Publications, **232**, 289–293. © The Geological Society of London, 2004.

Fig. 2. Plaintiff's helmet with original face shield (visor) attached.

Fig. 3. Helmet equipped with only one snap on either side, in which case a visor can be easily raised when necessary.

(2)   The plaintiff's wife also stated that her husband had removed the visor from its package and installed it on the helmet on the morning of the accident. The visor was knocked from the helmet by the force of the accident and was retained by the plaintiff's attorney (still unattached) until requested by the defence's expert witness.

(3)   The visor was attached to the helmet by two snaps on either side of the helmet (Fig. 2). In spite of the fact that the plaintiff admitted he had been driving motorcycles for nearly 4 years, when asked in his pre-trial deposition if the shield was designed to pivot upwards, he answered: 'I thought it flipped up.' While it is possible that a novice might not realize that there is a difference between the mobility of a visor attached to a helmet by two snaps on each side and one attached by only one snap (Fig. 3), it was difficult to believe that someone with 4 years' experience of motorcycles could be that uninformed.

As the suit that was filed against the local dealer and the national manufacturer was of the magnitude of $10 million, the local dealer immediately engaged a professional law firm to prepare for his trial defence. His attorney subsequently contacted professionals whom he hoped could support the dealer's claim that the visor had not been purchased in his store and had been acquired elsewhere.

Laboratory analysis

Visor curvature

On inspecting the visor, the defence's expert witness immediately suspected that there was an apparent misrepresentation of the facts because of the pronounced curvature of the visor. If the visor had only been attached the morning of the accident, and had been retained (unattached) by the plaintiff's attorneys following the accident, then it seemed unlikely that the visor acquired its significant curvature by being attached to the helmet for only a few hours. The witness then requested the local dealer to provide him with a dozen visors from his on-hand stock for testing. The dealer was adamant that these visors were the only visors that he had in stock and that they had been purchased directly from the national manufacturer. Each of the visors was packaged in a sealed, plastic bag and was supplied essentially flat (Fig. 4).

Each visor was affixed to the helmet for different periods of time, and a new visor was used for each new time period. It is obvious from Figure 5 that the curvature present on the plaintiff's visor could not have resulted from only a few hours of attachment. To duplicate that degree of curvature, attachment was required for nearly 3 months.

Refractive index of visor material

Analysis was carried out by simply scraping the edge of each visor and then determining the indices of refraction of the particles obtained from each visor. Standard Cargille® immersion oils were used and refractive indices were obtained for three wavelengths of light generated using a Bausch and Lomb

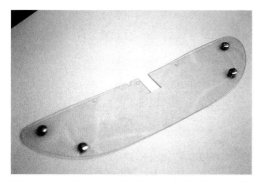

Fig. 4. A new visor as received from a dealer (the 'notch' is the sample removed for X-ray and scanning electron microscope examination).

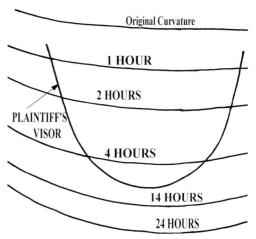

Fig. 5. Visor curvature resulting from different attachment times to the helmet.

monochromator. An interpolated value for light of 589 nm wavelength was then determined for each. The results are shown in Figure 6. The index of refraction for the plaintiff's visor (1.4936) was clearly much higher than that from a visor obtained from the local dealer (1.4775), thereby indicating that the visors differed in both mineralogical and chemical composition.

Crystallography of visor material

Confirmation that the crystal structure of the two visors was different was apparent on X-ray diffractograms of samples cut from each of the visors (Fig. 7). Broadband scattering was present on both (typical for most plastics), but also present on the pattern were distinctive peaks. The plaintiff's visor had only one low-angle peak, which occurred at 6.5 Å. The dealer's visor, in contrast, yielded peaks at 13.5 and 4.4 Å. This gave positive evidence that the visors were crystallographically different.

Scanning electron microscopy (SEM) of visor

The visors were then examined with an SEM equipped with an energy dispersive, X-ray fluorescence system. Corroboration that the two visors were also different chemically was clearly apparent on X-ray fluorescence spectra that were generated (Fig. 8). Distinct peaks indicating the presence of both chlorine and phosphorus were obtained from the plaintiff's visor whereas the dealer's visor, constructed from butyrate, lacked these two elements.

Additional evidence that supported the belief that the plaintiff's visor had been purchased some time before the accident took place was also obtained from the SEM analysis. The plaintiff's visor was

found to have not only numerous scratches on its exterior surface, but also a number of scratches on the interior surface (Fig. 9). While some of the exterior scratches may well have occurred during the accident, it is difficult to reconcile the abrasion and wear and tear on the inside of the visor with its alleged negligible use before the accident.

A number of the scratches on the inside appeared to be the result of debris (dust, sand, silt etc.) being dragged along the inside surface when the visor was raised and lowered on numerous occasions (the visor can be raised if one snap on each side is first released). Increased magnification of the SEM was used to identify the particulate material from the interior scratches. Some of the debris was found to be feldspar (Fig. 10). The accident took place within 80 km of the Gulf of Mexico, in an area totally within the outcrop belt of the Plio-Pleistocene Citronelle Formation. Many mineralogical analyses have been performed on this unit since it was first mapped in the early years of the past century, but no feldspar has ever been reported. The Citronelle Formation is a fluvial deposit whose provenance can be traced to erosion from older, deeply weathered formations of the Gulf Coastal Plain Province of the south-eastern USA. The rigorous weathering that occurred in this region during the Late Neogene and pre-Nebraskan Pleistocene resulted in removal of all but the most resistant heavy mineral and light mineral species. The presence of feldspar embedded in some of the scratches was therefore conclusive proof that this visor had, at some time in the past, journeyed at least 240 km north of the accident site (the nearest location of feldspathic sediments).

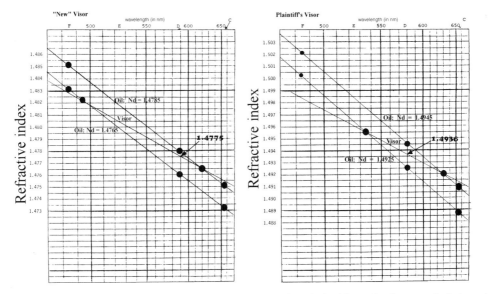

Fig. 6. Refractive index determination for samples taken from a new visor obtained from a local motorcycle dealer and from the plaintiff's visor.

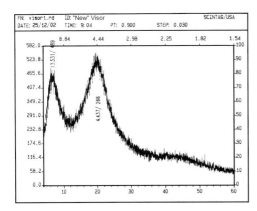

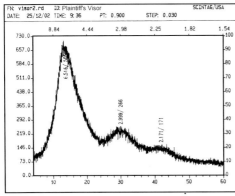

Fig. 7. X-ray diffractograms comparing the new visor with the plaintiff's visor.

Conclusions

The judicial system of most democratic nations guarantees a trial-by-jury system. Both sides are then permitted to present evidence for consideration by the jury. It is then presumed that the 'weight of evidence' will result in a decision in favour of the deserving party. The suit filed by the plaintiff in this case included two defendants: the local motorcycle dealer and the national motorcycle manufacturer. The suit claimed that both were remiss in omitting a warning that: (1) the visor was not an 'anti-fog' face shield; and (2) that the visor could be raised only if one snap on either side were released.

The evidence described in this paper was compiled by the expert witness working for an attorney who represented the local dealer. This attorney, as required under 'discovery' procedures in the USA before trial, made his expert's conclusions available to the plaintiff's attorneys. Those attorneys, after reviewing the witness's report, dismissed the local dealer from the suit 'with prejudice' (meaning that no further involvement of the local dealer would be forthcoming).

The evidence was not, however, passed on to the law firm representing the national motorcycle manufacturer, who chose not to contest the allegation that the two visors were different, but instead went to court and attempted to argue solely that the witness should have realized that the visor could not be raised and that the visor was not marketed as an 'anti-fog' face shield. No science was therefore presented at the trial and the jury's verdict for the plaintiff resulted in an award of $5 million.

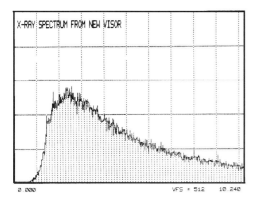

Fig. 9. Scanning electron microscope image showing numerous scratches on the inside of the plaintiff's visor.

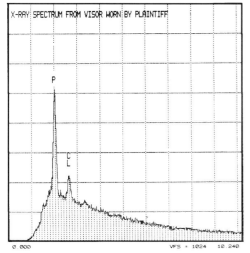

Fig. 8. Energy dispersive scan of samples from the new helmet visor and from the plaintiff's visor.

Fig. 10. Scanning electron microscope image of feldspar embedded in scratches on the inside of the plaintiff's visor.

Soil as significant evidence in a sexual assault/attempted homicide case

MARIANNE STAM

California Department of Justice, Riverside Criminalistics Laboratory, 7425 Mission Boulevard, Riverside, CA 92509, USA (e-mail: marianne.stam@doj.ca.gov)

Abstract: In October 1999, a woman and her two small children were assaulted by a male acquaintance on the banks of the New River in the Imperial Valley desert region of California, USA, approximately 160 km east of San Diego. The area includes a number of agricultural fields, old lake deposits, sand dunes and the Salton Sea, a large saline inland sea. The victims escaped from the suspect and hid for 30 h in the New River, thus diminishing the possibility of obtaining physiological fluids as evidence for DNA testing. Wet and soiled clothing and shoes were collected from a suspect's residence within hours of the assault being reported. The suspect denied any contact with the victims or with the crime scene. Analyses of the soil on the suspect's clothing and shoes, and of crime scene soil using microscopical methods and X-ray diffraction, indicated that the soil from the crime scene and soil on the suspect's clothing and shoes were similar. Predominant wind directions and observations of the soil distribution at the crime scene added significance to the observed similarities between the soil samples at the crime scene and the soil on the suspect's clothing and shoes.

In October 1999, a woman and her two small children were kidnapped by a male acquaintance and taken to the New River in the Imperial Valley near El Centro, California, USA (Fig. 1), where the woman was raped and her children were physically assaulted. The area is a desert, approximately 160 km east of San Diego, and includes agricultural fields, sand dunes, deposits from an ancient lake and the Salton Sea, a large saline inland sea. The victims escaped and hid in the river for 30 h. A suspect was

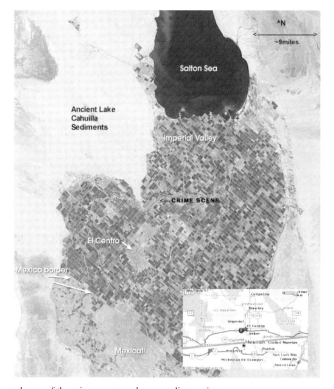

Fig. 1. Satellite view and map of the crime scene and surrounding region.

From: PYE, K. & CROFT, D. J. (eds) 2004. *Forensic Geoscience: Principles, Techniques and Applications.* Geological Society, London, Special Publications, **232**, 295–299. © The Geological Society of London, 2004.

Table 1. *Colour and content of soil samples from the crime scene and the suspect*

Sample	Munsell colour	Contents
Crime scene	10YR 6/3	Angular to subrounded mineral grains
	Pale brown	White snail shells
Suspect	10YR 6/3	Angular to subrounded mineral grains
	Pale brown	White snail shells

apprehended at his residence within hours of the incident being reported. Investigators found wet and soiled clothing and shoes at his residence. Upon questioning by investigators, the suspect denied any contact with the victims or with the crime scene. Police collected the clothing and shoes from the suspect and submitted them to the crime laboratory for examinations, along with soil samples from the crime scene, and a sexual assault evidence collection kit from the adult victim.

All sexual assault evidence tested negative. Soil evidence, therefore, was examined to help link the suspect to the crime scene, and to corroborate the victims' story.

Materials and methods

Soil collected from the suspect's shoes and clothing, and from the crime scene, was weighed and examined visually and with an Olympus SZ40 stereomicroscope to identify similarities in colour, content and grain morphologies. The samples were sieved and the silt size portions of soil from the suspect's right shoe and from the crime scene were separated into light and heavy mineral fractions, using bromoform. These fractions were mounted in 1.54 and 1.65 refractive index oils, respectively, for analysis using a Leica Orthoplan polarizing light microscope. The light and heavy minerals were then grouped into broad mineral categories and the heavy minerals were further grouped by colour. Each category of minerals was grouped by abundance, using estimates of the mineral counts in each field of view at ×63 magnification. In addition, the silt size fractions of the right shoe and crime scene samples were analysed using X-ray diffraction.

Results and discussion

Soil from the suspect's clothing and shoes, and from the crime scene, was macroscopically similar in colour, content and grain morphologies (Table 1).

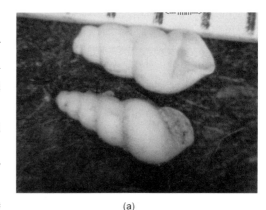

(a)

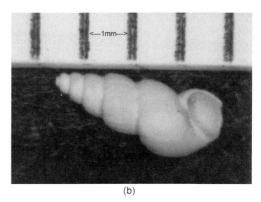

(b)

Fig. 2. Snail shells from: (**a**) suspect's shoe, and (**b**) crime scene.

Each sample contained small white shells, which were identified as being from a freshwater snail species, *Tryonia protea* (Fig. 2) (McClean pers. comm. 2000). Fossil shells of this species were deposited in Late Pleistocene Lake Cahuilla, a natural lake whose name in the Cahuilla Indian language means 'little shell' (McClean pers. comm. 2000). This lake filled an area approximately 160 km long and 80 km wide during the last Ice Age (between 10 ka and 50 ka) that includes the present-day Salton Sea. Consequently, the *Tryonia protea* shells are widely distributed throughout the Imperial Valley and are therefore less significant for linking the suspect directly to the crime scene.

Polarized light microscopy and X-ray diffraction revealed similar light and heavy minerals in the soil samples from the crime scene and the samples from the suspect's right shoe (Figs 3–6).

The significance of the soil similarities was determined by visiting the crime scene to look at the distribution of the soil. The crime scene is adjacent to the New River, which cuts through at least three different levels of river terraces in a northeast–

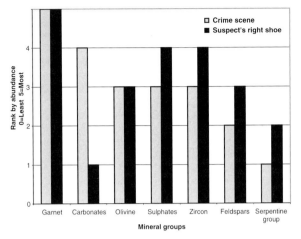

Fig. 3. Colourless heavy minerals ranked by abundances in soil from crime scene and from suspect's shoe.

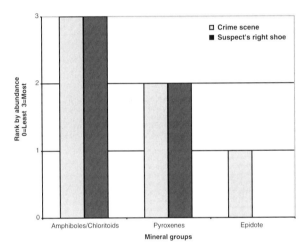

Fig. 4. Green heavy minerals ranked by abundance in soil from crime scene and from suspect's shoe.

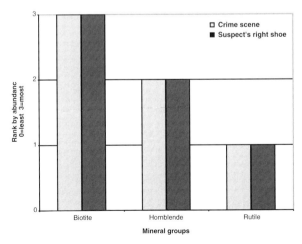

Fig. 5. Brown heavy minerals ranked by abundance in soil from crime scene and from suspect's shoe.

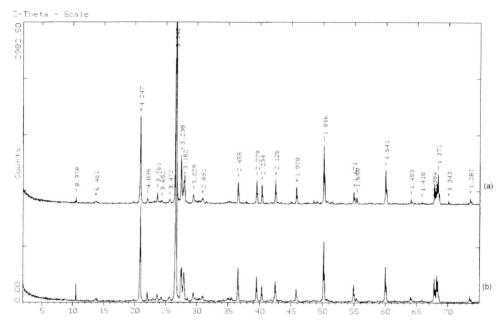

Fig. 6. X-ray diffraction traces for soil samples from: (**a**) suspect's right shoe; and (**b**) crime scene.

Fig. 7. The New River looking southwest.

Fig. 8. Crime scene site where victim and suspect struggled; looking southwest.

Fig. 9. White deposits above light brown river deposits; looking northwest.

Fig. 10. View of white deposits from north side of river looking south towards the crime scene.

southwest direction (Fig. 7). The adult victim identified an area of the river along the southern embankment where she and the suspect struggled during the assault. In this area, a distinct white sandy deposit overlies light brown silt terrace deposits (Figs 8 & 9). An approximately 5 km traverse along the southern and northern embankments revealed that similar white sandy deposits could only be observed on the southern side of the river, including the area that the victim identified (Fig. 10). These white, sandy deposits contained mostly quartz and some of the *Tryonia protea* snail shells. They were found not only along the southern embankment of the New River, but also in depressions on top of the terraces at the crime scene. This distribution, their content and the predominant, strong winds from the west at the crime scene suggested that the white sandy deposits could be windblown sediments of ancient Lake Cahuilla, which lies to the west (Fig. 1).

The soil reference samples collected from the crime scene site identified by the victim and the soil from the suspect's shoes and clothing were both similar, and appeared to represent mixtures of the white sandy deposits and the light brown silt deposits. Soil samples from the northern embankment of the river contained only light brown silt deposits and were not similar to the mixture of soil on the clothing and shoes or to sediments at the crime scene.

Conclusion

The similarities between the soil on the suspect's clothing and shoes and soil at the crime scene, combined with the limited distribution of similar soils at the scene, helped to corroborate the victims' story and to place the suspect in the vicinity of the crime scene.

The soil evidence, in addition to the victims' testimony, resulted in the conviction of the suspect.

I would like to thank R. Graham and K. Rose of the University of California, Riverside, for their assistance with the X-ray diffraction analysis and interpretation. I also would like to thank J. McClean of the Natural History Museum of Los Angeles for his assistance with the identification of the snail shells.

The nature of, and approaches to, teaching forensic geoscience on forensic science and earth science courses

C. W. LEE

School of Applied Sciences, University of Glamorgan, Treforest, Pontypridd CF37 1DL, UK
(e-mail: cwlee@glam.ac.uk)

Abstract: Whereas geoscience depends in part on the classical nomological method of the environmental sciences, it is also distinguished by a discrete set of logical procedures that lend themselves to forensic analysis. It has been argued that, as geology is a derivative science, it only partially lives up to the classical model of scientific reasoning. It does, however, provide a model of scientific reasoning based on interpretive techniques and its historical nature. Reasoning in geoscience offers a method that is applicable to the uncertainties and complexities of real-life situations because we are seldom in possession of all the data that we would like in order to make an unbiased or objective decision. The 'geological' method, in which we fill the gaps in our knowledge with interpretation and reasonable assumptions, is exemplified by the pragmatism of geological and geomorphological analysis and is herein considered analogous to that undertaken by forensic practitioners. A problem-based learning situation based on a past geological/forensic case study is included to illustrate the approach advocated.

Throughout the history of mankind, people have used the Earth in an attempt to resolve the conflicts of science by observing natural events and interpreting them. The relationship between geoscience and its forensic application (in its widest sense) has been in existence for thousands of years (Junger 1996). Forensic geoscience is now an increasingly accepted and commonly used part of the criminal and civil justice system (see this volume) and has been used in formal case formulation since the late 1800s. (Murray & Tedrow 1975). Most cases involve the analysis of earth materials that occur naturally or are synthesized from them, for example bricks and other building materials. An understanding of the process through which these materials are distributed is also important as objects of forensic interest (e.g. cadavers) often behave as 'clasts' controlled by fluid processes. (Haglund 1993; Bassett & Manheim 2002, Carniel *et al.* 2002.)

Forensic geoscientists have, or know of others who have, an understanding of minerals; igneous, sedimentary and metamorphic rocks; of fossils in both their macroscopic forms and, especially, microscopic forms such as pollen/spores, diatoms etc.; and of soils which may or may not contain components of all of the above. The sedimentary components of soils may be further analysed by grain size distribution and associated grain statistics together with heavy mineral associations.

(Smale & Trueman 1969; Junger 1996; Lee *et al.* 2002), all which can be used in case analysis. The colour of soils has also been shown to be of paramount importance in many cases but processed in a way that is somewhat different to that carried out by a general soil scientist.

In addition forensic earth science relationships are expressed by a knowledge of the spatial distribution offered by topographical, geological, geophysical and soil maps. Aerial photographs and specialized remote sensing methods are increasingly used and computer driven geographical information system analysis is now used in perpertrator profiling (Rossmo 2000). The analysis of specialized British Geological Survey (BGS) maps indicating abandonment and working plans of exploited minerals are important, especially those that leave publicly accessible voids for the dumping of contraband in its various guises (e.g. the Tooze case in 2000, where a suspect firearm was found in nearby 'old workings') and also for the understanding of voids that mysteriously appear overnight in residential areas built over various deposits subject to dissolution (Cooper & Waltham 1999). Maps and regional analyses that bring together 2-D and 3-D data may give an indication of permeability pathways for gases that explode some way from their point of origin (e.g. the Loscoe Landfill explosion).

What also makes geoscience somewhat different from other sciences is its historical basis that links an understanding of the concept of time with the sequential deposition of earth materials. In a case brought to court in 1908, the law of superposition was illustrated by material on the bottom of a boot that established a temporal and geographical framework for the perpetrator's movements. Similar procedures (microstratigraphy) take place when vehicle mudguards and other components are analysed for stratified deposits. Stratigraphic analysis has also proved beneficial in the elucidation of recent mass graves from the Balkans conflict (see Hanson 2004),

From: PYE, K. & CROFT, D. J. (eds) 2004. *Forensic Geoscience: Principles, Techniques and Applications.* Geological Society, London, Special Publications, **232**, 301–333. © The Geological Society of London, 2004.

and the factors governing the preservation of flesh (human or otherwise) in poorly oxygenated acidic peat environments (Fenning & Donnelly 2004) or mummified form (Honigschabl *et al.* 2002) are classic palaeontological topics.

Geophysical techniques also play an increasingly important role in forensic detection as a non-destructive method that may indicate an anomaly for further analysis. These vary through electrical, sound and heat recording techniques with ground-penetrating radar receiving much publicity through the Fred and Rosemary West case in 1994. Interestingly from an educational point of view, it is often the reasons why these techniques do not work in various terrains which is the most informative.

As in most branches of science, geoscience is a subject with endless branches and levels of sophistication. Before geoscience evidence can be collected, however, the investigators must be aware of its presence and importance.

Investigators need to recognize the discriminating power that is inherent in earth materials and the experts who exist to advise on the level of discrimination and uniqueness of the evidence. An investigator must have enough geoscience training to recognize the potential value of the earth material evidence and collect it in an appropriate way. Considerable effort must be placed on an appropriate method of sampling and avoidance of contamination. It is also important that the forensic geoscientist should have the knowledge to suggest which methods of analysis should be used without destroying the sample if further, repeat analysis is needed. Statistical sampling techniques, even on very small samples, are required to save a representation sample for further analysis. In addition a forensic geoscientist must be aware of, and able to, access any databases that are available for comparison with the evidence. Such databases are available from many university departments and commercial companies. Any sort of 'fingerprint' geological or otherwise, is only as good as the data base within which you can compare it.

Geological and forensic reasoning

During the course of teaching forensic geoscience at the University of Glamorgan, the annual course-monitoring procedure has suggested that students feel that the way in which we reason in geoscience is very similar to the way in which we would set out to solve a crime, and that we (geoscientists) think differently from the other more theoretical scientists (mostly chemists) on the award.

It is an often held assumption that geology is a derivative science (Bucher 1941; Schumm 1991) and geological reasoning has been thought to lack a distinctive methodology of its own, using a few 'rules of thumb' that are guided by the use of mathematics and the application of the laws of chemistry and physics to geological phenomena.

Geoscience is also seen to have many drawbacks caused by the incompleteness of data (gaps and poor resolution of the stratigraphic record) and by the lack of experimental control that is often possible in the laboratory based science. In addition the time required for geological processes to take place may make direct observation difficult or impossible. Geoscience is therefore perhaps less than ideal when compared with physics (most properly classical mechanics), which is the science that exemplifies the true nature of science as certain, precise and predictive, and gives us the basic model for understanding the nature of science and of knowledge in general.

However Frodeman (1995) argues that geoscience is not merely an 'applied and imprecise science' trying hard to achieve the degree of resolution and predictability of physics; instead the challenges and difficulties inherent in geological reasoning have caused geoscientists to develop their own variety of reasoning techniques which comprise a combination of logical procedures that are shared with the experimental sciences and those more typical of the humanities (Gould 1987, 1989). Geoscience differs from the theoretical sciences in that it is not necessarily concerned with the generation and testing of universal laws. These laws are generally taken for granted but are used to find and test particular geoscience problems, (Collinson & Thompson 1989). As such, geoscience often offers a better model than physics for the understanding of the nature of reasoning within the sciences and within everyday life (Frodeman 1995).

Where therefore does a distinctive methodology of geoscientific reasoning originate? Part of the problem possibly is that the philosophy of science is not generally practised by those experienced in doing science, or if so the science is generally assumed to be experimental/theoretical physics. Long ago Fairchild (1904) exclaimed that 'geologists have been too generous in allowing other people to make their philosophy for them' and, as Mackin (1963) explains, 'the best and highest use of the brains of our youngsters is the working out of cause and effect relations in geological systems'. The complex reasoning processes of geoscience are first manifest in the writings of Gilbert (1886, 1896), Chamberlin (1904), and Davis (1926) and Johnson (1933). These methodologies of are still seriously by earth scientists (Kitts 1973,1982), and although they have been criticized in modern philosophical terms (Blewett 1993; Rhoads & Thorn 1994), many citations of their methods are still occurring in the geological literature (Van Andel 1994; Webster & Yaalon 1994).

Together with these early protagonists was the philosopher and geologist Charles. S. Peirce of the US Coast Survey, who studied under Agassiz at Harvard. While with the US Coast Survey his work consisted of geodesy, involving detailed measurements of gravity through pendulum experiments. His standing at the time is indicated by the fact that he was placed fourteenth for distinction in mathematics on the 1903 all-American list (Cattel 1933). Peirce is also recognized as probably the greatest American Philosopher and is considered by Sir Karl Popper to be one of the greatest philosophers of all time (Popper 1959). Peirce is constantly referred to by Nordby (2000), with his idea of abduction providing the method of reasoning that takes an investigator from the presented signs to their probable explanation.

With his pragmatic approach, Pierce published a famous series of six papers between 1877 and 1878. These provided an in-depth account of his scientific reasoning and his concept of 'hypothesis' was established.

Using the classical logical device of syllogism Peirce (1878b) illustrates the basis in argument for the various forms of inference. He considered analytical reasoning, or deduction, as the application of general rules to specific cases. Deduction is a tool of science but offers nothing new that is not already known. Induction (synthetic reasoning) reverses the deductive syllogism but has received extensive comment in the modern philosophy of science, (the so-called 'fallacy of induction'), and Popper (1959) argued that induction cannot prove a scientific theory. Owing to the problems of induction (see Hawley 2002), Peirce (1878b) suggested a third form of inference: the making of an hypothesis, or abduction. His forms of inference were therefore as follows:

(1)   Analytical reasoning (deduction);
(2)   Synthetic reasoning 1 (induction);
(3)   Synthetic reasoning 2 (hypothesis, abduction).

Induction is seen as the process whereby we conclude that facts, similar to observed facts, are true in cases not examined, i.e. reasoning from particulars to a general law. By hypothesis (abduction) we conclude the existence of a fact quite different from anything observed, from which, according to known laws, something observed would necessarily result, i.e. from effect to case. In other words induction classifies, hypothesis explains.

Peirce (1878b) concludes his discussion of scientific inference by using his scheme to classify the sciences as purely inductive (e.g. chemistry), theoretical (e.g. physics), and hypothetical (e.g. geology).

As long ago as 1886 Gilbert presented this logic in the geological literature when he stated 'It is the province of research to discover the antecedents of phenomena. This is done with the aid of hypothesis'. However, this was not the mainstream philosophy of the day (Baker 1996). Gilbert (1896) states: 'When an investigator, (having under consideration a fact or groups of facts whose origin or case is unknown), seeks to discover their origin, his first step is to make a guess. In other words he frames a hypothesis and then proceeds to test the hypothesis. This method, based solely on observed facts (signs) is limited by imperfect observation and however widely accepted, and useful its conclusion, none is so sure that it cannot be called into question by a newly discovered fact'. In fact Peirce (Debrock 1992) defined sciences as an activity exercised by people who are passionately committed to the principle that they should be ready to abandon/amend any hypothesis that is contradicted by the facts. Peirce named this principle 'fallisbilism', a critical part of the pragmatism of the 1890s.

Gilbert's scientific method of 1896 can be summarized as follows:

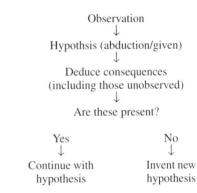

This procedure is not too dissimilar from a modern line of forensic enquiry (Nordby 2000).

As explained by Davis (1972) Peirce argued that all new knowledge arises from a kind of 'synthetic thinking' involving the continuous actions of comparing, connecting and organizing thoughts and perception. Each action is dominated by a succession of reasoning processes, respectively abduction, deduction and induction. Abduction is the creative hypothesis building, in which 'We find some curious circumstances which would be explained by the supposition that it was the case of a certain general rule, and thereupon adopt that supposition'. With induction, by comparison, 'We generalise from a number of cases in which something is true and infer that the same thing is true of a whole class'. As deduction is analytical, it yields no new knowledge. As can be seen, Peirce's view of deduction and induction follow classical logic but Pierce believed that hypothesis is an inferential proposition of similar logical status to deduction or induction and, as every induction follows from an abduction, all new knowledge arises from abduction.

Nordby (2000), suggests 'that in forensic analysis the hypothesis is not a guess but a bet. Whereas guessing is blind, and riddled with doubt, betting is not guessing but sighted and filled with belief. Both are uncertain, but informed betting is methodical, guessing is merely desperate'. As Pierce says, abduction is the process of proposing some explanation that is likely in itself but that must be tested before anyone can be fully justified in accepting it. Abduction is Nordby's informed betting. Induction on the other hand, involves observing some sign and noting the frequency of its association with another, for example every observed basalt is black, so probably all basalts are black (the more the better). Induction is statistical; indeed, statistics and probability, which quantify uncertainty, were developed around various logical problems of induction. However what explains the association? Why are basalts black? Induction without abduction explains nothing and, in other words, lacks imagination. Sometimes abductions may appear like deductions because they prove to be correct and so any doubts about their wisdom is resolved.

Abductions presume a great deal, but testing their presumptions provides great insight into the problem at hand. Abductions add something new that is not present in the original evidence. Deductions, on the other hand, do not amplify but draw out something that is already before us in the original observation. Good abductions also lessen doubt by best explaining the surprising sign in context

Sherlock Holmes, in Conan Doyle's *The Sign of Four* says: 'I never guess'. He should have added: 'I merely abduct, Watson'. Geoscience (and I would suggest forensic science) affords numerous examples of Peirce's 'method' of science, which has been termed by Baker (1996) as 'the critical philosophy of common sense'. The elements of this tradition include an affinity for fieldwork, a humility before the 'facts' of nature, a continuing effort 'to discriminate the phenomena observed from the observers inference in regard to them', a propensity to pose hypotheses and a willingness to abandon them when their consequences are contradicted by reality.

Forensic reasoning integrates abduction, deduction and induction into a unified process, much like that done in geological reasoning. As Peirce says: 'All ideas of science come to it by the way of abduction. Abduction consists of studying the facts and then devising a theory to explain them'. The hypotheses (tentative explanations) do not become 'less scientific' because they fail to invoke established laws of nature, or 'more scientific' because they appeal to covering laws. The scientific status of an explanation is dependent on its connection with the laws of nature.

Abduction depends upon explanation, and explaining the result depends upon finding mechanisms, manners and causes. Abduction therefore enables certain predictions as they uncover conjectures to be proved or disproved by future testing.

According to Nordby (2000) abductions constitute the heart of 'reasoning backward analytically', a fundamental of forensic analysis and something we do in geoscience as a matter of course.

However, the formulation and testing of a single hypothesis (Beveridge 1957, 1980) should be avoided. The 'method of multiple working hypotheses' (Chamberlin 1890, 1897) allows the formulation of as many hypotheses as possible, which are then selectively eliminated or combined to develop an explanation of the phenomenon. One hypothesis may dominate the thinking of an investigator and most importantly, if proved to be wrong, may mean that evidence for other hypotheses may not have been collected or may have become contaminated and inappropriate. The multiple hypotheses may then be developed in a sequential mode, with one hypothesis following another as weaknesses are found, or in a parallel mode with a number of hypotheses being developed and tested simultaneously (Schumm 1991). Closely related to this method of multiple working hypotheses is the process of differential diagnosis in medicine (Harvey & Bordley 1970), and there is evidence for the use of this process in medical investigation somewhat before Gilbert and Chamberlain (Semmelweis 1861). It is an interesting conjecture that this medical analysis may have had an effect on the teaching of Joseph Bell MD on whom Conan Doyle based his famous detective. This logic survives in the daily casework of detectives, forensic scientists, medical examiners and geoscientists as it helps them to recognize clues sort out evidence from coincidence, distinguish truth from falsehood and evaluate guilt or innocence (Nordby 2000).

Forensic science is also similar to geoscience in that it is all about observation. Reading signs depends on appreciating the small points, the details that are apparent to those schooled in observation but not to the novice, the incautious or the hasty. Observation can and should be taught to students.

However, observation itself may lead to potential sources of error. Seeing is a physical photochemical excitation that produces a neurological experience, whereas observation is an experience governed by prior knowledge, beliefs, values and the goals and purposes for looking in the first place. In other words, investigators see the same thing but they may interpret what they see differently and reason from the same observations to support different interpretations (Nordby 1992). Observation clearly influences interpretation and inference. Observations, however, differ subtly because each observer brings to the experience different background information, habits and theories that, in turn, supply different contexts for their observations. Therefore, if backgrounds and experience differ, investigators may see the same thing but interpret it differently. Experimental work indicates that elements in our experience do not cluster into visual patterns at random (gestalt psychology) but are governed by the 'laws of grouping' (Rock *et al.* 1990). The mind may complete or fail to complete a visual pattern by supplying data that is expected (but not actually present), or omitting data that is present but not expected. One may literally see things that are not there and fail to see things that are present. Nothing, perhaps, is self-evident unless you are looking for it. It is extremely important therefore that geoscience/forensic science students learn to 'observe' and be aware of the evidence offered at outcrop/scene of crime, as inappropriate expectation-laden observation can lead to observational error, leading to interpretive errors and finally errors of inference. As Schumm (1991) explains, we tend to see what we are trained to see and it is important that we are aware of the fact.

Teaching and learning

Based on the previous discussion this author believes that forensic geoscience is a vehicle for effective learning, especially if the most prominent educational process of constructionism is used. In constructionism it is thought that through experience we develop general conceptions that are models of reality. Learning occurs when these concepts are amended. We learn by matching new knowledge and understanding into, with or beyond pre-existing understanding. Learning is not so much about adding more knowledge but about transforming pre-existing knowledge.

Exercises in forensic geoscience have proved beneficial in the introduction of problem-based learning (PBL) techniques (Lee 2000) to undergraduates. PBL has grown dramatically since the 1970s and is being increasingly seen as a means of managing the knowledge explosion as standard curricula can no longer expand to cope with the demands (Savin-Baden 2000). As a result, students involved with PBL exercises are being equipped to 'manage knowledge' rather than being expected to have assimilated it all before entering employment.

It is possible to trace the origin of PBL to Socrates, who presented students with problems, so that, through questioning, he was able to help them explore their assumptions and the inadequacies of the proffered solution. Since these early times,

Dewey (1938) advocated the idea that knowledge is not something that is reliable and changeless, but is something that is an activity, a process of finding out. Dewey's pragmatic approach argued that knowledge was bound up with activity and he opposed theories that considered knowledge to be independent of its role in problem-solving inquiry. He emphasised 'learning by doing'.

PBL was popularized in the 1960s as a result of research by Barrows and Tamblyn (1980) into the reasoning abilities of medical students. They wished to develop in medical students the ability to relate their knowledge to the problems with which their patients presented (there are analogies here with the differential diagnosis method already mentioned), something they found that few medical students could do well. Their research indicated that there was a clear difference between problem-solving learning and learning in ways in which problem scenarios (such as scenes of crime) encouraged students to engage themselves in the learning process.

PBL is different from problem-solving learning as the focus of the information is centred on 'problem scenarios' and students work in groups or teams to solve or manage the situation. They are not required to solve a pre-determined series of 'correct answers' but are expected to engage with the complex situation presented. PBL also helps students to see that there are not always straightforward answers to problem scenarios and that not all problems are solvable (Lee 2001). Learning and life takes place in contexts that affect the kinds of solutions that are available and possible. Essential to the PBL method is that an individuals prior knowledge is in itself not sufficient for them to solve it. Students must decide what information they need to learn, or skills they need to gain, in order to manage the problem effectively and provide a solution. This has proved very beneficial on a forensic geosciences module where relatively advanced coal maceral and microlithotypes analyses, together with X-ray fluorescence and inductively coupled plasma mass spectrometry data for major oxides and minor and trace elements in coal, are studied to solve a coal slurry outburst event that caused millions of pounds worth of damage, (Hower *et al.* 2000). This type of data may not be studied even in a full geology degree.

The processes that an individual or group proceed through to achieve their problem solution has been detailed elsewhere, (Blumhof *et al.* 2001) but essentially the procedures assist the students to learn how to learn (Savin-Baden 2001). Students are encouraged to adopt a 'deep' approach to learning as research has shown (Norman & Schmidt 1992; & Schmidt 1993) that students using PBL techniques retain knowledge for longer, understand the topic better, produce more logical and coherent work and make the most connections between different topics.

Exercises in forensic geoscience encourage this deep learning through the relevance of the data, as real-life cases and issues are presented. It also encourages a positive working environment with a working interaction between peers and instructors as it requires an individual to put their hypotheses into words. Theses are then challenged, clarified and added to the argument. Forensic geoscience actively engages the individual by encouraging the generation of questions, identifies prior knowledge /areas of ignorance and applies the new knowledge to the problem. This is of special interest where many students on modular degrees fail to avoid the compartmentalization of knowledge and do not establish the links between different topics.

Conclusions

In this account I have argued that, while geoscience depends in part on the classic deductive-nomological method of the environmental sciences, geoscience is distinguished by a discrete set of logical procedures that lend themselves to forensic analysis. It has been argued elsewhere (Bucher 1941; Scriven 1959; Van Bemmelen, 1961; Pantin 1968) that geology is a derivative science and only partially lives up to the classic model of scientific reasoning as exemplified by physics (classical mechanics), but it does provide a model of a different type of scientific reasoning based on interpretive techniques (hermeneutics) and its historical nature (Frodeman 1995).

Unfortunately, 'scientific' reasoning is portrayed as providing infallible answers which are often (inevitably) very difficult to obtain. Reasoning in geoscience offers a method that is more applicable to the uncertainties and complexities of real-life situations as we are seldom in possession of all the data we would like to make a decision. It is also not always clear that the data we possess are unbiased or objective. We are therefore forced into filling the gaps in our knowledge with interpretation and reasonable assumptions (guesses, bets, hypotheses and abductions), that we hope will be subsequently confirmed. The 'geological' method of analysis, as exemplified by the pragmatism of Peirce, is therefore considered as analogous to that undertaken by forensic practitioners (Nordby 2000), and can be taught on forensic courses through problem based techniques. The nature of the information used also makes any exercise in forensic geoscience interesting, often compelling, to students on traditional geoscience courses as it certainly applies the subject and focuses the mind on the importance of the earth material analysis that may literally be the difference between life and death. Many basic geological principles and concepts can be communicated in this way.

With the increasing media coverage it is not surprising that, over the last several years, there has been a tremendous increase in forensic studies. In 1999 there were some 12 courses listed in the UK's Universities and College Admission System (UCAS) handbook with forensic in the title. In 2003 there were well over 200. Forensic geoscience has a part to play in many of these courses and can make a significant contribution to the overall education of a forensic scientist . In addition I would recommend forensic geoscience as a suitable vehicle for the illustration of many earth science principles and concepts on traditional earth and environmental science awards.

The author is grateful to the BGS (NERC) for permission to reproduce the figures and tables of the forensic case study and particularly to G. Williams who so readily made his work available. Thanks are also extended to two unknown referees who provided constructive criticism and N. Rossiter for continued support.

Appendix 1

A bungalow explodes: Forensic case study demonstrating problem based learning (PBL)

Introduction and background

At 06.30 on 24 March 1986 a bungalow at 51 Clarke Avenue, Loscoe, some 16 km north of Derby, UK (see Fig. 1), was completely destroyed by an explosion when the central heating switched on automatically. The three occupants, although badly injured, were lucky to escape with their lives. The day itself was overcast and a very deep atmospheric depression passed over the area with an associated barometric pressure drop of 0.04 bar (4 kPa).

Immediately after the explosion, gas samples were taken from the collapsed basement and were found to contain methane and carbon dioxide in ratios between 20–65% methane to 16–57% carbon dioxide.

The bungalow itself was underlain by a sequence of coals, mudstones, siltstones and sandstones of Carboniferous age (Fig. 2). Most of the strata is impermeable but the sandstones have an average porosity of 18–25% and a natural permeability of 600 mD (millidarcy). These deposits have been worked commercially since 1885 by opencast, shallow and deep mining methods.

The geological survey map of the area (1963) is shown in Figure 3 and a geological cross-section in Figure 4. Eight coal seams were worked from beneath the area from 1885 to the 1960s but this ceased when the shallowest seam was removed. Old records state that, in the oldest mines, naked flame

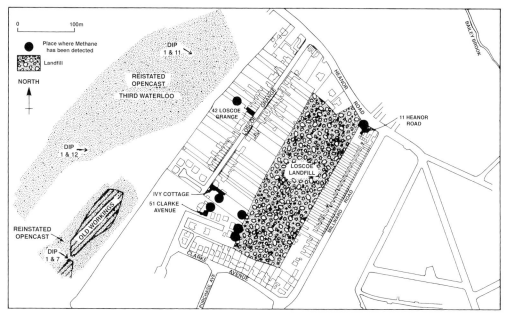

Fig 1. Site plan of the landfill and area surrounding 51 Clarke Avenue, Loscoe, Derby, UK. After Williams & Aitkenhead 1991, by permission of the BGS. (© NERC All rights reserved. IPR/43–52c).

lanterns were used for illumination. No deep shafts are recorded but some seams are so shallow that they may have been worked by drifts or adits to the north-west, where the rocks can be seen to dip regionally to the southwest. The majority of these shallow/surface workings have long become overgrown or reclaimed by agriculture. The working of one seam, the Roof Soft Seam, produced a zone of permanently extended or stretched strata as a result of the differential subsidence from below. This stretching increased the permeability of the rocks above (especially the sandstones), by the widening and extending of any pre-existing joints or fissures. The surface expression of this zone is shown in Figure 3.

On top of the bedrock (beneath the topsoil) is a thin deposit of Pleistocene Head deposits. The soil itself consists of clay, silt and sand. Both sequences tend to behave impermeably. In some cases, house foundations and trenches for service ducts and pipes are cut to bedrock level (0.9–3.0 m down). A number of wells and pumps are also shown on old maps.

To the southeast of the bungalow, the Loscoe brickpit had been worked for brick clay, stone and coal from before 1879. A planning consent issued in 1966 stipulated that this void should be backfilled with 'agreed' material, but the housing development was completed by 1973 when consent was given to deposit inert waste. In 1977 a waste disposal licence was granted under the provision of the Control of Pollution Act (1974) to deposit 50 tonnes of domestic (putrescible) waste per day. This was in disregard

of Government guidelines that stated that no houses should lie within 200 m of a landfill site. Tipping continued until 1982, and in 1984 the site was covered by permeable material. In 1985/6 the site was capped by a layer of impermeable clay to prevent water ingress and leachate production. This was effective as a positive pressure of 0.03 bar (3 kPa) was measured in the landfill during the drilling that took place subsequent to the explosion.

Recent history

In 1983 a pear tree in the garden of 51 Clarke Avenue began to die. Subsequently the soil became warm, dried out and crumbled. Other areas of the lawn died. Problems occurred after the lawn was returfed. Similar problems and unpleasant smells occurred in the garden of Ivy Cottage. At 42 Loscoe Grange, the occupier dug a hole 0.5 m deep and an unpleasant 'sewer' like smell was detected, together with a 'rumbling noise and a warm mould-like growth'. The gas was below the lower explosive limit for methane and no carbon monoxide was recorded. British Coal was contacted and installed a standpipe with a flame trap to allow the gas to vent harmlessly to the atmosphere. Gas analysis indicated 35% methane and 65% carbon dioxide. In 1983 smells were reported at 13a Heanor Road (Fig. 1). but no methane was detected. In 1984, smells and low concentrations of methane were detected at 44 Clarke

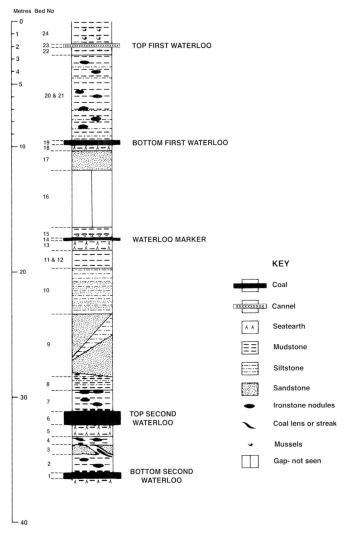

Fig 2. Stratigraphic log of Loscoe brickpit. After Williams & Aitkenhead 1991, by permission of the BGS (© NERC All rights reserved. IPR/43–52c).

Avenue. Investigations by the East Midlands Gas Board suggested that it was not mains gas because carbon dioxide levels were too high and ethane (normally present in mains gas at 3–4% by volume, with a methane/ethane ratio of about 25:1), was not recorded. Traces of methane and carbon dioxide were also detected at Purchase Avenue in 1985 and 1986. Initially, this soil heating and distressed vegetation was thought to be some sort of underground fire (perhaps a burning coal seam), but carbon monoxide levels, usually associated with coal burning in a limited supply of oxygen, were low. A bore-hole at Ivy Cottage showed a decrease in soil temperature with depth from 21 °C at depths of

0–0.5 m. below ground level (bgl) to 18 °C at 2.27 m bgl. This was accompanied by an increase in methane composition from 2% at the surface to 33.4% at 2.27 m bgl. Gas samples taken at 1.65 m bgl in sandstone contained 29.6% nitrogen. At 42 Loscoe Grange, gas at 3.0 m.bgl in a sandstone horizon contained 58% methane and 39% carbon dioxide. In the distressed soil areas the soil bacterium *Pseudomonas methanica* was identified. This has an affinity for methane and oxidizes it exothermically, with the production of water vapour and carbon dioxide. The heat causes net water loss and desiccation causes shrinkage and cracking at the surface, giving a direct route for methane venting.

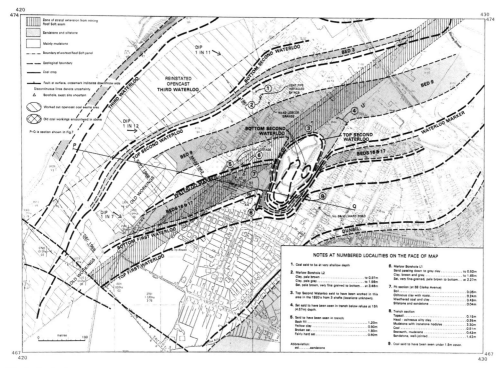

Fig 3. Geological plan of the Loscoe area. After Williams & Aitkenhead 1991, by permission of the BGS (© NERC All rights reserved. IPR/43–52c).

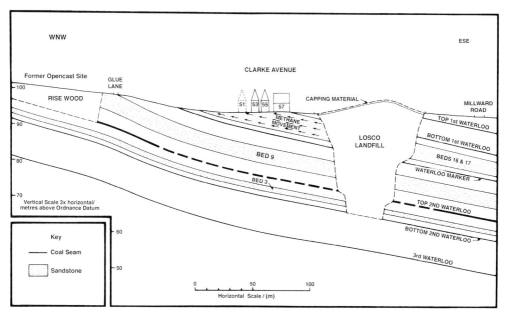

Fig 4. Geological cross-section through Loscoe landfill. (see Fig. 3 for line of section). After Williams & Aitkenhead 1991, by permission of the BGS (© NERC All rights reserved. IPR/43–52c).

Table 1. *Composition of methane-containing gases (% vol.) After Williams & Aitkenhead 1991, by permission of the BGS (© NERC. All rights reserved. IPR/43–52c)*

Source	CH_4	C_2H_6	C_3H_8	C_4H_{10}	C_{2+}	CO_2	CO	N_2	O_2
Landfill	20–65					16–57		0.5–37	<0.3
Coal									
seam	80–95	8	4			0.2–6		2–9	
drainage	22–95	3	1			0.5–6	0–10	1–61	
Anaerobic digestor	62–75					18–38		0–6	
Natural gas									
mains	94	3.2	0.6	0.2		0.5		1.2	
general	49–99	0.7–16	0.4–7.9	0.1–3.4	0–39	0–9.5		0.1–22	
'wet'	17–97	6.4	5.3	2.6	2.1–80				
'dry'	57–98	2.0	0.6	0.3	0.1–15			4.7	0.9
Marsh gas	11–88							3–69	
Glacial drift	45–97				0.8–1.4	0.2–8		1.6–54	
Deep marine biogenic	96–99				0–3				
Estuary/lake mud									
freshwater	3–86					0.3–13		16–94	
saltwater	55–79					2–13			

Table 2. *Summary of gas analyses from Loscoe. (After Williams & Aitkenhead 1991, by permission of the BGS (© NERC All rights reserved. IPR/43-52c)*

Location/date of sample	H_2	O_2	N_2	CO_2	CH_4	C_2H_6	CO
Loscoe Landfill							
16/4/86	1.1	0.3	11.7	33	53.9	ND	ND
Probe D, 51 Clarke Ave							
29/3/86	–	16.3	57.6	12.6	13.5	–	
10/3/86		1.09	3.05	32.8	30.1	32.9	ND
Standpipe, 42 Loscoe Grange							
28/8/85	–	–	–	36	64	–	
3–14/3/86	–	8.0	23.5	25.8	42.7	1	
3–14/3/86	–	1.8	5.2	34.5	58.5	–	
10/3/86	1.1	1.06	4.0	38.1	55.6	–	ND
23/10/86	ND	trace	2.0	40.1	59.7	267.8	
23/10/86	ND	trace	1.4	39.0	59.6	252.8	
Probe B, Ivy Cottage							
29/3/86	–	11.9	35.0	19.0	34.0	–	

All analyses are expressed in % by volume, except for C_2H_6 which is in p.p.m. ND, not detected.

Determining the gas component

A useful way of determining 'landfill' methane (modern source) and from the Coal Measures (ancient source) is to determine its 14_C content. Methane produced by the biodegration of recent organic waste material, paper, wood, garden refuse, sewage etc. reflects an amount of 14_C that is related to the present, relatively high concentration in the atmosphere, while 14_C from ancient geological sources has long since decayed away. (See Tables 1 & 2.)

The methane separated from the carbon dioxide at the standpipe at Loscoe Grange contained significant quantities of 14_C.

The problem

You are required to solve the problem by formulating multiple hypotheses, some of which may be discounted sooner than others. The following questions will help you to do this.

1. What caused the explosion? Coal gas? Landfill gas? Ordinance? Natural gas? Marsh gas? Meteorite impact? Mains gas? Sewer gas? Terrorist outrage? Does the gas responsible have a component that is specific (unique) to its particular source?
2. Can a permeability pathway be established from the suspected origin of the gas to the bungalow? (Use the maps and cross-section.)
3. Can a driving force be established to allow the gas to move from its origin to the bungalow? (A diffusion or pressure gradient is needed.) Note: Diffusion is unlikely, as calculations based on diffusion rates through a porous medium suggest an interval of 23 years to achieve a 5% methane content by movement through 90 m of sandstone. How long would it take to achieve a source composition of 60% methane and 40% carbon dioxide sampled at the bungalow.
4. What ignited the gas?
5. What evidence suggests the underlying coal seams were not gassy?
6. What are the explosive limits for methane?
7. Why was a mains gas explosion discounted?
8. What is the source of the nitrogen in the borehole at Ivy Cottage?
9. Why could a burning coal seam be discounted?
10. Why should the local ground hotspots be episodic?

Now present your case, stating who/what is to blame for the explosion. What, as a forensic geoscientist, would you suggest were the lessons learned from this case?

References

BARROWS, H. S. & TAMBLYN, R. M. 1980. *Problem-based Learning: An Approach to Medical Education*. New York, Springer.

BAKER, V. C. 1996. The pragmatic roots of American Quaternary geology and geomorphology. *Geomorphology*, **16**, 197–215.

BASSETT, H. E. & MANHEIM, M. H. 2002. Fluvial transport of human remains in the Lower Mississippi River. *Journal of Forensic Science*, **47**, 719–724.

BEVERIDGE, W. I. B. 1957. *The Art of Scientific Investigation*. 3rd edition. Norton, New York.

BEVERIDGE, W. I. B. 1980. The Seeds of Discovery. Norton, New York.

BLEWETT, W.L. 1993. Description, analysis and critique of the method of multiple working hypotheses. *Journal of Geological Education*, **41**, 254–259.

BLUMHOF, J., HALL, M., & HONEYBONE, A. 2001. Using problem-based learning to develop graduate skills. *In*: GASKIN, S. & KING, H. (eds) *Case Studies in Problem-Based Learning (PBL) from Geography, Earth and Environmental Sciences*. *Planet* (LTSN) Special Edition, **2**, 6–9.

BROWN, A. G., SMITH, A. & ELMHURST, O. 2002. The com-

bined use of pollen and soil analysis in a search and subsequent murder investigation. *Journal of Forensic Science*, **47**, 614–618.

BUCHER, W. H. 1941. The nature of geological inquiry and the training required for it. *New York A.I.M.E. Technical Publication*, **1377**, 6.

CARNIEL, S., URMGIESSER, G., SCLAVO, M., KANTHA, L. H. & MONTI, S. 2002. Tracking the drift of a human body in the coastal ocean using numerical prediction models of the oceanic, atmosphere and wave conditions. *Science and Justice*, **42**, 143–151.

CATTEL, J. M. 1933. The distribution of American men of Science in 1932. *In*: CATTEL, J.M. & CATTEL, J. (eds) *American Men of Science*. Science Press, New York, 1261–1278.

CHAMBERLIN, T. C. 1890. The method of multiple working hypotheses. *Science*, **15**, 92–96. [Reprinted 1965. *Science*, **148**, 754–759.]

CHAMBERLIN, T. C. 1897. The method of multiple working hypotheses. *Journal of Geology*, **5**, 837–848. [Reprinted 1944. *Scientific Monthly*, **59**, 356–362.]

CHAMBERLIN, T. C. 1904. The methods of earth sciences. *Popular Science Monitor*, **66**, 66–75.

COLLINSON, J. D. & THOMPSON, D. B. 1989. *Sedimentary Structures*. 2nd edition. Chapman and Hall, London.

COOPER, A. H. & WALTHAM, A. C. 1999. Subsidence caused by gypsum dissolution at Ripon, North Yorkshire. *Quarterly Journal of Engineering Geology*, **32**, 305–310.

DAVIS, W.M. 1926. The value of outrageous geological hypotheses. *Science*, **63**, 463–468.

DAVIS, W. H. 1972. *Pierce's Epistenology*. Martinus Nijhoff, The Netherlands, 163pp.

DEBROCK, G. 1992. *Peirce: A philosopher for the 21st century*. Part 1. *Introduction*. Transactions of Charles S. Peirce Society, **28**.

DEWEY, J. 1938. *Experience and Education*. Collier and Kappa Delta Pi, New York.

FAIRCHILD, H. leR. 1904. Geology under the planetismal hypothesis of Earth's origin. *Geological Society of America Bulletin*, **15**, 243–266.

FENNING, P. J. & DONNELLY, L. J. 2004. Geophysical techniques for forensic investigation. *In*: PYE, K. & CROFT, D. J. (eds) *Forensic Geoscience: Principles, Techniques and Applications*. Geological Society, London, Special Publications, **232**, 11–20.

FRODEMAN, R. 1995. Geological reasoning: geology as an interpretive and historical science. *Geological Society of America Bulletin*. **107**, 960–968.

GILBERT, G. K. 1886. The inculcation of scientific method by example. *American Journal of Science*, **31**, 284–299.

GILBERT, G. K. 1896. The origin of hypothesis, illustrated by a discussion of a topographical problem. *Science*, **3**, 1–13.

GLEICK, J. 1992. *Genius: The life and science of Richard Feynman*. Pantheon Books, New York.

GOULD, S. J. 1987. *Time's Arrow, Time's Cycle. Myth and Metaphor in the Discovery of Geological Time*. Harvard University Press, Cambridge.

GOULD, S. J. 1989. *Wonderful Life. The Burgess Shale and the Nature of History*. Norton, New York.

HAGLUND, W. D. 1993. Disappearance of soft tissue and the disarticulation of human remains from aqueous environments. *Journal of Forensic Science*, **38**, 806–815.

HANSON, I. D. 2004. The importance of stratigraphy in forensic investigation. *In*: PYE, K. & CROFT, D. J. (eds) *Forensic Geoscience: Principles, Techniques and Applications*. Geological Society, London, Special Publications, **232**, 39–47.

HARVEY, A. M. & BORDLEY, J. 1970. *Differential Diagnosis*. W. B. Saunders, Philadelphia.

HAWLEY, D. 2002. Building conceptual understanding in young scientists. *Journal of Geoscience Education*, **50**, 363–371.

HONIGSCHNABL, S., SCHADEN, E. *ET AL*. 2002. Discovery of decomposed and mummified corpses in domestic settings. *Journal of Forensic Science*, **47**, 837–842.

HORROCKS, M., COULSON, S. A. & WALSH K. A. J. 1998. Forensic palynology: variation in the pollen content of soil surface samples. *Journal of Forensic Science*, **43**, 320–323.

HOWER, J.C., SCHRAM, W. H. & THOMAS, G.A. 2000. Forensic petrology and geochemistry: tracking the source of a coal slurry spill, Lee County, Virginia. *International Journal of Coal Geology*, **44**, 101–108.

JOHNSON, D. 1933. Role of analysis in scientific investigation. *Geological Society America Bulletin*, **44**, 461–493.

JUNGER, E. P. 1996. Assessing the unique characteristics of close proximity soil samples: Just how useful is soil evidence? *Journal of Forensic Science*, 41, 27–34.

KITTS, D. B. 1973. Grove Karl Gilbert and the concept of 'hypothesis' in late nineteen-century geology. *In*: GIERE, R. N. & WESTFALL, R. S. (eds) *Foundations of Scientific Method: The Nineteen Century*. Indiana University Press, Bloomington, 259–274.

KITTS, D. B. 1982. The logic of discovery in geology. *Earth Science History*, **1**, 1–6.

LEE, B. D., WILLIAMSON, T. N., & GRAHAM, R. C. 2002. Identification of stolen rare palm trees by soil morphological and mineralogical properties. *Journal of Forensic Science*, **47**, 190–194.

LEE, C. W. 2000. A problem-based learning exercise in forensic geology. *In*: THOMAS, N. & KING, H. (eds) *Staff Resource Book, Earth Science Learning and Teaching in Higher Education*. National Subject Centre for Geography, Earth and Environmental Science, 8–10.

LEE, C.W. 2001. Problem based learning. *In*: GASKIN, S. & KING, H. (eds) *Case Studies in Problem-Based Learning (PBL) from Geography, Earth and Environmental Sciences*. *Planet* (LTSN), Special Edition, **2**, 10.

MACKIN, J. H. 1963. Rational and empirical methods of investigation in geology. *In*: ALBRITTON, C. C. (ed.), *The Fabric of Geology*. Freeman, Cooper, Stanford, 135–163.

MURRAY, R. C. & TEDROW, J. C. F. 1975. *Forensic Geology, Earth Sciences and Criminal Investigation*. Rutgers University Press, New Jersey

MURRAY, R. C. & TEDROW, J. C. F. 1992. *Forensic Geology*. Prentice Hall, New Jersey

NORDBY, J.J. 1992. Can we believe what we see, if we see what we believe? Expert disagreement. *Journal of Forensic Science*, 37, 1115–1124.

NORDBY, J. J. 2000. *Dead Reckoning. The Art of Forensic Detection*. C. R. C. Press, Baton Racon.

NORMAN, G. & SCHMIDT, H. 1992. The psychological basis of problem based learning: a review of evidence. *Academic Medicine*, **67**, 557–574.

PANTIN, C.F.A. 1968. *The Relations Between the Sciences*. Cambridge University Press, Cambridge.

PEIRCE, C. S. 1877. The fixation of belief. *Popular Science Monitor*, **12**, 1–15.

PEIRCE, C. S. 1878a. How to make our ideas clear. *Popular Science Monitor*, **12**, 286–302.

PEIRCE, C. S. 1878b. Deduction, induction and hypothesis. *Popular Science Monitor*, **13**, 470–482.

PEIRCE, C. S. 1878c. The probability of induction. *Popular Science Monitor*, **12**, 705–718.

POPPER, K.R. 1959. *The Logic of Scientific Discovery*. Basic Books, New York.

RHOADS, B.L. & THORN, C.E. 1994. Contemporary philosophical perspectives on physical geography with emphasis on geomorphology. *Geographical Review*, **84**, 90–101.

ROBERTS, E. 2003. Predicting weather effects predates weather forecasts. *Marine Scientist*, **3**, 18–21.

ROBERTS, R.M. 1989. *Serendipity: Accidental Discoveries in Science*. John Wiley, New York.

ROCK, T., IRVIN, J. & PALMER, S. 1990. The legacy of Gestallt Psychology. *Scientific American*, Dec. 84–90.

ROSSMO, D.K. 2000. *Geographical Profiling*. CRC Press, New York.

SABINE, P.A. 1991. Geologists at war: a forensic investigation in the field of war-time diplomacy. *Proceedings of the Geological Association*, **102**, 139–43.

SAVIN-BADEN, M. 2000. *Problem-Based Learning in Higher Education: Untold Stories*. Open University Press/SRHE, Buckingham.

SAVIN-BADEN, M. 2001. The problem-based learning landscape. *In*: GASKIN, S. & KING, H. (eds) *Case Studies in Problem-Based learning (PBL) from Geography, Earth and Environmental Sciences*. *Planet* (LTSN), Special Edition, **2**, 4–6.

SCHMIDT, H. 1993. Foundations of problem-based learning: some explanatory notes. *Medical Education*, **27**, 424–428.

SCHUMM, S. 1991. *To interpret the Earth: Ten Ways to be Wrong*. Cambridge University Press, Cambridge.

SCRIVEN, M. 1959. Explanation and prediction in evolutionary theory. *Science*, **130**, 477–482.

SEMMELWEIS, I. P. 1861. The etiology, the concept and prophylaxis of childbed fever. *Medical Classics*, **5**, 350–773. [Translated by F. Murphy 1941.]

SHUIRNAN G. & SLOSSON, J.E. 1992. *Forensic Engineering (Environmental Case Histories for Civil Engineers and Geologists)*. Academic Press, New York and London.

SMALE, D. & TRUEMAN, N.A. 1969. Heavy mineral studies as evidence in a murder case in outback Australia. *Journal of Forensic Science*, **9**, 3–4.

VAN ANDEL, T. 1994. The method of multiple working hypotheses and the agony of choice. *Terra Nova*, **6**, 222–223.

VAN BEMMELEN, R.W. 1961. The scientific character of geology. *Journal of Geology*, **69**, 453–463.

WATSON, R.A. 1969. Explanation and prediction in geology. *Journal of Geology*, **77**, 488–494.

WEBSTER, R. & YAALON, D.H. 1994. The research paper: an informal guide to authors. *Catena*, **21**, 3–11.

WILLIAMS, G.M. & AITKINHEAD, N. 1991. Lessons from Loscoe: the uncontrolled migration of landfill gas. *Quarterly Journal of Engineering Geology*, 24, 191–207.

Index

Page numbers in *italic* refer to figures, page numbers in **bold** refer to tables